Learn. Compete. Innovate.

Student, Parent, Teacher Resources

Improve Academic Performance

- Mathematics, Science, and English Language Arts standards
- STEM integration
- Standardized Test Practice
- Reading Strategies

Build Project Skills

- Chapter Technology Labs
- Unit projects
- TSA competitive event prep
- Safety skills

Innovate Your World

- Standards for Technological Literacy
- EcoTech for green technology
- Exploring Careers in Technology

Learn Anytime, Anywhere

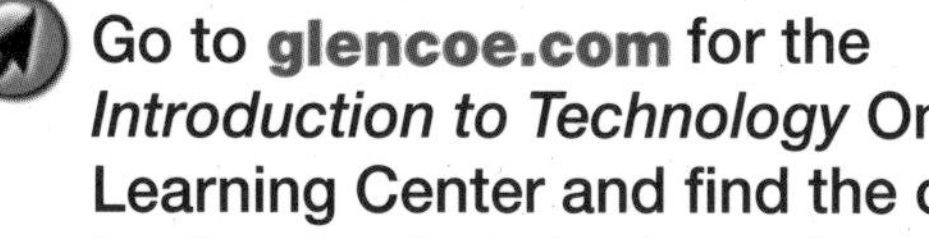

Go to **glencoe.com** for the *Introduction to Technology* Online Learning Center and find the online textbook, study tools, and projects

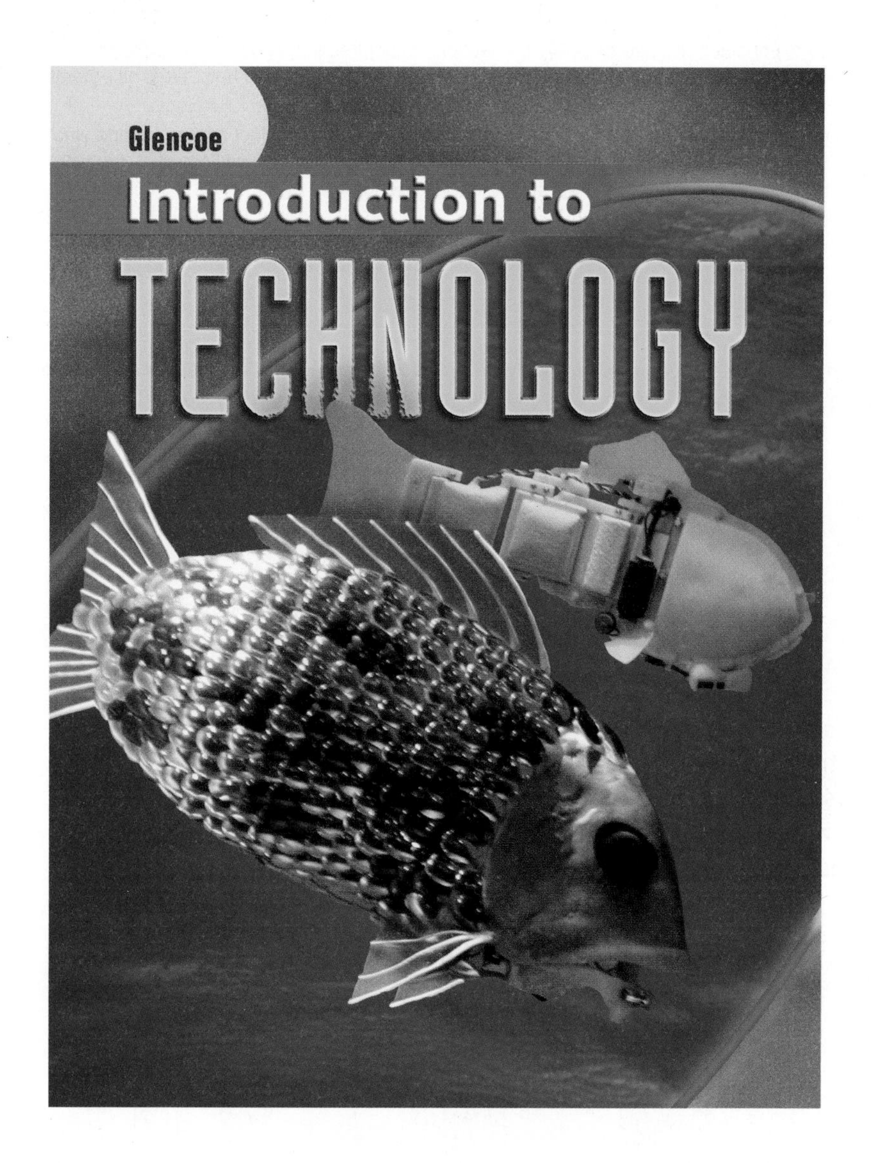

Alan J. Pierce

Dennis Karwatka

SAFETY NOTICE

The reader is expressly advised to consider and use all safety precautions described in this book or that might also be indicated by undertaking the activities described herein. In addition, common sense should be exercised to help avoid all potential hazards.

Publisher and Authors assume no responsibility for the activities of the reader or for the subject matter experts who prepared this book. Publisher and Authors make no representation or warranties of any kind, including but not limited to, the warranties of fitness for particular purpose or merchantability, nor for any implied warranties related thereto, or otherwise. Publisher and Authors will not be liable for damages of any types, including any consequential, special, or exemplary damages resulting, in whole or in part, from reader's use or reliance upon the information, instructions, warnings, or other matter contained in this book.

Notice: Information on featured companies, organizations, and their products and services is included for educational purposes only and does not present or imply endorsement of the *Introduction to Technology* program.

On the Cover

Aquatic Robotics The fish you see on the cover of *Introduction to Technology* is actually a robot. It is a real-life technology tool called a "robo-carp" that swims with real fish in the London Aquarium. Engineered by scientists at the University of Essex in England, this fish uses artificial intelligence and an infrared sensor in its mouth to detect and avoid obstacles. It was created to sniff out underwater mines, leaky oil pipes, and more.

Glencoe

The McGraw·Hill Companies

Printed in the United States of America.

Send all inquiries to:
Glencoe/McGraw-Hill
4400 Easton Commons
Columbus, OH 43219

ISBN: 978-0-07-879785-9 (Student Edition)
MHID: 0-07-879785-3 (Student Edition)

ISBN: 978-0-07-890711-1 (Teacher Annotated Edition)
MHID: 0-07-890711-X (Teacher Annotated Edition)

4 5 6 7 8 9 RJE/LEH 13 12 11 10

About the Authors

Dr. Alan J. Pierce has served as an educator at the elementary, middle, and college levels for more than 40 years. Dr. Pierce has authored numerous articles in the field of education, co-authored curriculum projects, and served as a technical consultant on a children's technology book. He is an editor and writer of the middle school Agricultural Biology curriculum for the NSF-funded TECHknow project at North Carolina State University. Dr. Pierce created the "Technology Today" column for *Tech Directions* magazine and has been writing this column since 1995.

Dennis Karwatka was a professor in the Department of Industrial Education and Technology at Morehead State University. He taught for many years in the high school Upward Bound Program. He has written two middle/high school textbooks, five technical history books, and many articles. Dennis Karwatka created the monthly "Technology's Past" column in *Tech Directions* and has been writing that column since 1980. He is a Registered Professional Engineer with experience in the Apollo lunar landing program and jet engine development.

Contributing Writer

Thomas Shown
Industrial Technology & Human Services
North Carolina Department of Public Instruction
Raleigh, North Carolina
Technology Students Association
Member of Competition Regulations Committee

Our McGraw-Hill Partner

BusinessWeek is the leading global resource for ground-breaking business news and news analysis that offers essential insight into the real world of business, technology, education, and other areas. *BusinessWeek* is the world's most widely read business magazine in print and online with more than 8 million weekly readers.

Reviewers and Advisors

Educational Reviewers

We wish to acknowledge the contributions of the following reviewers:

Lane Beard
St. James Parish Career and Technology Center
Lutcher, Louisiana

Courtney Bullock
Walter Reed Middle School
Studio City, California

LaMarr Brooks
Pendleton High School
Pendleton, South Carolina

Lewis H. Chappelear
James Monroe High School
North Hills, California

Lisa K. Coleman
Gautier High School
Gautier, Mississippi

Eric Elder
Warren East Middle School
Bowling Green, Kentucky

Michael Fontaine
Murdock Middle High School
Winchendon, Massachusetts

Ed Hamilton
Berryhill High School
Tulsa, Oklahoma

Lauren Withers Olson
Biloxi Junior High
Biloxi, Mississippi

Tim Schultz
Lafayette Tecumseh Junior High School
Lafayette, Indiana

Trade and Technology Industry Advisory Board

Johan Gallo
Manager of Human Resources
Government and Agency Affairs
Bridgestone-Firestone
Laguna Hills, California

Cheryl Horn
Corporate Manager, Community Relations
Northrop Grumman Corporation

Erin Kuhlman
Vice President, Corporate Relations
Parsons, Inc.
Pasadena, California

Donna Laquidara-Carr
National Training Manager
McGraw-Hill Construction
Lexington, Massachusetts

Floyd McWilliams
Executive Vice President
American Design Drafting Association
Program Director, Drafting
Anthem College
Phoenix, Arizona

Olen Parker
Executive Director
American Design Drafting Association
Nashville, Tennessee

Robert Ratliff
President
The Manufacturing Institute
Cumming, Georgia

Joe Vela
Associate Principal
AEDIS Architecture & Planning
San Jose, California

Table of Contents

Table of Contents

Reading Strategies

In each section, look for these reading strategies:

- Before You Read
- Graphic Organizer
- Reading Check
- As You Read
- After You Read

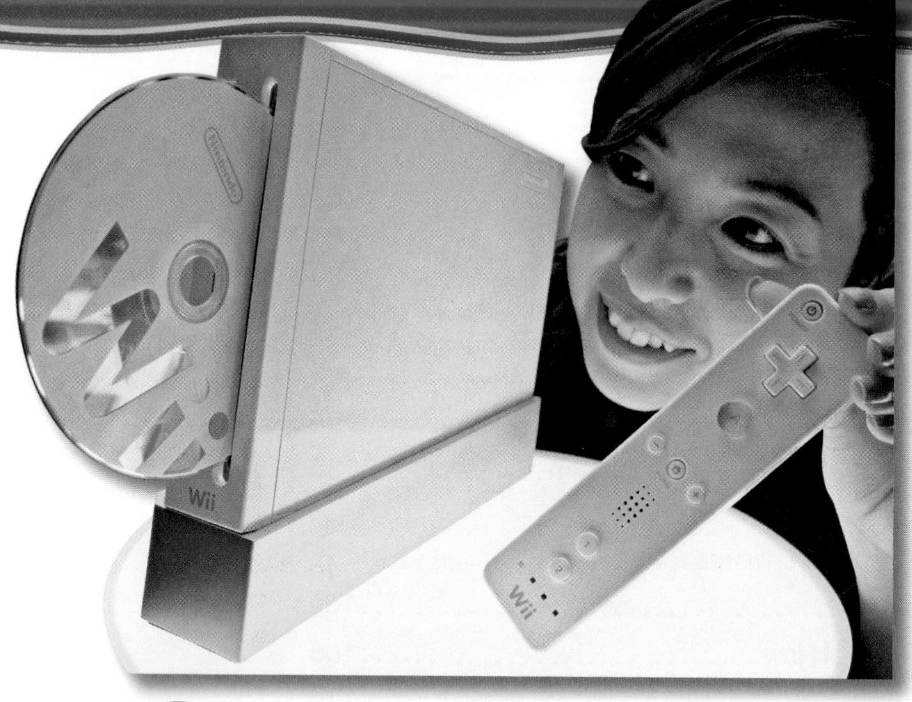

Reading Strategies

In each section, look for these reading strategies:

- Before You Read
- Graphic Organizer
- Reading Check
- As You Read
- After You Read

Unit 4 Biotechnologies 286

Unit 5 Manufacturing Technologies 336

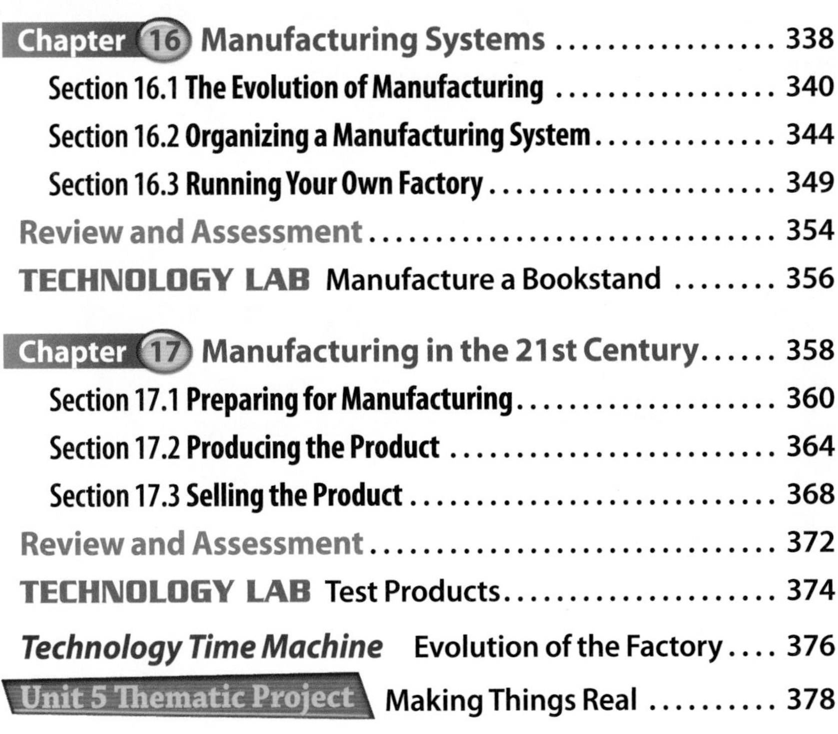

Table of Contents

Unit 6 Construction Technologies 380

Table of Contents

Welcome to Introduction to Technology

Explore Your Textbook

Technology is everywhere. Did you know that you use technology each time you ride in a car, talk on a phone, watch TV, play computer games, or read a book? Start exploring your textbook so you can understand, use, design, and even make technology!

Units

Introduction to Technology has seven units. Each unit contains two to six chapters. The units explore all the major types of technology.

Unit Opener

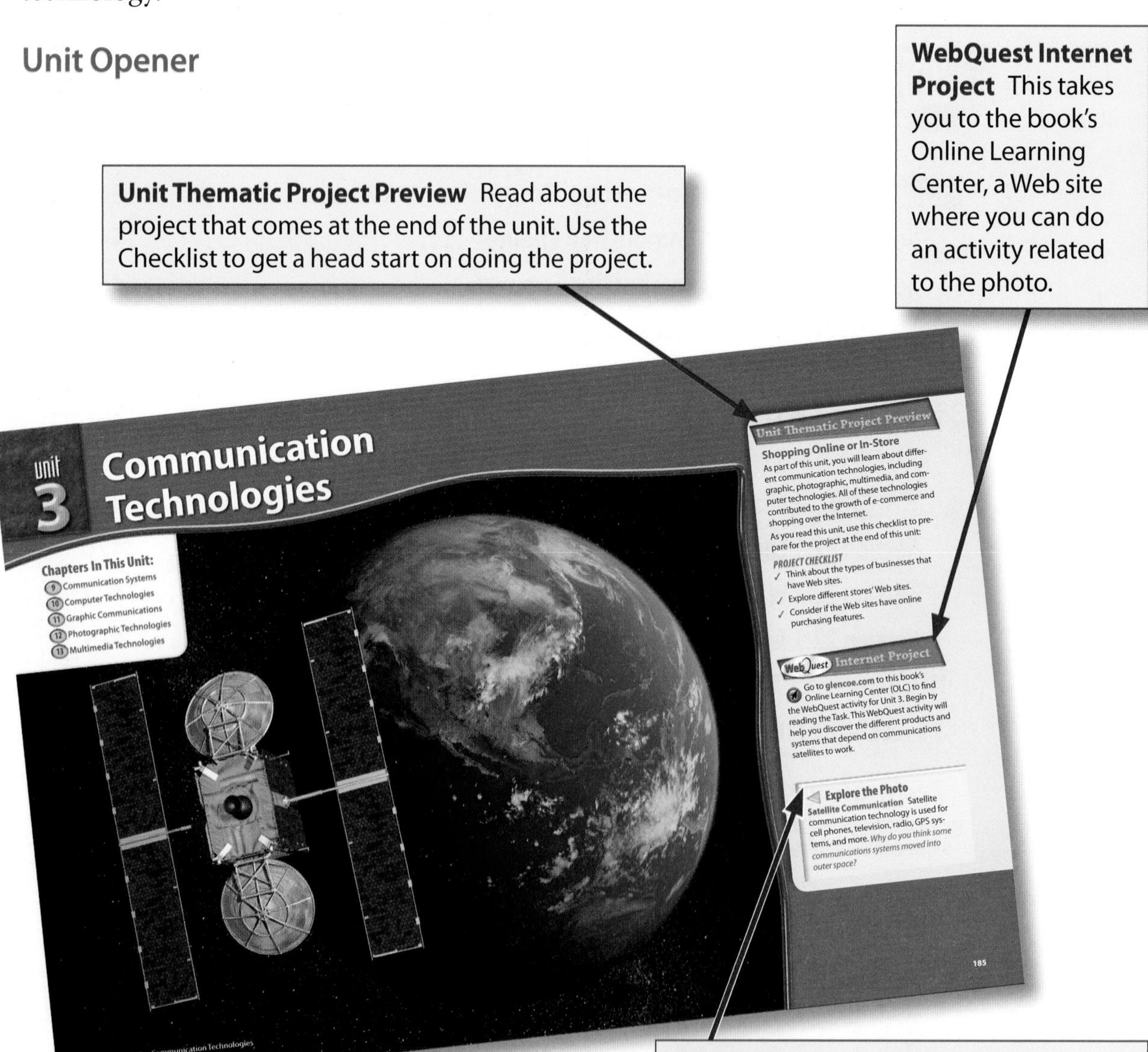

Unit Close

Each unit ends with a fun feature and a project-based activity that will give you a chance to apply what you learn.

Technology Time Machine

Play this game to learn about the history of technology that we use today.

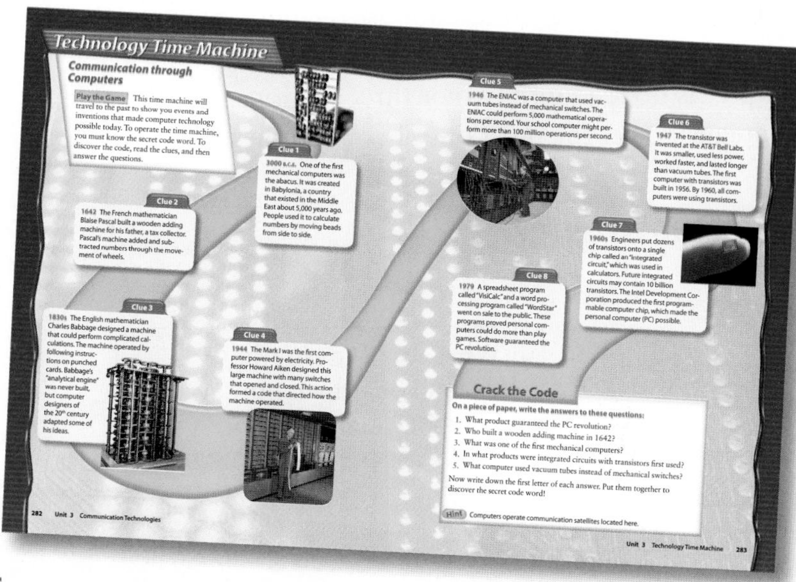

Technology Time Machine

Communication through Computers

Play the Game This time machine will travel to the past to show you events and inventions that made computer technology possible today. To operate the time machine, you must know the secret code word. To discover the code, read the clues, and then answer the questions.

Clue 1
3000 B.C.E. One of the first mechanical computers was the abacus. It was created in Babylonia, a country that existed in the Middle East about 5,000 years ago. People used it to calculate numbers by moving beads from side to side.

Clue 2
1642 The French mathematician Blaise Pascal built a wooden adding machine for his father, a tax collector. Pascal's machine added and subtracted numbers through the movement of wheels.

Clue 3
1830s The English mathematician Charles Babbage designed a machine that could perform complicated calculations. The machine operated by following instructions on punched cards. Babbage's "analytical engine" was never built, but computer designers of the 20th century adapted some of his ideas.

Clue 4
1944 The Mark I was the first computer powered by electricity. Professor Howard Aiken designed this large machine with many switches that opened and closed. This action formed a code that directed how the machine operated.

Clue 5
1946 The ENIAC was a computer that used vacuum tubes instead of mechanical switches. The ENIAC could perform 5,000 mathematical operations per second. Your school computer might perform more than 100 million operations per second.

Clue 6
1947 The transistor was invented at the AT&T Bell Labs. It was smaller, used less power, worked faster, and lasted longer than vacuum tubes. The first computer with transistors was built in 1956. By 1960, all computers were using transistors.

Clue 7
1960s Engineers put dozens of transistors onto a single chip called an "integrated circuit," which was used in calculators. Future integrated circuits may contain 10 billion transistors. The Intel Development Corporation produced the first programmable computer chip, which made the personal computer (PC) possible.

Clue 8
1979 A spreadsheet program called "VisiCalc" and a word processing program called "WordStar" went on sale to the public. These programs proved personal computers could do more than play games. Software guaranteed the PC revolution.

Crack the Code

On a piece of paper, write the answers to these questions:

1. What product guaranteed the PC revolution?
2. Who built a wooden adding machine in 1642?
3. What was one of the first mechanical computers?
4. In what products were integrated circuits with transistors first used?
5. What computer used vacuum tubes instead of mechanical switches?

Now write down the first letter of each answer. Put them together to discover the secret code word!

Hint Computers operate communication satellites located here.

282 Unit 3 Communication Technologies

Unit 3 Technology Time Machine 283

The Unit Thematic Project

Each Thematic Project begins with an introduction. Complete five steps to evaluate your resources, conduct research, create a project, present your findings, and evaluate your report.

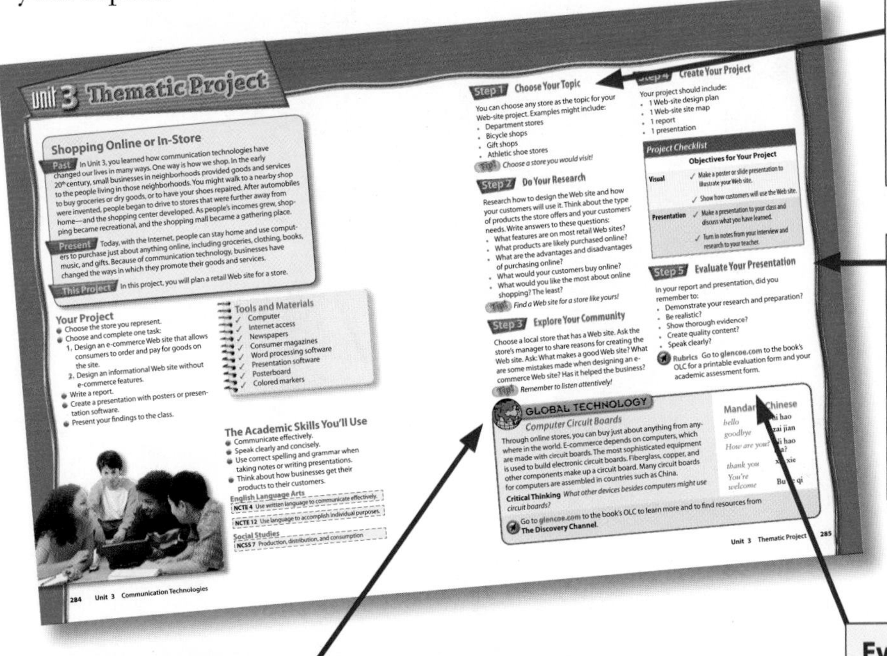

Unit 3 Thematic Project

Shopping Online or In-Store

Past In Unit 3, you learned how communication technologies have changed our lives in many ways. One way is how we shop. In the early 20th century, small businesses in neighborhoods provided goods and services to the people living in those neighborhoods. You might walk to a nearby shop to buy groceries or dry goods, or to have your shoes repaired. After automobiles were invented, people began to drive to stores that were further away from home—and the shopping center developed. As people's incomes grew, shopping became recreational, and the shopping mall became a gathering place.

Present Today, with the Internet, people can stay home and use computers to purchase just about anything online, including groceries, clothing, books, music, and gifts. Because of communication technology, businesses have changed the ways in which they promote their goods and services.

This Project In this project, you will plan a retail Web site for a store.

Your Project

- Choose the store you represent.
- Choose and complete one task:
  1. Design an e-commerce Web site that allows consumers to order and pay for goods on the site.
  2. Design an informational Web site without e-commerce features.
- Write a report.
- Create a presentation with posters or presentation software.
- Present your findings to the class.

Tools and Materials

- ✓ Computer
- ✓ Internet access
- ✓ Newspapers
- ✓ Consumer magazines
- ✓ Word processing software
- ✓ Presentation software
- ✓ Posterboard
- ✓ Colored markers

The Academic Skills You'll Use

- Communicate effectively.
- Speak clearly and concisely.
- Use correct spelling and grammar when taking notes or writing presentations.
- Think about how businesses get their products to their customers.

English Language Arts
NCTE 4 Use written language to communicate effectively.
NCTE 12 Use language to accomplish individual purposes.

Social Studies
NCSS 7 Production, distribution, and consumption

Step 1 Choose Your Topic

You can choose any store as the topic for your Web-site project. Examples might include:

- Department stores
- Bicycle shops
- Gift shops
- Athletic shoe stores

Tip! *Choose a store you would visit!*

Step 2 Do Your Research

Research how to design the Web site and how your customers will use it. Think about the type of products the store offers and your customers' needs. Write answers to these questions:

- What features are on most retail Web sites?
- What products are likely purchased online?
- What are the advantages and disadvantages of purchasing online?
- What would your customers buy online?
- What would you like the most about online shopping? The least?

Tip! *Find a Web site for a store like yours!*

Step 3 Explore Your Community

Choose a local store that has a Web site. Ask the store's manager to share reasons for creating the Web site. Ask: What makes a good Web site? What are some mistakes made when designing an e-commerce Web site? Has it helped the business?

Tip! *Remember to listen attentively!*

Step 4 Create Your Project

Your project should include:

- 1 Web-site design plan
- 1 Web-site site map
- 1 report
- 1 presentation

Project Checklist

	Objectives for Your Project
Visual	✓ Make a poster or slide presentation to illustrate your Web site.
	✓ Show how customers will use the Web site.
Presentation	✓ Make a presentation to your class and discuss what you have learned.
	✓ Turn in notes from your interview and research to your teacher.

Step 5 Evaluate Your Presentation

In your report and presentation, did you remember to:

- Demonstrate your research and preparation?
- Be realistic?
- Show thorough evidence?
- Create quality content?
- Speak clearly?

Rubrics Go to glencoe.com to the book's OLC for a printable evaluation form and your academic assessment form.

GLOBAL TECHNOLOGY

Computer Circuit Boards

Through online stores, you can buy just about anything from anywhere in the world. E-commerce depends on computers, which are made with circuit boards. The most sophisticated equipment is used to build electronic circuit boards. Fiberglass, copper, and other components make up a circuit board. Many circuit boards for computers are assembled in countries such as China.

Critical Thinking *What other devices besides computers might use circuit boards?*

Go to glencoe.com to the book's OLC to learn more and to find resources from The Discovery Channel.

Mandar	Chinese
hello	i hao
goodbye	zai jian
How are you?	i hao a?
thank you	x xie
You're welcome	Bu qi

284 Unit 3 Communication Technologies

Unit 3 Thematic Project 285

Choose Your Topic Step 1 helps you focus on a topic or project that you would like to complete.

Research, Connect, Report Steps 2, 3, and 4 guide you through researching procedures, as you connect to the real world and report your findings.

Global Technology At the end of each project, discover how your project relates to the real world and another country. Learn new words in a world language!

Evaluate Your Presentation Step 5 takes you to the Online Learning Center to download a rubric for evaluating your project.

Discover, Learn, Apply

Chapters

The chapters of *Introduction to Technology* are organized around subjects that relate to all kinds of technology. Each chapter is divided into two or three sections. Use the reading strategies to get the most out of your studies.

Chapter Opener

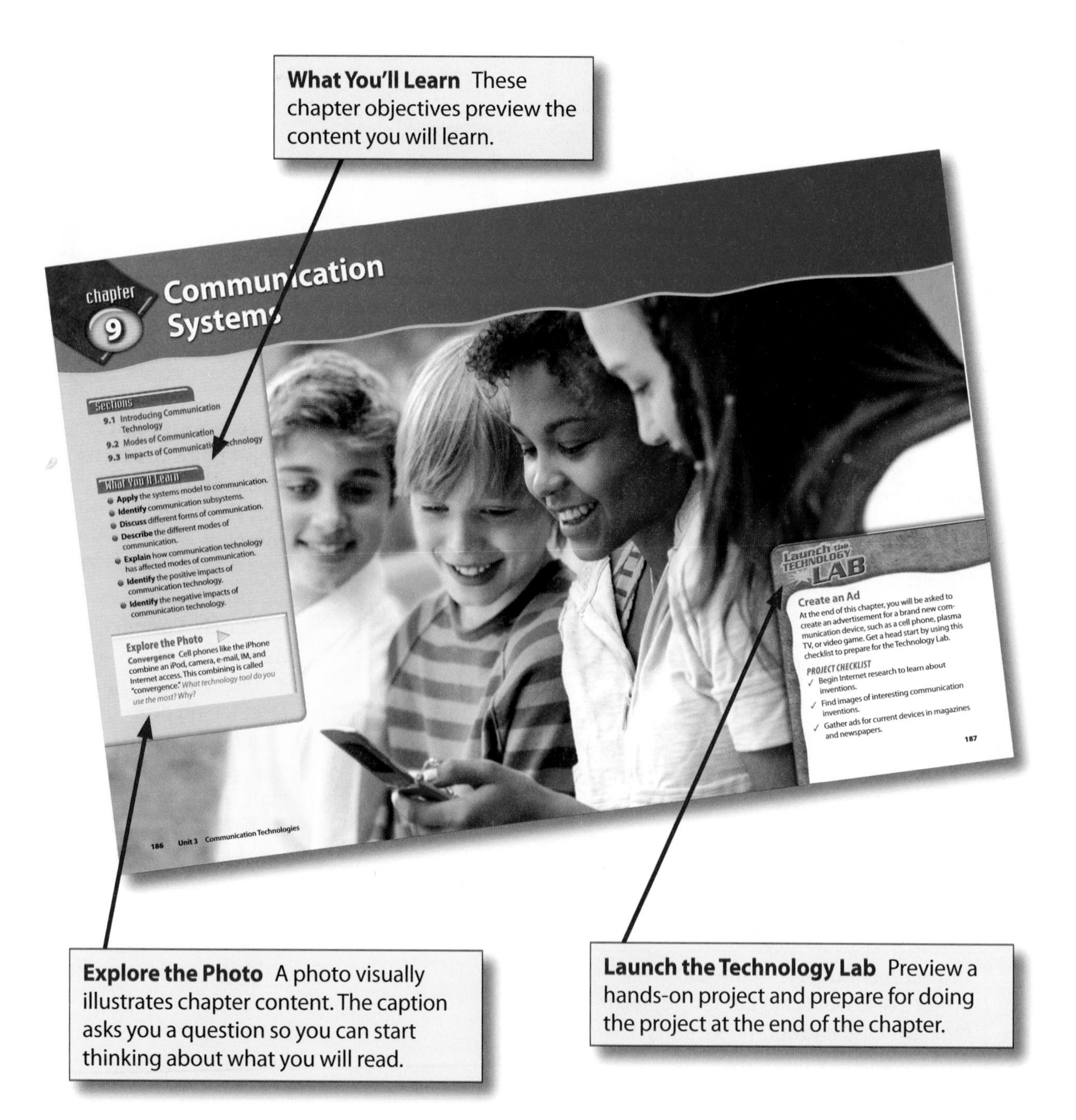

Practice Your Skills

Chapter Close—The Technology Lab

At the end of every chapter, you can create your own technology project. You might design an ad, build a model vehicle, or even make a robot!

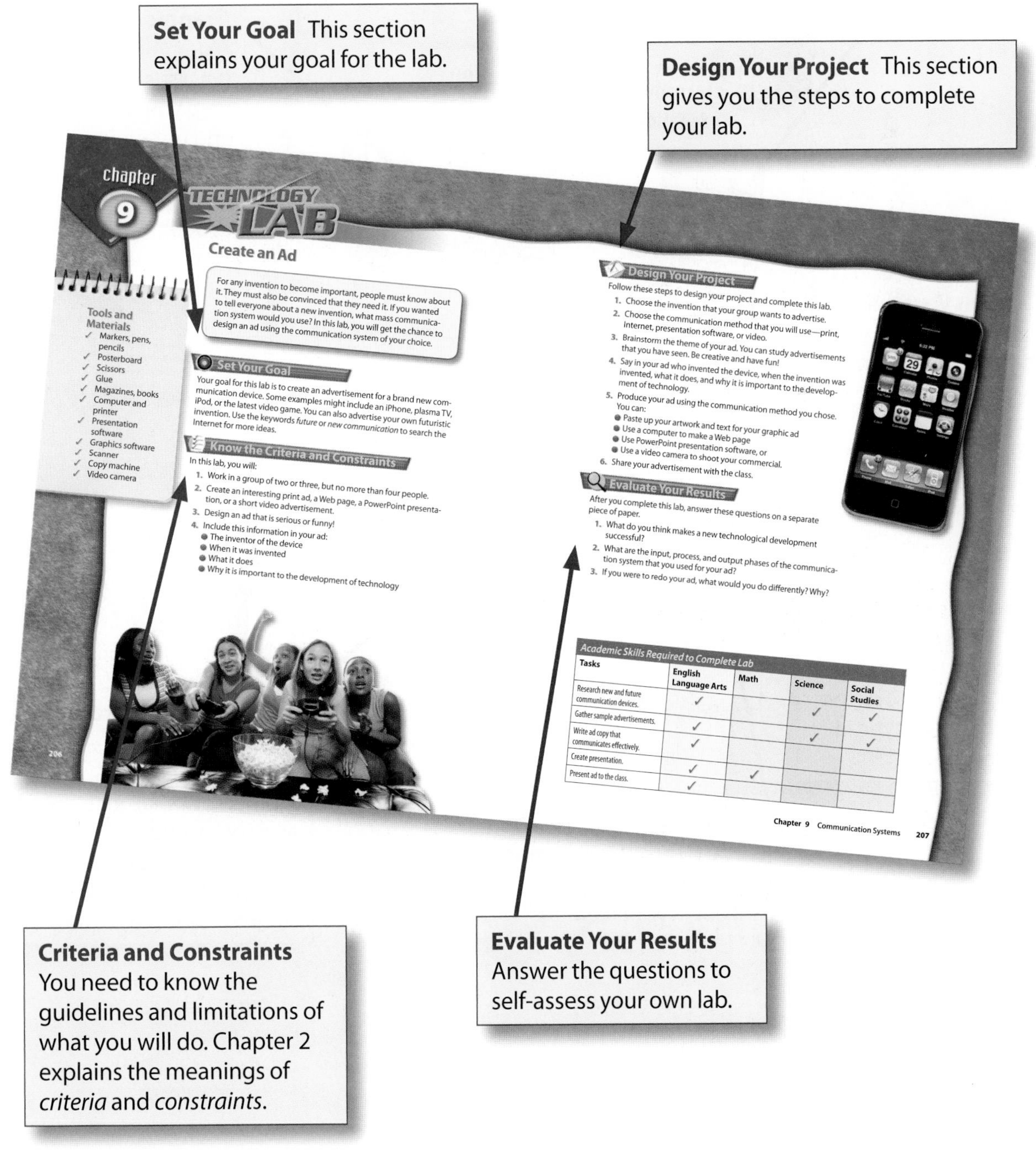

chapter 9

TECHNOLOGY LAB

Create an Ad

For any invention to become important, people must know about it. They must also be convinced that they need it. If you wanted to tell everyone about a new invention, what mass communication system would you use? In this lab, you will get the chance to design an ad using the communication system of your choice.

Tools and Materials
- ✓ Markers, pens, pencils
- ✓ Posterboard
- ✓ Scissors
- ✓ Glue
- ✓ Magazines, books
- ✓ Computer and printer
- ✓ Presentation software
- ✓ Graphics software
- ✓ Scanner
- ✓ Copy machine
- ✓ Video camera

Set Your Goal

Your goal for this lab is to create an advertisement for a brand new communication device. Some examples might include an iPhone, plasma TV, iPod, or the latest video game. You can also advertise your own futuristic invention. Use the keywords *future* or *new communication* to search the Internet for more ideas.

Know the Criteria and Constraints

In this lab, you will:

1. Work in a group of two or three, but no more than four people.
2. Create an interesting print ad, a Web page, a PowerPoint presentation, or a short video advertisement.
3. Design an ad that is serious or funny!
4. Include this information in your ad:
   - The inventor of the device
   - When it was invented
   - What it does
   - Why it is important to the development of technology

206

Design Your Project

Follow these steps to design your project and complete this lab.

1. Choose the invention that your group wants to advertise.
2. Choose the communication method that you will use—print, Internet, presentation software, or video.
3. Brainstorm the theme of your ad. You can study advertisements that you have seen. Be creative and have fun!
4. Say in your ad who invented the device, when the invention was invented, what it does, and why it is important to the development of technology.
5. Produce your ad using the communication method you chose. You can:
   - Paste up your artwork and text for your graphic ad
   - Use a computer to make a Web page
   - Use PowerPoint presentation software, or
   - Use a video camera to shoot your commercial.
6. Share your advertisement with the class.

Evaluate Your Results

After you complete this lab, answer these questions on a separate piece of paper.

1. What do you think makes a new technological development successful?
2. What are the input, process, and output phases of the communication system that you used for your ad?
3. If you were to redo your ad, what would you do differently? Why?

Academic Skills Required to Complete Lab

Tasks	English Language Arts	Math	Science	Social Studies
Research new and future communication devices.	✓		✓	✓
Gather sample advertisements.	✓		✓	✓
Write ad copy that communicates effectively.	✓			
Create presentation.	✓	✓		
Present ad to the class.	✓			

Chapter 9 Communication Systems 207

Set Your Goal This section explains your goal for the lab.

Design Your Project This section gives you the steps to complete your lab.

Criteria and Constraints You need to know the guidelines and limitations of what you will do. Chapter 2 explains the meanings of *criteria* and *constraints*.

Evaluate Your Results Answer the questions to self-assess your own lab.

Section Opener

At the beginning of each section, check out the Reading Guide. Develop your reading and comprehension skills as you preview the content, vocabulary, standards, and main ideas for that section.

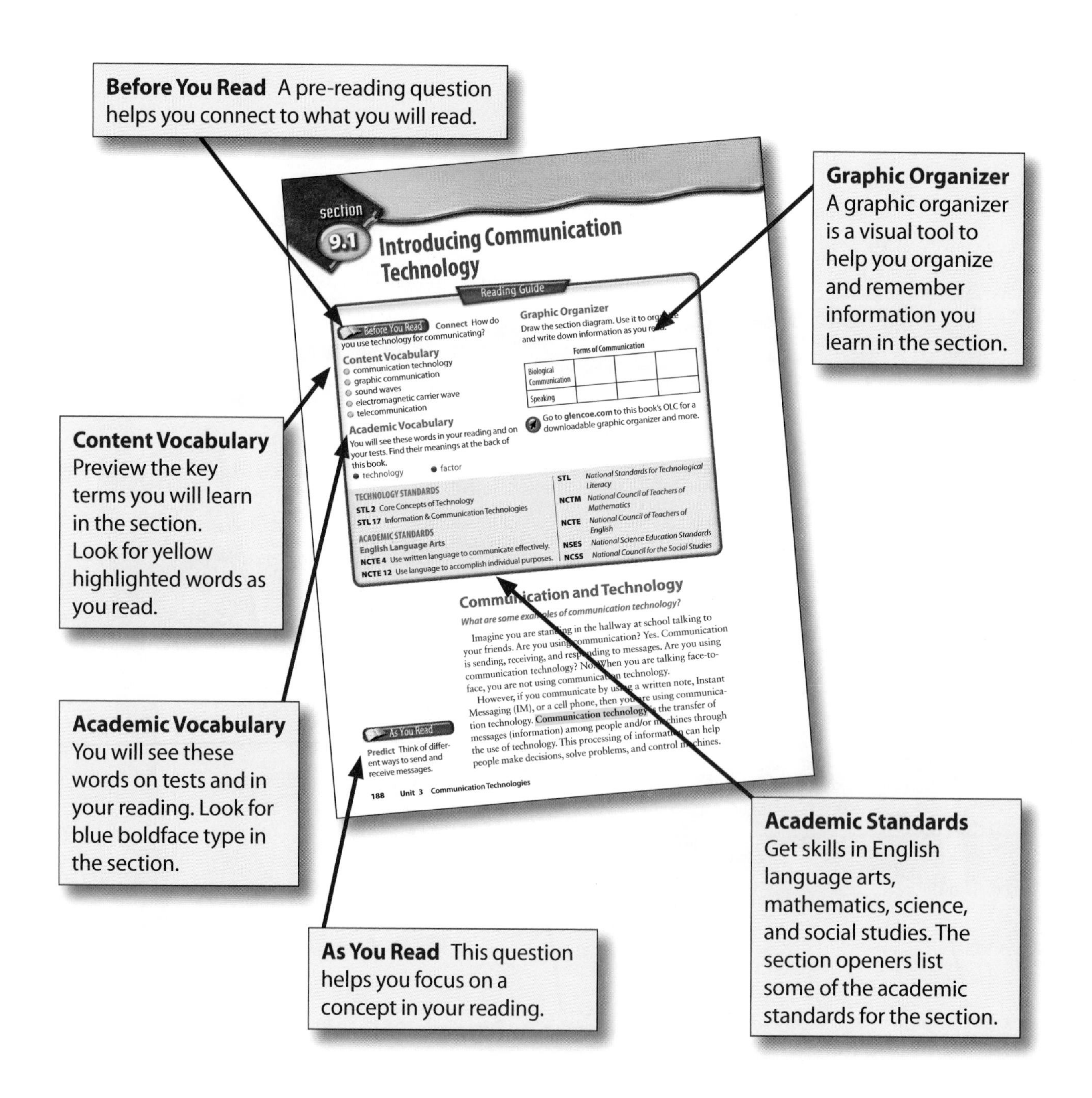

Features

Interesting features introduce basic technology topics in each chapter. You will also learn about people of the past and present. These features help you see how what you read relates to the real world.

Margin Features

Academic Connections See how technology can link to almost every subject—mathematics, science, English language arts, or social studies.

Academic Connections
Science

Now Hear This! You have your own built-in sound receivers—your ears. Sound waves striking your eardrums make them vibrate.

Apply Find an illustration of the ear in an encyclopedia at school. Make a drawing showing the main parts of the ear. Label the parts and functions.

STEM

EcoTech Technology can be "green"! This feature shows you how technology can be friendly to our environment. **Try This** gives you ideas that can help.

EcoTech

Treeless Paper

Making paper from trees can use a lot of energy and toxic chemicals. But paper can also be made from hemp, bamboo, or kenaf—a plant that grows quickly and uses eco-friendly chemicals for production.

Try This To save paper, write on both sides of your paper, use scrap paper—and recycle it.

Imagine This Learn about amazing innovations in technology for today and tomorrow.

Imagine This...

Computers You Can Wear

Imagine wearing a computer that is a lightweight, voice-activated box with a headset and eyepiece. When you look into the eyepiece, you see a computer screen that appears as if it is a few feet away. Researchers at NASA are working on a model called a Wearable Augmented Reality Prototype (WARP) for astronauts to wear so their hands are free for other tasks. *How might wearing your computer help you in everyday communication?*

Go to **glencoe.com** to this book's OLC for answers and to learn more about NASA and technology.

***BusinessWeek* Tech News** This feature appears in a chapter once in every unit. It reports special news on technology.

BusinessWeek Tech NEWS

Building a Super Cell Phone

The newest wave of tech entrepreneurs is transforming our mobile phones into personal computers. "The most common digital device in the world is the cell phone," says Motricity's chief technology officer. Young entrepreneurs are looking to social networking, Internet video, and online photo-sharing. They extend those applications by putting them on cell phones.

Critical Thinking *What modes of communication would apply to a super cell phone? Why?*

Go to **glencoe.com** to this book's OLC read more about this news.

To the Student

In-Chapter Features

Tech Stars This feature profiles an innovator in a technology field that relates to the chapter.

Tech Stars

Tim Berners-Lee
Inventor of the World Wide Web

When Tim Berners-Lee was a teenager in England, he built a computer using a soldering iron and an old television. Later, he worked as a computer programmer at CERN, a physics laboratory in Geneva, Switzerland. There, he wrote a program with Robert Cailliau for storing information by using "random associations." This idea led to "hyperlinking," which is how the World Wide Web works.

Tim said the World Wide Web should be like a "global hypertext project." It should help people work together through a "web" of documents. The documents could be opened from anywhere on the Web.

The WC3 In 1994, Tim set up the World Wide Web Consortium (WC3) at the Massachusetts Institute of Technology (MIT). The group manages Web development around the world. Of course, the WC3 has its own Web site at w3.org.

English Language Arts/Writing Write an article for your school newspaper about spam and how it affects using the Web.

Go to **glencoe.com** to this book's OLC to learn about young innovators in technology.

Ethics in Action This feature lets you look at the ethical questions and choices that face the world of technology today.

Ethics in Action

Misinformation on the Net

Mistakes, deliberate falsehoods, and personal opinion make their way into newspapers, books, the Internet, and magazines everyday. Both ethical and unethical people use communication media.

Just the Facts How can you protect yourself from misinformation? Stay informed using many sources. For example, during an election read or listen to what each side says about the candidates. Try to gather all the facts.

English Language Arts/Writing

Comparing News Select a news event and compare the coverage that it receives in a newspaper, a TV news broadcast, and a news Web site.

1. Use a spreadsheet to chart and compare how each covers the story.
2. Write a paragraph summarizing your conclusions.

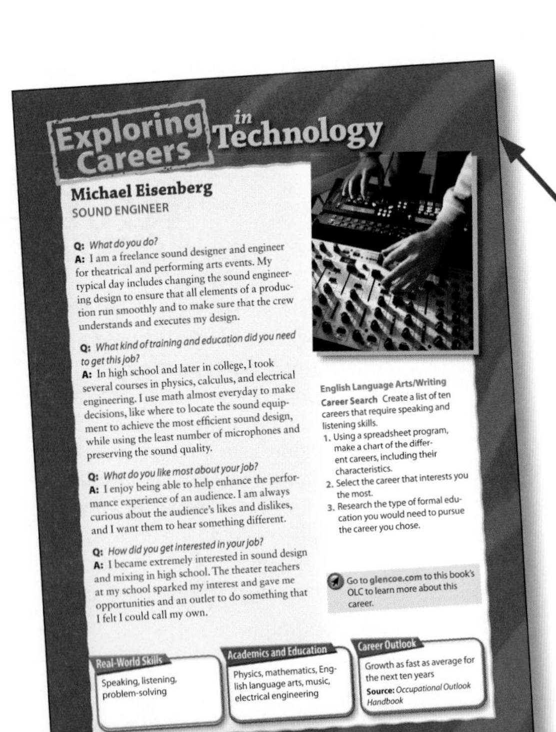

Exploring Careers in Technology

Michael Eisenberg
SOUND ENGINEER

Q: *What do you do?*
A: I am a freelance sound designer and engineer for theatrical and performing arts events. My typical day includes changing the sound engineering design to ensure that all elements of a production run smoothly and to make sure that the crew understands and executes my design.

Q: *What kind of training and education did you need to get this job?*
A: In high school and later in college, I took several courses in physics, calculus, and electrical engineering. I use math almost everyday to make decisions, like where to locate the sound equipment to achieve the most efficient sound design, while using the least number of microphones and preserving the sound quality.

Q: *What do you like most about your job?*
A: I enjoy being able to help enhance the performance experience of an audience. I am always curious about the audience's likes and dislikes, and I want them to hear something different.

Q: *How did you get interested in your job?*
A: I became extremely interested in sound design and mixing in high school. The theater teachers at my school sparked my interest and gave me opportunities and an outlet to do something that I felt I could call my own.

English Language Arts/Writing
Career Search Create a list of ten careers that require speaking and listening skills.
1. Using a spreadsheet program, make a chart of the different careers, including their characteristics.
2. Select the career that interests you the most.
3. Research the type of formal education you would need to pursue the career you chose.

Go to **glencoe.com** to this book's OLC to learn more about this career.

Real-World Skills	Academics and Education	Career Outlook
Speaking, listening, problem-solving	Physics, mathematics, English language arts, music, electrical engineering	Growth as fast as average for the next ten years **Source:** *Occupational Outlook Handbook*

Chapter 9 Communication Systems 203

Exploring Careers in Technology Read an interview with a real-world person working in technology. Learn about the tasks of the job, the skills and education needed, and the future of the career.

Assessment

Assessment is an important part of learning. Knowing what you have learned is a good way to find out what you need to study.

Section Assessment

Each section ends with a review and two academic activities.

After You Read: Self-Check
Check your comprehension with vocabulary and concept questions. Connect content to academics with **Academic Skills** questions.

section 9.1 assessment

After You Read **Self-Check**
1. Name the factors are involved in message design.
2. List the six purposes of a message.
3. Define telecommunication.

Think
4. Explain why smoke signals would be classified as a form of telecommunication.

Practice Academic Skills

English Language Arts/Writing
5. Until the late 1950s, many people shared telephone lines, or party lines. Only one party could use the phone at a time, because you might hear another person talking. Write a paragraph comparing today's social networking technologies to ones your parents used at your age.

STEM **Mathematics**
6. Shanda wants to create an Internet ad for China and Thailand and has to calculate the combined population of the two countries. If China has 1.306 billion people and Thailand has 65 million, what is the total population?

Math Concept **Representing Large Numbers** When you add larger numbers, make sure that they are expressed in the same form.
1. Express the number for China's population, 1.306 billion, as 1,306 million.
2. Add 1,306 million to 65 million to find the total combined population.

For help, go to glencoe.com to this book's OLC and find the Math Handbook.

Chapter Review and Assessment

The Chapter Review and Assessment contains a variety of review questions and activities.

Chapter Summary These summaries restate the main concepts of each section.

Real-World Skills STEM-based technology and real-world activities give you the chance to practice your skills and apply what you have learned.

Academic Skills These activities connect STEM-based mathematics and science as well as English language arts and social studies with technology concepts. **Math Concept** shows ways to solve the mathematics problems.

Review Vocabulary The activity focuses on technology terms and words you should know for reading and tests.

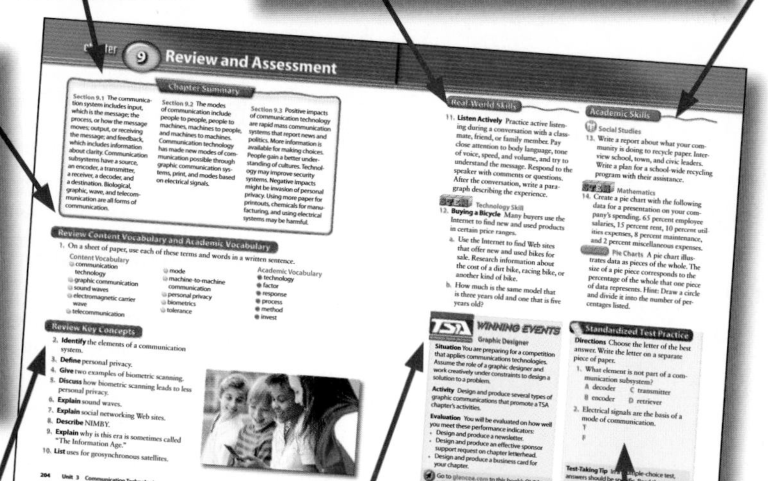

Review Key Concept Practice your skills and reinforce your understanding of key concepts in the chapter.

TSA Winning Events Prepare for competitive events as you role-play situations from real TSA events.

Standardized Test Practice Prepare for standardized tests through multiple-choice and true/false questions. The **Test-Taking Tip** gives you advice on taking tests.

Learn, Compete, Innovate!

TSA and Competitive Events

The Technology Student Association (TSA) promotes personal growth, leadership, and opportunities in technology, innovation, design, and engineering. TSA members use and apply science, technology, engineering, and mathematics concepts by participating in co-curricular activities, competitive events, and related programs.

TSA Winning Events

Participating in TSA and other competitive events is a fun way to apply and test your technology knowledge. Get to know other students as you work on exciting technology projects. Use the role-play exercise at the end of each chapter to increase your competitive advantage.

For more information about TSA competitive events, visit www.tsaweb.org.

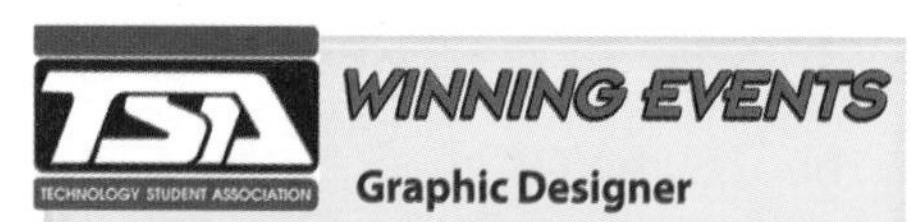

Graphic Designer

Situation You are preparing for a competition that applies communications technologies. Assume the role of a graphic designer and work creatively under constraints to design a solution to a problem.

Activity Design and produce several types of graphic communications that promote a TSA chapter's activities.

Evaluation You will be evaluated on how well you meet these performance indicators:

- Design and produce a newsletter.
- Design and produce an effective sponsor support request on chapter letterhead.
- Design and produce a business card for your chapter.

Go to **glencoe.com** to this book's OLC for information about TSA events.

What If English Is Not Your First Language?

The English Language Learner

Today's classrooms offer terrific opportunities to learn many skills, including language skills. You might be a part of a classroom in which ten or more countries are represented. You and your classmates may actually speak different native languages. You can use this multicultural environment to practice speaking, writing, and listening skills. Practicing these skills with your teacher and your classmates will help you to communicate effectively in your other classes and in your future career.

Highlight Vocabulary Check the section *Reading Guide* for vocabulary words that are not familiar. Write down these words and look them up in the glossaries at the back of the book.

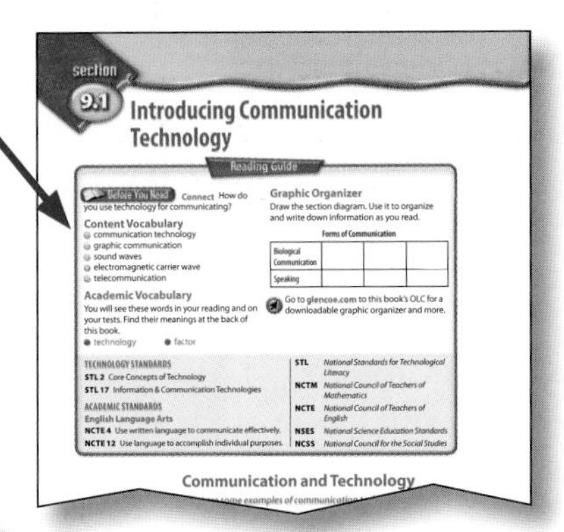

Learn through Pictures Look at the photos and figures carefully. Make a connection between what you have seen in the real world and what you see in the picture. If something is not familiar, ask your teacher or a classmate to explain it.

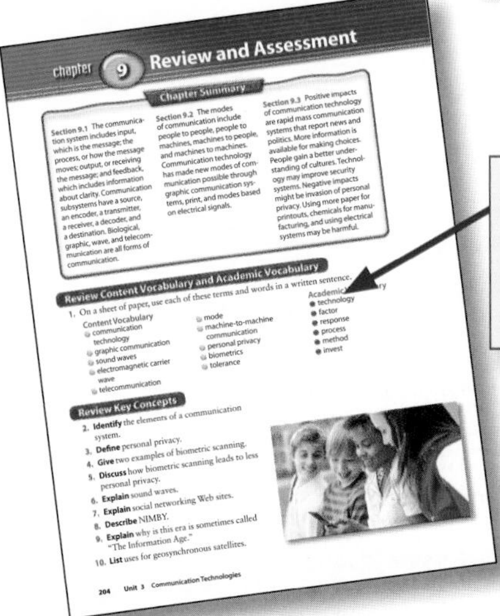

Practice Communication Study with a classmate. At the end of each chapter, review the chapter together. Start by choosing two vocabulary words from the list. Take turns defining them. Who can find the words in the chapter?

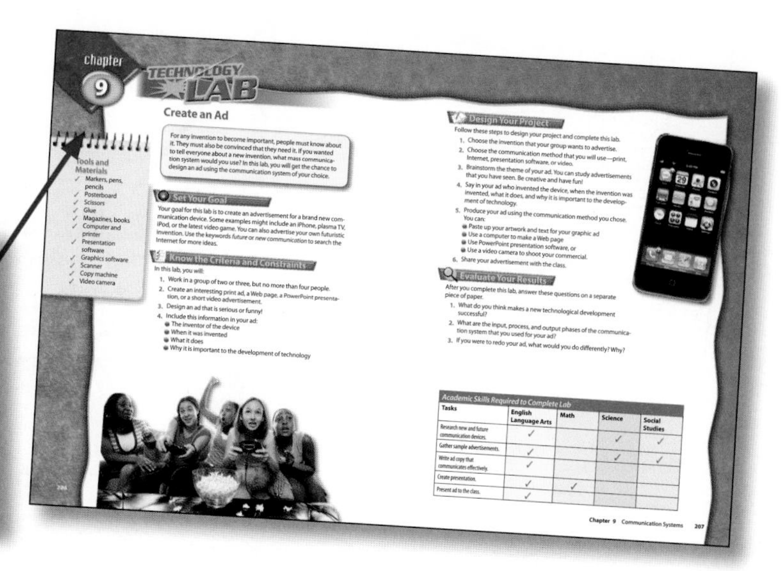

Connect to Your World Show what you have learned by doing projects like the *Technology Lab* or the *Unit Thematic Project*. When you must choose a topic, think of your culture, experiences, and skills. Then select a topic that interests you. Relate your project to your world!

Reach for Excellence in Math, Science, and Technology

Welcome to the World of Engineering

Important organizations like NASA, the space agency, and the National Science Foundation helped launch STEM learning. STEM stands for Science, Technology, Engineering, and Mathematics. STEM is also a nationwide initiative, or plan, developed for students just like you. STEM promotes learning those subjects to prepare you for success in tomorrow's world. Studying STEM subjects can lead you to exciting careers. In fact, STEM applies to most career fields today—from aircraft engineers, forensic scientists, and architects to firefighters, game developers, and auto mechanics.

In this book, you will find many Science, Technology, Engineering, and Mathematics activities in every chapter.

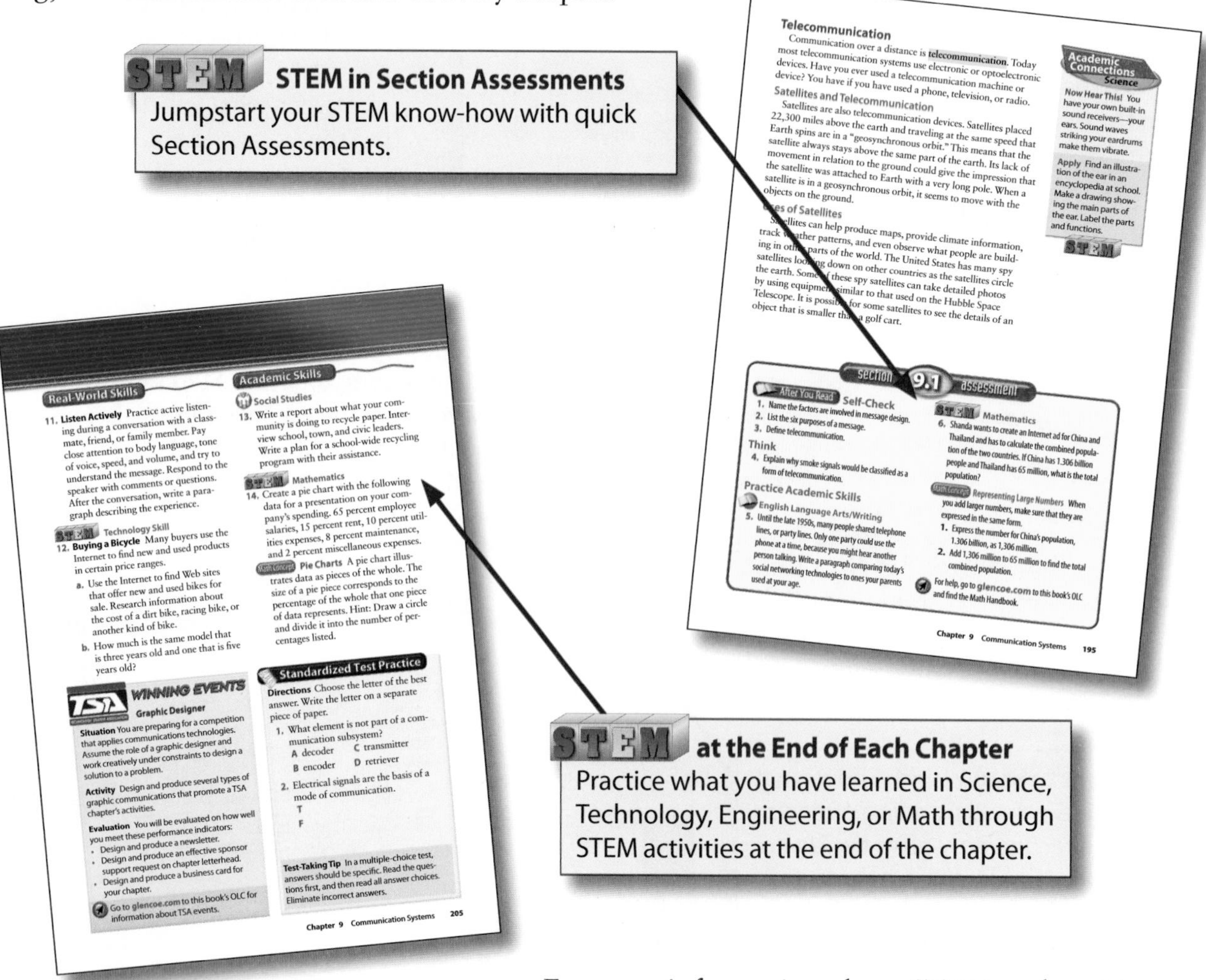

Telecommunication

Communication over a distance is telecommunication. Today most telecommunication systems use electronic or optoelectronic devices. Have you ever used a telecommunication machine or device? You have if you have used a phone, television, or radio.

Satellites and Telecommunication

Satellites are also telecommunication devices. Satellites placed 22,300 miles above the earth and traveling at the same speed that Earth spins are in a "geosynchronous orbit." This means that the satellite always stays above the same part of the earth. Its lack of movement in relation to the ground could give the impression that the satellite was attached to Earth with a very long pole. When a satellite is in a geosynchronous orbit, it seems to move with the objects on the ground.

Academic Connections Science

Now Hear This! You have your own built-in sound receivers—your ears. Sound waves striking your eardrums make them vibrate.

Apply Find an illustration of the ear in an encyclopedia at school. Make a drawing showing the main parts of the ear. Label the parts and functions.

Section 9.1 Assessment

After You Read Self-Check

1. Name the factors are involved in message design.
2. List the six purposes of a message.
3. Define telecommunication.

Think

4. Explain why smoke signals would be classified as a form of telecommunication.

Practice Academic Skills

English Language Arts/Writing

5. Until the late 1950s, many people shared telephone lines, or party lines. Only one party could use the phone at a time, because you might hear another person talking. Write a paragraph comparing today's social networking technologies to ones your parents used at your age.

STEM Mathematics

6. Shanda wants to create an Internet ad for China and Thailand and has to calculate the combined population of the two countries. If China has 1.306 billion people and Thailand has 65 million, what is the total population?

Representing Large Numbers When you add larger numbers, make sure that they are expressed in the same form.

1. Express the number for China's population, 1.306 billion, as 1,306 million.
2. Add 1,306 million to 65 million to find the total combined population.

For help, go to glencoe.com to this book's OLC and find the Math Handbook.

Chapter 9 Communication Systems 195

Real-World Skills

11. Listen Actively Practice active listening during a conversation with a classmate, friend, or family member. Pay close attention to body language, tone of voice, speed, and volume, and try to understand the message. Respond to the speaker with comments or questions. After the conversation, write a paragraph describing the experience.

STEM Technology Skill

12. Buying a Bicycle Many buyers use the Internet to find new and used products in certain price ranges.
a. Use the Internet to find Web sites that offer new and used bikes for sale. Research information about the cost of a dirt bike, racing bike, or another kind of bike.
b. How much is the same model that is three years old and one that is five years old?

TSA WINNING EVENTS

Graphic Designer

Situation You are preparing for a competition that applies communications technologies. Assume the role of a graphic designer and work creatively under constraints to design a solution to a problem.

Activity Design and produce several types of graphic communications that promote a TSA chapter's activities.

Evaluation You will be evaluated on how well you meet these performance indicators:
- Design and produce a newsletter.
- Design and produce an effective sponsor support request on chapter letterhead.
- Design and produce a business card for your chapter.

Go to glencoe.com to this book's OLC for information about TSA events.

Academic Skills

Social Studies

13. Write a report about what your community is doing to recycle paper. Interview school, town, and civic leaders. Write a plan for a school-wide recycling program with their assistance.

STEM Mathematics

14. Create a pie chart with the following data for a presentation on your company's spending. 65 percent employee salaries, 15 percent rent, 10 percent utilities expenses, 8 percent maintenance, and 2 percent miscellaneous expenses.

Pie Charts A pie chart illustrates data as pieces of the whole. The size of a pie piece corresponds to the percentage of the whole that one piece of data represents. Hint: Draw a circle and divide it into the number of percentages listed.

Standardized Test Practice

Directions Choose the letter of the best answer. Write the letter on a separate piece of paper.

1. What element is not part of a communication subsystem?
A decoder C transmitter
B encoder D retriever

2. Electrical signals are the basis of a mode of communication.
T
F

Test-Taking Tip In a multiple-choice test, answers should be specific. Read the questions first, and then read all answer choices. Eliminate incorrect answers.

Chapter 9 Communication Systems 205

For more information about STEM and exciting STEM careers, visit www.stemedcoalition.org.

What Are Standards?

Standards for Learning

Introduction to Technology was written to meet technology standards, so you will have a better understanding of technology and its impact on our world.

Technology Standards

Each year we depend more and more on technology. For that reason, understanding technology is very important for everyone. Years ago the International Technology Education Association (ITEA) created the Technology for All Americans Project. This project developed standards for what students should be able to do and know to be technologically literate. These standards are called "Standards for Technological Literacy" (STL). They are also called "Technology Content Standards."

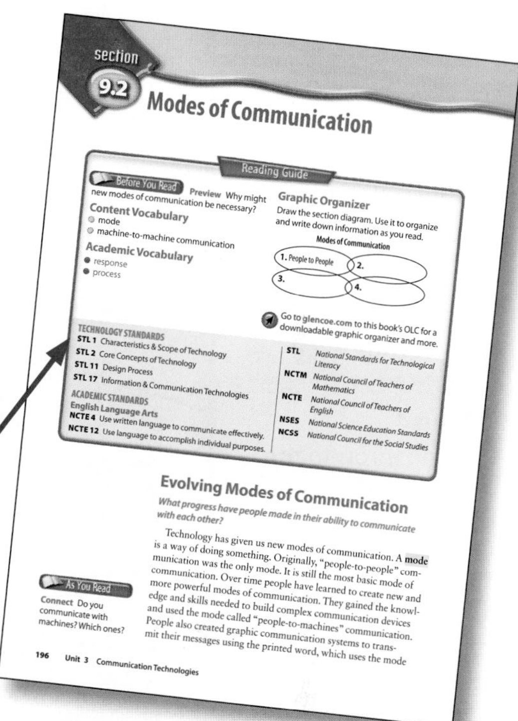

section 9.2

Modes of Communication

Reading Guide

Before You Read Preview Why might new modes of communication be necessary?

Content Vocabulary
- mode
- machine-to-machine communication

Academic Vocabulary
- response
- process

Graphic Organizer
Draw the section diagram. Use it to organize and write down information as you read.

Modes of Communication: 1. People to People, 2., 3., 4.

Go to glencoe.com to this book's OLC for a downloadable graphic organizer and more.

TECHNOLOGY STANDARDS
STL 1 Characteristics & Scope of Technology
STL 2 Core Concepts of Technology
STL 11 Design Process
STL 17 Information & Communication Technologies

ACADEMIC STANDARDS
English Language Arts
NCTE 4 Use written language to communicate effectively.
NCTE 12 Use language to accomplish individual purposes.

STL National Standards for Technological Literacy
NCTM National Council of Teachers of Mathematics
NCTE National Council of Teachers of English
NSES National Science Education Standards
NCSS National Council for the Social Studies

Evolving Modes of Communication

What progress have people made in their ability to communicate with each other?

Technology has given us new modes of communication. A **mode** is a way of doing something. Originally, "people-to-people" communication was the only mode. It is still the most basic mode of communication. Over time people have learned to create new and more powerful modes of communication. They gained the knowledge and skills needed to build complex communication devices and used the mode called "people-to-machines" communication. People also created graphic communication systems to transmit their messages using the printed word, which uses the mode

As You Read Connect Do you communicate with machines? Which ones?

196 Unit 3 Communication Technologies

Technology Standards Each chapter of this book is divided into sections. In the Reading Guide, there is a list of the STLs covered in that section.

Academic Standards

The English Language Arts, Mathematics, Science, and Social Studies standards are listed under the "Academic Standards" in the Reading Guide. You will practice these academic skills as you move through the chapter.

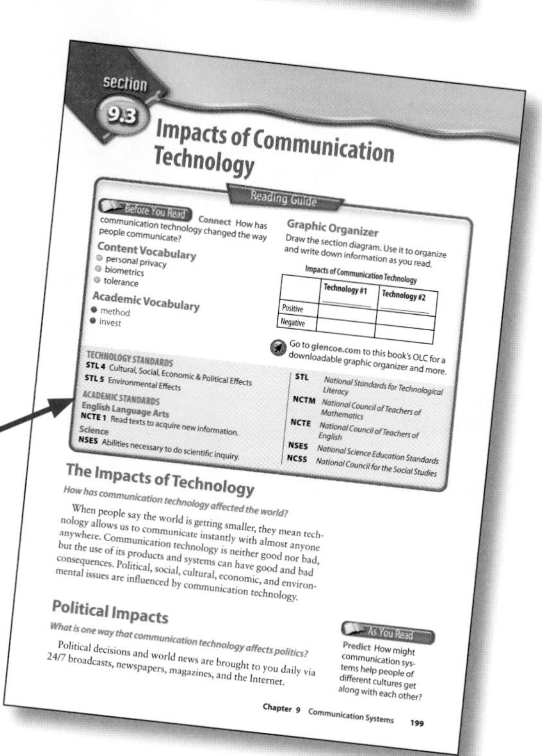

section 9.3

Impacts of Communication Technology

Reading Guide

Before You Read Connect How has communication technology changed the way people communicate?

Content Vocabulary
- personal privacy
- biometrics
- tolerance

Academic Vocabulary
- method
- invest

Graphic Organizer
Draw the section diagram. Use it to organize and write down information as you read.

Impacts of Communication Technology

	Technology #1	Technology #2
Positive		
Negative		

Go to glencoe.com to this book's OLC for a downloadable graphic organizer and more.

TECHNOLOGY STANDARDS
STL 4 Cultural, Social, Economic & Political Effects
STL 5 Environmental Effects

ACADEMIC STANDARDS
English Language Arts
NCTE 1 Read texts to acquire new information.
Science
NSES Abilities necessary to do scientific inquiry.

STL National Standards for Technological Literacy
NCTM National Council of Teachers of Mathematics
NCTE National Council of Teachers of English
NSES National Science Education Standards
NCSS National Council for the Social Studies

The Impacts of Technology

How has communication technology affected the world?

When people say the world is getting smaller, they mean technology allows us to communicate instantly with almost anyone anywhere. Communication technology is neither good nor bad, but the use of its products and systems can have good and bad consequences. Political, social, cultural, economic, and environmental issues are influenced by communication technology.

Political Impacts

What is one way that communication technology affects politics?

Political decisions and world news are brought to you daily via 24/7 broadcasts, newspapers, magazines, and the Internet.

As You Read Predict How might communication systems help people of different cultures get along with each other?

Chapter 9 Communication Systems 199

Academic Standards The Reading Guide provides a list of the Academic Standards covered in that section.

How to Access the OLC Student Center

Follow these steps to get to all the textbook resources at the *Introduction to Technology* Online Learning Center.

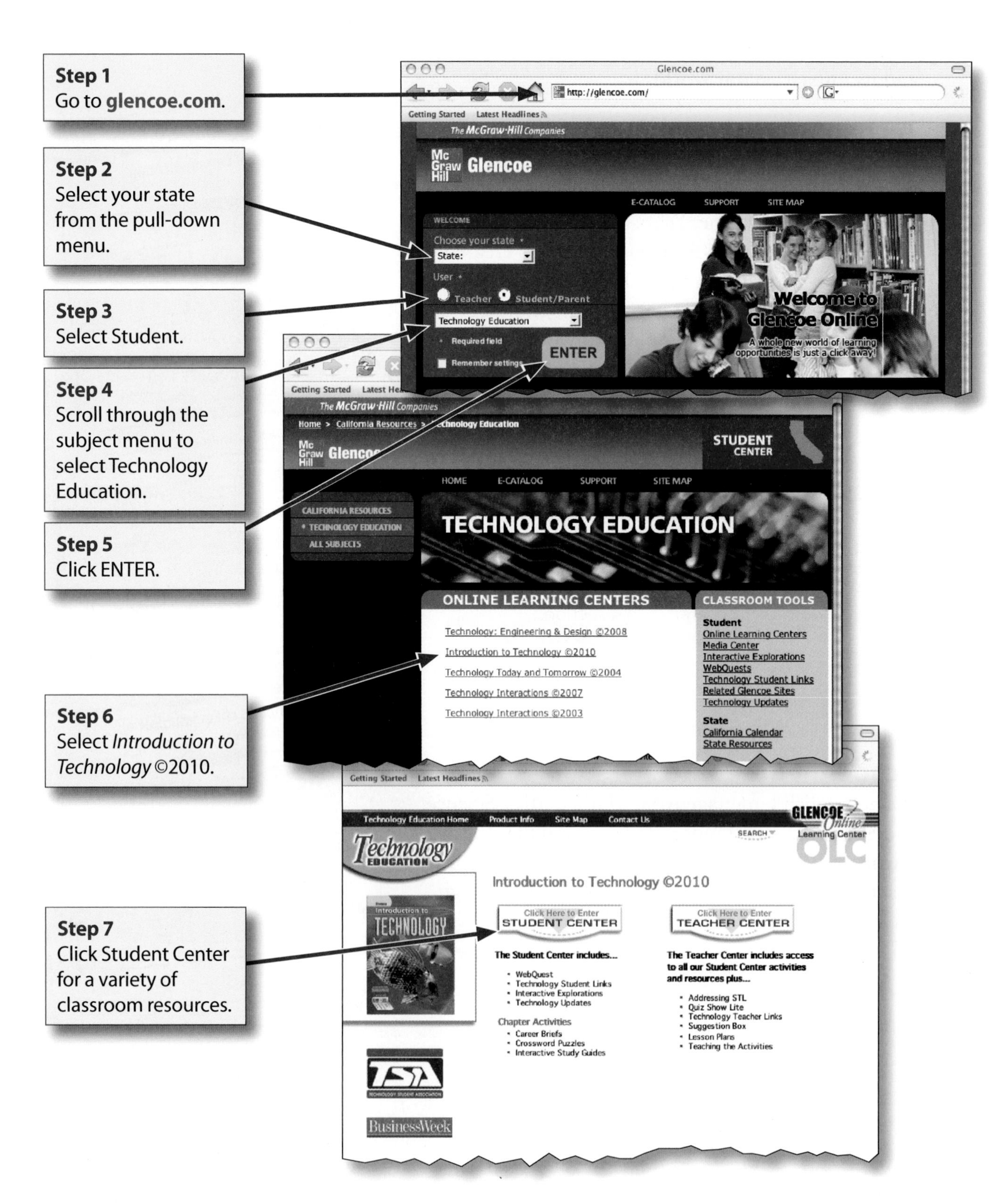

Treasure Hunt

Introduction to Technology **contains a wealth of information. Knowing where to find information in your textbook will help you get the most out of your study time. Come on, let's discover your textbook!**

1. How many chapters are in the book? How many units?
2. What part of the textbook will show you where printing is taught?
3. Where do you find the WebQuest Internet Project?
4. What topic is featured in the Unit 3 Technology Time Machine?
5. If you need help with one of the math applications, where would you look?
6. Where can you find the definitions of *gravure printing* and *hardware*?
7. What skill do you practice in Chapter 10, Section 10.2 Assessment?
8. Where do you find a Unit Thematic Project?

Reading Skills Handbook

Reading: What's in It for You?

What role does reading play in your life? The possibilities are countless. Are you on a sports team? Perhaps you like to read about the latest news and sports. Are you enrolled in an English class, an algebra class, or a science or technology class? Then your assignments require a lot of reading.

Improving or Fine-Tuning Your Reading Skills Will:

- Improve your grades
- Allow you to read faster and more efficiently
- Improve your study skills
- Help you remember more information accurately
- Improve your writing

The Reading Process

Good reading skills build on one another, overlap, and spiral around in much the same way that a winding staircase goes around and around while leading you to a higher place. This handbook is designed to help you find and use the tools you will need **before, during,** and **after** reading.

Strategies You Can Use

- Identify, understand, and learn new words.
- Understand why you read.
- Take a quick look at the whole text.
- Try to predict what you are about to read.
- Take breaks while you read and ask yourself questions about the text.
- Take notes.
- Keep thinking about what will come next.
- Summarize.

Vocabulary Development

Word identification and vocabulary skills are the building blocks of the reading and the writing process. By learning to use a variety of strategies to build your word skills and vocabulary, you will become a stronger reader.

Use Context to Determine Meaning

The best way to expand and extend your vocabulary is to read widely, listen carefully, and participate in a rich variety of discussions. When reading on your own, though, you can often figure out the meanings of new words by looking at their **context,** the other words and sentences that surround them.

Tips for Using Context

Look for clues such as:

A synonym or an explanation of the unknown word in the sentence:
*Elise's shop specialized in **millinery, or hats for women.***

A reference to what the word is or is not like:
*An **archaeologist, like a historian, deals with the past.***

A general topic associated with the word:
*The **cooking** teacher discussed the best way to braise meat.*

A description or action associated with the word:
*He used the **shovel** to **dig up** the garden.*

Predict a Possible Meaning

Another way to find the meaning of a word is to take the word apart. If you understand the meaning of the **base,** or **root,** part of a word, and also know the meanings of key syllables added either to the beginning or end of the base word, then you can usually figure out what the word means.

Word Origins Since Latin, Greek, and Anglo-Saxon roots are the basis for much of our English vocabulary, having some background in languages can be a useful vocabulary tool. For example, *astronomy* comes from the Greek root *astro,* which means "relating to the stars." *Stellar* also refers to stars, but its origin is Latin. Knowing root words in other languages can help you determine meanings, derivations, and spellings in English.

Prefixes and Suffixes A prefix is a word part that can be added to the beginning of a word. For example, the prefix *semi* means "half" or "partial," so *semicircle* means "half a circle." A suffix is a word part added to the end of a word. Adding a suffix can change a word's part of speech.

Using Dictionaries A dictionary provides the meaning or meanings of a word. Look at a dictionary entry to see what other information it provides.

Thesauruses and Specialized Reference Books A thesaurus provides synonyms and often antonyms. A synonym is a word that means the same thing as the word you are using. Check the exact definition of the listed words in a print or online dictionary before you use a thesaurus.

Glossaries Many textbooks contain condensed dictionaries that provide an alphabetical listing of words used in the text and their definitions.

Recognize Word Meanings across Subjects Have you learned a new word in one class, and then noticed it in your reading for other subjects? The word might not mean exactly the same thing in each class, but you can use the meaning you already know to help you understand what it means in another subject area. For example:

Math After multiplying the two numbers, explain how you found the **product.**

Science One **product** of photosynthesis is oxygen.

Economics The Gross National **Product (GNP)** is the total dollar value of goods and services produced by a nation.

Understanding What You Read

Reading comprehension means understanding—deriving meaning from—what you have read. Using a variety of strategies can help you improve your comprehension and make reading more interesting and more fun.

Read for a Reason

To get the greatest benefit from what you read, you should **establish a purpose for reading.** In school, you have many reasons for reading. Some of them are:

- To learn and understand new information
- To find specific information
- To review before a test
- To complete an assignment
- To prepare (research) before you write

As your reading skills improve, you will notice that you apply different strategies to fit the different purposes for reading. For example, if you are reading for entertainment, you might read quickly, but if you read to gather information or follow directions, you might read more slowly, take notes, construct a graphic organizer, or reread sections of text.

Draw on Personal Background

Drawing on personal background may also be called activating prior knowledge. Before you start reading a text, ask yourself questions like these:

- What have I heard or read about this topic?
- Do I have any personal experience relating to this topic?

Using a KWL Chart A KWL chart is a good device for organizing information you gather before, during, and after reading. In the first column, list what you already **know,** then list what you **want** to know in the middle column. Use the third column when you review and you assess what you **learned.** You can also add more columns to record places where you found information and places where you can look for more information.

K (What I already know)	W (What I want to know)	L (What I have learned)

Adjust Your Reading Speed Your reading speed is a key factor in how well you understand what you are reading. You will need to adjust your speed depending on your reading purpose.

Scanning means running your eyes quickly over the material to look for words or phrases. Scan when you need a specific piece of information.

Skimming means reading a passage quickly to find its main idea or to get an overview. Skim a text when you preview to determine what the material is about.

Reading for detail involves careful reading while paying attention to text structure and monitoring your understanding. Read for detail when you are learning concepts, following complicated directions, or preparing to analyze a text.

Techniques to Understand and Remember What You Read

Preview

Before beginning a selection, it is helpful to **preview** what you are about to read.

Previewing Strategies

- Read the title, headings, and subheadings of the selection.
- Look at the illustrations and notice how the text is organized.
- Skim the selection: Take a glance at the whole thing.
- Decide what the main idea might be.
- Predict what a selection will be about.

Predict

Have you ever read a mystery, decided who committed the crime, and then changed your mind as more clues were revealed? You were adjusting your predictions. Did you smile when you found out you guessed the murderer? You were verifying your predictions.

As you read, take educated guesses about story events and outcomes; that is, **make predictions** before and during reading. This will help you focus your attention on the text, and it will improve your understanding.

Determine the Main Idea

When you look for the **main idea**, you are looking for the most important statement in a text. Depending on what kind of text you are reading, the main idea can be located at the very beginning (news stories in newspaper or a magazine) or at the end (scientific research document). Ask yourself:

- What is each sentence about?
- Is there one sentence that is more important than all the others?
- What idea do details support or point out?

Taking Notes

Cornell Note-Taking System There are many methods for note taking. The **Cornell Note-Taking System** is a well-known method that can help you organize what you read. To the right is a note-taking chart based on the Cornell Note-Taking System.

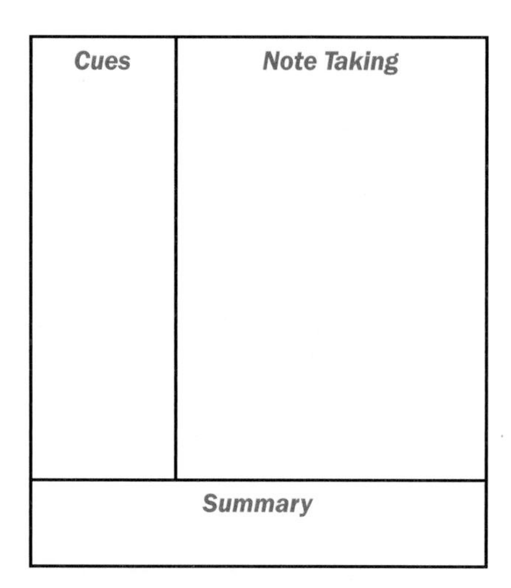

Graphic organizers Using a graphic organizer to retell content in a visual representation will help you remember and retain content. You might make a **chart** or **diagram,** organizing what you have read. Here are some examples of graphic organizers:

Venn diagrams: When mapping out a comparison-and-contrast text structure, you can use a Venn diagram. The outer portions of the circles will show how two characters, ideas, or items contrast, or are different, and the overlapping part will compare two things, or show how they are similar.

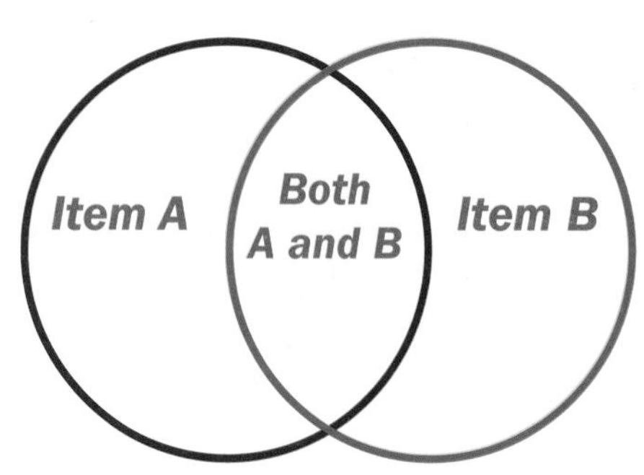

Flow charts: To help you track the sequence of events, or cause and effect, use a flow chart. Arrange ideas or events in their logical, sequential order. Then draw arrows between your ideas to indicate how one idea or event flows into another.

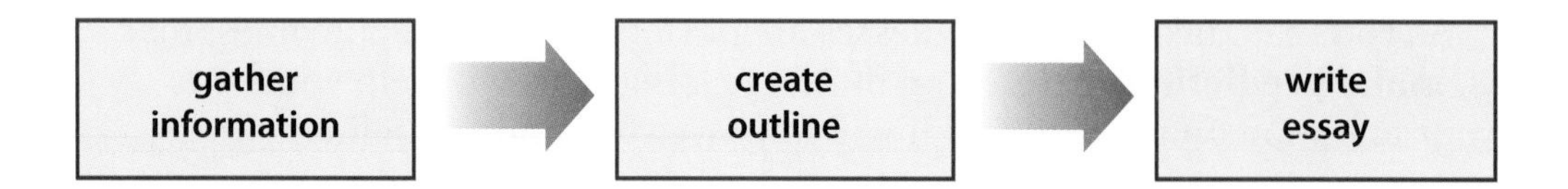

Visualize

Try to form a mental picture of scenes, characters, and events as you read. Use the details and descriptions the author gives you. If you can **visualize** what you read, it will be more interesting, and you will remember it better.

Question

Ask yourself questions about the text while you read. Ask yourself about the importance of the sentences, how they relate to one another, if you understand what you just read, and what you think is going to come next.

Clarify

If you feel you do not understand meaning (through questioning), try these techniques:

What to Do When You Do Not Understand

- **Reread confusing parts of the text.**
- **Diagram (chart) relationships between chunks of text, ideas, and sentences.**
- **Look up unfamiliar words.**
- **Talk out the text to yourself.**
- **Read the passage once more.**

Review

Take time to stop and review what you have read. Use your note-taking tools (graphic organizers or Cornell notes charts). Also, review and consider your KWL chart.

Monitor Your Comprehension

Continue to check your understanding by using the following two strategies:

Summarize Pause and tell yourself the main ideas of the text and the key supporting details. Try to answer the following questions: Who? What? When? Where? Why? How?

Paraphrase Pause, close the book, and try to retell what you have just read in your own words. It might help to pretend you are explaining the text to someone who has not read it and does not know the material.

Understanding Text Structure

Good writers do not just put together sentences and paragraphs, they organize their writing with a specific purpose in mind. That organization is called "text structure." When you understand and follow the structure of a text, it is easier to remember the information you are reading. There are many ways text may be structured. Watch for **signal words**. They will help you follow the text's organization (also, remember to use these techniques when you write).

Compare and Contrast

This structure shows similarities and differences between people, things, and ideas. This is often used to demonstrate that things that seem alike are really different, or vice versa.

Signal words: similarly, more, less, on the one hand / on the other hand, in contrast, but, however

Cause and Effect

Writers used the cause and effect structure to explore the reasons for something happening and to examine the results or consequences of events.

Signal words: so, because, as a result, therefore, for the following reasons

Problem and Solution

When writers organize text around the question "how?" they state a problem and suggest solutions.

Signal words: how, help, problem, obstruction, overcome, difficulty, need, attempt, have to, must

Sequence

Sequencing tells you the order in which to consider thoughts or facts. Examples of sequencing are:

Chronological order refers to the order in which events take place.

Signal words: first, next, then, finally

Spatial order describes the organization of things in space (to describe a room, for example).

Signal words: above, below, behind, next to

Order of importance lists things or thoughts from the most important to the least important (or the other way around).

Signal words: principal, central, main, important, fundamental

Reading for Meaning

It is important to think about what you are reading to get the most information out of a text, to understand the consequences of what the text says, to remember the content, and to form your own opinion about what the content means.

Interpret

Interpreting is asking yourself, "What is the writer really saying?" and then using what you already know to answer that question.

Infer

Writers do not always state exactly everything they want you to understand. By providing clues and details, they sometimes imply certain information. An **inference** involves using your reason and experience to develop the idea on your own, based on what an author implies or suggests. What is most important when drawing inferences is to be sure that you have accurately based your guesses on supporting details from the text. If you cannot point to a place in the selection to help back up your inference, you may need to rethink your guess.

Draw Conclusions

A conclusion is a general statement you can make and explain with reasoning or with supporting details from a text. If you read a story describing a sport in which five players bounce a ball and throw it through a high hoop, you may conclude that the sport is basketball.

Analyze

To understand persuasive nonfiction (a text that discusses facts and opinions to arrive at a conclusion), you need to analyze statements and examples to see if they support the main idea. To understand an informational text (a text, such as a textbook, that gives you information, not opinions), you need to keep track of how the ideas are organized to find the main points.

Hint: Use your graphic organizers and notes charts.

Distinguish Facts and Opinions

This is one of the most important reading skills you can learn. A fact is a statement that can be proven. An opinion is what the writer believes. A writer may support opinions with facts, but an opinion cannot be proven. For example:

Fact: California produces fruit and other agricultural products.

Opinion: California produces the best fruit and other agricultural products.

Evaluate

Would you take seriously an article on nuclear fission if you knew it was written by a comedic actor? If you need to rely on accurate information, you need to find out who wrote what you are reading and why. Where did the writer get information? Is the information one-sided? Can you verify the information?

Reading for Research

You will need to **read actively** in order to research a topic. You might also need to generate an interesting, relevant, and researchable **question** on your own and locate appropriate print and nonprint information from a wide variety of sources. Then you will need to **categorize** that information, evaluate it, and **organize** it in a new way in order to produce a research project for a specific audience. Finally, **draw conclusions** about your original research question. These conclusions may lead you to other areas for further inquiry.

Unit 1 The Nature of Technology

Chapters In This Unit:

Unit Thematic Project Preview

Redesigning an Invention

As part of this unit, you will learn about the concepts, tools, and materials of technology. Products are created through design and problem solving, using drawings and prototypes. Knowing about the nature of technology and its history can lead to new inventions and innovations.

As you read this unit, use this checklist to prepare for the project at the end of this unit:

PROJECT CHECKLIST

✓ Think about some simple tools or inventions you use every day.

✓ Ask your teacher what equipment you will need to do this lab.

✓ Find out where you can get an application for a patent.

WebQuest Internet Project

Go to **glencoe.com** to this book's Online Learning Center (OLC) to find the WebQuest activity for Unit 1. Begin by reading the Task. This WebQuest activity will help you learn about the different types of plastics used for technology and how they can be recycled.

Explore the Photo

The Wide World of Technology You probably know that computers, cell phones, and rockets are all technology. Some technologies depend on satellite technology. *How might satellite dishes relate to products you use?*

chapter 1

Why Study Technology?

Sections

What You'll Learn

- **Define** technology.
- **Identify** reasons for studying technology.
- **Explain** the advantages of being technologically literate.
- **Name** the workers who do technology.
- **Describe** how science, engineering, and technology are linked.
- **Explain** how teens have contributed to technology.
- **Discuss** how technology changes.
- **Describe** the influence of democracy on technology in the United States.

Explore the Photo

Build a Better Robot Technology helps build robots. Building small industrial robots is a lot of fun, and some students can enter robot competitions for prizes. *Besides getting the chance to work on fun projects, what are some other reasons for studying technology?*

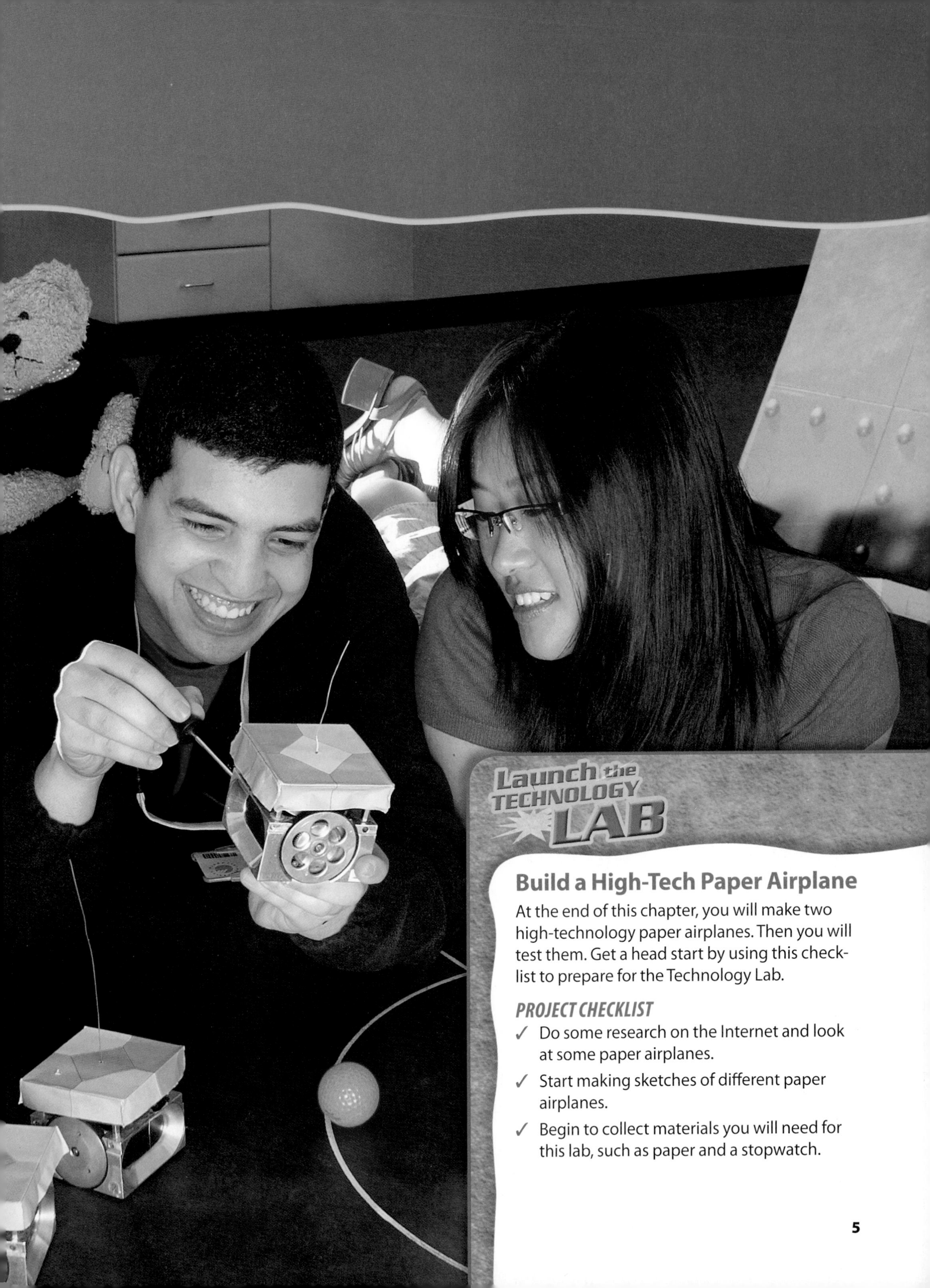

Launch the TECHNOLOGY LAB

Build a High-Tech Paper Airplane

At the end of this chapter, you will make two high-technology paper airplanes. Then you will test them. Get a head start by using this checklist to prepare for the Technology Lab.

PROJECT CHECKLIST

- ✓ Do some research on the Internet and look at some paper airplanes.
- ✓ Start making sketches of different paper airplanes.
- ✓ Begin to collect materials you will need for this lab, such as paper and a stopwatch.

section 1.1

Technology and You

Reading Guide

Preview How would you define the word *technology*?

Content Vocabulary

- technology
- technologically literate

Academic Vocabulary

You will see these words in your reading and on your tests. Find their meanings at the back of this book.

- project
- device

Graphic Organizer

Draw the section diagram. Use it to organize and write down information as you read.

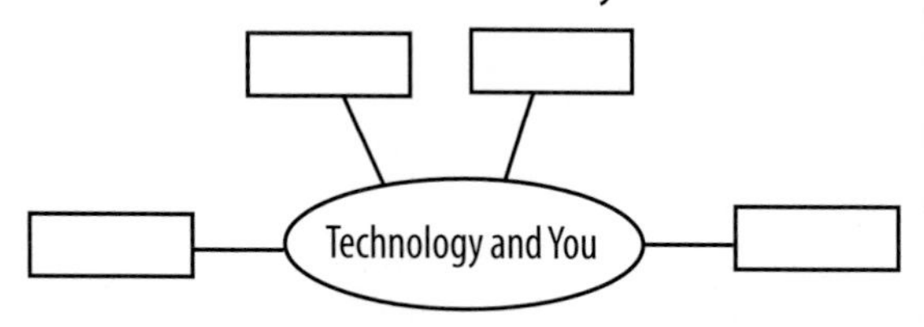

Go to **glencoe.com** to this book's OLC for a downloadable graphic organizer and more.

TECHNOLOGY STANDARDS

STL 4 Cultural, Social, Economic & Political Effects

STL 6 Role of Society

STL 17 Information & Communication Technologies

ACADEMIC STANDARDS

Science

NSES F Science and technology in society

Social Studies

NCSS 8 Science, Technology, and Society

STL *National Standards for Technological Literacy*

NCTM *National Council of Teachers of Mathematics*

NCTE *National Council of Teachers of English*

NSES *National Science Education Standards*

NCSS *National Council for the Social Studies*

Enjoying Technology

***What is the origin of the word* technology?**

Technology means different things to different people. A physician might think of technology as a way to produce a new medicine. To a space engineer, it might mean making better rocket engines. Technology is so widespread that it is part of everyone's life.

The word *technology* comes from the Greek word *techne*, which means "art." You might think that art means only paintings or sculpture. But the Greeks believed an artist could make useful products from natural materials such as trees, rocks, and plants.

In this book you will learn *about* technology, inventing, and how things work. You will also learn to *do* technology—and you may learn to *enjoy* technology.

As You Read

Connect What kinds of technology do you use every day?

Technology and Society

What are the different kinds of technology?

Technology is the practical use of human knowledge to extend human abilities and to satisfy human needs and wants. Technology provides us with most things we use in our society. It can be broken down into six general groups or types of technologies.

1. **Energy and power**—This type of technology deals with the electricity and power that makes things run. We can generate electricity and power from many sources today.
2. **Biotechnology**—This type of technology is based on biology, or the science of living things. Breakfast cereals, medicines, and bionic arms and legs are products of biotechnology.
3. **Communication**—This type of technology includes the use of cell phones, iPods, DVD recorders, and similar items.
4. **Manufacturing**—This type of technology includes items that were made or processed in a factory. Practically everything you use is a product of manufacturing technology.
5. **Construction**—This type of technology deals with building houses, bridges, skyscrapers, and other **projects** such as playgrounds and monuments.
6. **Transportation**—This type of technology deals with moving people or products in cars, ships, airplanes, trains, or other vehicles.

All of this technology has improved over the years. When your grandparents were your age, they may have watched black-and-white television. Today you might watch your favorite shows on the Internet or on a high-definition television. Think of how cell phones have changed. Houses have also changed, and so have the foods and entertainment that you enjoy. What other changes in technology can you identify?

Define What is technology?

Seeing the Light Today technologists use lasers in many fields that involve communication technology, manufacturing technology, and biotechnology. *Why do you think this student is wearing goggles?*

Tech Stars

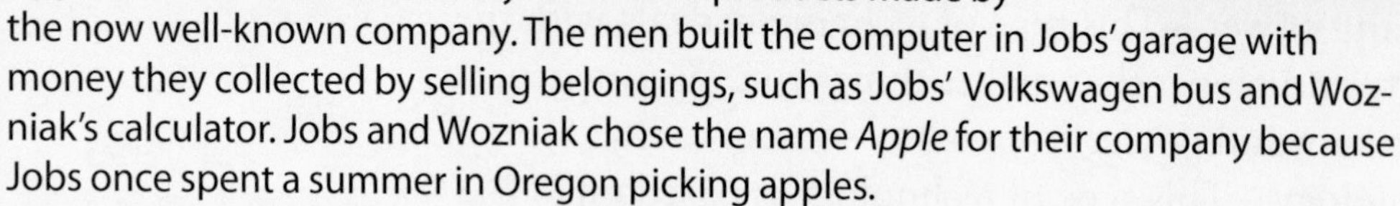

Steve Jobs and Steve Wozniak

Founders of Apple Computers

In 1977, Steve Jobs and Steve Wozniak designed and built the world's first commercially successful personal computer, the Apple II. It was the first of many innovative products made by the now well-known company. The men built the computer in Jobs' garage with money they collected by selling belongings, such as Jobs' Volkswagen bus and Wozniak's calculator. Jobs and Wozniak chose the name *Apple* for their company because Jobs once spent a summer in Oregon picking apples.

iPod® Nation By 1985, Apple Computers had become a large, successful company. Jobs and Wozniak left the company to pursue other interests. Jobs rejoined Apple in 1996 to rebuild the brand. Apple is now a leader in the digital music world with iPod and iTunes®.

English Language Arts/Writing *Write a short essay identifying types of technology you think Apple uses, and explain your choices.*

Go to **glencoe.com** to this book's OLC to learn about young innovators in technology.

Why We Study Technology

What school subjects are related to technology?

Why should you study technology? That question is easy to answer. Technology is fun, rewarding, and exciting. It is fun because you get to work with your hands. It is rewarding because you get to see the results of your work. Each day brings new ideas and new challenges, which make technology exciting.

Studying technology will also help you develop your problem-solving skills. You can learn to identify a problem and come up with a solution.

You will also find that technology is related to other subjects that you study in school such as:

- Mathematics
- Science
- Social studies
- English language arts
- Art

You might enjoy all of your classes more after you begin to see the relationships between technology and other subjects.

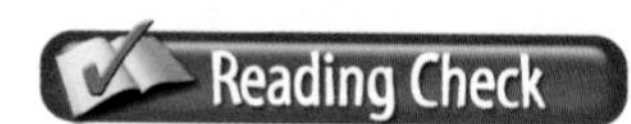

List What are some reasons for studying technology?

Being Technologically Literate

Do you know what "technologically literate" means?

Technology is often in the news. A journalist might report on a particular electrical power plant, a food additive, or a safety device on an automobile. It is important that you understand the importance of technology. Every day in many ways, technology affects the lives of people around the world.

For example, automobile air bags have saved many lives in collisions. However, they inflate so quickly that they have caused injury and death in some cases. As a result, the federal government allows car owners to install an on/off switch for the air bags. Do you think the government should let people do this? To answer this question, you first need to know something about the technology being discussed.

Does working on a small engine sound difficult to you? With a basic understanding of technology, it might not be as hard as you think. Being **technologically literate** means understanding technology and feeling comfortable with it. Sometimes there is no one correct answer to a problem. You need to think about and evaluate each situation, and then make a decision. When you can do this, you will be technologically literate.

A Green World

Workplaces, businesses, and public places are adopting sustainable and "green" practices that are friendly to our environment. Laws are requiring this, and many consumers and employees are demanding it. Saving energy and using resources wisely can save the environment and money, too.

Try This Think of five ways to make your school "green." For example, use energy-saving light bulbs.

section 1.1 assessment

After You Read Self-Check

1. Name two communication products.
2. Name two transportation products.
3. Explain what it means to be technologically literate.

Think

4. Discuss how you think cars today are different from cars made 50 years ago. How did society's demands and values influence these differences?

Practice Academic Skills

English Language Arts/Writing

5. Walter Chrysler, the man who started the Chrysler car company, once said, "Someday I'd like to show a poet how it feels to design and build a railroad locomotive." Imagine designing a new technology product that changes the world. Imagine inventing that new product, and write a poem about how that would make you feel.

STEM Mathematics

6. Kevin did research for a paper at the library. It took him 25 minutes to drive to the library, 35 minutes to determine what books would help him, 15 minutes to find the books, and 2 hours to collect the information. For a similar paper, he did his research online. It took him 1 hour and 50 minutes to gather the information he needed. How much time did he save by doing research online?

Math Concept **Time** When you are adding or subtracting time, work with minutes and hours separately.

1. Total the individual times to determine how many minutes it took.
2. Remember that 60 minutes equals one hour.

For help, go to **glencoe.com** to this book's OLC and find the Math Handbook.

section 1.2

Making Technology Happen

Reading Guide

Preview How does technology involve people?

Content Vocabulary

- science
- engineering

Academic Vocabulary

- automatic
- survey

Graphic Organizer

Draw the section diagram. Use it to organize and write down information as you read.

Science, Technology, and Engineering

1. Science	2. Technology	3. Engineering
	Makes things happen	

Go to **glencoe.com** to this book's OLC for a downloadable graphic organizer and more.

TECHNOLOGY STANDARDS

STL 3 Relationships & Connections

STL 7 Influence on History

STL 19 Manufacturing Technologies

ACADEMIC STANDARDS

Science

NSES E Understandings about science and technology

Social Studies

NCSS 4 Individual Development and Identity

STL *National Standards for Technological Literacy*

NCTM *National Council of Teachers of Mathematics*

NCTE *National Council of Teachers of English*

NSES *National Science Education Standards*

NCSS *National Council for the Social Studies*

The People Who Do Technology

Who are the people who create and use technology?

More than anything else, technology involves people. People make things happen. People apply the technology. Long ago workers in technology called themselves "artisans." That word is based on the word *art*, just like *artist*. An artisan is a highly skilled worker or craftsperson. Two modern names for people who work in technology are *technician* and *technologist*. You can use either word to describe a person who works in technology.

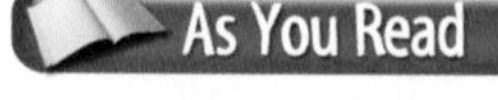

Predict Think of different words that might be related to the word *technology*.

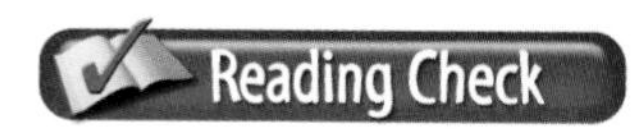

Identify What is another word for technologist?

Science, Technology, and Engineering

How are the areas of science, technology, and engineering related?

You may notice that the words *science* and *technology* are often used together. Your dictionary probably uses the word *science* in the definition of the word *technology*. Although science and technology are related, they are not the same. **Science** explains *how* things happen. Technology *makes* things happen.

Scientists helped to create electronic microchips. Technologists used those microchips to make digital cameras, automatic sprinklers, MP3 players, and other electronic devices. Scientists discovered lasers before technologists built laser-operated medical equipment.

Engineering

Another important profession is **engineering**. It often fits between science and technology. Using their knowledge of science and mathematics, engineers determine *how to make* things. For example, chemical engineers who work with mechanical engineers design machines that produce plastics and other materials. Technologists build the products that the engineers design.

Technology Today and Tomorrow

In every chapter, you will discover exciting innovations in technology today. The technology of tomorrow may cause important changes in our lives. *What kind of new inventions can you imagine for communication of the future?*

Go to **glencoe.com** to this book's OLC for answers and to learn more about the technology of tomorrow.

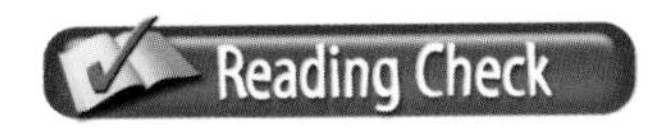

Distinguish What is the difference between science and engineering?

Teens and Technology

How have teenagers contributed to the development of technology?

Believe it or not, teens have made important contributions to technology. Have you ever heard of George Westinghouse? He started a company that still uses his last name. He is known for inventing improved brakes for trains. However, when he was only 19 years old, Westinghouse patented a new type of steam engine. It was not successful, but he was on his way.

A Young Technologist As a young man, George Washington surveyed millions of acres of land in Virginia. He used a tool called a circumference set. *Why do you think a surveyor is a technologist?*

Academic Connections
Social Studies

An Inventive President U.S. presidents are known for many things. One president was an inventor who patented his invention. A patent is a government document granting a person the right to produce or sell his or her invention. No one else may copy it.

Apply Go to the Internet to research past presidents. Identify the president who was granted a patent and name his invention.

George Washington, the first president of the United States, was a self-taught surveyor long before he became a military leader in the Revolutionary War. During the 1700s, Washington assisted in surveying 5 million acres for the largest landowner in the state of Virginia. He also helped plan the city of Alexandria, Virginia. He was appointed Official Surveyor of Culpeper County in Virginia when he was only 17 years old.

At age 14, Elmer Sperry invented a swiveling headlamp for trains so that an engineer could see around curves while the train was traveling along the railroad tracks. Although the headlamp was not successful, the gyrocompass he later invented was remarkable. Sperry's gyrocompass is still used in all ship and airplane guidance systems.

Good inventions are still being created by young people today. Students at Hampshire College in Massachusetts created the "Grease Car," which turns used cooking oil into fuel. Other students invented a hand braking system for wheelchairs and a scooter-bicycle combination.

Would you like to be an inventor? Ask your teacher for more information about organizations like the National Collegiate Inventors and Innovators Alliance as well as other groups dedicated to young innovators.

section 1.2 assessment

After You Read Self-Check

1. Name two words that describe a person who works in technology.
2. Explain the difference between science and technology.
3. Identify who designs a gear—a scientist, an engineer, or a technologist. Explain why.

Think

4. Find an engineer who works in your community. Briefly describe what this person does for a living.

Practice Academic Skills

STEM Science

5. Manufacture a useful product out of raw materials. Present the product to your class, explaining the materials you used to make it and its purpose. For example:
   - Tree branches to make small wooden items
   - Rocks to make jewelry or other decorative items
   - Dyed sand to make paintings, as used by Native Americans
   - Grass or straw to make woven hot pads

STEM Mathematics

6. Roberta is designing a car that will run on solar energy. The flywheel of the car needs to have a circumference of 62.8 inches. What is the diameter of the flywheel?

Math Concept **Determining Measurement** When working with geometric shapes, be sure to use the correct formulas.

1. Remember, the circumference of a circle is equal to its diameter times *pi*.
2. Use 3.14 to represent *pi*.

For help, go to **glencoe.com** to this book's OLC and find the Math Handbook.

section

1.3 How Technology Changes

Reading Guide

Before You Read **Preview** How has technology changed to meet people's needs?

Content Vocabulary

- nanotechnology
- machine tool

Academic Vocabulary

- predict
- tradition

Graphic Organizer

Draw the section diagram. Use it to organize and write down information as you read.

U.S. Technology Products

1. ______	4. ______
2. ______	5. ______
3. ______	6. ______

Go to **glencoe.com** to this book's OLC for a downloadable graphic organizer and more.

TECHNOLOGY STANDARDS

STL 3 Relationships & Connections
STL 4 Cultural, Social, Economic & Political Effects
STL 6 Role of Society
STL 7 Influence on History

ACADEMIC STANDARDS

Science

NSES 8 Content Standard G History of Science

English Language Arts

NCTE 1 Read text to acquire new information.

STL *National Standards for Technological Literacy*
NCTM *National Council of Teachers of Mathematics*
NCTE *National Council of Teachers of English*
NSES *National Science Education Standards*
NCSS *National Council for the Social Studies*

Technology and History

What examples of today's technology are based on past technology?

Throughout history, technology has helped change societies and cultures. It has influenced politics and economies. In turn, technology itself has been influenced by changes in civilization. See **Figure 1.1** on page 14.

As You Read

Analyze How does technology of the past lead to technology of the future?

Building on the Past

In everything we do, we build on the efforts of people who came before us. Isaac Newton was a famous British scientist during the 1700s who investigated the motion of the planets.

Figure 1.1 The History of Technology

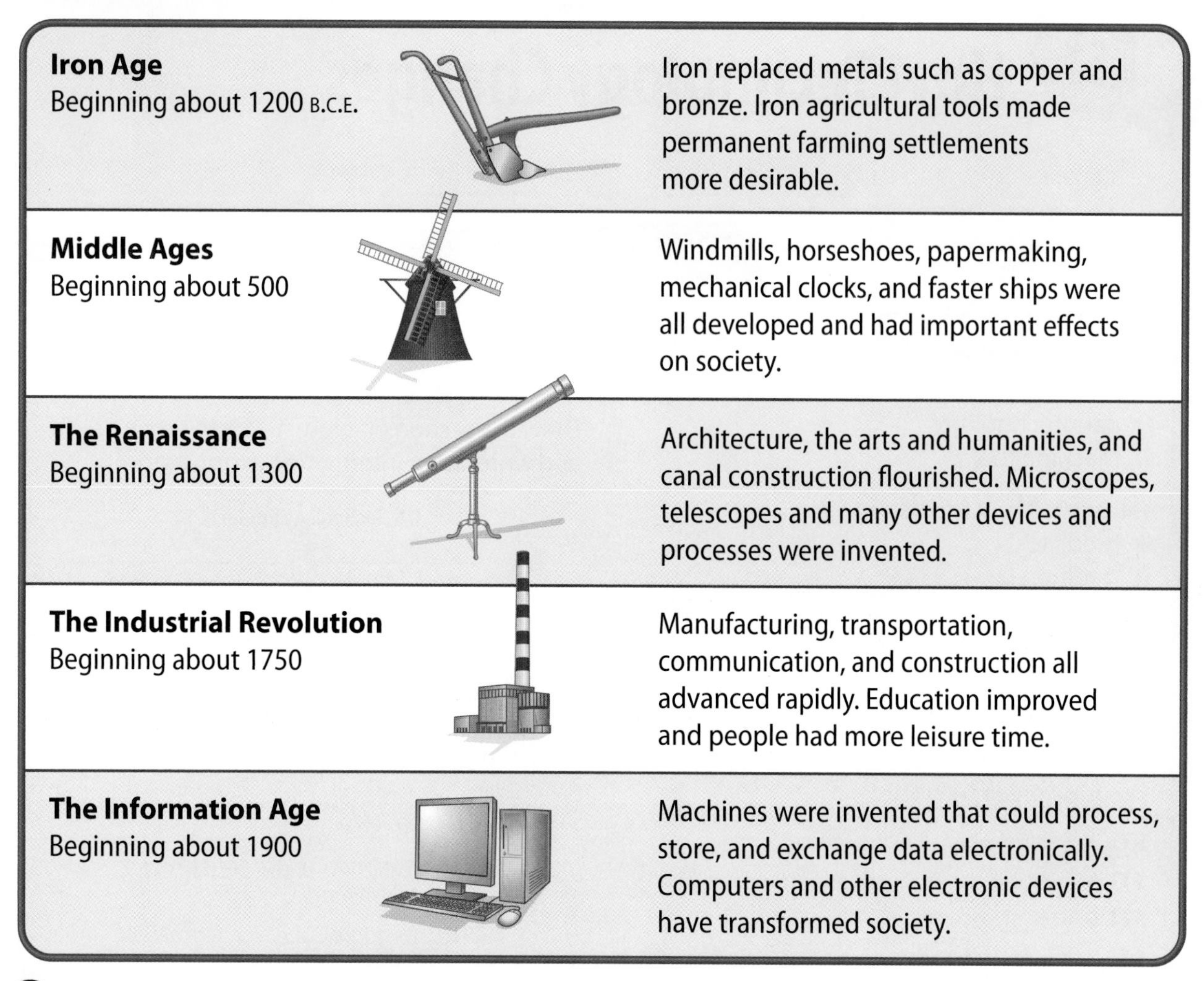

Through the Ages Technology has been evolving since the beginning of time. *What might be some more examples of Information Age technology?*

Newton said, "If I have seen further [than others], it is by standing upon the shoulders of giants." He meant that his accomplishments were based on the earlier work of other people. In technology, those people are the artisans, technicians, engineers, and scientists who came before us.

The first Boeing 707 airplanes were designed by people who used slide rules to make calculations. Today's airplanes are designed by people who use pocket calculators and computers. Technology moves forward by adapting, so that each new product is an improvement over an existing product.

The Evolution of Technology

Technology is continually evolving, which means it is changing and growing. Thomas Edison invented the phonograph, but he knew nothing about tape recordings. Compact disc recordings replaced tapes. Digital files are replacing compact discs.

For example, a new technology you might have heard about is **nanotechnology**. It is the science of working with the atoms or molecules of materials to develop very small machines. The term comes from the word *nano*, which means "one billionth." Years ago a famous physicist, Richard Feynman, predicted that we would be able to build a machine so small that it would be the size of just a few thousand atoms. That is very small. His prediction came true. Some tiny machines have parts so small that you need a microscope to see them!

Technology is changing more rapidly today than ever before. See **Figure 1.2** on page 16. One reason for this is better communication. Centuries ago news traveled slowly. Today the Internet allows information to travel around the world in seconds.

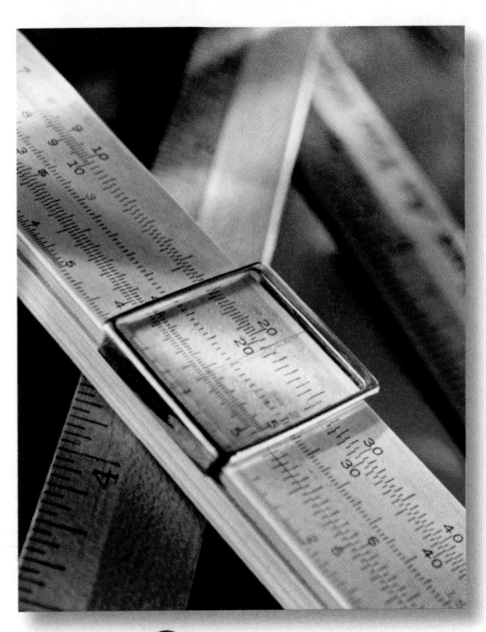

Changing Rules The engineers who built the first Boeing 707 airplanes used slide rules like these to make calculations. Today's engineers use computers and small calculators. *Is this an example of evolution in technology? Why or why not?*

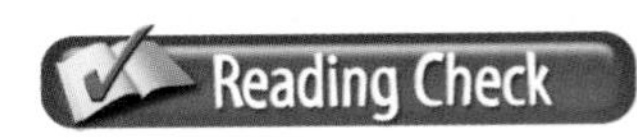

Reading Check **Summarize** What are some ways in which technology changes?

Technology in the United States

What happened when technology met democracy?

Since the 1700s, the United States has developed a reputation as a place where an intelligent and energetic person could be successful. Americans have developed new ideas about work, community, and success. Under democracy, Americans are basically free to try different ways of doing things. This system has encouraged technical advances and new businesses.

Ethics in Action

Making the Grade Online

Many teachers are now posting grades on school Web sites. Parents seem to like the idea. Online grades make it easier for parents and students to keep track of progress. They can check the students' most recent grades at work or at home.

"A" for Effort But online grading programs have drawbacks. One problem is hackers. They might erase or change grades. Another problem is privacy. If your grades are posted online, other students might be able to see them.

English Language Arts/Writing

Do Unto Others Yuri's teacher, Mr. Santiago, posts grades online. To protect the class's privacy, Mr. Santiago uses ID numbers instead of names.

1. By accident, Yuri learns another student's ID number. On the Web site, Yuri sees that the other student got a low grade.
2. What should Yuri do and why? Write your answer in one paragraph.

Figure 1.2 What's Next?

Edison's cylindrical records adapted to . . .	Flat records adapted to . . .	Cassette tapes adapted to . . .	Compact disc recordings adapted to . . .	MP3 players

Changing Tunes Voice and music recording media have evolved from Edison's records to today's MP3s. *What kind of recording technology do you think will be popular ten years from now?*

After the United States began manufacturing machine tools, Americans developed many other technological products. **Machine tools** are machines for shaping and finishing metals and other materials. These products include electronic computers, industrial robots, liquid-fueled rockets, reliable suspension bridges, photocopy machines, diesel engines for locomotives, electronic television, the metal-framed skyscraper, and the practical helicopter.

The heritage of American technology is very rich. You have benefited from and have inherited a powerful technological tradition. You can help to continue that tradition as you study and use technology.

section 1.3 assessment

After You Read Self-Check

1. Identify Isaac Newton.
2. Name the type of product the United States used to first enter the arena of technology.
3. List five products that were originally developed in the United States.

Think

4. Give an example from your own life of how you have built on the efforts of someone else.

Practice Academic Skills

English Language Arts/Writing

5. Choose and investigate an interesting career in technology. Write a short essay describing what you like about it.

Social Studies

6. Technology has played an important role in society throughout history to make things easier, safer, and more efficient. Choose a technology product you use today. Use the Internet and/or your library and research its history. Find out when and why it was invented. What was used before it was invented? Write a few paragraphs describing what you find.

Exploring Careers in Technology

Aaron Stewart-Ahn

FILMMAKER

Q: *What do you do?*

A: I am a filmmaker who works on music videos, documentaries on musicians, and now commercials. I hope to make feature films soon.

Q: *What kind of training and education did you need to get this job?*

A: I attended a post-graduate film school. However, today's filmmaking technology has become very accessible to everyone. You can buy enough equipment to become a filmmaker for less than the cost of attending an expensive film school. I suggest young filmmakers get an education in the liberal arts—literature, philosophy, history—so they can develop stories and ideas to share.

Q: *What do you like about your job?*

A: I like everything about the process of filmmaking. I like working with other creative people, the unexpected happy accidents, the way film involves all the performing arts and techniques. But it takes hard work to get the next job, and you must sometimes work long hours.

Q: *How did you get interested in your job?*

A: My parents would take me to movies at a very young age. Then I started to play with toys by lining up action figures the way a director sets up a shot. My mother told me what a director did when I was five. I said, "I want to do that." Despite the ups and downs of my life, I never gave up.

English Language Arts/Writing

Script Writing Write a script for a short film or just a scene in a film. Write dialogue (talking) and describe scenery, and any action.

1. Using a word processor, write a script including dialogue, action, camera direction, etc.
2. Make a "storyboard" of your script. This means: Draw on paper or use a graphics program to show how you picture your script.
3. Display your storyboard and read your script to the class. Or if you have a presentation software program, combine your script and images as a slideshow to show the class.

Go to **glencoe.com** to this book's OLC to learn more about this career.

Real-World Skills

Writing, speaking, listening, problem-solving

Academics and Education

English language arts, social studies, mathematics

Career Outlook

Growth as fast as average for the next ten years

Source: *Occupational Outlook Handbook*

chapter 1 Review and Assessment

Chapter Summary

Section 1.1 Technology is the practical use of human knowledge to extend human abilities and to satisfy human needs and wants. It involves turning natural items into useful products. We study technology because it is fun, rewarding, and exciting. If you are technologically literate, you are comfortable with technology. You can evaluate each situation and make informed decisions.

Section 1.2 The areas of science, technology, and engineering are related but different. Scientists try to explain how things happen. Engineers figure out how to make things. Technologists make things by operating machines and assembling parts. All of these people work together to create and produce the products we need. Teenagers can also contribute to the development of technology.

Section 1.3 We build on the efforts of people who came before us. Technology advances by adapting, so each new product is an improvement over an existing one. New inventions are almost always based on earlier accomplishments. One reason technology has thrived in the United States is because people feel free under democracy to try different ways of doing things.

Review Content Vocabulary and Academic Vocabulary

1. On a sheet of paper, use each of these terms and words in a written sentence.

Content Vocabulary	Academic Vocabulary
○ technology	● project
○ technologically literate	● device
○ science	● automatic
○ engineering	● survey
○ nanotechnology	● predict
○ machine tool	● tradition

Review Key Concepts

2. **Define** technology.
3. **List** reasons for studying technology.
4. **Explain** the advantages of being technologically literate.
5. **Identify** the workers who do technology.
6. **Discuss** the connection between science, engineering, and technology.
7. **Explain** how teens have contributed to technology.
8. **Explain** how technology changes.
9. **Describe** how democracy in the United States has influenced technology.
10. **Identify** the technology you use in your life.

Real-World Skills

11. Observation Practice your observation skills by looking for technology at home or in the classroom. Write a summary of what you see, describing why the technology is needed and how your life would be different without it.

STEM Technology Skill

12. Spreadsheets Many companies use spreadsheet software to organize data or information. Spreadsheets make comparing data fast and easy.

a. Research the cost of a product at several stores. Use spreadsheet software and input what you find. Include information like the price, cost of transportation, and mailing costs.

b. Compare the costs. What is the least expensive option? Consider all costs.

Academic Skills

Social Studies

13. Write a report describing three products developed in the United States. Discuss why each product was developed at the time in this country. Discuss political, social, and economic factors at the time.

STEM Mathematics

14. Clarence rides his bicycle to school every day. The diameter of his front tire is 24 inches. His school is 2,500 feet away. How many times will his front wheel rotate on his way to school?

Math Concept **Measurement** The distance around a circle is called the "circumference." When a wheel rotates, it covers a distance equal to its circumference as it moves on the ground. The circumference of a circle is equal to its diameter multiplied by *pi*, which is 3.14.

WINNING EVENTS

Student Representative

Situation You and three team members have been elected student representatives for your school. You must present arguments to your school board about two issues:

1. Access and use of cell phones by students during school hours
2. Access and use of the Internet by students

Activity Work together to prepare arguments for and against the use of these technologies. Present to your classmates. Two members will argue pros, and two will argue cons.

Evaluation You will be evaluated on how well you meet these performance indicators:

- Knowledge of the subject
- Quality of your argument
- Stage presence—quality of voice, poise

Go to **glencoe.com** to this book's OLC for information about TSA events.

Standardized Test Practice

Directions Choose the letter of the best answer. Write the letter on a separate piece of paper.

1. What is the definition of the term *technologically literate*?
   - **A** Reading about technology
   - **B** Being able to read manuals
   - **C** The practical use of human knowledge to extend ability
   - **D** Being informed about technology and feeling comfortable using it

2. Science makes things happen, and technology explains how things happen.
   T **F**

Test-Taking Tip When you are taking a test, do not rush yourself. Read the entire question and look for key words. Focus on your work.

chapter 1

TECHNOLOGY LAB

Build a High-Tech Paper Airplane

You probably already know how to fold a simple paper airplane. The ordinary pointed-nose style has been around for a long time. However, you may not have had the opportunity to fold a high-technology paper airplane (HTPA). An HTPA requires careful planning and folding.

In 1967, *Scientific American* magazine held its first International Paper Airplane Competition. There were 12,000 entries from 28 countries. This lab is based on activities from the competition.

Tools and Materials

- ✓ Paper
- ✓ Pencil
- ✓ Scissors
- ✓ Colored markers
- ✓ Ruler
- ✓ Yardstick, meter-stick, or tape measure
- ✓ Stopwatch
- ✓ Paper clips
- ✓ Tape

Set Your Goal

Your goal is to make two different HTPAs, and then find out which one flies farther and stays up longer. The ordinary pointed-nose style flies about 15 feet and can stay up for about four seconds. One HTPA winner in the *Scientific American* competition stayed in the air for 10.2 seconds. The winner for distance flew 91 feet. Different conditions can affect a paper airplane's abilities. For example, a breeze from an open window or a heating vent could help or hurt a flight.

Know the Criteria and Constraints

In this lab, you will use a simple *systems technique.* A systems technique breaks down a complex project into basic elements. Here are the basic elements for each HTPA construction:

1. Select an HTPA design.
2. Draw the plans.
3. Construct the HTPA.
4. Operate and fly the HTPA.
5. Collect flight information.
6. Evaluate the flight information.

F4F Wildcat

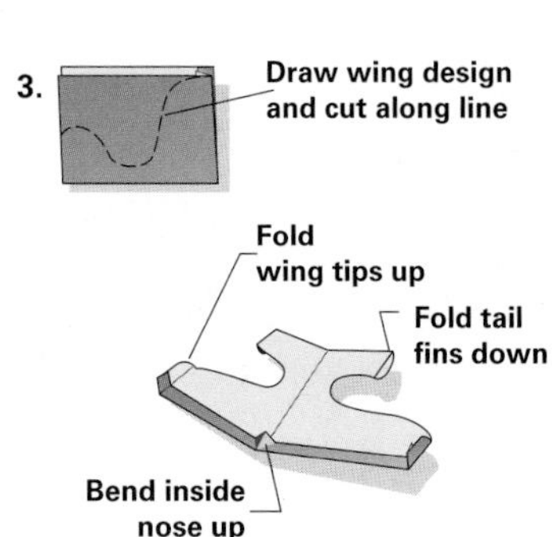

Design Your Project

Follow these steps to design your project and complete this lab.

1. Select a paper airplane design from the three plans in your textbook.
2. Carefully draw the plans for your design on a sheet of paper.

3. Fold the airplane according to the plans. Fold sharp edges.
4. Decorate your airplane using colored markers so that you can easily identify it.
5. Fly your HTPA three times.
6. Use the stopwatch to measure the length of each flight in seconds.
7. Use a measuring device such as a tape measure to determine the straight-line distance from the point where the HTPA was launched to the point where it stopped.
8. If your plane does not fly well at first, try placing a very small weight at the nose. You can use a piece of tape or small paper clip for a weight.
9. Repeat the preceding steps with another design.

Mark VB Spitfire

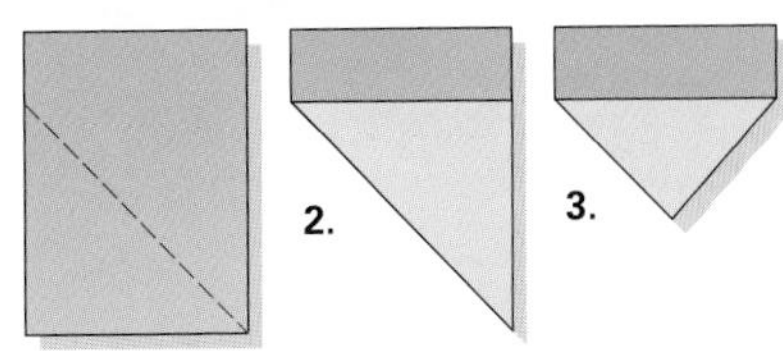

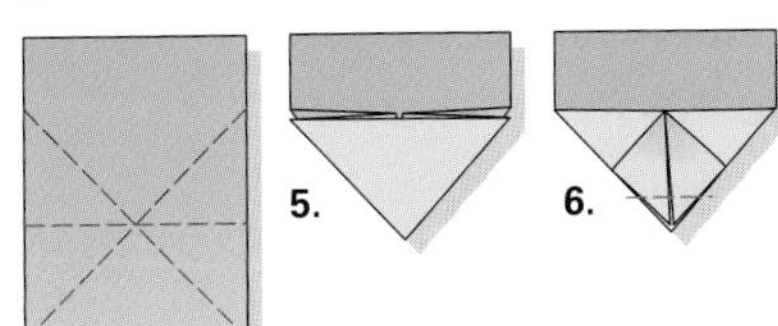

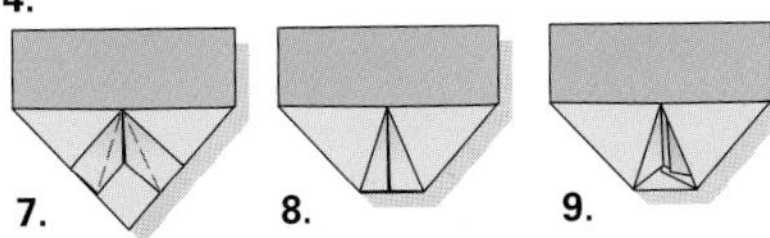

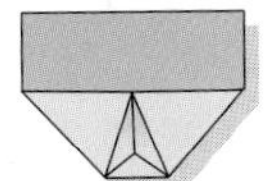

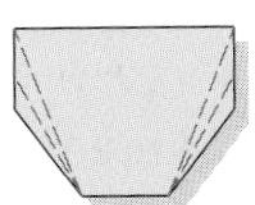

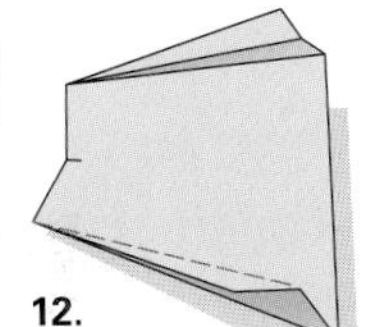

Evaluate Your Results

After you create, fly, and measure two HTPAs, answer these questions on a separate piece of paper.

1. Which HTPA flew farther and stayed up longer? Why?
2. Which HTPA was more fun to fly? Why?

P47 Thunderbolt

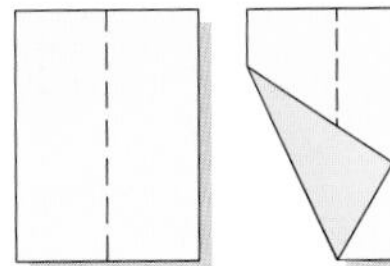

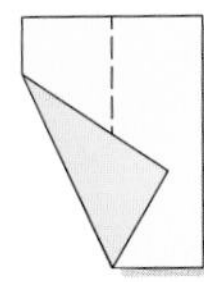

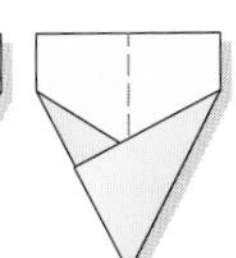

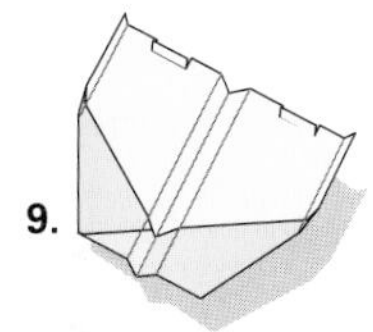

Academic Skills Required to Complete Lab				
Tasks	**English Language Arts**	**Math**	**Science**	**Social Studies**
Research performance of the different planes you create.	✓	✓	✓	
Gather data of time in air and distance planes fly.	✓	✓	✓	
Write results and evaluation.	✓	✓	✓	
Create presentation.	✓			
Present process, results, and evaluation to the class.	✓			

chapter 2

Concepts of Technology

Sections

What You'll Learn

- **Identify** the seven technology resources.
- **Explain** how skills and creativity lead to new inventions.
- **Describe** the six simple machines.
- **Explain** systems and subsystems.
- **Discuss** the difference between open- and closed-loop systems.
- **Explain** how systems relate to technology.
- **Describe** how requirements affect the development of products and systems.
- **Define** criteria and constraints, optimization, and maintenance and control.
- **Identify** tradeoffs during product development.
- **Explain** the positive and negative effects of technology.

Explore the Photo

Early Technology Our ancestors used tools and technology that were innovations during their time. *Would you be able to survive in their time by using the resources that they used? Why or why not?*

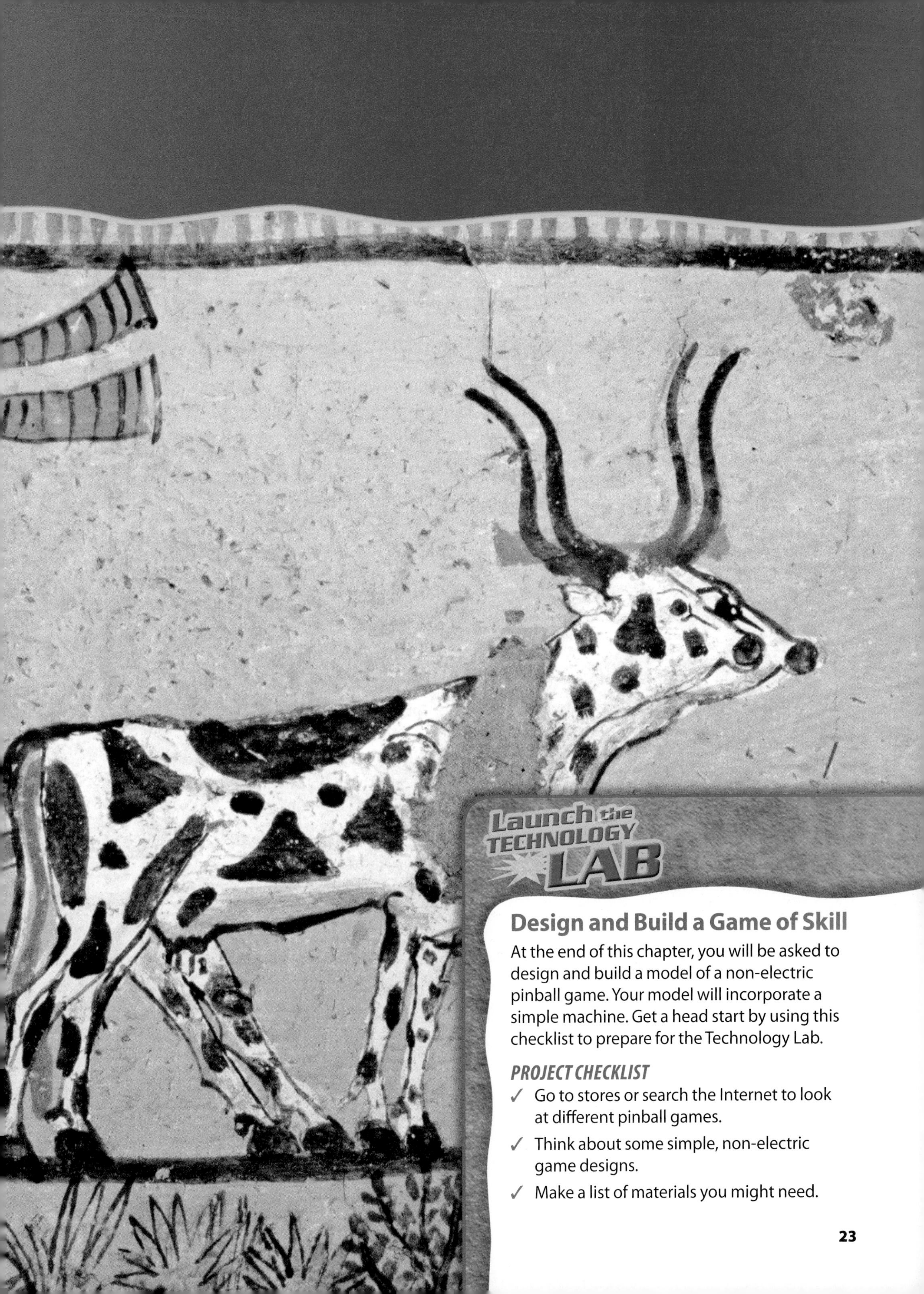

Design and Build a Game of Skill

At the end of this chapter, you will be asked to design and build a model of a non-electric pinball game. Your model will incorporate a simple machine. Get a head start by using this checklist to prepare for the Technology Lab.

PROJECT CHECKLIST

- ✓ Go to stores or search the Internet to look at different pinball games.
- ✓ Think about some simple, non-electric game designs.
- ✓ Make a list of materials you might need.

section

2.1 Technology Resources

Reading Guide

 Before You Read **Connect** What tools have you used? Why did you need to use them?

Content Vocabulary

- resource
- skill
- tool
- primary tool
- machine
- capital

Academic Vocabulary

You will see these words in your reading and on your tests. Find their meanings at the back of this book.

- recall
- labor

Graphic Organizer

Draw the section diagram. Use it to organize and write down information as you read.

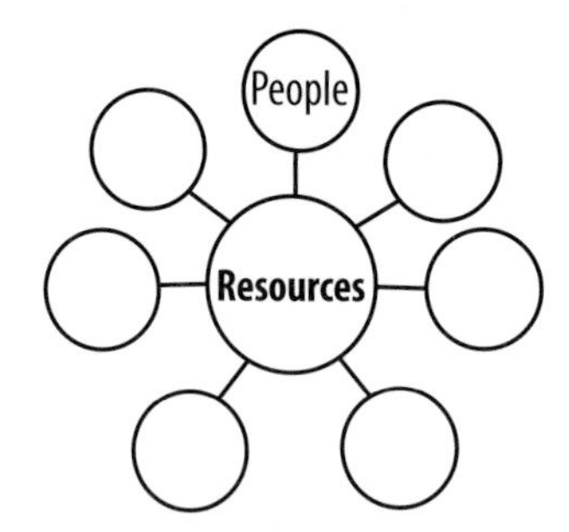

 Go to **glencoe.com** to this book's OLC for a downloadable graphic organizer and more.

TECHNOLOGY STANDARDS

STL 1 Characteristics & Scope of Technology

STL 4 Cultural, Social, Economic & Political Effects

STL 7 Influence on History

STL 19 Manufacturing Technologies

ACADEMIC STANDARDS

Science

NSES Content Standard F Science and technology in society

Mathematics

NCTM Geometry Characteristics and properties of two- and three-dimensional geometric shapes

STL *National Standards for Technological Literacy*

NCTM *National Council of Teachers of Mathematics*

NCTE *National Council of Teachers of English*

NSES *National Science Education Standards*

NCSS *National Council for the Social Studies*

Technology Past and Present

What would you need to create technology?

 As You Read

Identify What are some simple machines?

Technology requires knowledge, skill, raw materials, tools, and energy to create products and services. People today use the same resources used in ancient days to develop new technology. Our ancestors knew very little about technology. However, they were able to use their limited knowledge and their hands to form raw materials like stones into useful tools. Their tools were simple by our standards. Our technology will probably seem simple to people of the 23rd century.

In ancient times technology allowed humans to survive. Early weapons made people the *hunters* instead of the *hunted*. By the 1600s, telescopes and microscopes allowed people to see what was once invisible. Today communication technology allows us to hear over great distances by using telephones and satellites. Computers give us the ability to recall details and solve problems in seconds.

To create new technology today, people use the same resources that they used long ago. A **resource** is something that gives help or aid to a system. The seven technology resources include:

- People
- Information
- Tools and machines
- Capital
- Time
- Materials
- Energy

My Surgeon Is a Robot

Imagine a tiny robot, the size of a small caterpillar. It was developed in 2007 at the Robotics Institute at Carnegie Mellon University. "HeartLander," the robot, enters the body through a small incision, using front and rear suction cups to move. The robot can attach leads for a pacemaker and place a needle to inject heart medicine. *Why have doctors tested this robot on pigs before using it on people?*

Go to **glencoe.com** to this book's OLC for answers and to learn more about robotic surgery.

People and Technology

Why must technology be created by people?

All past and present technology was developed by people who used their creativity and imaginations to find new solutions to existing problems. Machines in the future might design and create new technology without human intervention. Until then people will remain necessary for the development of new technology. However, robots can perform repetitive jobs.

Can you define the term *skill*? **Skill** is an ability you develop when you combine knowledge and practice in order to perform an activity well. Our ancient ancestors developed the necessary skills to convert their ideas into real products, systems, or processes. The skills that they developed in the past continue to be the foundation for the skills that you will develop in the future.

People have learned to create new tools and pass their inventions on to future generations. Each generation can benefit from the accomplishments of the past.

Old News This cell phone is already out of date. *What two features, found on new cell phones, were not available when this phone was first sold?*

People are also the users of the products that their technology has built. Between the designer and the user, there are many jobs performed by people. People build the tools and machines, set up the factories, run the machines, and package and ship the products. Other people work in the service area of technology. They sell, install, and repair these products.

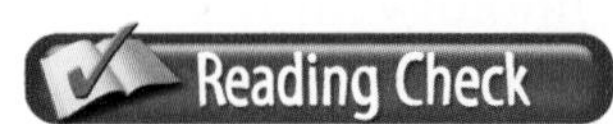

Define What is a skill?

Information

Why do we need information to create technology?

Information can lead to the knowledge, learning, scholarship, understanding, and wisdom needed to create technology. We use information, skill, and natural resources to meet our needs and wants. If a chimpanzee takes a branch (natural resource) and moves an object into its reach (skill), it is using technology to get food (need).

When our early ancestors used a stick to gather food, they used elementary technology similar to that used by a chimp. This basic tool was refined by each generation and passed down to us. People learned (gained knowledge) that a stone attached to the stick improved its performance. Others learned that the reaching stick could also be used as a weapon.

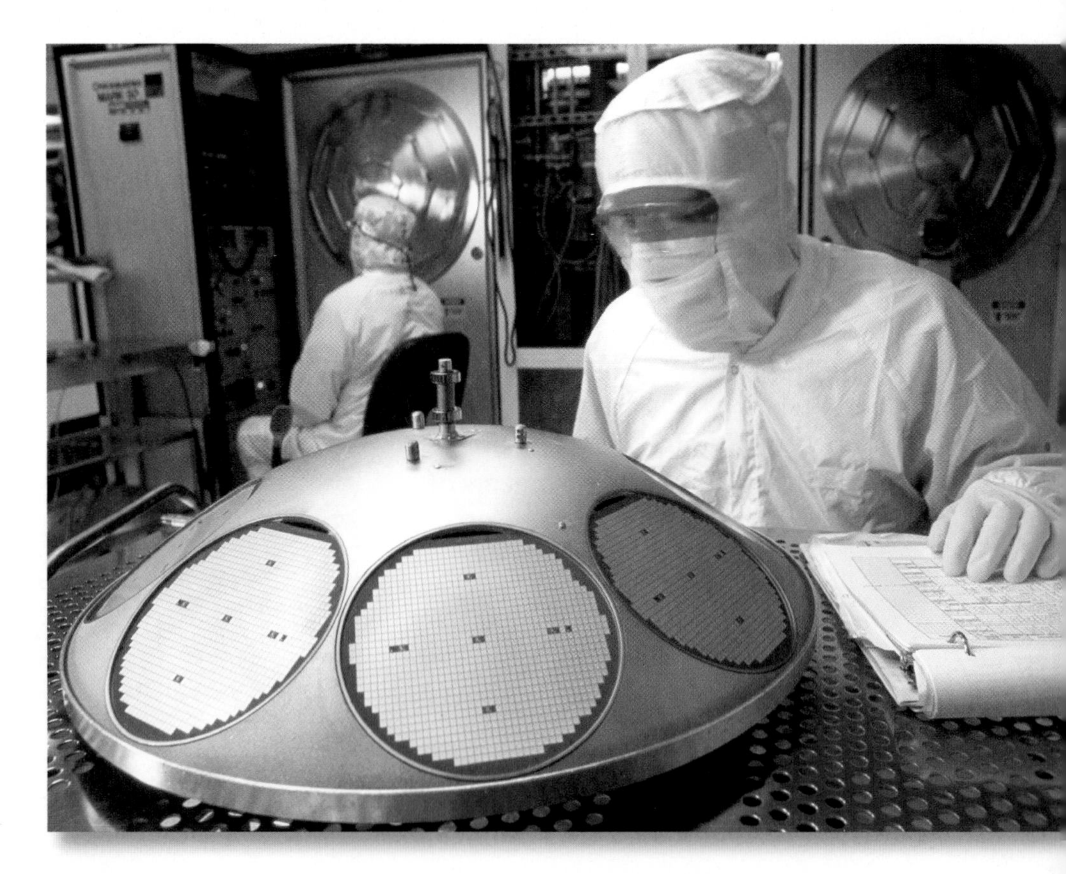

Tech Savvy The technological world needs skilled people to create components for technology, such as these computer chips. *Can skilled workers who perform a repetitive job be replaced by robots?*

Dr. Fujio Masuoka

Inventor of Flash Memory

Dr. Fujio Masuoka invented flash memory technology in 1984 while working at Toshiba in Japan. Flash memory uses small flash chips to store information in many types of electronics—computers, cell phones, automobiles, and MP3 players. Flash memory is considered the most important semiconductor innovation of its time.

Masuoka invented a product worth more than $76 billion. However, because he invented flash memory while working for Toshiba, he did not own the patent.

A Faster Flash Masuoka left Toshiba in 1994 and became a college professor. In 2004, he sued Toshiba and won a settlement worth more than $750,000. Masuoka is still inventing, working on a semiconductor that could increase the speed of a flash chip. Masuoka is also securing his invention with the U.S. Patent Office.

English Language Arts/Writing Write an article for your school newspaper describing any products you use that utilize flash memory technology.

 Go to **glencoe.com** in this book's OLC to learn about young innovators in technology.

Tools and Machines

Why are tools and machines needed to develop technology?

People consider all devices that help them perform their jobs as "tools of their trade." A **tool** is a device that increases our ability to do work. If learning is your work, then pens, pencils, and books are your tools.

Primary Tools

The first tools were all handheld tools and muscle-powered tools. These **primary tools** are basic handheld tools that increase a person's ability to hold, cut, drill, bend, and hammer materials. People used these early mechanical tools to construct things that met human wants and needs at that time. These primary tools were also used to make other tools. Without these tools, more complex technology would never have developed.

Machines

Machines are often referred to as tools. A **machine** is a tool with a power system that takes advantage of certain scientific laws that make the tool work better. All mechanical power systems use one or more of the six simple machines to change direction, speed, or force. (See **Figure 2.1** on page 28.) Complex machines can use a combination of the simple machines.

Figure 2.1 Simple Machines

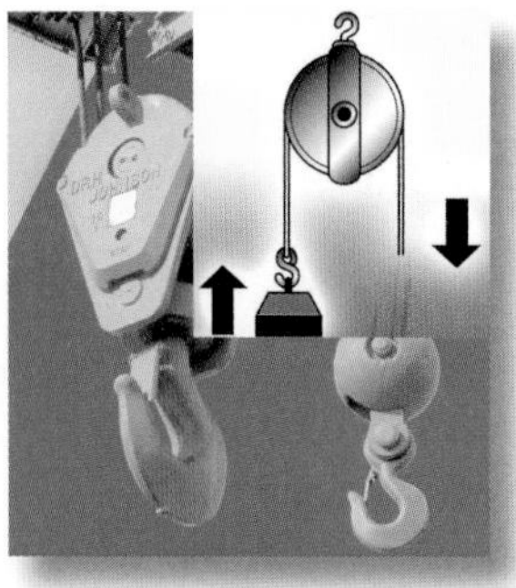

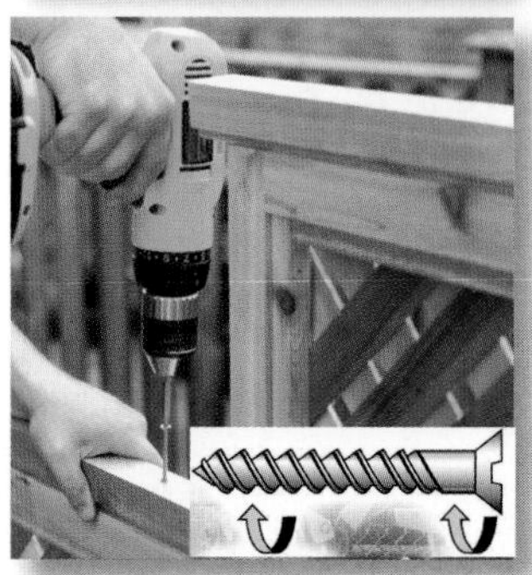

The Basics Simple machines can be incorporated into complex modern products and machines. *Can you identify one item you have used for each machine pictured?*

1. **Wheel and Axle** The best-known simple machine is the wheel. It is round and connected to an axle, which is the center shaft. Gears and cams are related to the wheel and axle. The gear is a wheel with teeth around its circumference, or outer rim. The teeth allow gears to mesh (fit together) without any chance of slipping. Your bicycle has a gear that you turn by pushing your feet on the pedals. This gear meshes with the chain that meshes with the gear that drives your rear wheel. The cam uses the principle of the wheel with the principle of the inclined plane. Most cams look like wheels that are not perfectly round.
2. **Pulley** The pulley uses the principle of the wheel in combination with a rope or chain to lift heavy objects. In a one-pulley system, the full weight of the object can be lifted by pulling the rope. In a two-pulley system, the object feels as if it weighs one-half its actual weight.
3. **Lever** The lever is a bar that turns on a fixed point and allows you to lift something heavy. You have probably played on a seesaw. This playground toy consists of a long board that is fastened securely at its middle so each end can move up and down. If a heavier person sat closer to the middle of the board, a lighter person could easily push up that person by sitting on one end of the seesaw.
4. **Inclined Plane** The inclined plane is a ramp, or angled board, that makes it easier to raise things by rolling them uphill. Cars driving into parking garages move along an inclined plane upward to the next level of the garage.

5. **Wedge** The wedge is a small inclined plane used to spread things apart. Its shape transforms downward movement into a force that separates things. The axe is a wedge on a stick. Scissors are two wedges joined together. The plow is one of the most important wedge-shaped tools.
6. **Screw** The screw is actually an inclined plane that runs around a metal rod. Notice how a ramp in a parking garage looks like a giant screw.

Not all machines have mechanical power systems. Some machines use electronic power systems. The computer is an electronic machine. Its power transfer system has no moving parts. It works by pushing electrons through a conducting material.

Other machines are biological. You are a perfect example of a biological machine. Today scientists are turning cells into machines that can manufacture needed chemicals.

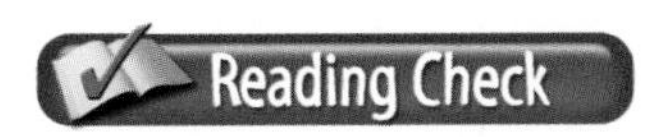

Identify What are simple machines?

Capital and Technology

Why do inventors need capital for technology inventions?

Capital is money, credit, or property—or accumulated wealth. At the dawn of technology, inventors probably created their tools without financial help from others. They did not need capital.

However, later inventors could not get all the necessary tools and materials without financial assistance. To trade for or purchase tools, materials, and **labor**, they had to use some form of capital. The importance of capital for inventing grew over time.

Capital from investors and companies can buy resources. Today a team approach is used to develop most new ideas. Corporations hire experts and obtain materials, tools, information, and skilled and creative people to develop useful products.

However, spending great sums of money does not guarantee success. An independent inventor might create the next invention that will become a multibillion-dollar business.

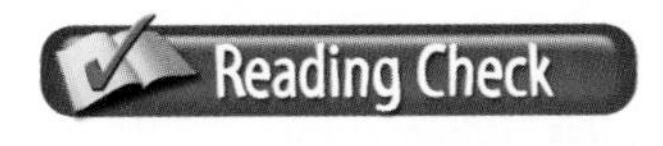

Explain Why has capital not always been needed for the development of technology?

Time and Technology

How does time affect the development of technology?

Everything takes time to develop. Also, people are paid for the time they work. Products developed by human labor are usually more expensive than products made by machines.

Most food recipes require the mixing, stirring, heating, or cooling of the contents for a specific amount of time. Whether you are making a cake, building a car, or designing a new product, results will take shape over time.

Materials and Technology

Could we develop technology without materials?

Materials are needed to create the products and processes of technology. People create *new* materials by combining or refining natural resources in ways not done by nature. Material resources can be classified according to how they were formed. Materials can be raw, processed, manufactured, or synthetic.

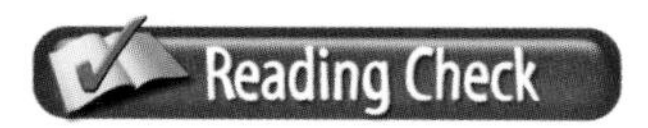

Identify What are the categories of materials?

Energy and Technology

Why must we use energy to create technology?

After a hard workout playing sports, you might feel you have run out of energy. Your muscles use a great deal of energy to perform the tasks you do daily. Even when you are at rest, you use energy to breathe, think, and pump blood through your body.

Energy is also the source of power that runs our technological systems. There are many sources of energy. These sources may be natural or synthetic. Chapter 7 discusses more about energy.

section 2.1 assessment

Self-Check

1. Name seven resources to create new technology.
2. Explain why people need skills and imagination to create new technology.
3. Identify the six simple machines.

Think

4. Describe how we differ from early humans.

Practice Academic Skills

English Language Arts/Writing

5. Build an object using either raw materials, processed materials, manufactured materials, or synthetic materials. Write a step-by-step instruction page on how you made the object.

STEM Mathematics

6. Dani is moving into her new apartment. She places heavy things on a cart so she can wheel them into the building using the ramp at the entrance. The ramp is 12 feet long, 5 feet high. What distance will Dani cover when pushing the cart on the ramp?

Math Concept **Geometric Formulas** When you are doing geometric calculation, be sure to use the correct formulas.

1. To help determine a solution, draw a picture.
2. Use the Pythagorean equation ($a^2 + b^2 = c^2$) to determine the unknown length of a side of a triangle.

section 2.2

Technology Systems

Reading Guide

Before You Read **Preview** What is the purpose of a system?

Content Vocabulary

- system
- subsystem
- open-loop system
- input
- process
- output
- closed-loop system
- feedback

Academic Vocabulary

- distinct
- achieve

Graphic Organizer

Draw the section diagram. Use it to organize and write down information as you read.

Closed-Loop Systems

Output

Go to **glencoe.com** to this book's OLC for a downloadable graphic organizer and more.

TECHNOLOGY STANDARDS

STL 2 Core Concepts of Technology

STL 11 Design Process

ACADEMIC STANDARDS

Science

NSES Content Standard F Science and technology in society

English Language Arts

NCTE 1 Read texts to acquire new information.

STL *National Standards for Technological Literacy*

NCTM *National Council of Teachers of Mathematics*

NCTE *National Council of Teachers of English*

NSES *National Science Education Standards*

NCSS *National Council for the Social Studies*

Understanding Systems

Why are systems so important for technology?

A system is an organized way of doing something. A **system** is made of parts that work together to complete a task. Systems are one of the building blocks of technology.

Subsystems are smaller systems that exist within larger systems. A subsystem cannot usually function properly without its surroundings. The jet engine is one of many subsystems of an airplane. However, some systems can be both. The airplane is a distinct system, but it is also a subsystem of a transportation system.

As You Read

Compare What is the difference between input and output?

Diagramming Systems

Why do people use diagrams when they make plans?

Football coaches often diagram plays to help team members understand what they are going to do during a game. Technology uses a method of diagramming, originally developed by engineers, that helps people understand how any system operates. A diagram shows how one part of a system relates to the other parts. This same diagram can also help people organize plans for new ideas.

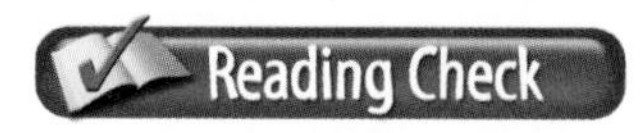

Name Who first developed the method of diagramming used for technology?

Open-Loop Systems

Why is this system called "open-loop"?

When a system has no way to measure or control its product, the system is called an **open-loop system**. Old-fashioned bathtubs, stoves, and traffic lights are all examples of open-loop systems. These devices cannot shut down by themselves. A bathtub can overflow. A stove will stay on and burn food. A red traffic light automatically goes on even when there is no cross traffic.

The open-loop system includes three parts: input, process, and output. In **Figure 2.2**, **input** includes the resources, ideas, and activities that determine what we need to accomplish. For example, suppose you want to run for school president. You decide to make campaign posters and buttons. All the steps that lead up to the idea of creating these posters and buttons are part of input.

Figure 2.2 Open-Loop System

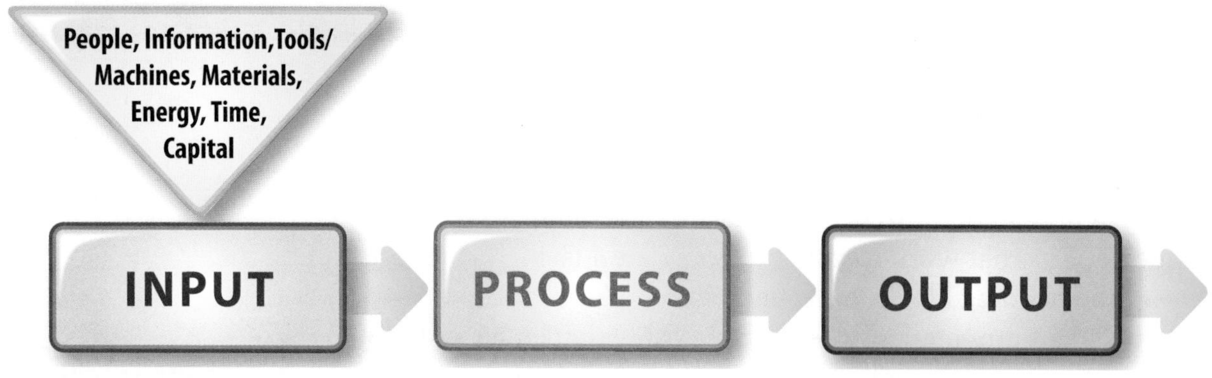

People, Information, Tools/ Machines, Materials, Energy, Time, Capital

INPUT → PROCESS → OUTPUT →

Good Planning People often graph systems to focus their attention on a particular project. *If you were making buttons for your school team, what information would you place in each section of your open-looped system?*

Figure 2.3 Closed-Loop System

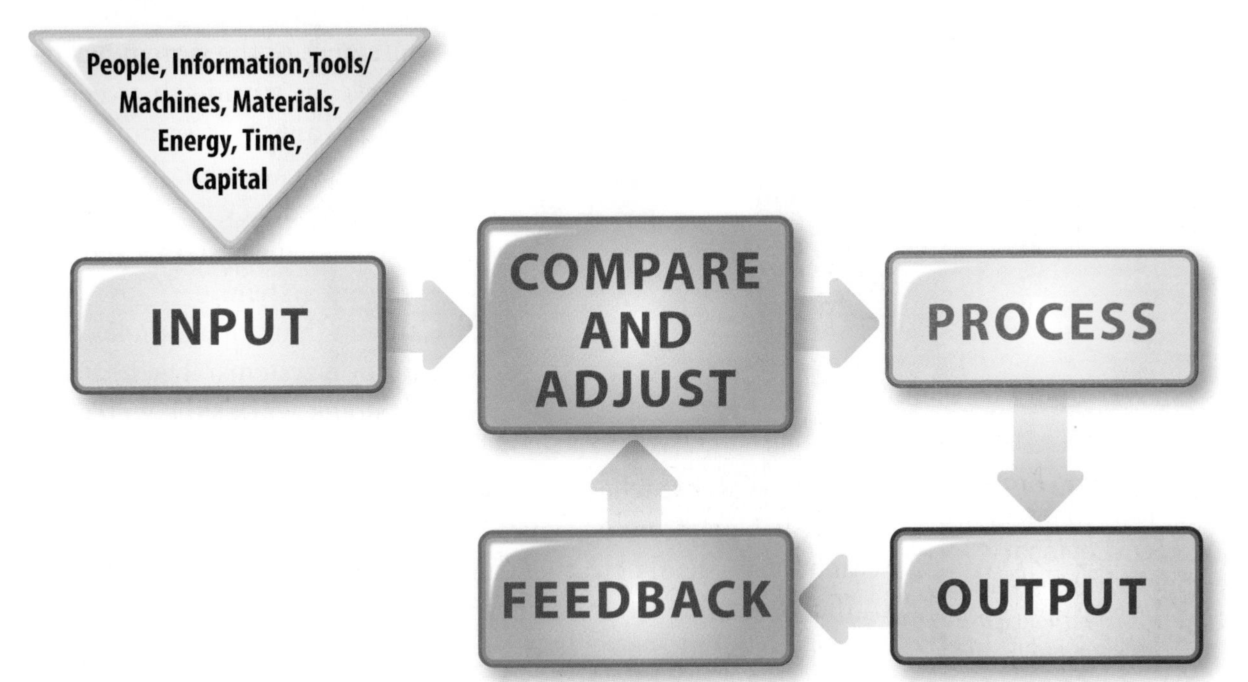

Feedback A closed-loop system adds a method of measuring the effectiveness of the system. *Why is it important to know if a system is effective?*

The **process** is the conversion of ideas or activities into products by using machines and labor. The process of our system diagram includes designing and making your buttons and posters.

Output is simply what the system produces or achieves. Your posters and buttons would be the output of your election planning. The three parts of an open-loop system diagram contain an idea (input), which leads to an action (process), which leads to an outcome (output).

Can an open-loop system measure effectiveness? Could you tell if your buttons and posters accurately communicated your message to the other students? Probably not. How could you measure the effectiveness of your buttons and posters? How would you add a controlling device to regulate a traffic light?

Define What is a process?

Closed-Loop Systems

A **closed-loop system** is an open-loop system with an added feature that provides you with up-to-date information about your end product (output). See **Figure 2.3**. If students did not like your posters, what would you do? You might change them.

Swimming Around These fish cannot survive in this environment without a water filtration system. *What makes this water filter/heater in this aquarium part of a closed-loop system?*

Feedback is the part of the system that measures and controls the outcomes of the system. Feedback serves as a bridge between what you want to do (input) and what you are actually doing (process). Feedback "closes" the loop to make the system a stable closed-loop system.

Can you think of examples of closed-loop systems? The heater in a fish tank warms the water in the tank. The heater shuts off when the water reaches the right temperature. If it did not shut off, the fish might not survive. A traffic light at an intersection with built-in metal detectors can stay green for lanes that have traffic. It remains red for empty lanes.

Feedback is also used to open doors in some public buildings. In one system, a motion detector placed above the door senses movement in the same way that a police radar system can detect a speeding car.

Some complex technological systems have many layers of feedback and control. Systems can be connected. Some buildings have computer systems that control the elevators, escalators, heating, cooling, and lighting for the entire structure. The independent systems work together as parts of a larger system.

section 2.2 assessment

After You Read Self-Check

1. Describe systems and subsystems.
2. Explain why people diagram plans.
3. Define open- and closed-loop systems.

Think

4. What might be the simplest operating system? Explain why.

Practice Academic Skills

English Language Arts/Writing

5. Construct two LEGO® objects. Make one of them an example of an open-loop system. Make the other an example of a closed-loop system. Then write two paragraphs explaining why each one represents either system.

STEM Mathematics

6. Jorge is interested in buying a new computer. The system he wants will cost $2,500. He has $1,375 saved so far. At his part-time job, he makes $175 a week. He is able to save $125 per week for the purchase. How many weeks will it take him to save enough money?

Math Concept **Multi-Step Problems** Writing an equation can help you solve a problem that has multiple steps.

1. The first step is to determine how much Jorge still needs to save.
2. The next step is to determine how many weeks of work it will take him to save that amount.

section 2.3

Developing a System or Product

Reading Guide

Before You Read **Connect** What is important to you when you buy a new product?

Content Vocabulary

- criteria
- constraints
- optimization
- trade-off
- Beanstalk Principle

Academic Vocabulary

- enhance
- impact

Graphic Organizer

Draw the section diagram. Use it to organize and write down information as you read.

Impacts and Effect of Technology

	Technology #1	Technology #2
	CAR	
Positive	Provides transportation from place to place.	
Negative		

Go to **glencoe.com** to this book's OLC for a downloadable graphic organizer and more.

TECHNOLOGY STANDARDS

STL 2 Core Concepts of Technology

STL 6 Role of Society

STL 8 Attributes of Design

STL 10 Troubleshooting & Problem Solving

ACADEMIC STANDARDS

English Language Arts

NCTE 1 Read texts to acquire new information.

Science

NSES Content Standard A Abilities necessary to do scientific inquiry

STL *National Standards for Technological Literacy*

NCTM *National Council of Teachers of Mathematics*

NCTE *National Council of Teachers of English*

NSES *National Science Education Standards*

NCSS *National Council for the Social Studies*

Product Development

How do companies make people want to buy a new product?

All products are designed and built to meet the needs or wants of the people who will buy the products. The most expensive advertising campaign can fail if the advertised product is not what customers want and need. Products must be safe, attractive, useful, and reliable. They must sell at a price consumers will pay.

Some new products are developed before consumers have said they want them. For example, consumers were not aware of personal video players or MP3 players. Public relations and marketing companies built consumer interest in these products.

As You Read

Define What do the words *criteria* and *constraints* mean?

Academic Connections
Social Studies

Events Influence Technology Historical events, even tragic ones, influence technology. Several new innovations resulted after the attack on the World Trade Center on September 11, 2001. One was a tiny camera that can be worn by search-and-rescue dogs so they can find victims of disasters.

Apply Research and write a paragraph about other inventions and innovations resulting from disasters.

Criteria and Constraints

Why should you know the meanings of criteria *and* constraints?

Product designers decide which features a product must have. **Criteria** are requirements or specifications for a product to be successful. They might include more than requirements related to a product being useful and practical. Consumers might demand a certain level of performance, special features, a designer label, or a very low price. Designers look for a winning combination.

The first cell phones weighed two pounds. They were the size of a brick and provided only a half-hour of talk time. These cell phones were successful because, for the first time, they met the criteria of allowing people to make wireless telephone calls while on the go.

Early cell phones had many engineering and infrastructure **constraints**, which are limits on a product's design. These contraints were eliminated as the phones became popular. New cell sites were added very quickly; electronics shrunk in size; battery power was enhanced; and talk time and features skyrocketed.

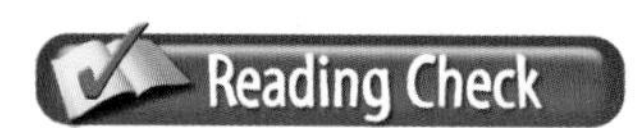

Identify Who decides product features?

Optimization and Trade-Offs

Why should price be just one consideration when designing a new product?

A designer's goal is to create the best system, product, or process by using all of the best tools, materials, and processes available. In technology, we call this **optimization**. You *optimize* your product by making the most of its positive features while reducing its negative features.

Changing Times Cell phones keep shrinking in physical size even though features such as cameras, MP3 players, and IM keyboards are added. *What new feature would you add to a cell phone?*

Ethics in Action

Voting in the 21st Century

During the 2000 election, paper ballots confused some voters who accidentally voted for the wrong candidate. Now many states are using electronic voting machines. Some officials think they are more reliable and easier to use.

Every Vote Counts However, hackers, software glitches, or power outages could erase or alter electronic votes. Some voting machines do not keep paper records, and so officials could not do a recount of lost votes.

English Language Arts/Writing

Rock the Vote Supporters of electronic voting say it helps people with disabilities to vote. Critics say the machines are expensive and can drop votes. Both electronic voting and paper ballots have pros and cons.

1. Do Internet or library research on paper ballots and electronic voting machines.
2. Consider the pros and cons, and trade-offs. Then write a paragraph discussing the method you think should be used and why.

However, building the best product also involves trade-offs. A **trade-off** is a compromise—you give up one thing in order to gain something else. Many trade-offs involve cost. You choose components that will function at the most reasonable price. You choose the least expensive materials and processes. A completely safe car would be expensive and might not sell.

As a designer, you would need to ask: Are the materials readily available? Are they the best choice? Will there be waste? All these considerations may involve trade-offs.

Maintenance and Control

How do modern control systems maintain performance?

You probably know the story of "Jack and the Beanstalk." Jack's beanstalk grew much larger than normal. The **Beanstalk Principle** is the rule that states systems, processes, and products should not grow beyond an optimal or ideal size. If this principle is ignored, you can expect system failures. The larger and more complex a system is, the harder it is to keep it working properly and to control the way it functions.

Systems require maintenance to keep them working. Maintenance is basic care and upkeep. To ensure proper maintenance, many products, such as cars and computers, have control elements to watch over their systems and report any problems.

On the other hand, other products may not be worth maintaining. It may be less expensive or more convenient just to replace them. This happens quite often today, and so our society has been called the "throw-away society."

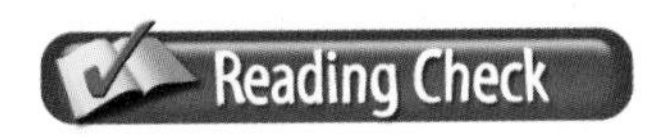

Explain How does the Beanstalk Principle relate to technology?

EcoTech

Investing in the Future

Internet technology allows people to find socially responsible investments. There are stock funds that focus completely on these kinds of investments. You can also research a company and find out if its products benefit the environment and society.

Try This To support companies that benefit the environment and people, do an Internet search with the key words "green companies" and "socially responsible companies." Find out about their products and decide if you would use them.

Impacts of Technology

Should we develop a new technology if it might have some negative effects?

As you know, surprises can be good or bad. In the same way, the unexpected effects of technology can be positive or negative. Past technological inventions and innovations have extended our human capabilities. An invention is a new product. An innovation is a change made in an existing product. However, unexpected effects can hurt the very people that the technology was designed to serve.

For example, the internal combustion engine, the automobile, and our system of roads and highways have given us the ability to travel in comfort. However, some of these technologies have also polluted our atmosphere and may have helped cause global warming. Technologists continue to work at reducing these negative impacts.

Technology Literacy

Energy and power, information and communication, biotechnical and medical, manufacturing, construction, and transportation technologies are all discussed in this book. As you explore each area, look for the positive as well as the negative impacts of the particular technology. To become truly technologically literate, you must learn how to weigh one effect against the other effect.

section 2.3 assessment

Self-Check

1. Explain criteria and constraints.
2. Define optimization.
3. Discuss how trade-offs relate to technology.

Think

4. Explain why technologists must consider cost when they plan to manufacture a product.

Practice Academic Skills

English Language Arts/Writing

5. Working in teams of three or four, choose one communications, clothing, or personal product made by several competitors. Compare criteria, constraints, optimization, maintenance, and control (if applicable) that different manufacturers may have considered. Present your findings in a written report and chart.

STEM Science

6. Imagine or choose a technology product that would help the environment in some way. It could be something that would be used to reduce waste, help the recycling process, or eliminate using non-renewable resources. Write a few paragraphs describing the purpose of the product, how it would be produced, and how you feel it could benefit the environment.

Exploring Careers in Technology

Jeff Briggs

IMAGE ARCHIVIST

Q: *What do you do?*

A: I am an image archivist for a major film and television studio. We have assets going back 100 years. I am responsible for locating, organizing, and preserving still images from motion pictures. Still images are like photographs. Once we find the best material from a film, it gets scanned and placed into our digital asset system, where employees and clients can use it.

Q: *What kind of training and education did you need to get this job?*

A: Although I have a bachelor's degree, it is in a field that's not related to the film industry. The image archive was created 5 years after I began working in the general studio archive. I gained knowledge from one of my coworkers, who had been a studio photo editor for 20 years.

Q: *What do you like most about your job?*

A: All my life I have loved movies of all kinds. Being around historic photo material is a thrill for me. Even the lesser-known titles are fascinating to explore. It's not just film history; it's the entire history of the 20th century and beyond.

Q: *How did you get interested in your job?*

A: After college I looked for work in the entertainment industry. I wanted to apply my knowledge of film history to a job. A studio archive was the best match. My career has proved interesting from day one!

English Language Arts/Writing

Start an Archive Create a form that you can use to file, archive, and organize information.

1. Choose three movies and collect the same basic information about each of them.
2. Use a spreadsheet or a file-making program and enter the information into a document.
3. Share your method and results with other students and identify what type of system you are using and why.

Go to **glencoe.com** to this book's OLC to learn more about this career.

Real-World Skills

Speaking, listening, problem-solving, organization

Academics and Education

English language arts, social studies, mathematics

Career Outlook

Growth as fast as average for the next ten years

Source: *Occupational Outlook Handbook*

chapter 2 Review and Assessment

Chapter Summary

Section 2.1 Technology developed because people had ideas they turned into useful devices. The seven resources of technology include people, information, tools and machines, capital, time, materials, and energy. Companies buy resources needed to create new technology. They hire people with the knowledge and skill to make products.

Section 2.2 Technology has produced many systems and subsystems. A car engine is a subsystem of the automobile. An automobile is a complete system, but it is also a subsystem of our transportation system. In an open-loop system, input is the information, ideas, and activities needed to plan for production. Process is the construction stage. Output is what the system produces.

Section 2.3 Products are designed to meet the needs and wants of consumers. Marketing firms use advertising to create desire for new products. Product designers determine features, considering criteria and constraints. Trade-offs may be needed to turn a design into a new product that can be built at a reasonable cost. To protect people and the environment, we must monitor new technology to determine negative outcomes.

Review Content Vocabulary and Academic Vocabulary

1. On a sheet of paper, use each of these terms and words in a written sentence.

Concept Vocabulary
- resource
- skill
- tool
- primary tool
- machine
- capital
- system
- subsystem
- open-loop system
- input
- process
- output
- closed-loop system
- feedback
- criteria
- constraints
- optimization
- trade-off
- Beanstalk Principle

Academic Vocabulary
- recall
- labor
- distinct
- achieve
- enhance
- impact

Review Key Concepts

2. **Identify** seven technology resources.
3. **Explain** how skills and creativity lead to new inventions.
4. **Describe** the six simple machines.
5. **Describe** systems and subsystems.
6. **Compare** open- and closed-loop systems.
7. **Explain** how systems play a part in technology.
8. **Describe** how requirements affect product development.
9. **Define** criteria and constraints.
10. **Explain** how technology has positive and negative effects.

Real-World Skills

11. **Conceptualize** Choose a mechanical system you use regularly. Determine which simple machine is part of the system. Write a summary of the system and how the simple machine is used.

Technology Skill

12. **Researching Raw Materials** Raw materials are used to create the products of technology. Some materials can cause harm to the environment. However, safer materials can be substituted.
    a. Use the Internet to research some of the products and processes that use petroleum in their production.
    b. Is petroleum harmful in these processes? How is it harmful? Is there another material that could replace petroleum?

Academic Skills

Social Studies

13. Choose one country where students might benefit from a low-cost computer. Research the government, education system, and teens there. Find companies that make low-cost computers. Present your findings to the class.

STEM Mathematics

14. Chad bought a new MP3 player for $145.50. His friend just bought a laptop computer for nine times the amount Chad paid for his MP3 player, minus $109.75. How much was the computer?

Math Concept **Equations** A word problem can be easier to solve as an equation. Key words tell what operation to use. Words such as *times, less than,* and *goes into* are key words. *Times* means to multiply. *Less than* means to subtract. *Goes into* means to divide.

WINNING EVENTS

Technical Writer

Situation You are to research and write about a technology and its positive and negative impacts on people, the economy, and the environment. Gather information from the library, Internet, and experts.

Activity Research and write a report on one of these topics:

- The Internet: Opportunities and Threats
- The Increasing World Population and the Role of Technology

Evaluation You will be evaluated on how well you meet these performance indicators:

- Understanding of the topic
- Quality of writing
- Quality of research

Go to **glencoe.com** to this book's OLC for information about TSA events.

Standardized Test Practice

Directions Choose the letter of the best answer. Write the letter on a separate piece of paper.

1. If a CD player costs $89.50, how much would 12 players cost?
   A $890.50 **C** $1,800.50
   B $1,074.00 **D** $912.00
2. An open-loop system has no way of measuring its product.
   T
   F

Test-Taking Tip When you first receive your test, do a quick review of the entire test so that you know how to budget your time.

chapter 2

TECHNOLOGY LAB

Design and Build a Game of Skill

Companies spend a lot of money to determine what consumers want to buy, including toys, electronic gadgets, cars and trucks, home appliances, clothing, sporting goods, and business supplies. Companies may hire people to test early versions of their products before deciding how the final product will look. In this lab, you will get the chance to create a game of skill.

Tools and Materials

- ✓ Paper
- ✓ Ruler
- ✓ Glue
- ✓ Wood screws
- ✓ Masonite® board
- ✓ Rubber bands
- ✓ Electric drill press
- ✓ Speed bores
- ✓ Woodworking vises
- ✓ Hand woodworking tools
- ✓ Large steel ball bearings
- ✓ Solid foam plastic shapes
- ✓ Markers
- ✓ Clay
- ✓ Dowels
- ✓ Scroll saw
- ✓ Nails
- ✓ Wood

Set Your Goal

Your goal is to design and build a model of a non-electric game for the Arcadian Pinball Machine Company, which has decided to design and market a new game of skill. The illustration shows two sample solutions to meet this challenge. All workers must pass a safety test before using any tools or machines to construct this game machine.

Know the Criteria and Constraints

In this lab, you will:

1. Work in a group of two or three, but no more than four people.
2. Create a sketch of your design, build a model, and meet these requirements:
   - The game must consist of a board on which a ball or puck will roll, slide, or drop.
   - It must include at least one simple machine.
   - The game board cannot be larger than 1 foot by 2 feet.
   - The game should have some obstacles to overcome. Give extra points for reaching a more difficult area of the board.
   - A player's final score should reflect his or her level of skill.
3. Present your game to the class.

SAFETY

Reminder

In this lab, be sure to follow appropriate safety procedures and rules so that you and your classmates do not get hurt.

Design Your Project

Follow these steps to design your project and complete this lab.

1. Identify the simple machine principles used in the games in the illustrations.
2. Discuss ideas for a game board, controllers, obstacles, and method of scoring.
3. Develop rough sketches of all good ideas. Then choose the design to construct.
4. List all the materials that you will need to construct the game.
5. Call in an outside consultant (your teacher) to determine if your design can be produced with equipment and materials that you have.
6. Select construction materials for your model.
7. Construct your model and test it. Does it hold your attention? Is it a game of skill?
8. Present your solution to the class. Ask for feedback.
9. As a class, pick the best solution or brainstorm how to combine a number of ideas into a super game.

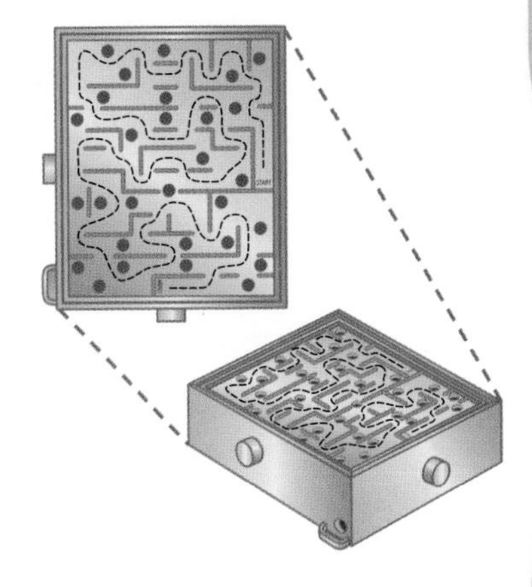

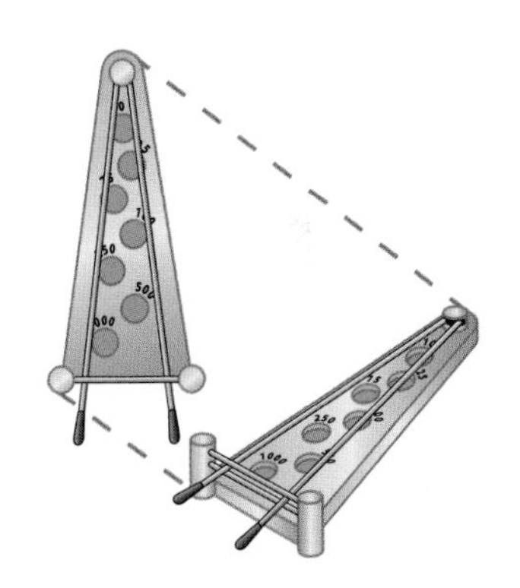

Evaluate Your Results

After you complete this lab, answer these questions on a separate piece of paper.

1. What feedback did your design team get?
2. What part of this activity did you enjoy the most?
3. Did you learn anything that will be useful outside your technology lab? Explain.
4. What was the most difficult part of this activity? What would you do differently?

Academic Skills Required to Complete Lab

Tasks	English Language Arts	Math	Science	Social Studies
Research possibilities for game.	✓		✓	
Plan game in small group.	✓	✓	✓	
Construct and test game.		✓	✓	
Write evaluation.	✓		✓	
Present game to class.	✓		✓	

chapter 3

Processes, Tools, and Materials of Technology

Sections

3.1 Technology Processes
3.2 Tools and Machines
3.3 Engineering Materials

What You'll Learn

- **Describe** separating, forming, combining, conditioning, and finishing processes.
- **Explain** how the different processes are used.
- **Describe** the purpose of several hand tools.
- **Describe** the purpose of several portable power tools.
- **Explain** the importance of safety when using tools and machines.
- **Identify** some basic properties of materials.
- **Name** the common engineering materials.

Explore the Photo 

Using the Right Tools Carpenters use tools to process materials. Processing helps build or create a house, building, or any product. *Can you think of an item that is made without processing?*

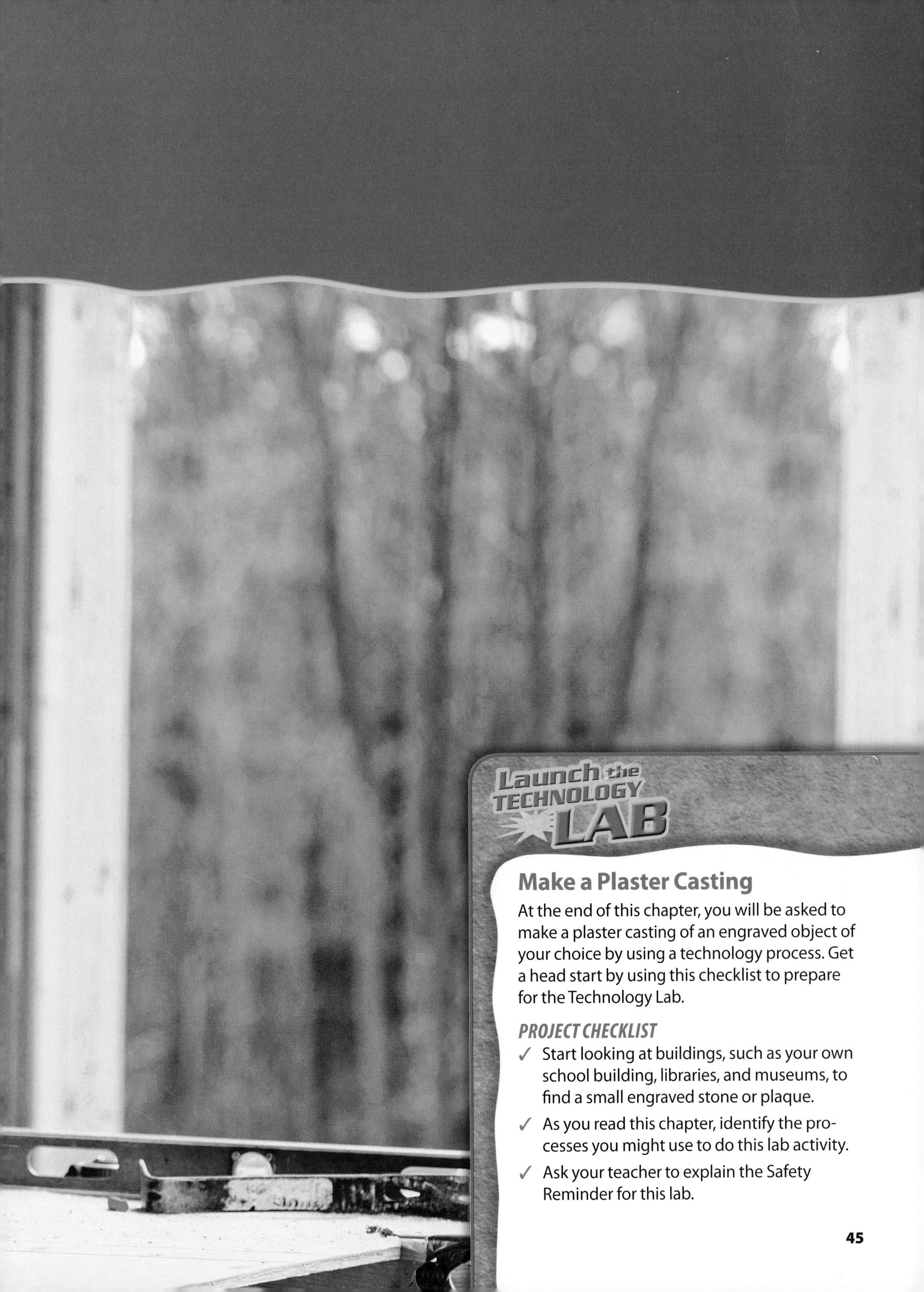

Make a Plaster Casting

At the end of this chapter, you will be asked to make a plaster casting of an engraved object of your choice by using a technology process. Get a head start by using this checklist to prepare for the Technology Lab.

PROJECT CHECKLIST

✓ Start looking at buildings, such as your own school building, libraries, and museums, to find a small engraved stone or plaque.

✓ As you read this chapter, identify the processes you might use to do this lab activity.

✓ Ask your teacher to explain the Safety Reminder for this lab.

section

3.1 Technology Processes

Reading Guide

Before You Read **Preview** How are products made?

Content Vocabulary

- separating
- forming
- combining
- conditioning
- finishing

Academic Vocabulary

You will see these words in your reading and on your tests. Find their meanings at the back of this book.

- technique
- similar

Graphic Organizer

Draw a section diagram. Use it to write and organize information as you read.

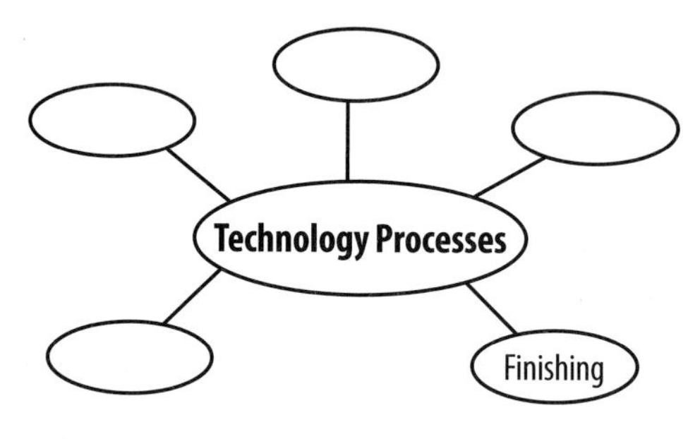

Go to **glencoe.com** to this book's OLC for a downloadable graphic organizer and more.

TECHNOLOGY STANDARDS

STL 9 Engineering Design

STL 19 Manufacturing Technologies

ACADEMIC STANDARDS

Science

NSES Content Standard E Understandings about science and technology

Mathematics

NCTM Measurement Understand measurable attributes of objects and the units, systems, and processes of measurement.

STL *National Standards for Technological Literacy*

NCTM *National Council of Teachers of Mathematics*

NCTE *National Council of Teachers of English*

NSES *National Science Education Standards*

NCSS *National Council for the Social Studies*

Tools and Progress

What were some early tools?

A tool is an instrument that increases your ability to do work. The development of tools moved from simple to complex. The first tools were hand tools and muscle-powered. Cave dwellers used them for hunting and gathering—a tree limb became a club. With tools, people changed materials they found in nature.

The first machine tool was created when an inventor attached a mechanical power system to a hand tool. In every time period, people have created new tools, new materials, and new products.

As You Read

Identify What is the purpose of separating materials?

Using Technology Processes

How do processes change materials?

Whether products are simple or complex, people use different processes to make them. The steps or operations that are used change materials in some way. For example, to make furniture, wood must be cut, shaped, fastened, and finished. Most products have materials that need to be cut and shaped, but not all products have parts that must be fastened and finished. Each product is made with processes that are right for the specific material and the desired result.

Separating

How are materials separated?

Separating is removing pieces of a material. One separating process you are probably familiar with is sawing. When you saw a board to make it the size you want, you divide it into at least two pieces. Small pieces in the form of sawdust are removed as well. Almost all separating can be done with hand tools or electrically powered tools. Separating processes include drilling, sawing, grinding, turning, milling, planing, and shaping. See **Figure 3.1**.

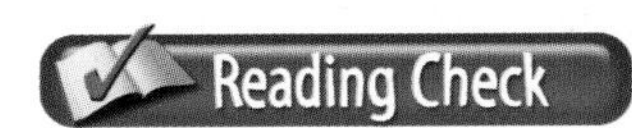

Recall What are the separating processes?

Futuristic Fibers

Imagine special fibers and fabrics that are engineering materials. Designers are developing a fabric vest that becomes a personal airbag during a crash. The D-Air® vest is controlled by a small computer sewn in the vest. Other futuristic materials include shirt fabric that can monitor heart patients, balloon fabric strong enough to lift a building, and steel fibers knitted into heat-producing blankets. *Do research on the Internet to find out more about the D-Air vest. Write about your findings.*

Go to **glencoe.com** to this book's OLC for answers and to learn about smart fibers.

Figure 3.1 Separating Processes

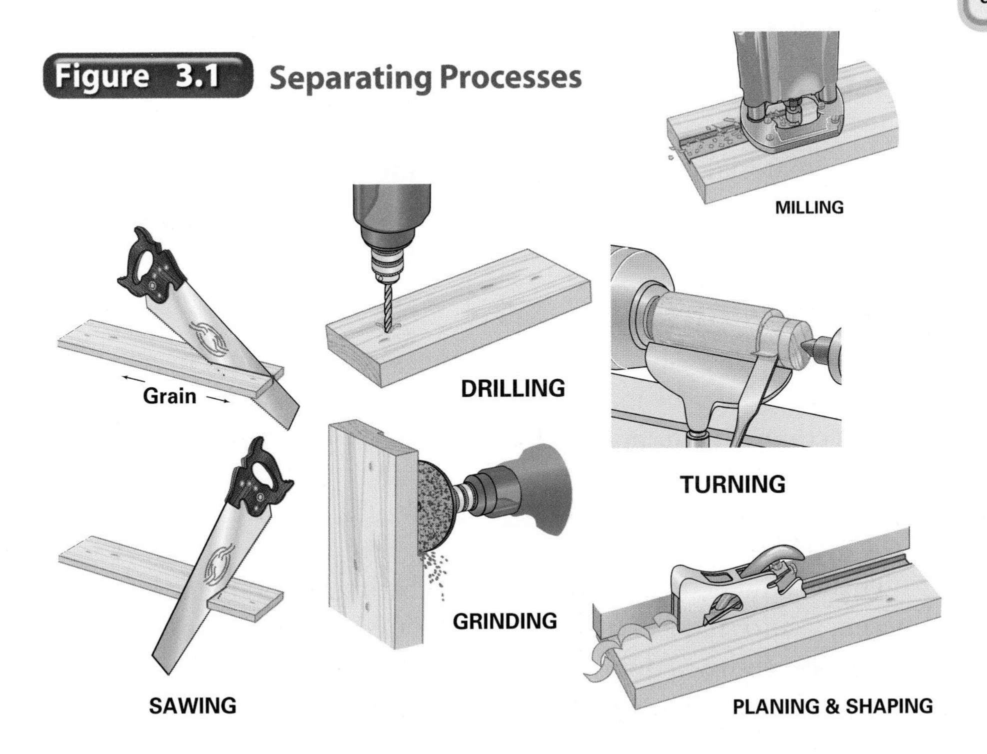

Ways to Separate Either handheld or electric tools can separate materials. *What separating operation might use just a handheld tool? Explain.*

Figure 3.2 Casting

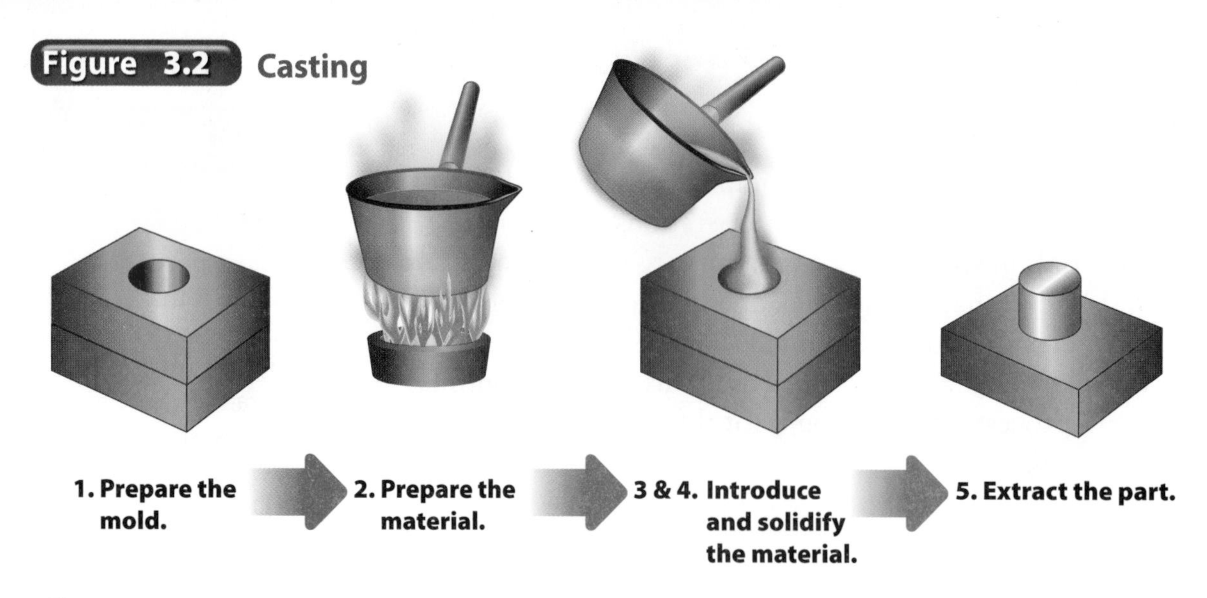

Liquid to Solid Molten metal, glass, clay, or plastic material takes on the shape of the mold as it cools. *What items do you use at home or school that might have been made by casting?*

The Forming Process

What are some different ways to form materials?

Forming is a process that changes the shape of materials. If you have ever used your hands to mold clay, then you have *formed* a material. Forming can be done in several ways.

Bending

In bending, material is formed by forcing part of it to move into a different position. This type of forming is commonly used with metal. However, wood can also be bent into different shapes by using heat and moisture.

Casting

In casting, a liquid material is poured into a mold. As it hardens, the material takes on the shape of the mold. The liquid can be a molten (hot and melted) metal, glass, plastic, or liquid clay. See **Figure 3.2**.

Compression

In compression, a flat material is pressed into a mold by a strong force, and the material takes on the shape of the mold. This is commonly done with metal sheets to form them into things such as car doors.

Forging

The technique of shaping metal by heating it, and then hammering it into shape is called "forging." Old-time blacksmiths formed horseshoes, door hinges, and other items by forging. Modern forging is done with huge and powerful machines.

Extruding

In extruding, softened material is squeezed through a small opening. You "extrude" toothpaste from a tube when you apply it to your toothbrush. Pipe and wire are some common items that are formed by extrusion.

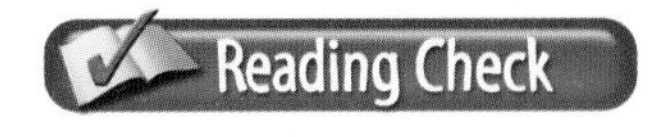

Define What does *forming* mean?

The Combining Process

How do people join materials together?

Joining several parts together to make a finished product is called **combining**. For example, a wooden pencil is made out of four different materials that are joined together: a wooden barrel, pencil graphite, a rubber eraser, and a metal ferrule, which holds the eraser to the barrel. Combining also occurs when materials are mixed together. For example, paint and cake mixes are both made of combined materials.

Mechanical Fastening

If you have ever nailed two pieces of wood together or tightened a screw, then you have done mechanical fastening. Mechanical fasteners are small pieces of metal or plastic that hold parts together. Examples of **similar** fasteners include nails, staples, wood screws, nuts and bolts, pins, and rivets. See **Figure 3.3**. Some fasteners hold parts together permanently, while others allow materials to be taken apart when needed.

Figure 3.3 Mechanical Fasteners

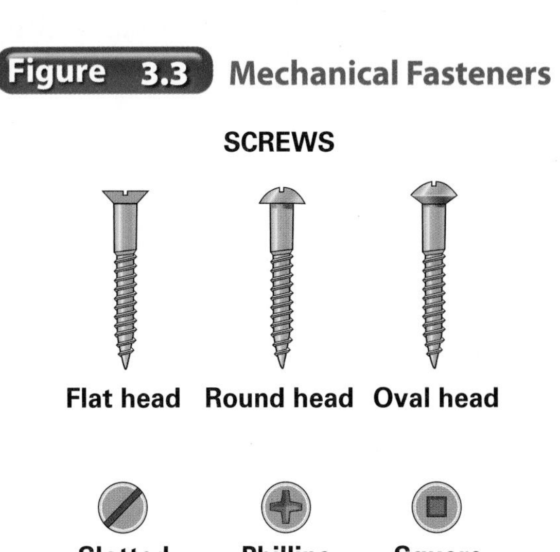

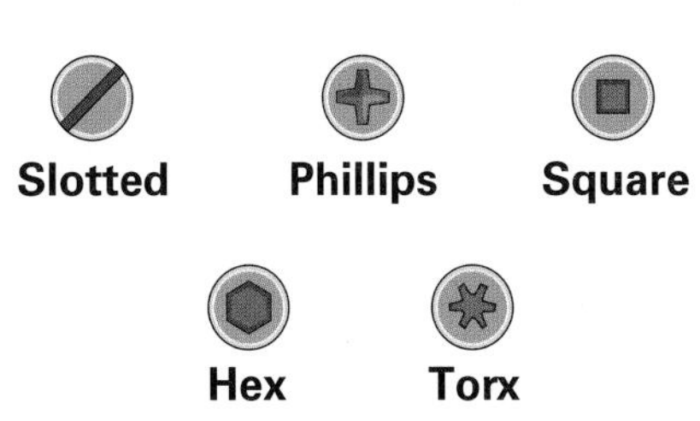

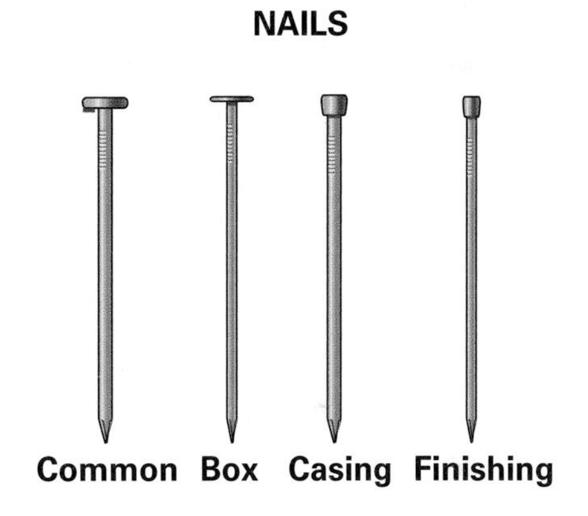

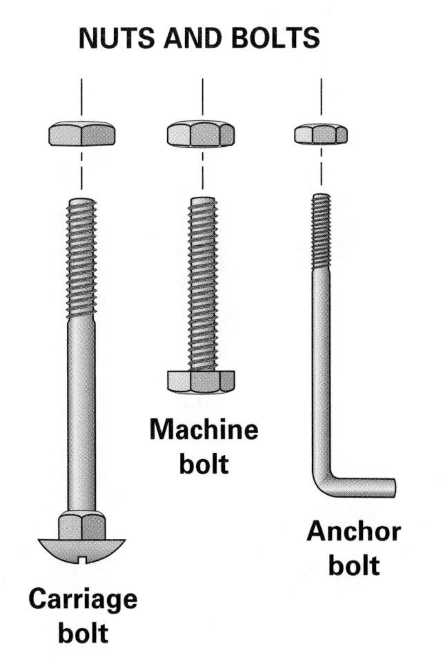

Building Supplies Screws, nails, nuts, and bolts are all mechanical fasteners. *Are all of these fasteners used for items that can be taken apart? Why or why not?*

Heat Fastening

When some materials are heated, they soften, melt, and flow into each other. This common way to combine metal parts is called "welding." A special burning gas or an electrical current heats the material until the parts flow together.

Plastic parts can also be combined using heat. Plastic requires much less heat than metal requires. If you look carefully at some plastic items, you might be able to see a spot where heat was used to melt one plastic part in order to attach it to another part.

Gluing

One of the easiest ways to join parts is to use glue. Glue is called an "adhesive" because it makes one part of an item adhere, or stick, to another part. Adhesives form a film on the surfaces being joined. The film adheres to both surfaces, which holds the parts together. White glues are used for wood. Epoxy or polyester resin (pronounced REZZ-in) can hold metal and ceramic parts together. Super glues can hold some plastics together. However, glue does not work on all plastics.

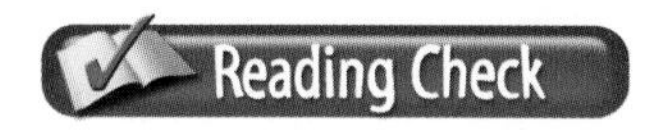

Recall What are the different methods of fastening?

Conditioning

How does conditioning change a material?

Conditioning is done to change the inner structure of a material. When you bake cookies, for example, you are using a conditioning process. The oven's heat changes the dough from a clay-like mass into a light, crispy cookie.

Materials are conditioned to improve their performance. Heat is used to harden ceramics and some metals. After a steel piece is formed into a certain shape, heat and chemicals can help harden its surface. This allows the item to last longer.

Creating with Fire The tremendous heat generated during welding softens a material until the parts can meld together. *Heat fastening only works with some materials. Why?*

Other conditioning processes may be used to soften materials. Leather, for example, is softened before it is made into shoes. Still other processes relieve stresses and strains caused by heat and forming methods. Some metals, for example, become brittle and crack if they are not treated with a conditioning process to improve performance.

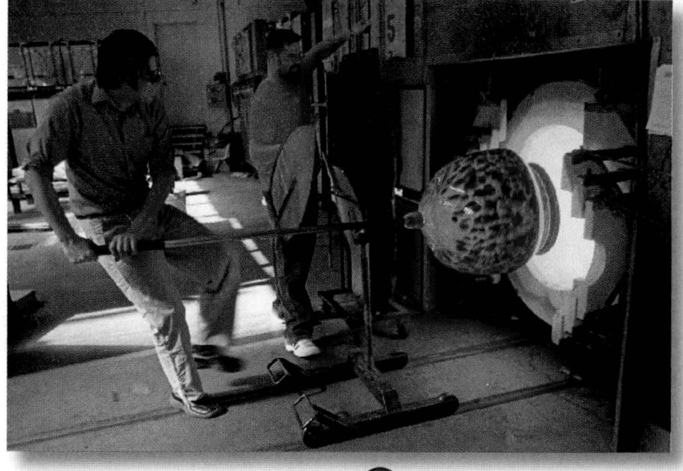

Processing Art Glass makers often reheat the glass object. Reheating removes stresses and strains built up in the piece during the forming process. *What might happen if those stresses and strains were not removed?*

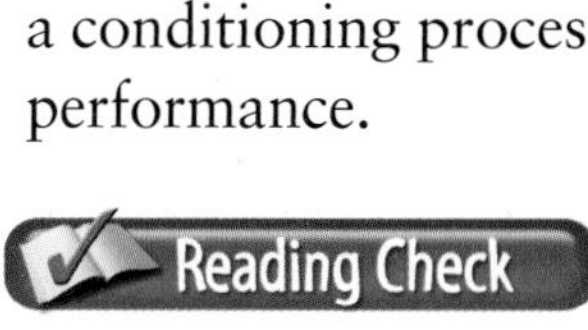

Identify What is an example of conditioning?

Finishing

What is finishing?

Finishing is the last step in making a product. The purpose of finishing is to improve the product's appearance. It can be done in several different ways. Finishing can be simply smoothing and polishing the surface of the product. Other finishing methods use coatings, such as paints, clear finishes, or plastic.

section 3.1 assessment

Self-Check

1. Compare the processes of forging and casting.
2. Name the processes used to shape wire, shape a car door, glue furniture, and paint a bike.
3. Identify five examples of mechanical fasteners.

Think

4. Explain why a screw might have more holding strength than a nail.

Practice Academic Skills

Social Studies/History

5. The original paper clip was patented around the year 1900. Do Internet or library research to prepare a one-page report on the history of the paper clip. Then try to form a better one. Take a standard paper clip, straighten it out, and bend it into another shape. Test your design. Present your report and demonstrate to the class.

STEM Mathematics

6. Drew works in a metal shop where he forges brass hinges for doors. Each hinge requires 5.8 ounces of brass. If Drew forges 80 hinges, how many pounds of brass will he use?

Math Concept **Unit Conversion** When problems require unit conversion, it is sometimes easier to do the unit conversion as the last step.

1. Use multiplication to figure out how many ounces of brass are used.
2. There are 16 ounces in 1 pound. Divide by 16 to convert ounces to pounds.

For help, go to **glencoe.com** to this book's OLC and find the Math Handbook.

section

3.2 Tools and Machines

Reading Guide

Before You Read **Connect** Why are safety rules important to know when you use tools and machines?

Content Vocabulary

- measuring tool
- hand tool
- portable electric tool

Academic Vocabulary

- ultimate
- attitude

Graphic Organizer

Draw a section diagram. Use it to write and organize information as you read.

Safety Rules

1. Work alone	**5.**
2.	**6.**
3.	**7.**
4.	**8.**

Go to **glencoe.com** to this book's OLC for a downloadable graphic organizer and more.

TECHNOLOGY STANDARDS

STL 12 Use & Maintenance

STL 17 Information & Communication Technologies

ACADEMIC STANDARDS

Science

NSES Content Standard E Understandings about science and technology

NSES Content Standard F Natural hazards

STL *National Standards for Technological Literacy*

NCTM *National Council of Teachers of Mathematics*

NCTE *National Council of Teachers of English*

NSES *National Science Education Standards*

NCSS *National Council for the Social Studies*

Uses of Tools

What tools and machines are commonly used in the technology lab?

Primary tools and machines are used in all areas of technology. The way they are used for processing materials determines their categories. Some tools are for measuring and laying out. Others are for holding, separating, combining, conditioning, or finishing. Tools and machines can also help diagnose and repair malfunctioning products. Tools and machines designed to cut materials will also cut you if you do not use them correctly. Your teacher will demonstrate the correct use of these tools and ask you to pass a safety test before you can use them.

As You Read

Connect What are some measuring tools that you use?

Measuring Tools and Machines

Is it possible to use a laser beam as a ruler?

Measuring tools help identify size, shape, weight, distance, density, and volume. Tools such as rulers measure materials directly. Others, such as a marking gauge and compass, help transfer measurements from one place to another. **Figure 3.4** shows several kinds of measuring tools. Measuring machines contain lasers or infrared beams to measure electronically.

Using Metric Measures

The metric system is based on units of ten and is, therefore, easier to work with than the measurement system used by most people in the United States. Many other countries use only the metric system. So their products, including those sold in the United States, are **ultimately** designed and built using metric measurements. To stay competitive, many U.S. industries and scientific institutions have adopted the metric system.

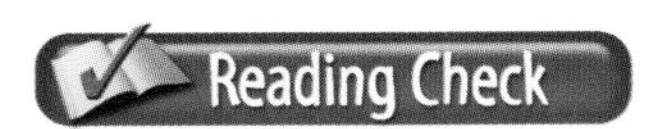

Name What are some measuring tools?

Academic Connections
Math

Calculating Metric Conversions The U.S. measurement system can be converted to metric measures, but the results are not always as accurate as results from using metric tools.

Apply Using the appendix at the back of this book, measure the length and width of this book using metric rules. Convert the measurements to U.S measurements. Use a U.S. ruler to measure the book and convert the measurements to metric measurements.

STEM

Figure 3.4 **Measuring Tools**

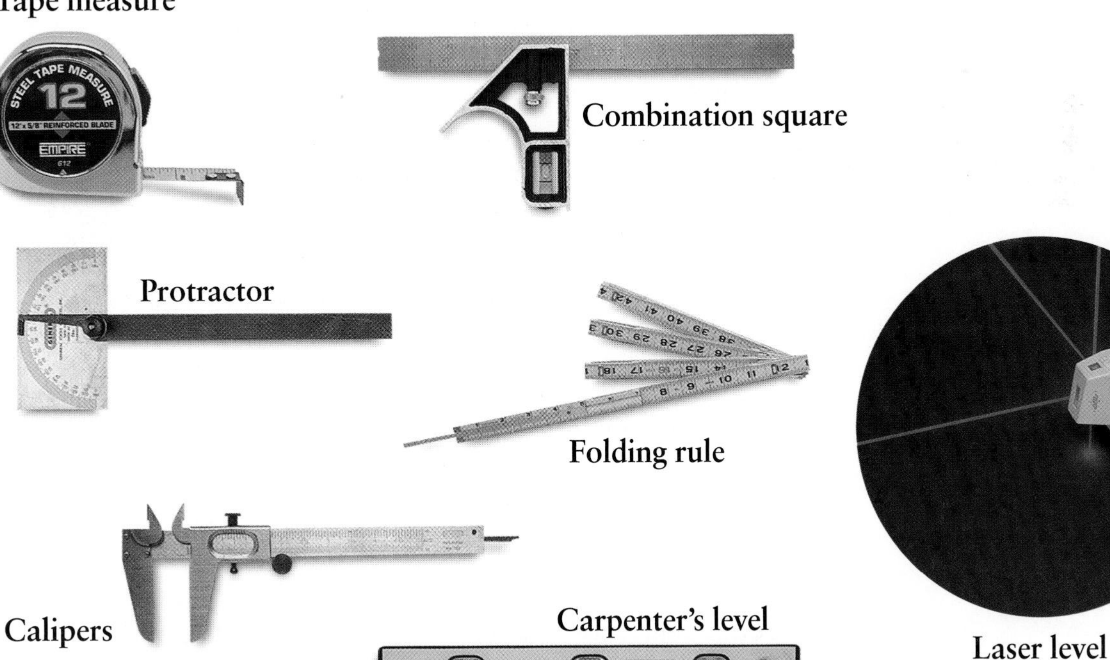

Measuring Accurately Builders and craftspeople cannot create quality products without accurate measuring tools. *What would happen if you used a ruler that had no markings?*

Figure 3.5 Holding Devices

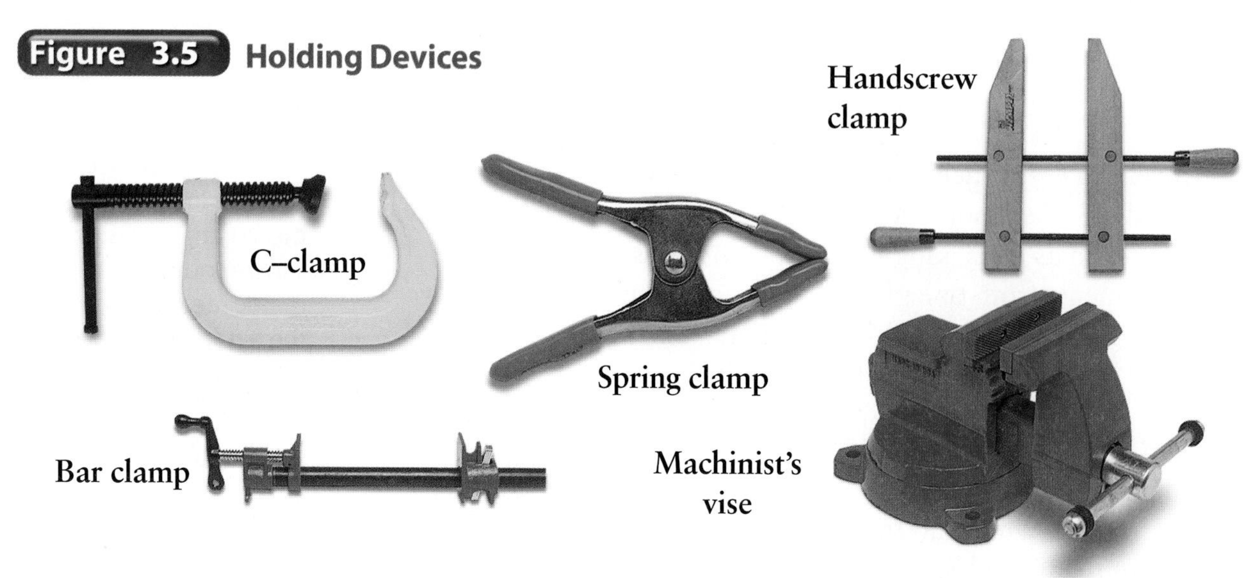

Hands-Free Holding devices come in many forms. *What are two reasons for using holding devices?*

Holding Devices

Why are holding devices needed?

If the material that you are cutting is not clamped in place, your risk of injury increases. To protect you and your workpiece when cutting, bending, drilling, or hammering, place the workpiece in a vise or clamp. **Figure 3.5** shows various holding devices.

Types of Tools

How do a hand tool and a portable electric tool differ?

A **hand tool** requires your muscle power to work. A **portable electric tool**, or power tool, is a small portable tool powered by electricity. Power equipment is usually faster and more efficient than hand tools. However, hand tools can be safer to use.

Your technology laboratory might also have some large machine tools that can also perform the same processes. **Figure 3.6** on pages 55–57 shows different types of tools.

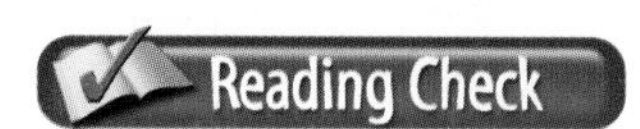

Recall What is the advantage of power tools?

Thinking about Safety

Why is safety instruction important?

Equipment, tools, materials, and activities determine the dangers of a situation. Therefore, safety rules around a swimming pool, gymnasium, and technology laboratory are different. Your teacher will provide you with general safety rules, fire safety rules, safety instruction, safety tests, and an emergency exit plan to keep you safe in your lab.

Figure 3.6 Tools

The Right Tool These are the tools of the trade for many types of technology. *What are three examples of portable electrical tools? Machine tools?*

Claw hammer—the most commonly used hammer. The curved claw provides leverage for pulling nails.

Ball-peen hammer—one face used to strike cold chisels and punches; the other face for shaping soft metal

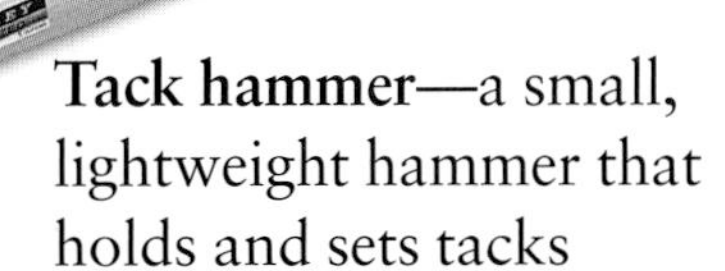

Tack hammer—a small, lightweight hammer that holds and sets tacks

Power brad nailer—drives and countersinks brads without marring the surface of the wood

Rubber mallet—used mainly for assembling projects

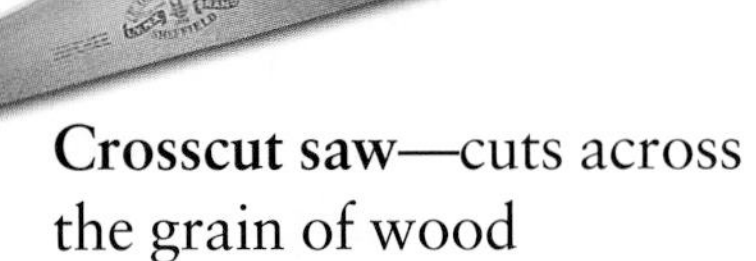

Crosscut saw—cuts across the grain of wood

Band saw—cuts curves and resaws stock to thinner sizes

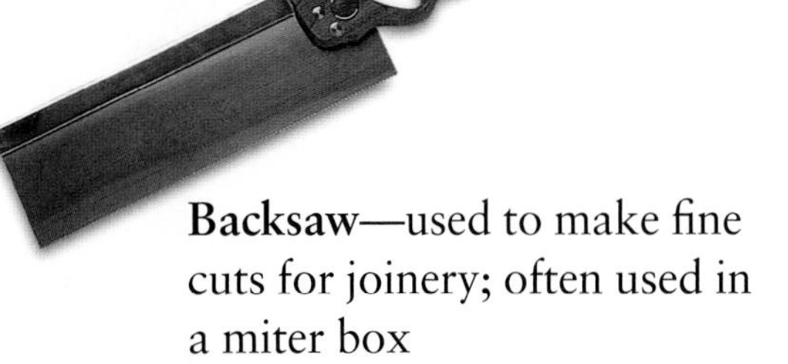

Backsaw—used to make fine cuts for joinery; often used in a miter box

Scroll saw—the best tool for intricate and accurate irregular curves

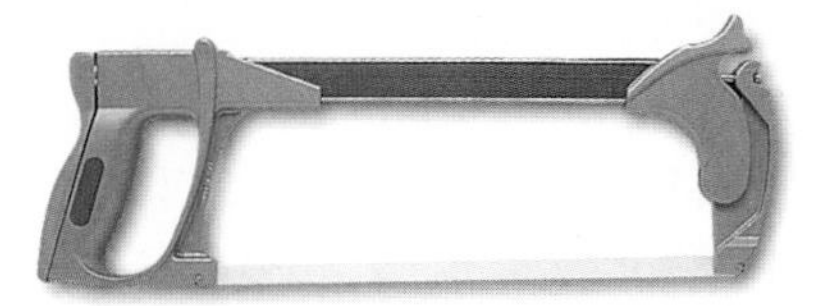

Hacksaw—cuts metal

Figure 3.6 continued

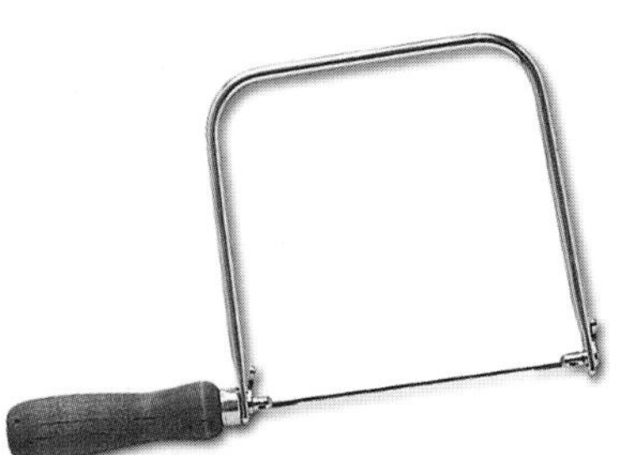

Coping saw—used to cut curves, scroll work, and molding as finishing trim

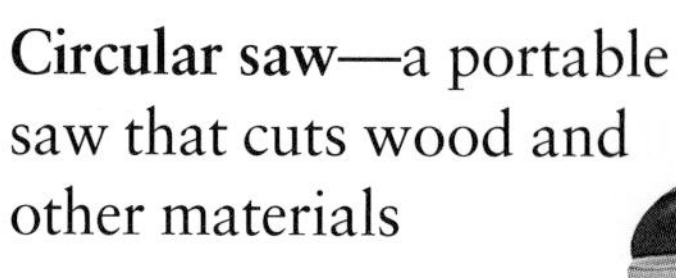

Circular saw—a portable saw that cuts wood and other materials

Brace—bores holes in wood by hand. Special auger bits must be used with the brace.

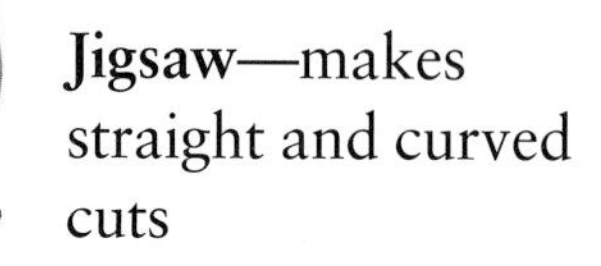

Jigsaw—makes straight and curved cuts

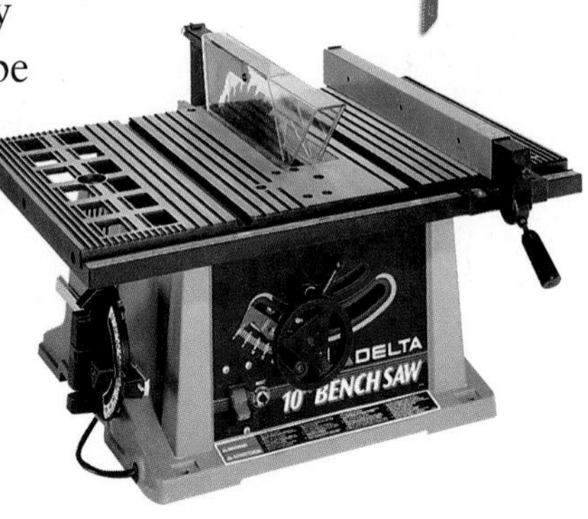

Table saw—the most commonly used saw. Its size is determined by the diameter of the largest blade it can use.

Twist bit—designed for wood. If you use it with metal, lubricate it with machine oil.

Spade bit—the long point makes it easy to place the hole exactly where you want it.

Countersink bit—drills a neat taper for the head of a wood screw

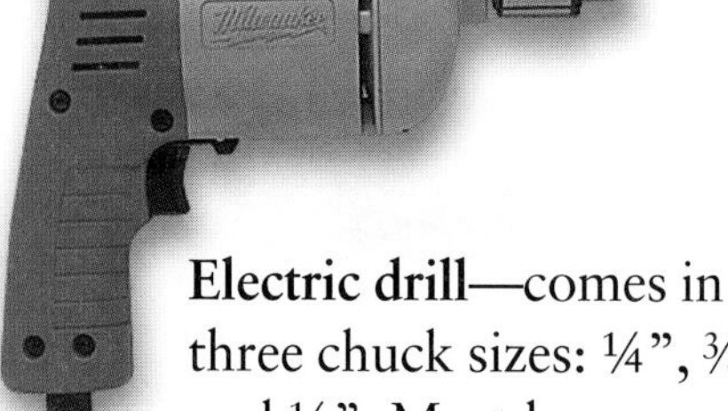

Electric drill—comes in three chuck sizes: ¼", ⅜", and ½". Most have a reverse drive and variable speed.

Hole saw—cuts large holes in wood, plastic, and thin metal

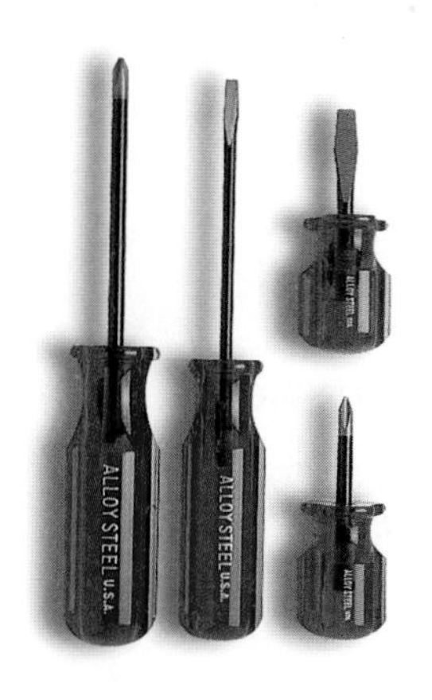

Screwdriver set—left: Phillips-head; middle: standard slotted; right: Phillips-head and standard stubby screwdrivers

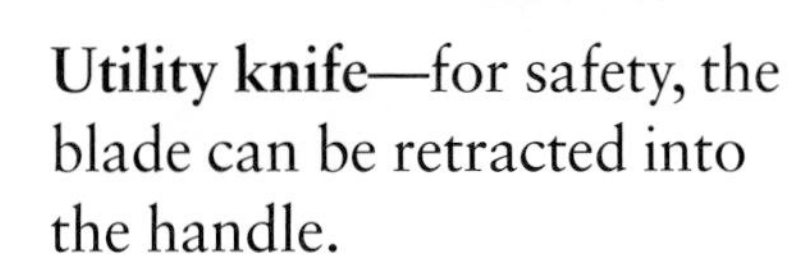

Utility knife—for safety, the blade can be retracted into the handle.

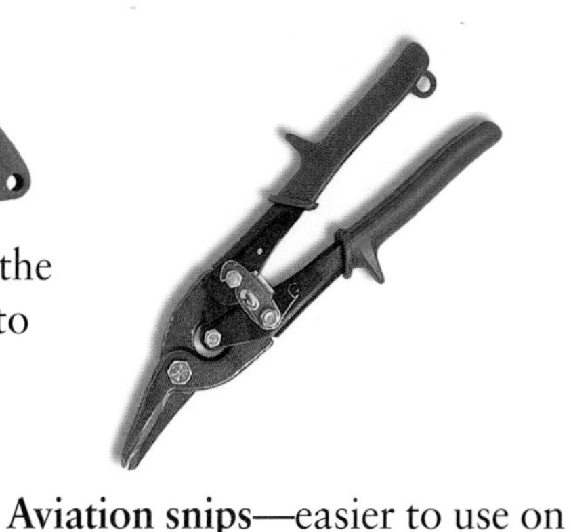

Aviation snips—easier to use on metal than tin snips; especially designed to make curved and straight cuts in metal

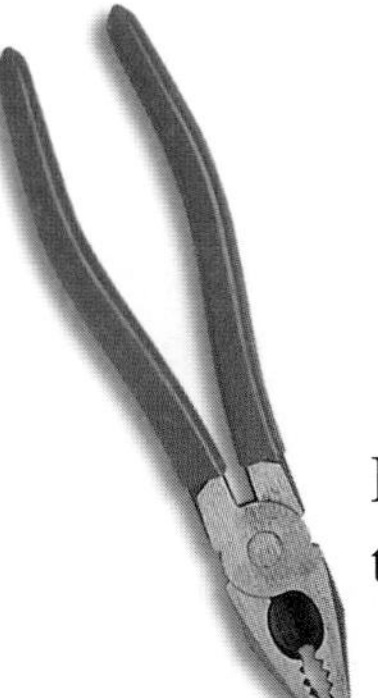

Lineman's pliers—mainly for twisting and cutting wire

Tin snips—used to make straight cuts in lightweight sheet metal

Straight-jaw locking pliers—clamps firmly to an object

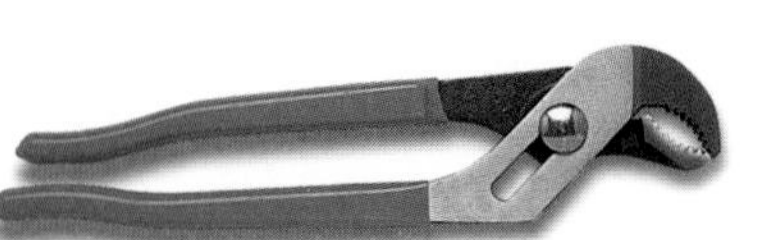

Groove-joint pliers—grips objects that are round, square, or hexagonal

Slip-joint pliers—has small and large teeth to grip objects. The jaw size can be expanded.

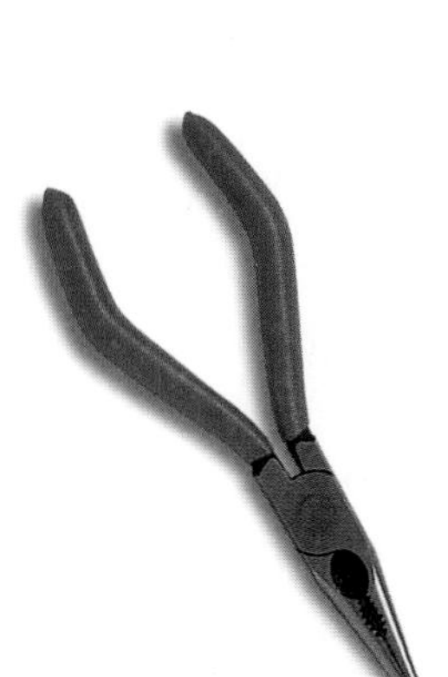

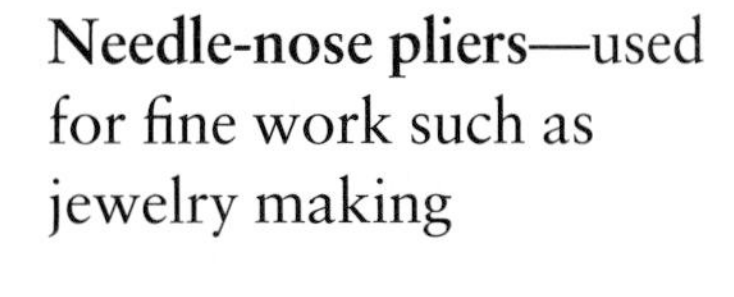

Needle-nose pliers—used for fine work such as jewelry making

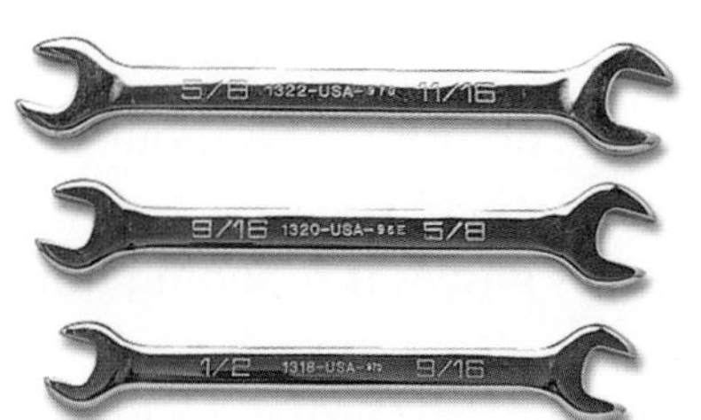

Open-end wrench—has a different size opening at each end

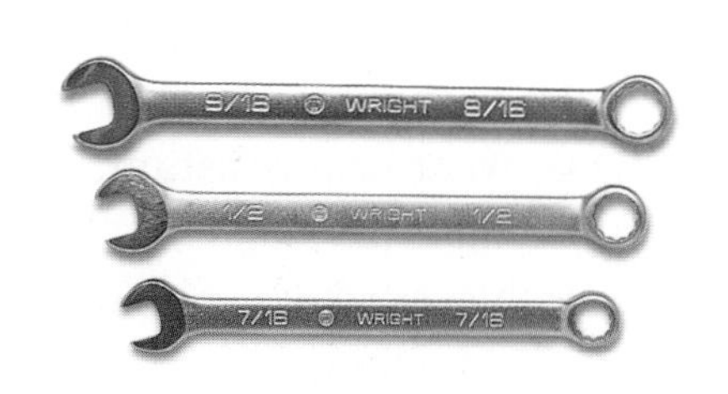

Combination wrench—has both box and open-end heads. Both ends are for the same size bolt.

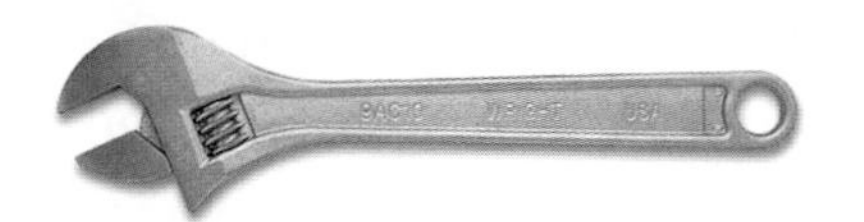

Adjustable wrench—can be used on a variety of bolts and nuts. It should be used only when a box- or open-end wrench is not available.

Figure 3.7 Colors for Safety

Red—Danger or emergency

Orange—Be on guard

Yellow—Watch out

White—Storage

Green—First aid

Blue—Information or caution

Symbolizing Safety When you see warning symbols and labels in these colors, pay attention to their meanings to maintain safety. *What color represents the most hazardous situations?*

Safety Precautions

In general, recognizing hazards is one way to avoid danger. Accidents usually occur because people are not aware of the dangers that exist around them. You can avoid accidents by having the right **attitude** and paying attention to what you are doing.

Six colors are used for signs and labels to indicate danger or other safety factors. See **Figure 3.7**. Be sure you know what they mean as you work in the lab.

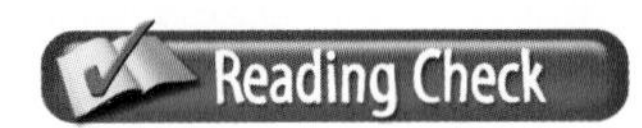

Recall Why do most accidents occur?

The Do's and Don'ts of Lab Safety

Use common sense and follow safety rules to make your experience in the lab enjoyable and safe. Here are a few basic rules:

- Protect your eyes by wearing proper eye protection.
- Wear a protective apron and roll up your sleeves.
- Never use equipment, tools, and materials unless your teacher has approved them.
- Never plug in or turn on an electrical device without your teacher's permission.
- Inform your teacher if you are injured.
- Inform your teacher if you find any broken, dull, or damaged tools or equipment.

Most injuries occur because people do not think about what they are doing. To avoid injury, follow these rules:

- Wear heat resistant, non-asbestos gloves if touching hot material.
- Never touch spinning rollers, which can pull your fingers into a machine.
- Never rest your fingers in areas where they can be pinched.
- Wear eye and clothing protection if using chemicals.
- Never wear loose clothing and jewelry near machines.
- Never use electric tools with broken wires or insulation.
- Never use tools that should be plugged into a three-prong plug in non-grounded wall outlets or extension cords.

The Safe Use of Hand Tools

Your teacher will show you the correct way to use hand tools to complete the activities in this book. Here are some reminders:

- Only use tools designed to do the particular job.
- Always cut *away* from yourself. Accidents happen when people cut *toward* themselves.
- Use sharp tools. A dull tool is more dangerous than a sharp one because people use more force to operate a dull tool, making it more likely to slip.
- Never use broken tools or tools without proper handles.

The Safe Use of Machines

The machines in your technology lab are not toys. They are designed to process materials. They cut, bend, and reshape what goes into them. To avoid injury, follow these rules at all times:

- Stay out of the safety area that surrounds a machine unless you are the operator.
- Never use any machine until your teacher has shown you how to operate it.
- Never use a machine until your teacher gives permission.
- Work alone. Be sure other people are clear of the area before you start any machine.
- Wear safety goggles.
- Watch what you are doing. Do not rush. Concentrate.
- If you have difficulty, turn off the machine. Request help.
- Never walk away from a machine that is running.

section 3.2 assessment

After You Read Self-Check

1. Name three tools used for separating.
2. Explain why a dull tool is more dangerous than a sharp one.
3. Discuss why you should wear safety glasses or goggles when using tools to cut materials.

Think

4. Explain why a hammer is considered a combining tool. Give examples.

Practice Academic Skills

English Language Arts/Reading

5. Read an instruction manual for a portable power tool to learn about using it safely. Make a presentation to the class about its safe use.

STEM Mathematics

6. Mike is a carpenter who makes display cases for the stores in his neighborhood. He cuts the pieces for the cases out of lumber that is 1 inch thick. He starts with a stack of wood that is 6 feet high. How high will the stack be, in inches, after he uses 23 pieces of wood?

Math Concept **Algebra** Use algebra to analyze change.

1. The basic equation for this problem is: Quantity before – Quantity used = End quantity
2. Remember that 1 foot is equal to 12 inches.

For help, go to **glencoe.com** to this book's OLC and find the Math Handbook.

section

3.3 Engineering Materials

Reading Guide

Before You Read **Preview** What could be the meaning of *sensory property*?

Content Vocabulary
- organic material
- inorganic material
- mechanical property
- sensory property
- alloy
- ceramic
- composite

Academic Vocabulary
- obtain
- create

Graphic Organizer
Draw a section diagram. Use it to write and organize information as you read.

Engineering Materials

1. Wood	2. ____	3. ____	4. ____	5. ____
Furniture				
Fuel				
Const. Mat'ls				
Paper				
Books				

Go to **glencoe.com** to this book's OLC for a downloadable graphic organizer and more.

TECHNOLOGY STANDARDS

STL 7 Influence on History

STL 11 Design Process

STL 19 Manufacturing Technologies

ACADEMIC STANDARDS

Science

NSES Content Standard B Properties and changes of properties in matter

NSES Content Standard F Science and technology in society

STL *National Standards for Technological Literacy*

NCTM *National Council of Teachers of Mathematics*

NCTE *National Council of Teachers of English*

NSES *National Science Education Standards*

NCSS *National Council for the Social Studies*

Engineering Materials

Why are engineering materials important?

Materials used to make products are called "engineering materials" or "production materials." They are the building blocks of our designed world. These materials must be found and processed before they are used. Materials such as oil and natural gas are obtained by drilling. Then they are processed in a refinery.

Natural materials are classified by how they originated. **Organic materials**, such as wood, leather, and cotton, come from living things. **Inorganic materials**, such as stone, metals, and ceramics, come from mineral deposits. They were never alive.

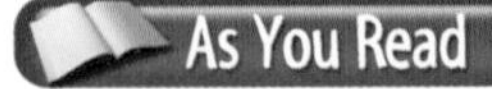
As You Read

Compare What are organic and inorganic materials?

Stanford Ovshinsky

Inventor, Engineer, Physicist

Stanford Ovshinsky was born in 1922. After high school, the self-taught engineer began working with amorphous materials, which do not have a definite crystalline structure. By the 1950s, he had created a new area of materials science.

Ovshinsky's discoveries led to electro-photography, printing, imaging, optical memory switching, and holographic information storage. His invention of a reversible optical memory disk was a prototype for the rewritable CD. Ovshinsky's work has also contributed to the electric car and solar energy.

Flat Screen Science Ovshinsky has earned 200 patents, which include patents for materials essential for photocopying and fax machines, and for flat panel liquid crystal displays (LCD) of computer monitors and flat screen televisions.

English Language Arts/Writing Research amorphous materials and write a few paragraphs describing them. Also identify products you own that may utilize them.

Go to **glencoe.com** in this book's OLC to learn about young innovators in technology.

Types of Properties

Engineers and designers check a material's properties to determine whether it is suitable to use for a particular project. The **mechanical properties** of materials are characteristics that determine how a material reacts to forces. There are four basic mechanical properties:

- **Strength—**The strength of a material is determined by how it withstands forces like tension, compression, shear, and torsion.
- **Elasticity—**Elasticity is a material's ability to stretch out of shape and return to its original shape.
- **Hardness—**This characteristic is determined by a material's ability to withstand scratches, dents, and cuts.
- **Fatigue—**This is the ability to resist bending and flexing.

Sensory properties are characteristics detected by our senses—color, texture, temperature, odor, flavor, and sound.

Chemical properties control how a material reacts to chemicals. Optical properties relate to how a material reacts to light. Thermal properties relate to how it reacts to heat or cold. Magnetic properties determine a material's reaction to magnetism.

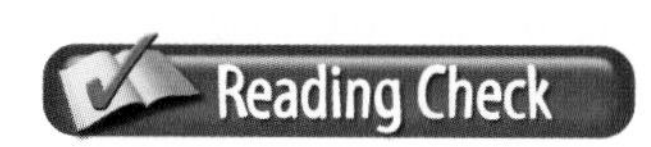

Identify What determines suitability of a material for a project?

Wood Categories

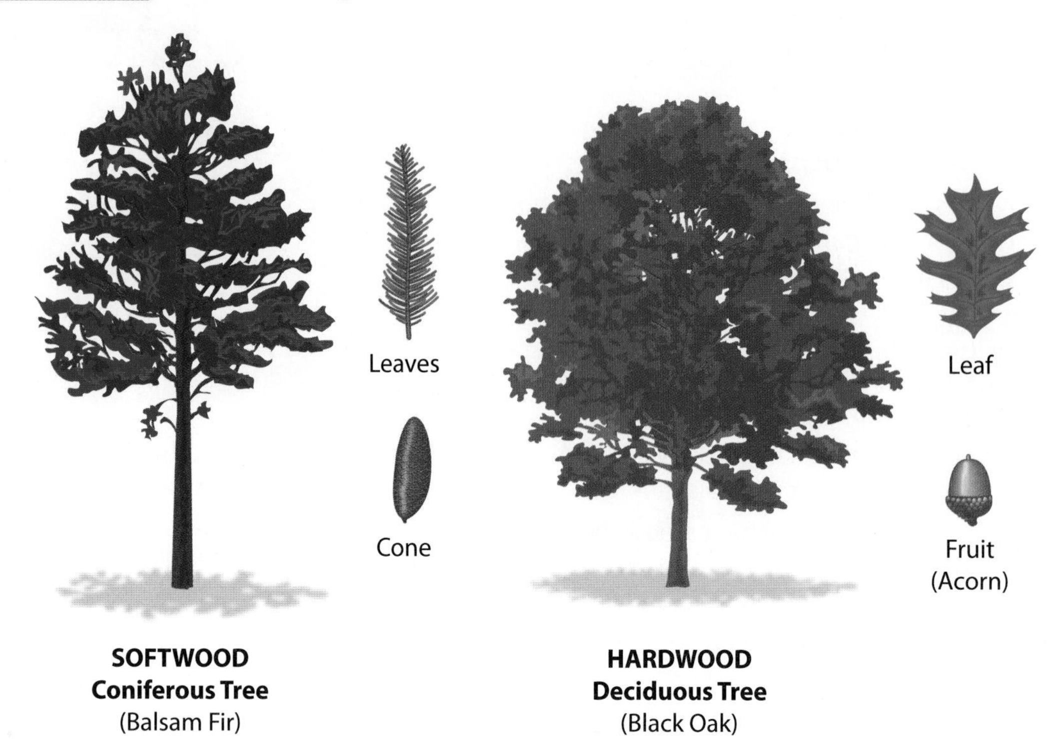

Softwood or Hardwood? Trees may be considered as either softwood or hardwood. *Which type of wood is usually used in fine furniture?*

Kinds of Materials

What materials are most often used to make products?

Our world is filled with many useful and unusual materials. However, some materials are most commonly used to make products—wood, plastics, metals, ceramics, and composites.

Wood

Trees fit into two categories—hardwoods and softwoods. If a tree bears cones and keeps its leaves all year long, it is a softwood tree. If a tree loses its leaves during cold or very dry seasons, the tree is a hardwood tree. See **Figure 3.8**.

Once harvested, trees are eventually turned into products such as furniture, fuel, construction materials, or paper. Logs are cut into lumber at a sawmill. The lumber is then seasoned to match normal humidity by air-drying the wood for at least a year or by drying it in an oven, or kiln. The lumber is then ready for additional processing.

Plastics

Plastics are synthetic materials, which means they are not found in nature. Many plastics come from petroleum. Some are produced from plants. Chemical processing can create plastics such as thermoplastics and thermosetting plastics.

- **Thermoplastic**—Thermoplastics are formed into products by using heat and pressure. When recycled, thermoplastic parts can be melted, formed, and made into new products. If recycled, thermoplastics are not as harmful to the environment as some other plastics. See **Figure 3.9**.
- **Thermosetting Plastic**—This type of plastic can be heated and formed into a product only once. Thermosetting plastics are difficult to recycle. They can be chopped up and mixed with other materials. Otherwise, they remain in landfills unchanged for centuries.

Figure 3.9 **Types of Recyclable Plastics**

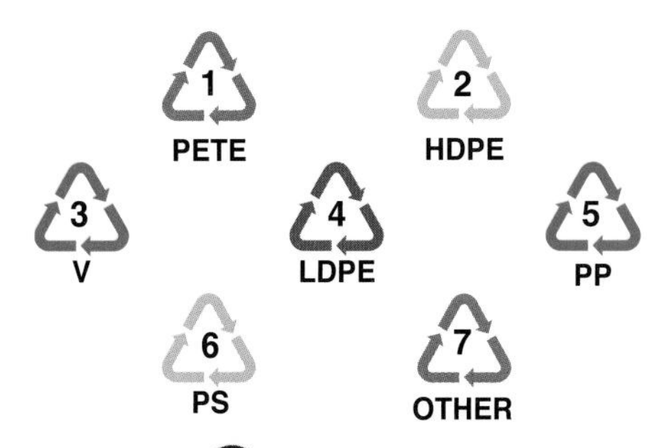

Identifying Plastics These symbols are used to identify recyclable plastics. Generally, only plastics labeled number 1 or number 2 are recycled. *Why should you know the difference between plastics?*

Metals

Metals are mined from natural rock deposits. Gold mines, uranium mines, and other mining facilities are set up for separating valuable metals from the rock. A mining operation might process tons of rock in order to produce a small quantity of pure metal. For example, the Mission open-pit copper mine in Arizona processes 2,000 pounds of rock to obtain 13 pounds of copper.

Metals are used directly or mixed with other metals or materials to create **alloys**. For example, by itself, copper is used to make copper wire and electronic components. It can be alloyed with tin to make bronze. Iron is alloyed with carbon and other materials to make types of steel. Other metals used in industry include aluminum, chromium, zinc, and lead.

Metals can be processed into many different shapes. Steel, for example, is made into beams for construction that are I-shaped, U-shaped, or L-shaped. Other metal shapes include a sheet, bar, rod, square, hexagon, tube, angle, channel, and octagon.

Ethics in Action

A Call to Recycle

Consumers buy and enjoy having new phones with the latest designs and features. As a result, Americans discard 125 million used cell phones each year. People throw them away even if they still work.

Bad Call Unfortunately, many electronic devices contain toxic substances that can harm the environment. So, electronic waste, or "e-waste," is becoming a major problem in the United States and other countries.

English Language Arts/Writing

Can You Hear Me Now? Many businesses recycle electronic devices. Do some research on the Internet about recycling.

1. Find out which businesses and organizations recycle cell phones.
2. Find out where to recycle or donate used cell phones in your area.
3. Design a poster that includes your findings and encourages recycling.

EcoTech

Environmental IT

How do you make a computer run "green"? Cleaner manufacturing, energy-efficient cooling systems, and recycling old parts help the environment. Using computer memory efficiently is another way.

Try This Back up files and use your disk space fully. Delete junk mail, spam, and unnecessary files in your e-mail inbox to gain more memory.

Ceramics

Ceramics are made from inorganic, mostly nonmetallic, minerals, such as clay, sand, or quartz. High temperatures fuse these minerals into useful products. The firing of clay and sand are ancient technologies.

The two oldest ceramic products, pottery and glass, are very different. Pottery and almost all other ceramics except glass are *thermosetting* materials. Once they are heated and formed, they can never be softened again. Glass, however, can be formed again and again by using heat. Ceramics are used to make sandpaper, pottery, dinnerware, bathroom fixtures, spark plugs, space shuttles, and a variety of other products.

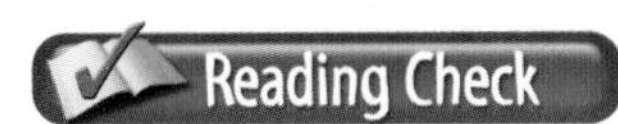

List What are the most used materials?

Composites

A **composite** is a material created by combining two or more materials to form a new material that is better than each of the original materials would have been. A composite's ingredients provide the correct physical properties, and a binder holds the materials together.

Composites are made with glue, resin, or epoxy binder to bond layers of wood or wood fibers, Kevlar®, or metal together. Concrete is a composite material that is made by mixing sand, gravel, and Portland cement. Most buildings could not be built without using concrete.

Section 3.3 Assessment

Self-Check

1. Name three organic materials that technology turns into products.
2. Explain the difference between a thermoplastic and a thermosetting plastic.
3. Define the words *softwood* and *hardwood*.

Think

4. Name three materials used to make this textbook.

Practice Academic Skills

English Language Arts/Writing

5. Investigate one of the material properties discussed in this chapter. Develop a chart, report, or display that shows what the property is all about.

Social Studies

6. The material used to make the products we use comes from all over the world. Choose a common household product or industrial product. Determine the materials from which the product is made. Research where these materials are found, and then write a short report on your discoveries.

Exploring Careers *in* Technology

Jess Clark

FORESTRY TECHNICIAN, REMOTE SENSING AGENT

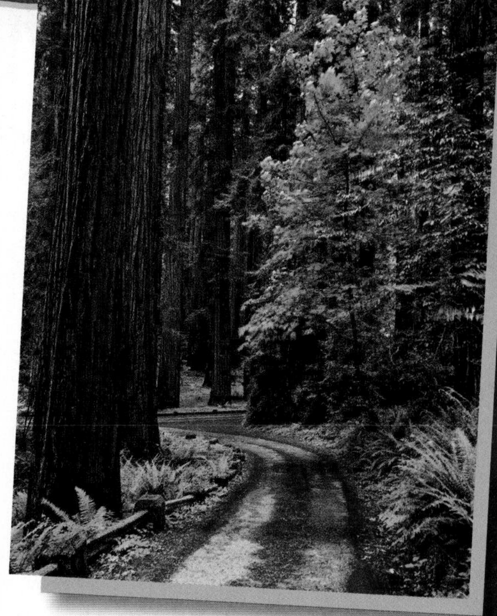

Q: *What do you do?*

A: I use remote sensing (satellite imagery) to create maps of wildfires that show the most severely burned areas. I highlight areas of greatest concern that could be affected by weather events after a fire. One of the biggest problems after a wildfire is the loss of natural vegetation. The ground is more likely to erode during the next rainstorm, and there is a danger of floods and mudslides.

Q: *What kind of training and education did you need to get this job?*

A: I have a degree in geographic information systems (GIS) and a Master of Science in geography, with an emphasis on remote sensing. I also interned with the U.S. Forest Service while I was in college. That government internship turned into a full-time job after I graduated. I worked on my master's degree while working full time.

Q: *What do you like most about your job?*

A: Mapping wildfires is exciting. I map fires all over the country and look at some pretty fascinating imagery. People in the field use my maps, which show the burn severity, to make fast decisions in emergency situations.

Q: *How did you get interested in your job?*

A: I loved geography in high school and was good on a computer. Remote sensing and GIS are perfect fits for me.

English Language Arts/Writing

Tools of Technology The tools that people use to do their jobs have advanced over time.

1. Identify the tools used by a forestry technician, remote sensing agent.
2. Research these tools and the tools that forestry technicians used in the past.
3. Write a short report on your findings.

Go to **glencoe.com** to this book's OLC to learn more about this career.

Real-World Skills

Speaking, listening, analyzing, problem-solving, flexibility

Academics and Education

Geography, computer technology, meteorology, cartography

Career Outlook

Growth as fast as average for the next ten years

Source: *Occupational Outlook Handbook*

chapter 3 Review and Assessment

Chapter Summary

Section 3.1 Separating is removing pieces of a material. Forming changes the shape of materials. In bending, material is formed by moving a part into a different position. In casting, a liquid material is poured into a mold. In compression, a flat material is pressed into a mold. Forging is shaping metal by heating and hammering. In extruding, softened material is squeezed through a small opening. Combining is joining several parts. Mechanical fasteners hold parts together. Conditioning changes the inner structure of a material to improve it. Finishing is the last step.

Section 3.2 Measuring tools are used to identify size, shape, weight, distance, density, and volume. Muscle power operates a hand tool. Electricity supplies the power to operate a portable electric tool. Before you begin a "hands-on" activity, stop and think seriously about safety.

Section 3.3 Mechanical properties of materials relate to how a material reacts to forces. Sensory properties are those we perceive with our senses. Lumber comes from trees. Plastics are synthetic. Metals that are mixed with other metals or other materials create alloys. Ceramics are made from minerals. Composites are two or more materials combined to form a new material.

Review Content Vocabulary and Academic Vocabulary

1. On a sheet of paper, use each of these terms and words in a written sentence.

Content Vocabulary
- separating
- forming
- combining
- conditioning
- finishing
- measuring tool
- hand tool
- portable electric tool
- organic material
- inorganic material
- mechanical property
- sensory property
- alloy
- ceramic
- composite

Academic Vocabulary
- technique
- similar
- ultimate
- attitude
- obtain
- create

Review Key Concepts

2. **Identify** the different technology processes.
3. **Explain** how the different processes are used.
4. **Describe** the purpose of two hand tools.
5. **Describe** the purpose of portable power tools.
6. **Discuss** how to prevent injury using tools.
7. **List** some basic properties of materials.
8. **Identify** common engineering materials.
9. **Explain** metric measurements.
10. **Explain** natural materials.

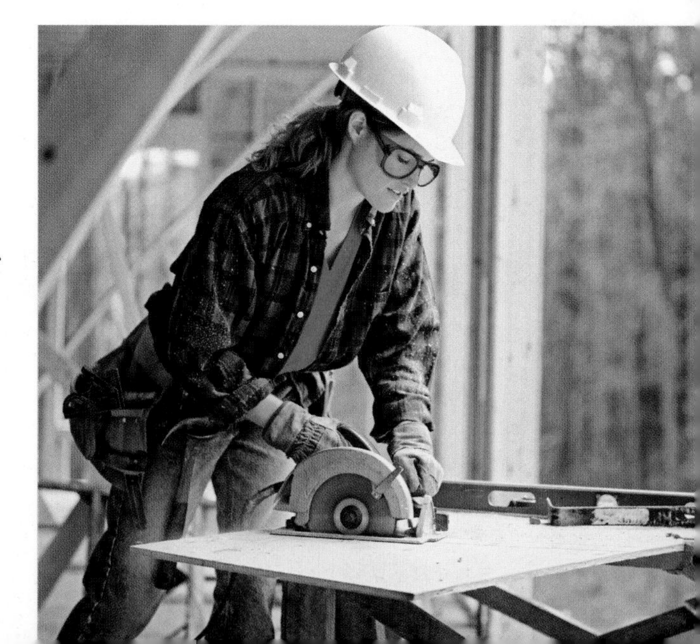

Real-World Skills

11. **Self-Assess** Think about accidents you have had in the past. What could you have changed about your behavior that might have prevented the accidents? Write a paragraph discussing your conclusions.

STEM Technology Skill

12. **Joining Processes** Glue is an adhesive that forms a film on surfaces that are to be joined. This film adheres to both surfaces being joined and holds the parts together. Different glues are used for different materials.
    a. Use the Internet to find a glue manufacturer that sells different types of glue and explains their uses.
    b. What kind of glue is used on wood? What kind is used on ceramics?

Academic Skills

Science

13. Many products are made from nonrenewable resources. Choose a product that uses such material. Research materials made of renewable resources that could be used to make the product. Find out why companies are not using them. Present your findings in class.

STEM Mathematics

14. Eric is a roofer in charge of laying down sheets of protective material before shingles are applied to a roof. The roof he is working on is 43 feet long and 19 feet wide. What is the area of the material needed for the roof?

Math Concept **Area** Surface area is found by multiplying the length times the width of a two-dimensional surface. Calculating surface area will result in an answer in square units.

WINNING EVENTS

Product Designer

Situation Your team is designing a product to sell to students. The prototype will be made of any materials available. Use discarded material from local businesses. Use tools and machines in your lab to make a prototype.

Activity Brainstorm things to make. Choose and make sketches and prototypes of the best ideas. Compare all prototypes by evaluating which would produce the greatest profit. You may survey classmates as research.

Evaluation The prototypes will be evaluated by these criteria:

- Product design—inventive, easy to make
- Marketable—students will want to buy it
- Profitable—team will make a profit

Go to **glencoe.com** to this book's OLC for information about TSA events.

Standardized Test Practice

Directions Choose the letter of the best answer. Write the letter on a separate piece of paper.

1. Which process is a forming process?
   A separating **C** casting
   B combining **D** conditioning
2. Organic materials come from mineral deposits.
   T
   F

Test-Taking Tip Space your studying out over a few days or weeks, and continually review class material. Do not wait until the night before to cram and learn everything at once.

chapter 3

TECHNOLOGY LAB

Make a Plaster Casting

Some buildings and monuments have engraved cornerstones that are inscribed with the date of construction or other information. You can cast a copy of these inscribed images after you make a mold from the original image.

Tools and Materials

- ✓ Heavy-duty aluminum foil
- ✓ Scissors
- ✓ Adhesive tape
- ✓ Foam shaving cream
- ✓ Thick, heavy-duty cardboard or lightweight wood larger than the image
- ✓ Plaster of Paris
- ✓ Bowl, stirring stick, and water for plaster mixture
- ✓ Paint or dark wax shoe polish

Set Your Goal

Your goal for this lab is to first locate an engraved image that you would like to copy. Then you will make a mold and cast a replica of the engraved image.

Know the Criteria and Constraints

In this lab, you will:

1. Determine if you want to copy an image that is part of private or government property.
2. Obtain permission from the owner or manager of the property.

Design Your Project

Follow these steps to design your project and complete this lab.

1. Find an engraved image that is fairly small, about 8 inches by 10 inches.
2. Go to the site of the engraving with a group.
   - Bring foil, scissors, tape, shaving cream, and cardboard.
   - Cut a piece of aluminum foil that includes a 3- to 4-inch border.
   - Center the foil over the image and carefully push it into the image with your fingers. Work slowly and carefully to avoid tearing the foil.

SAFETY

Reminder

In this lab, you will be using tools and materials. Be sure to follow appropriate safety procedures and rules so you and your classmates do not get hurt.

3. Create a mold of the engraving.
   - While one person holds the foil, another covers the piece with shaving cream.
   - Quickly place the cardboard or wood on the shaving cream, which acts as a light glue.
   - Carefully pull back on the cardboard to pull the foil from the building surface.

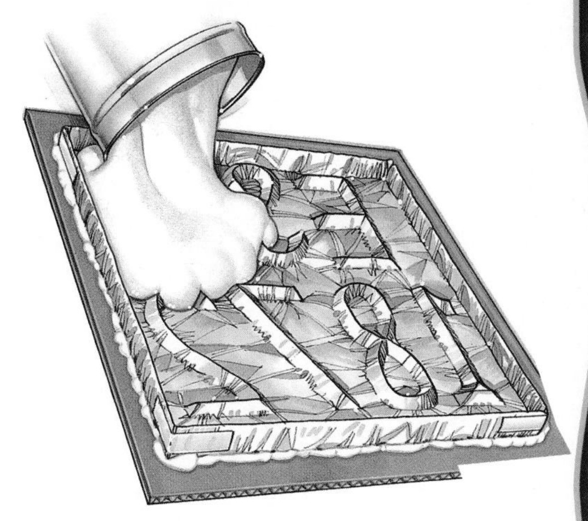

4. Return to your classroom.
   - Place the cardboard on a table with the foil facing up.
   - Fold up the edges of the foil, and double them over to form sides.
   - Strengthen the corners of the mold with tape.
5. Create the casting.
   - Mix the plaster of paris according to directions. Make enough to cover the mold.
   - Pour the plaster into your mold and shake the mold a little to help prevent air bubbles. It hardens in a short time, so work quickly.
   - After the casting hardens, peel away the aluminum foil. Your casting will be a brilliant white.
6. Paint or rub the background with dark wax shoe polish to make the message easier to read.
7. Display the casting in class with a description and photo of the original image.

Evaluate Your Results

After you complete the lab, answer these questions on a separate piece of paper.

1. Did your first casting turn out as well as you expected?
2. How long did the plaster remain workable before it hardened?
3. What properties did it have that made it suitable for this activity?

Note: If you wish, make another casting to improve your skill.

Academic Skills Required to Complete Lab

Tasks	English Language Arts	Math	Science	Social Studies
Research local sites to find engravings.		✓		✓
Make mold and plaster cast.		✓	✓	
Write evaluation of process.	✓		✓	
Create presentation of cast.	✓		✓	✓
Present to the class.	✓			

chapter 4

Design and Problem Solving

Sections

What You'll Learn

- **Explain** how creativity influences design.
- **Explain** the importance of engineering and appearance in design.
- **Discuss** ways to evaluate designs.
- **List** the six steps in problem solving.
- **Describe** an effective problem statement.

Explore the Photo

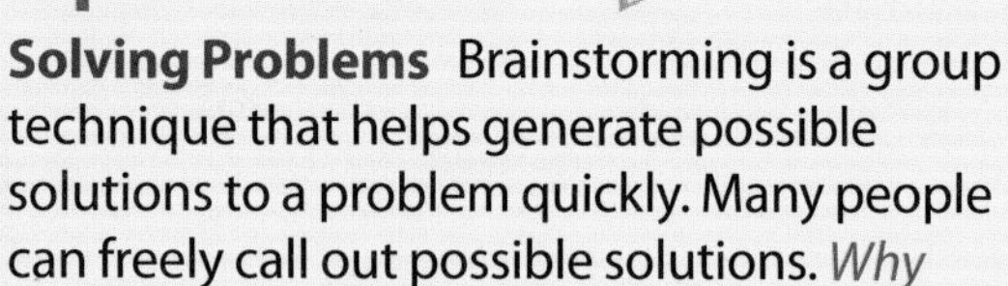

Solving Problems Brainstorming is a group technique that helps generate possible solutions to a problem quickly. Many people can freely call out possible solutions. *Why should you consider all ideas in a brainstorming session, including silly ones?*

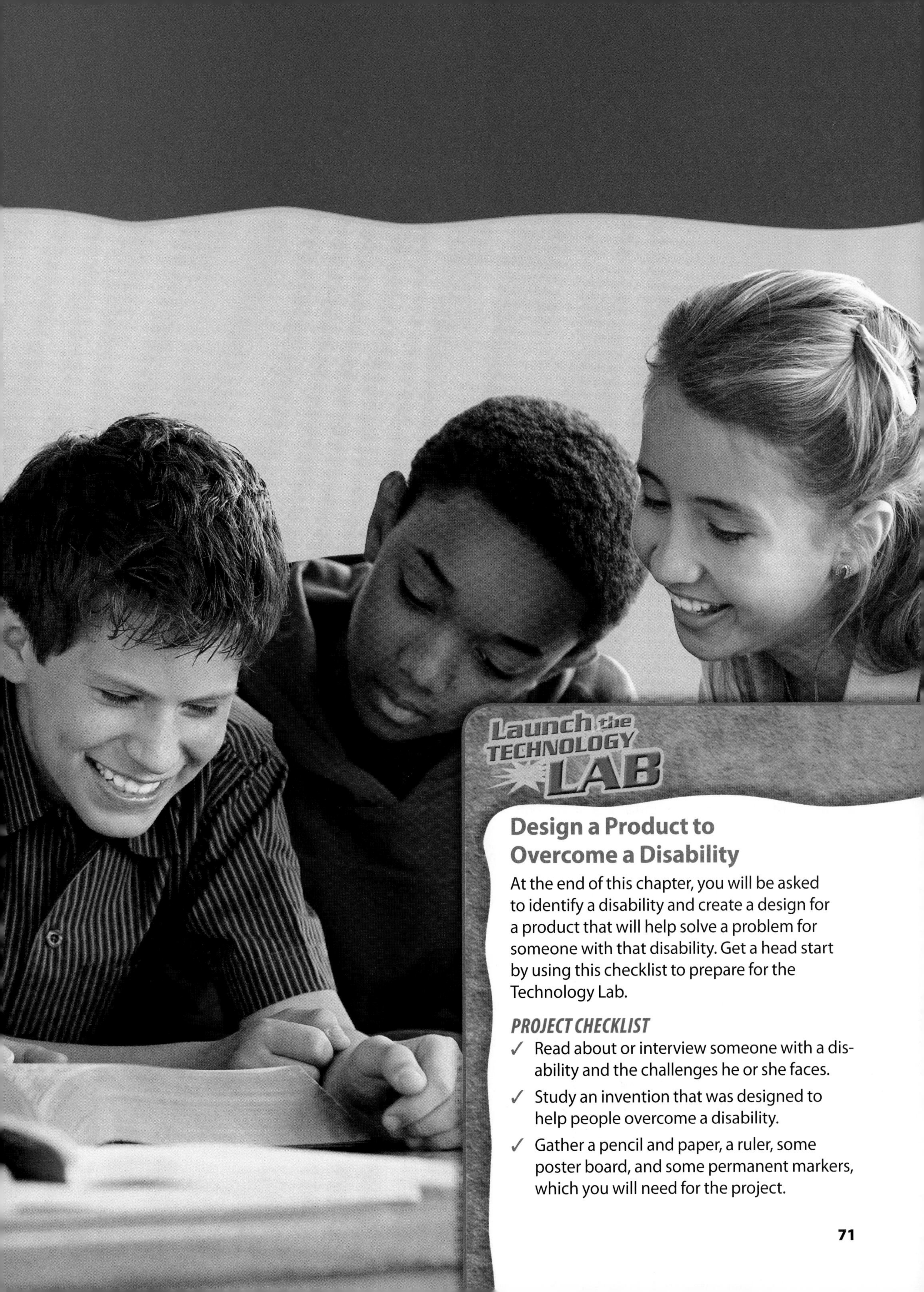

Design a Product to Overcome a Disability

At the end of this chapter, you will be asked to identify a disability and create a design for a product that will help solve a problem for someone with that disability. Get a head start by using this checklist to prepare for the Technology Lab.

PROJECT CHECKLIST

✓ Read about or interview someone with a disability and the challenges he or she faces.

✓ Study an invention that was designed to help people overcome a disability.

✓ Gather a pencil and paper, a ruler, some poster board, and some permanent markers, which you will need for the project.

section 4.1

The Design Process

Reading Guide

 **Before You Read** **Preview** What is the first step in the design process?

Content Vocabulary

- innovation
- invention
- human factors engineering
- balance
- proportion
- unity

Academic Vocabulary

You will see these words in your reading and on your tests. Find their meanings at the back of this book.

- theory
- range

Graphic Organizer

Draw the section diagram. Use it to organize and write down information as you read.

Principles of Design

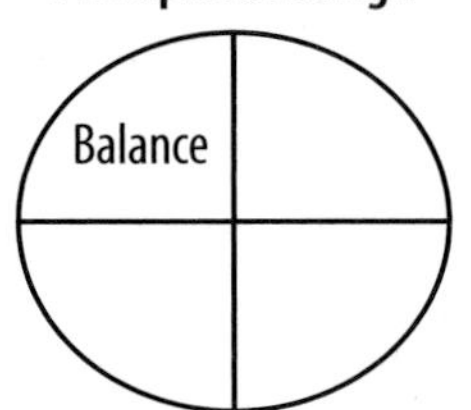

Go to **glencoe.com** to this book's OLC for a downloadable graphic organizer and more.

TECHNOLOGY STANDARDS

STL 1 Characteristics & Scope of Technology

STL 10 Troubleshooting & Problem Solving

ACADEMIC STANDARDS

Mathematics

NCTM Algebra Represent and analyze mathematical situations and structures using algebraic symbols.

Science

NSES Content Standard E Abilities of technological design

STL *National Standards for Technological Literacy*

NCTM *National Council of Teachers of Mathematics*

NCTE *National Council of Teachers of English*

NSES *National Science Education Standards*

NCSS *National Council for the Social Studies*

The Role of Creativity

How does creativity influence design?

As You Read

Connect What role does engineering play in the design of new products?

The design of all new products and processes begins with an idea. In 1837, John Deere was the first person to use a steel plow. Previously, farmers had used only wooden or iron plows. Deere must have done some creative thinking, which is the first step in the design process. Creativity often leads to ideas that are original and imaginative.

John Deere improved on the wooden plow. Modifying an existing product is **innovation**. Thomas Edison owned 1,093 patents, the most patents ever granted to one person by the U.S. Patent Office. Edison's favorite invention was the phonograph. No one before him had ever attempted to patent a similar recording device. **Invention** is turning ideas and imagination into new devices or systems.

A Earth-Friendly School Supplies These lunch bags are creative and reusable. They were designed to insulate food and drinks separately. *Why do these lunch bags come in different colors and patterns?*

The Role of Engineering

How is engineering similar to design?

Design influences how well things work. A cell phone with buttons that are slightly raised is easier to use than one with buttons that are flat or flush with the surface. Engineers make sure the product works well. They also make sure it is durable, reliable, and easy to maintain.

Engineers use mathematics and science to help calculate strength and other important characteristics of a design. Designers and engineers must be able to think in terms of principles and theories as well as objects.

Human Factors Engineering

Engineering also considers how a product relates to the human body. This application is called **human factors engineering**, or "ergonomics." Clothes are more comfortable, playgrounds are safer, and tools are easier to use when designers think about how these things will fit the people who use them.

Flexibility

Engineering can also determine a product's flexibility. The design may have to be flexible enough to meet the needs of many potential users. Some refrigerator doors can be hinged on either the left side or right side. Clothing marked "medium" must include a range of sizes of people. In addition, some software works with more than one computer operating system.

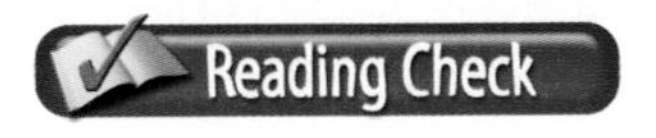

Discuss How does engineering influence the design of new products?

Ethics in Action

Green Design

Designers are trying to create products that do not harm the environment. They are building automobiles, electronic devices, and schools that use less energy and make less waste. This is called "green design."

You've Got Green The computer manufacturing company HP has already built a "green" computer. The rp5700 slim desktop PC contains no toxic materials. It is made so that it can be taken apart and recycled easily. For the computer's casing and packaging, HP used recycled materials.

English Language Arts/Writing

More or Less? You are designing a new computer. You could use materials that are good for the environment, but they will cost more.

1. You must choose to charge more for the computer or make less money.
2. Write a short paragraph explaining what you would do and why.

The Role of Appearance

What makes a product attractive?

The appearance of a product is usually important. Dinnerware, for example, is designed to be attractive, as are clothes, cars, and MP3 players. The function of all of those products is enhanced when they are attractive. An unattractive cell phone might work just as well as one with a sleek, high-tech design. However, the attractive one looks better and gives an impression of quality to most people.

We all have different ideas about what we find attractive. That is one reason why products come in a variety of designs. Designers follow certain principles that usually produce attractive and successful results. These design principles include balance, proportion, contrast, and unity.

- **Balance**—A sense of **balance** is achieved when different elements in a design are arranged to seem stable and steady. For example, you probably would not place all the furniture on one side of your living room. The arrangement would seem off balance.
- **Proportion**—Different parts of a design should be in proportion to one another. **Proportion** is the proper relation of sizes or quantities. Putting huge tractor wheels on a tiny car would create a problem of proportion.
- **Contrast**—Sometimes, differences in size or materials can deliberately call attention to one part of a design. This effect is called "contrast." For example, a designer might place a red collar on a white shirt for contrast.

- **Unity**—How do you know when everything in a design works together? The design has **unity**—which means function and form are carefully planned and are in harmony.

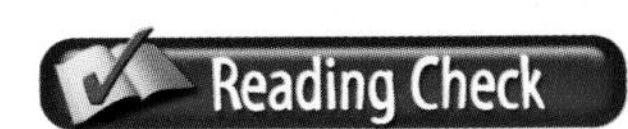

Explain What makes a product attractive?

Evaluating Design

How are designs evaluated?

During the design process, the design needs to be checked and critiqued. Sometimes basic ideas about a design are changed. For example, someone designing an energy-efficient light bulb may learn about a new technology that will make the original idea out of date. Designers are always refining and improving ideas.

After a new design is developed, engineers construct a model or prototype on which to test it. When an airplane company designs a new aircraft, the engineers build one for flight testing. A test pilot follows a carefully planned flight procedure. Engineers evaluate the tests, and the airplane is changed if necessary before manufacturing begins.

Today engineers use special computer programs to test many product designs. These programs analyze different aspects of the design, such as safety factors.

Packing Your Lunch

Did you know that one student's average school lunch generates about 67 pounds of waste a year? Plastic bags, plastic bottles, plastic forks and spoons, food packaging, and paper bags go in the trash and might not be recycled.

Try This Place your food in containers and lunch bags or boxes designed to be washed and reused. You can reuse plastic bottles and recycle them.

section 4.1 assessment

After You Read Self-Check

1. Explain why creativity is the first step in the design process.
2. Identify the influence engineering has on a design.
3. Name at least three principles that designers follow when considering appearance.

Think

4. Name some items with which you are familiar and that you think are well designed. Explain why. Name some you think are poorly designed. Explain why.

Practice Academic Skills

Science

5. Obtain a glider airplane made of balsa wood. Fly it several times to observe flight characteristics. Make a list of its criteria and constraints. Make some design changes by trimming a little wood from both ends of the wing and see how it flies. Do the same for the top of the vertical fin at the rear and the ends of the horizontal stabilizer at the rear. Keep trimming and flying. Write your observations and conclusions.

STEM Mathematics

6. Kyle is designing and constructing a small storage locker. The base needs to be 12.5 feet long by 8.6 feet wide. He needs the locker to hold 1,677 cubic feet of material. How high should he make the walls?

Math Concept **Algebra** It helps to construct equations to solve problems. Use variables such as x or y to represent unknown quantities.

1. The equation for volume is volume = length × width × height.
2. Use h to represent the height.

For help, go to **glencoe.com** to this book's OLC and find the Math Handbook.

section 4.2

Problem Solving

Reading Guide

Before You Read **Preview** How might problem solving relate to design?

Content Vocabulary

- problem statement
- brainstorming

Academic Vocabulary

- sequence
- evaluate

Graphic Organizer

Draw the section diagram. Use it to organize and write down information as you read.

Six Steps in Problem Solving

1. Define the Problem
2.
3.
4.
5.
6.

Go to **glencoe.com** to this book's OLC for a downloadable graphic organizer and more.

TECHNOLOGY STANDARDS

STL 8 Attributes of Design

STL 9 Engineering Design

STL 10 Troubleshooting & Problem Solving

ACADEMIC STANDARDS

Science

NSES Content Standard E Understandings about science and technology

English Language Arts

NCTE 7 Conduct research and gather, evaluate, and synthesize data to communicate discoveries.

STL *National Standards for Technological Literacy*

NCTM *National Council of Teachers of Mathematics*

NCTE *National Council of Teachers of English*

NSES *National Science Education Standards*

NCSS *National Council for the Social Studies*

Problem Solving Steps

How can you apply problem solving to designing?

Problem solving is a part of the design process. The steps in problem solving help designers arrive at a design that does the best job. These steps are: 1) defining the problem, 2) collecting information, 3) developing possible solutions, 4) selecting one solution, 5) putting the solution to work, and 6) evaluating the solution. You can perform these steps in sequence and repeat them.

Predict What is the purpose of a problem statement?

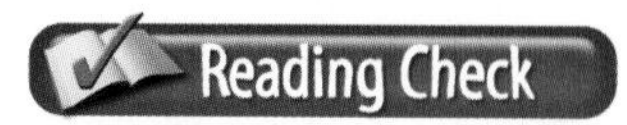

List What are the six steps in problem solving?

Define the Problem

What does it mean to "define the problem"?

Before you can solve a problem, you must identify it. Design problems are seldom clearly defined. However, even when the problem is easy to identify, it is important to define it.

The Problem Statement

Engineers and product designers create a problem statement. A **problem statement** clearly identifies the problem that the product will solve. Suppose your friends have asked you to design a bookcase for them. (See **Figure 4.1**.) To write a problem statement, you need to know something about the bookcase, such as its location, capacity, and cost.

Writing the Problem Statement

A problem statement for the bookcase might look like this: *Design a bookcase with two shelves that measures 28 inches tall, 33 inches wide, and 12 inches deep. It must be made from wood that is finished to match the living room furniture, which is made of walnut. Cost of materials must not be more than $75.*

A problem statement helps focus your thinking. A poorly worded problem statement results in wasted effort. "Design an improved animal carrier" is not an effective statement.

Figure 4.1 The Problem Solving Process

Define the problem

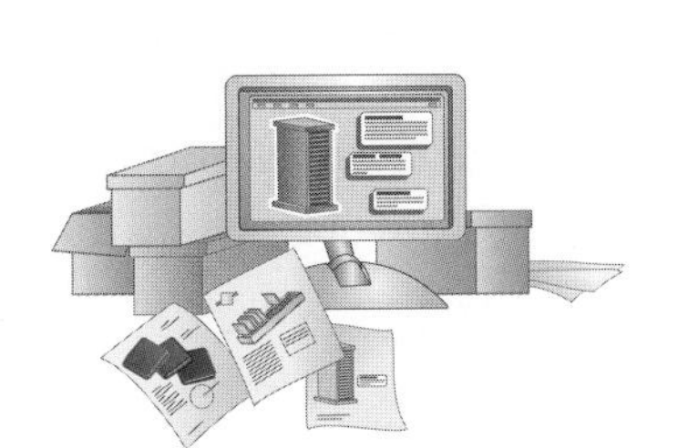

Collect information

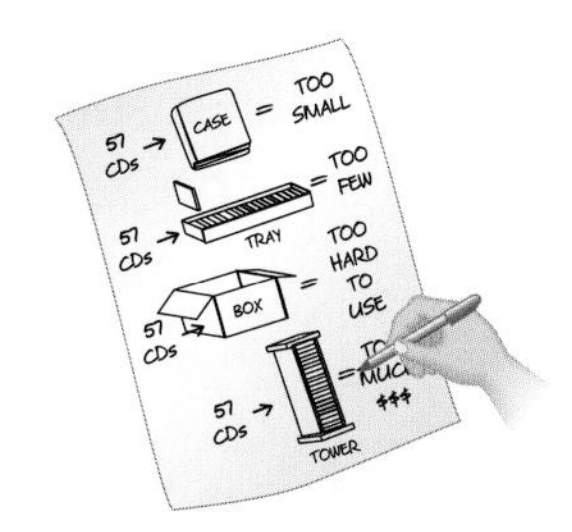

Develop possible solutions

Select a solution

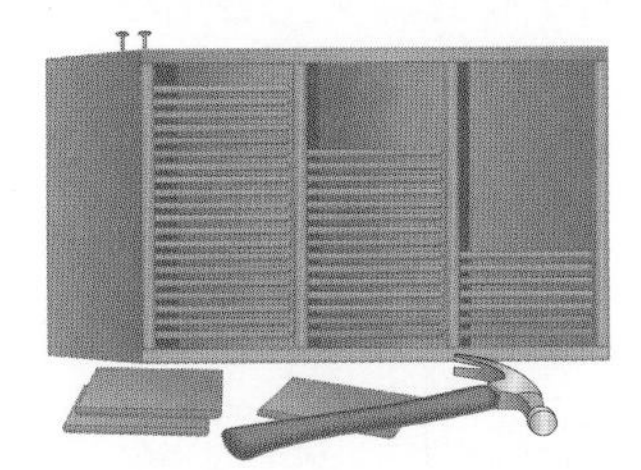

Put the solution to work

Evaluate the solution

Step by Step The steps in the problem solving process can be repeated as often as needed. *Why is it important to define the problem?*

Tune Your Sunglasses

Some sunglasses darken when you wear them in bright sunlight. Imagine sunglasses that change their color and tint at the touch of a button. Technologists at the University of Washington have used creative thinking to create these high-tech sunglasses. The lenses use voltage from a small battery to alter the color of special material. *What are some of the criteria and constraints for designing sunglasses?*

Go to **glencoe.com** to this book's OLC for answers and to learn more about high-tech materials.

Revising the Problem Statement

An improved problem statement for the animal carrier might be: *Design a plastic and wire-screen carrier for cats weighing up to ten pounds. It must have a latching door hinged at the side and a top-mounted handle.* (See **Figure 4.2**.)

Criteria and Constraints

Criteria are usually identified once the problem is defined. These criteria may be included in the problem statement. They are the requirements and specifications that help establish what designers must achieve. Designers must determine the constraints, or limits, on the design at the beginning, or they may waste time.

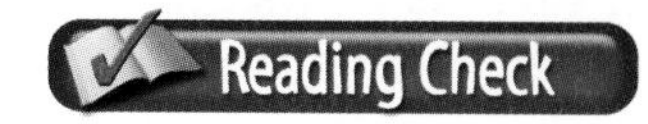

Explain How can you write a good problem statement?

Collect Information

Why should you find out about past solutions to problems?

If other people have already worked on a similar problem, look at what they have done. This could save you time and lead to a better solution. You also need to learn about all the factors associated with the problem. For example, you might want to test other animal carriers to learn which are most appealing to cats.

Develop Possible Solutions

Does every problem have one perfect solution?

Most problems have more than one possible solution. However, some solutions may work better than others. Some may be more practical, less costly, more efficient, or easier to produce. At this stage, it is time to develop the solutions that seem to have the most potential. Always try to explore at least two.

Figure 4.2 Defining the Problem

Animal Carriers Be clear when wording your problem statement. *What criteria might affect a fish carrier design?*

Brainstorming

Brainstorming is a group technique that helps you develop possible solutions to a problem very quickly. The group chooses a leader who states the problem. Many people in the group freely call out possible solutions. Someone writes down all the ideas, no matter how silly they may seem. No one criticizes or explains. Silly ideas often trigger other ideas that will work. Brainstorming can provide inspiration.

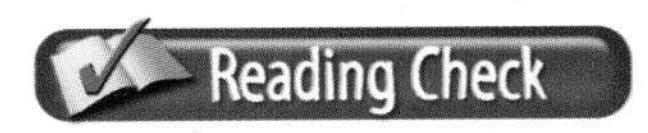

Summarize How does brainstorming help find solutions to problems?

Select One Solution

What should you consider when selecting a solution?

The most important factors to consider when selecting a solution are the criteria and constraints set down when defining the problem. Then each solution can be **evaluated** in terms of its advantages and disadvantages. For example, suppose two kinds of plastic are proposed for your animal carrier. Both meet all the requirements, but one is more attractive than the other. The appearance of the carrier may mean nothing to a cat, but it might mean something to the cat's owner.

Ernesto Blanco

Inventor, Engineer, Teacher

Ernesto Blanco was born in Cuba, where he majored in engineering at Havana University. He came to the United States in 1949 to work and to further his studies. Blanco is best known for inventing the stair-climbing wheelchair with retractable, spring-loaded spokes, and the electric Braille typewriter. Many of his inventions were designed to assist the physically challenged.

Back to School Blanco has taught at several universities. As a professor at the Massachusetts Institute of Technology, he is one of the school's most highly-respected creative thinkers and teachers.

English Language Arts/Writing Have you ever had an idea for an invention? Think of an invention and write a short essay about how it might work and who might use it.

Go to **glencoe.com** to this book's OLC to learn about young innovators in technology.

Academic Connections
Science

A Life-or-Death Problem On the *Apollo 13* spacecraft, a small tank of liquid oxygen exploded, damaging the craft. The astronauts lost oxygen and electric power. Workers at NASA had to help the three astronauts make repairs and get back to Earth by using only items they had on board. They found a solution by using problem solving.

Activity Do some research at the library on the *Apollo 13*. Determine how problem solving brought the astronauts home safely. Write a paragraph about your findings.

Put the Solution to Work

What is the best way to test a solution?

The best way to test a solution is to put it to work. If the solution is a process, you might carry it out. If you want to mass-produce animal carriers, you might set up work stations and do a test run. If the solution is a product, you would make a prototype and test it.

The best kind of test involves actual use. If you made a prototype cat carrier, you would want to test it with a real cat. Did the cat easily damage it? Did users find it unattractive? Every solution must be evaluated, and most designs must be refined.

Evaluate the Solution

Should you always stick with your first idea?

Did your product work the way you thought it would? Did it meet all the criteria and constraints? If you cannot identify the source of a problem, you can try troubleshooting. Troubleshooting helps identify the cause of a malfunction in a system. It is done through a step-by-step process.

Look for better solutions and use new knowledge gained from experimentation to create new and better solutions. Do not feel that you must stick to your first solution. Any results or processes must be communicated to team members, coworkers, or customers. If you choose a new solution, you will repeat the problem-solving steps.

section 4.2 assessment

After You Read Self-Check

1. Name the six steps in problem solving.
2. Describe a brainstorming session.
3. You test and adjust your solution. Identify which system you are using—open-loop or closed-loop.

Think

4. Identify what you think is the most important step in problem solving. Explain.

Practice Academic Skills

English Language Arts/Writing

5. Think of a tool or type of technology that you use on a regular or semi-regular basis. It could be something at home, work, or school. Write a complete problem statement for this item. Share your problem statement with others in your class. See how well it defines the problem that is solved by your chosen item.

STEM Science

6. As part of a small group, brainstorm ideas for a product that would make a bicycle more useful. Use the problem-solving process to design the product. Build a model and evaluate your solution.

Exploring Careers in Technology

Mike Bickford

COMPUTER-AIDED DESIGN (CAD) TECHNICIAN

Q: *What do you do?*

A: I use AutoCAD, which is a computer-aided design software that is used to produce drawings and designs. I work primarily with civil engineers, who are responsible for the design. Sometimes I convert paper drawings to 3D images on the computer. My company specializes in land development, as well as mapping for public and private subdivisions.

Q: *What kind of training and education did you need to get this job?*

A: When I was a freshman in high school, I took a drafting class and really enjoyed it. After that, I continued taking drafting classes every year, because I knew that's what I wanted to do. Also, taking math classes was very helpful. It laid the foundation for my advanced training. At ITT, I focused on CAD classes and learned to use various types of software. Once I got my associate's degree and started working, I learned more specific skills that apply to my day-to-day job.

Q: *What do you like most about your job?*

A: I like the fact that I am always learning and being challenged. For example, I am always learning new ways to use software to do my job better. Also, my work environment is very diverse, and I do not feel pinned down to one aspect of the job.

English Language Arts/Writing

Design a Survey You have been asked to design a new cereal. You will need to collect information about your classmates' cereal preferences.

1. Write and carry out a survey. Ask your classmates at least five questions about the kinds of cereal they like.
2. Graph your data by hand or by using graphing software.
3. Summarize your findings in a one paragraph conclusion and present them to your class.

Go to **glencoe.com** to this book's OLC to learn more about this career.

Real-World Skills

Problem-solving, communication, computer technology

Academics and Education

Mathematics, science, art

Career Outlook

Growth slower than average for the next ten years

Source: *Occupational Outlook Handbook*

chapter 4 Review and Assessment

Chapter Summary

Section 4.1 The design of new products begins with an idea in someone's mind. Invention is turning ideas into new devices. Innovation means modifying and improving existing ones. Engineering makes sure the product works well and is reliable. Designers follow certain principles: balance, proportion, contrast, and unity. Engineers test new designs on a model or prototype.

Section 4.2 Problem solving steps include defining the problem, collecting information, developing possible solutions, selecting one solution, putting the solution to work, and evaluating the solution. A problem statement clearly identifies the problem that a product will solve. The most important factors to consider when selecting a solution are the criteria and constraints.

Review Content Vocabulary and Academic Vocabulary

1. On a sheet of paper, use each of these terms and words in a written sentence.

Content Vocabulary
- innovation
- invention
- human factors engineering
- balance
- proportion
- unity
- problem statement
- brainstorming

Academic Vocabulary
- theory
- range
- sequence
- evaluate

Review Key Concepts

2. **Explain** how creativity influences design.
3. **Discuss** the importance of engineering and appearance in design.
4. **Discuss** ways to evaluate design.
5. **List** the six steps in problem solving.
6. **Write** an effective problem statement.
7. **Describe** what makes a product attractive.
8. **Discuss** how problem solving is applied to design.
9. **Explain** what a problem statement includes.
10. **Discuss** what is involved in finding solutions to problems.

Real-World Skills

11. Evaluating Design Find two tools at home that are designed for the same purpose. You might choose tools designed to open cans or stir sauces. Write a paragraph comparing the design of the tools and telling which you think is easier to use and why. Suggest any improvements.

Technology Skill

12. Cut Down a Tree Imagine there is a tree at school with a broken branch about to fall. You must cut it down.

- **a.** Use the Internet to research techniques for cutting down a tree. Consider different processes and tools.
- **b.** Write a plan for solving the problem. List the tools you will use and why.

Academic Skills

Science

13. Think of a technology item that you think has a design problem. Develop several possible solutions. Write a few paragraphs describing the possible solutions. Share what you write with another student for feedback.

Mathematics

14. Stacy's car used to get 25 miles per gallon of gas, but now it gets 22 mpg. If gas cost $2.75 a gallon, how much more would she pay for gas on a 132-mile trip?

Math Concept **Multi-Step Problems** Write down all the information a question gives you, and all of the things it is asking you to solve. Then go through the steps for each of those things.

WINNING EVENTS

Problem Solvers

Situation You are preparing for a team problem-solving competition. The problems must be solved within two hours. You must apply the following problem-solving strategies:

- Work well as a team and team member.
- Use simple hand tools safely and effectively.
- Identify the problem precisely.
- Know the problem's criteria and constraints.

Activity Work with a teammate to solve the following: Use 30' of fishing line, 1 large plastic cup, 3 balloons, 3 pins, 7 paper clips, 3 ping-pong balls, 3 sheets of 8½" × 11" paper, and white glue; design a device that will travel 15 feet; pop the inflated balloon; and deposit the balls into the cup.

Evaluation Your device will be evaluated by the criteria stated in the activity.

Go to **glencoe.com** to this book's OLC for information about TSA events.

Standardized Test Practice

Directions Choose the letter of the best answer. Write the letter on a separate piece of paper.

1. What is the average speed of a car that travels 140 miles in 3.5 hours?
   A 42 mph
   B 35 mph
   C 45 mph
   D 40 mph
2. A problem statement clearly defines a problem to be solved.
   T
   F

Test-Taking Tip Exercising for a few days before the test can help reduce stress.

chapter 4 TECHNOLOGY LAB

Design a Product to Overcome a Disability

People who live with disabilities face challenges. In the past a person with a physical disability often faced significant limits to what he or she could accomplish. Today people can more easily work with a disability by using new devices that advanced technologies have made possible.

Tools and Materials

- ✓ Pencil and paper
- ✓ Ruler
- ✓ Poster board
- ✓ Permanent markers

Set Your Goal

You and your team will identify a disability and produce the design for a product that will help solve a problem for someone with that disability. An example might include a device to help a blind person identify the color of clothing.

Know the Criteria and Constraints

In this lab, you will:

1. Design a product that safely aids a person with a physical disability.
2. Create a new invention or an improvement on an existing device.
3. Produce a poster or a simple model of your product. A poster must include a drawing of the product.
4. Turn in a problem statement and an explanation of your solution.

Design Your Project

Follow these steps to design your product and complete this lab.

1. Discuss some of the difficulties of daily living that people with disabilities encounter. Consider people who rely on wheelchairs or who have vision or hearing impairments. Do some research. If possible, interview someone who has a disability and learn about the obstacles that person must overcome.

2. Identify one problem to solve. Here are some suggestions. Be original and think of others.
   - A device to pick up small items from the floor for a person who uses a wheelchair.
   - A system to enable a visually impaired person to locate the entrance to a public building.
   - A system to awaken a hearing-impaired person in the morning.
3. Write a problem statement.
4. Brainstorm to arrive at a potential solution to the problem.
5. Select two or three potential solutions and develop them with sketches and written descriptions.
6. Review all the sketches and descriptions. As a team, select one idea that meets all the criteria and constraints in your problem statement. Make any adjustments that are necessary to refine it.
7. Create a poster with a drawing or simple model that shows how your solution will work. Your poster or model must include the problem statement and a brief explanation of the solution.
8. Present your solution to the class.

Evaluate Your Results

After you complete this lab, answer these questions on a separate piece of paper.

1. Why did your team decide to work on this particular design?
2. How did the problem statement help you as you worked?
3. How did each team member contribute when brainstorming?

Academic Skills Required to Complete Lab

Tasks	English Language Arts	Math	Science	Social Studies
Research the difficulties experienced by people with disabilities.	✓			✓
Write a problem statement.	✓		✓	
Brainstorm solutions and decide which to present.	✓		✓	
Create model or poster of device.	✓		✓	
Present solution to the class.	✓		✓	

chapter 5

From Drawings to Prototypes

Sections

- **5.1** The Graphic Language
- **5.2** Drafting Applications
- **5.3** Building Models and Prototypes

What You'll Learn

- **Compare** freehand sketching and technical drawing.
- **Describe** the alphabet of lines and drawing to scale.
- **Identify** the purpose of each kind of technical drawing.
- **List** examples of drafting applications.
- **Explain** the purpose of working drawings and schematic diagrams.
- **Explain** the purpose of models and prototypes.
- **Define** rapid prototyping.
- **Explain** the purpose of scientific and engineering visualization.

Explore the Photo 

Blueprint for Success This architectural drawing shows the plan for the Freedom Tower in New York City with its surrounding towers and the World Trade Center memorial. *Does this drawing help you visualize what the site might look like when construction is completed? Why?*

Launch the TECHNOLOGY LAB

Make a Back-Massage "Vehicle"

At the end of this chapter, you will be asked to draft and make a model of a back-massage machine. Get a head start by using this checklist to prepare for the Technology Lab.

PROJECT CHECKLIST

✓ Decide whether you will use a software program or hand sketching to create a design.

✓ Go to stores to find examples of novelty back massagers.

✓ Think of simple designs of small wooden cars or trucks.

section 5.1

The Graphic Language

Reading Guide

 Before You Read **Preview** What kinds of drawings are used for technology?

Content Vocabulary

- technical drawing
- drafting
- scale drawing
- multiview drawing
- dimension
- pictorial drawing
- isometric drawing
- oblique drawing
- perspective drawing
- section drawing
- CAD

Academic Vocabulary

- individual
- relationship

Graphic Organizer

Draw the section diagram. Use it to organize and write down information .

Pictorial Drawings	
1.	
2.	Oblique Drawing
3.	

 Go to **glencoe.com** to this book's OLC for a downloadable graphic organizer and more.

TECHNOLOGY STANDARDS

STL 8 Attributes of Design

STL 12 Use & Maintenance

STL 17 Information & Communication Technologies

ACADEMIC STANDARDS

Science

NSES Content Standard F Science and technology in society

Social Studies

NCSS 8 Science, Technology, and Society

STL *National Standards for Technological Literacy*

NCTM *National Council of Teachers of Mathematics*

NCTE *National Council of Teachers of English*

NSES *National Science Education Standards*

NCSS *National Council for the Social Studies*

Understanding Graphic Languages

What are some different ways to create drawings?

There is a saying: "A picture is worth a thousand words." This is why drawings are used to construct the machines, buildings, highways, products, and systems of our technological world. These drawings can be created by freehand sketching, **technical drawing**, and computer-aided design and drafting. You will see how the graphic language is the foundation of our designed world.

As You Read

Compare What is the difference between a technical drawing and an artistic illustration?

Drafting

Drafting refers to all the drawing techniques that are used to describe the size, shape, and structure of objects. The drawings are used by architects, designers, engineers, technicians, tradespeople, and many others to help people make things. Drawing techniques are also used in science and engineering to help people understand data. Drafting is also used to create images for video games and for movies.

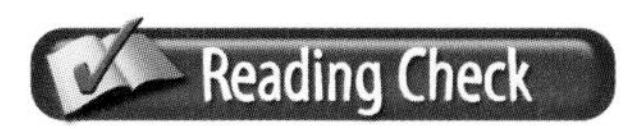

Explain What are graphic languages?

Freehand Sketching

How are freehand sketches used for technology?

Freehand sketching can be the first stage in the development of a drawing that will be used to construct a building, bridge, or automobile. Designers often make many freehand sketches of possible designs to get their ideas down on paper. They use the sketches to evaluate their designs and pick the best one.

Designers draw freehand sketches using only pencil and paper. To make your first sketching experience easier, try drawing on graph paper. See **Figure 5.1**. In the technology laboratory, you will use freehand sketching to show the size and shape of a project you plan to construct. Drawing is one of the simplest ways of communicating ideas to other people.

Academic Connections
Math

Mainly Metrics Drawings and models made for the international market must be made using the metric system of measurement. The United States is one of the only countries in the world that does not use the metric system as its standard.

Apply Select a drawing or model you have made for this course and convert the measurements to metrics. If you use CAD, you can do it automatically.

STEM

Figure 5.1 **Freehand Drawing**

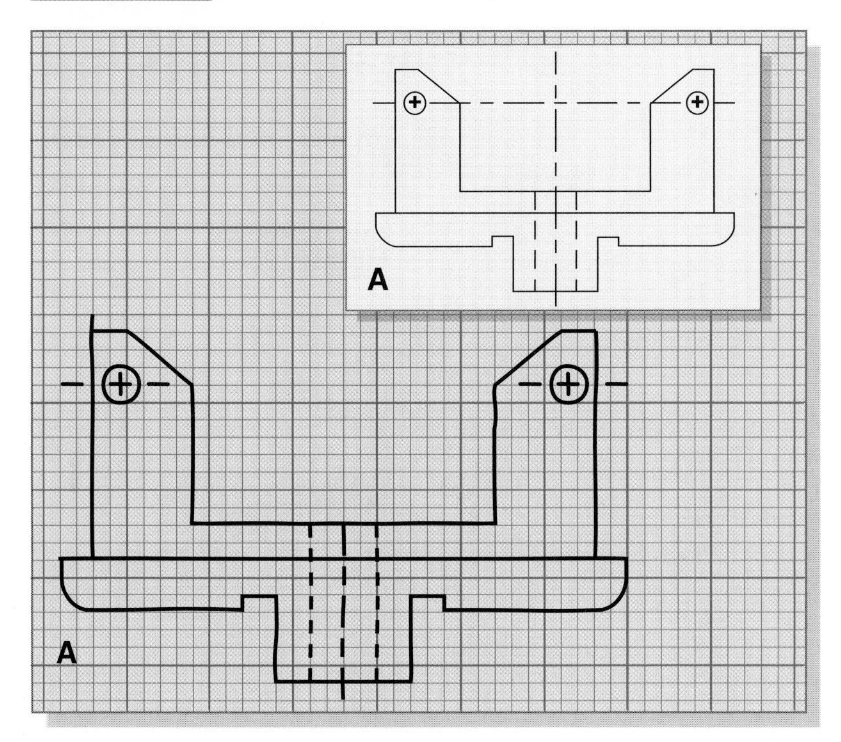

Precise Drawing This freehand sketch was drawn on graph paper. *If you had a choice, would you prefer freehand sketching on graph paper or blank paper? Explain why.*

Technical Drawing

How does a technical drawing differ from a freehand sketch?

Many simple projects can be made from freehand sketches. As the project becomes more complex, it becomes necessary to have more accurate plans. If the project consists of many parts that will be made by different people, plans must be very accurate.

Technical drawing, or mechanical drawing, uses mechanical or electronic tools to accurately show the size and shape of objects. When you produce a technical drawing, you may use pencils, pens, paper, a drawing board, rulers, triangles, compasses, and a computer with special software. These tools give the drafter the ability to draw perfectly straight, round, or curved lines.

Drawing Techniques

Drafting has rules about how to draw each type of line. These lines are "the alphabet of lines." See **Figure 5.2**.

Figure 5.2 The Alphabet of Lines

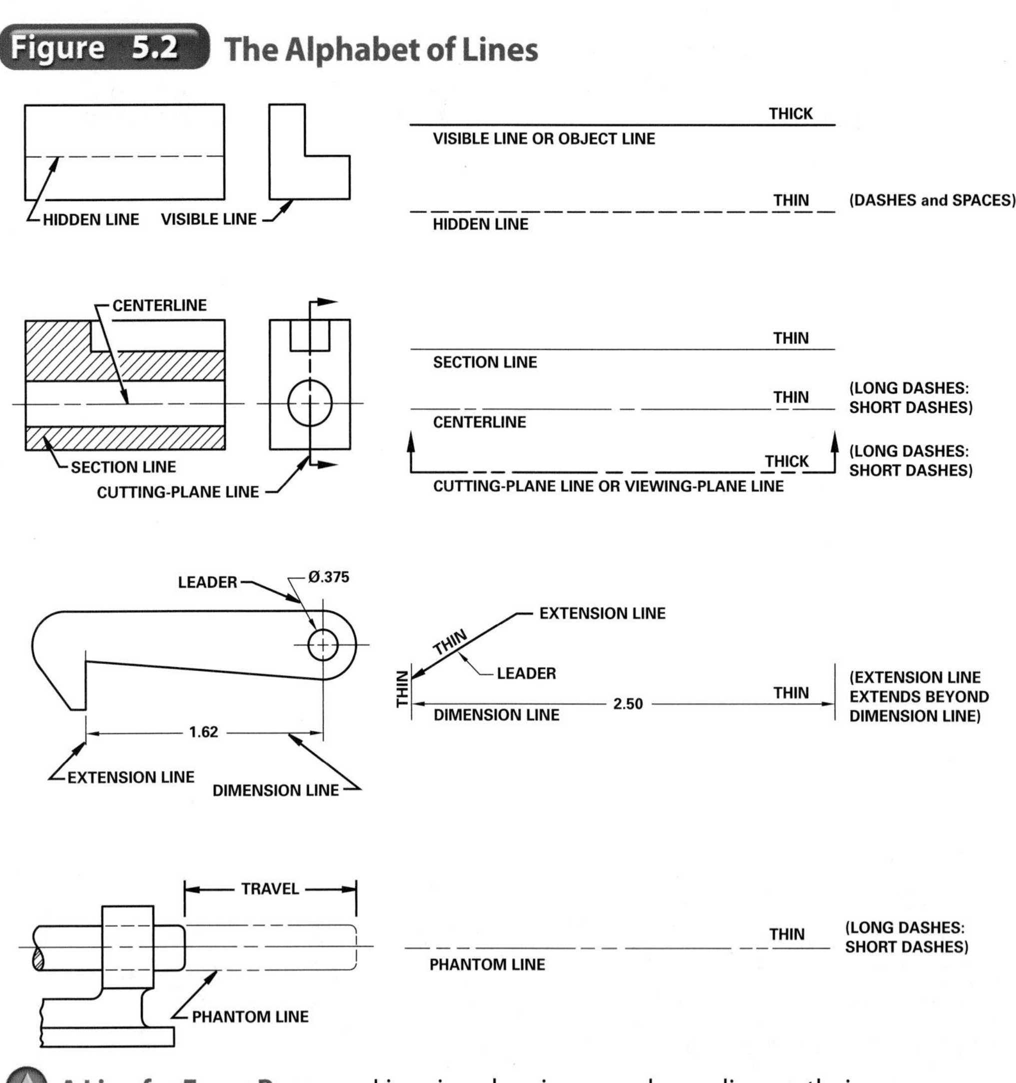

A Line for Every Purpose Lines in a drawing vary depending on their purpose. *Can you tell the difference between a hidden line and a visible line?*

Figure 5.3 Multiview Drawings

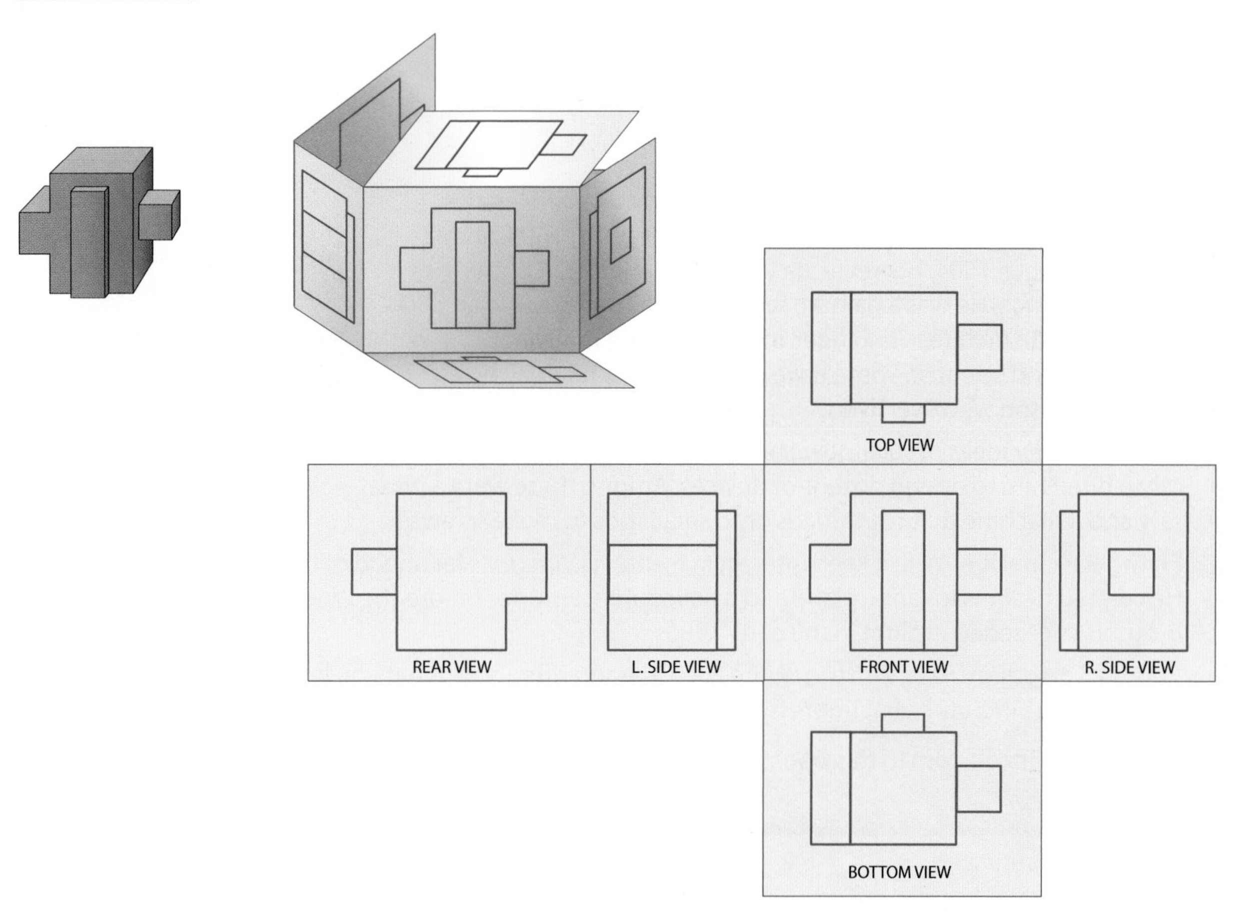

A Full-Shape Description A multiview drawing shows an object as if you took pictures of it through each side of a glass box. *Which of these views might not be necessary? Why?*

Can you imagine carrying a set of house plans that are as big as a house? To make them easy to carry and use, most technical drawings are drawn to scale. A **scale drawing** shows an object's actual shape drawn smaller or larger than its actual size. House plans might be drawn so one-quarter inch equals one foot.

Multiview Drawings

Imagine an object placed inside a square glass box. If you were to photograph this object through each side of the box, you would have six photographs that show each individual side of the object.

If you were to draw, rather than photograph, these different views of the object, you would have a **multiview drawing**.

Although six drawings can be produced, most objects can be explained by drawing the front, top, and right side. **Figure 5.3** shows a multiview drawing. A drawing needs only as many views as necessary to provide the full-shape description. When necessary, the drafter creates additional, detailed views.

Tech Stars

Leonardo da Vinci

Artist, Engineer, Inventor

Most people know Leonardo da Vinci as a famous Italian Renaissance artist. He painted *The Last Supper*, completed in 1498, and the *Mona Lisa* in 1504. However, da Vinci was a master of both arts and science. He was a painter, sculptor, writer, and musician. He was also an architect, engineer, inventor, mathematician, anatomist, and scientist. Some have called him the most diversely talented person who ever lived.

Da Vinci's early ideas of solar power and calculators were ingenious. The Duke of Milan hired him to design dozens of devices. Among these were a water-powered saw and a machine that cut threads onto small rods to make screws.

Flying Fascination With a keen interest in flying, da Vinci was far ahead of his time. He created designs for an airplane and a helicopter centuries before the Wright brothers succeeded in flight in the early 1900s.

English Language Arts/Writing Write a few paragraphs about what you know of Leonardo da Vinci and what impresses you about him. Do research if needed.

 Go to **glencoe.com** to this book's OLC to learn about young innovators in technology.

It is easy to place exact dimensions on a multiview drawing. **Dimensions** give the size of the object and the exact size and location of holes, cutouts, and other features.

When the drawings are finished, copies are given to people who check for drawing errors and design problems. If licenses, patents, or permits must be obtained, copies of the drawings are filed with the appropriate government agencies. Additional copies are sent to the construction site or factory where the finished product will be made.

Pictorial Drawings

It is sometimes hard to visualize what an object actually looks like from a multiview drawing. For this reason, drafters often include a **pictorial drawing**, which is a realistic-looking drawing of a three-dimensional (3D) object.

Isometric Drawings

In an **isometric drawing**, the object being drawn is rotated 30 degrees and tilted forward 30 degrees so three sides are shown. The surfaces of the object in **Figure 5.4** that face toward you are all equally distorted. An isometric drawing is the only pictorial drawing that can be measured along all three axes.

Oblique Drawings

In **oblique drawings** you see a perfect undistorted front view of the object that is combined with a distorted angle view of the other two sides. See Figure 5.4. again. Oblique drawings do not always present a clear picture of an entire object. Therefore, they are the least drawn pictorials. They are most often used for making furniture.

Figure 5.4 Pictorial Drawings

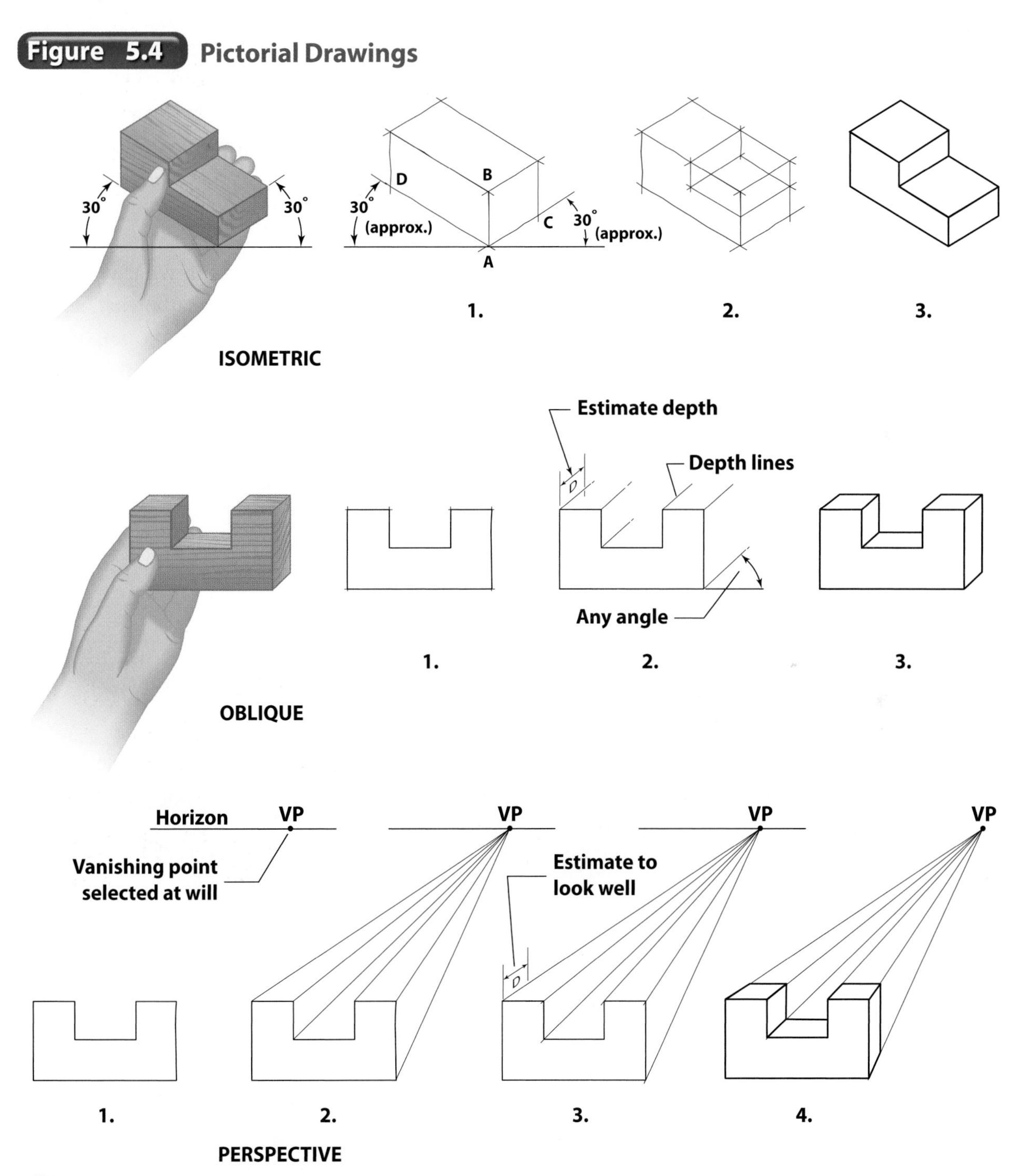

Three of a Kind Three kinds of pictorial drawings are isometric, oblique, and perspective. *Why is a perspective drawing the easiest one to understand?*

Perspective Drawings

In a **perspective drawing**, the object appears as it would in real life. See Figure 5.4 again. Here more distant parts of the object appear smaller, and parallel lines vanish at a distant point, commonly referred to as a *vanishing point*.

This is the hardest of the pictorial drawings to create. Yet, with certain objects, it is the easiest drawing to understand when viewing it.

Can you see how positioning of the object before it is drawn can affect your final drawing? Can you see why isometric, oblique, or perspective drawings are better suited for some objects than they are for other objects? Study the examples to see the relationships among the object, the drawing, and the observer.

Section Drawings

If you wanted to see what was inside a candy bar, what would you do? You would cut open the candy bar. In the same way, a drafter can show the insides of objects by cutting them open and drawing a view of the interior. Drawings that show the inside of an object are called **section drawings**. See **Figure 5.5**. These drawings are used when the insides of objects are complex.

Recall What is the purpose of section drawings?

Figure 5.5 **Section Drawings**

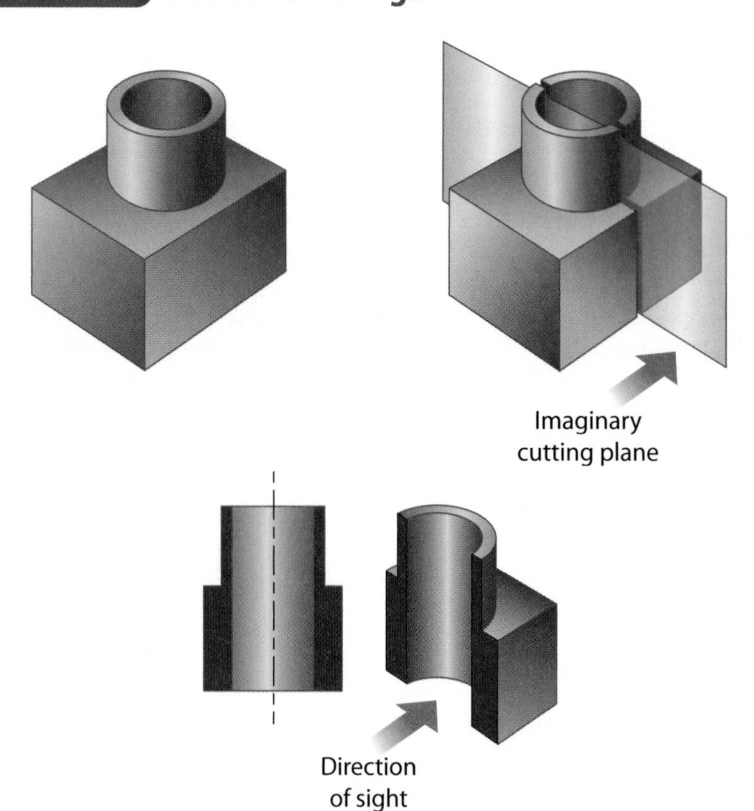

The Inside View Section drawings show the internal features of an object. *What is the purpose of the imaginary cutting plane?*

Computer-Aided Drafting and Design

How does CAD differ from traditional design and drafting?

Today most drafters use a computer instead of a drawing board, paper, T-square, and triangles to make a drawing. The use of computers to do design and drafting is known as **CAD** (computer-aided design).

The computer serves as a very powerful tool because it lets you change and test your design while you are still drawing it. For example, an automotive design can be tested for wind resistance right on the computer. Even an expensive CAD system, however, cannot turn poor designs or drawings into good ones.

With CAD, the drafter selects the type of lines to be drawn from an assortment of drawing tools. The drafter enters the location of the line into the computer by indicating the starting and ending points. The computer program also contains many pre-drawn symbols that the drafter wants to include. The drafter picks what is needed, and then cuts and pastes it into the drawing.

Computer-aided drafters can increase or decrease the size of their drawings. For fine detail work, they magnify an area of a drawing, add the details, and then reduce the area down to size.

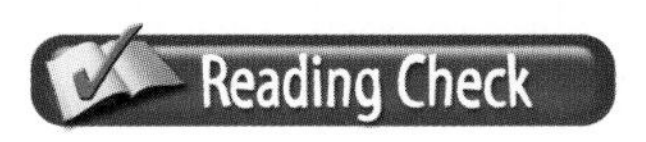

Explain What kind of testing can be done with CAD?

Screen Savers

Did you know that a screen saver displaying moving images causes your monitor to use as much electricity as it does in active use? A blank screen saver is only slightly better. Also, leaving your computer on overnight is less efficient than shutting it down and rebooting.

Try This Turn off your monitor when you are not using it to reduce energy consumption and carbon dioxide emissions.

Computerized Drawing Most technical drawings are now drawn using CAD. *Why do you think drafters prefer CAD over traditional mechanical drawing methods?*

When the drawing is finished, the computer will add the dimensions, draw the pictorial, and rotate the object so that you can see what the other side looks like. The computer can also send the drawing to a printer or plotter.

Modeling, Rendering, and Animation

Using a computer, a drafter can produce a three-dimensional (3D) model. The model is built by combing various shapes, lines and arcs to create a surface structure. There are three kinds of models in 3D CAD:

- Wireframe models
- Surface models
- Solid models

Wireframe Models

A wireframe model appears to be built out of wire mesh rather than solid surfaces, as illustrated in the picture below. You can use a wireframe to determine the amount of space between objects. Wireframes do not provide as much information as surface or solid models.

Surface Models

CAD surface models can give more information. The actual area of the object can be shown. Surface models can look like the shape of the object when viewed only one way.

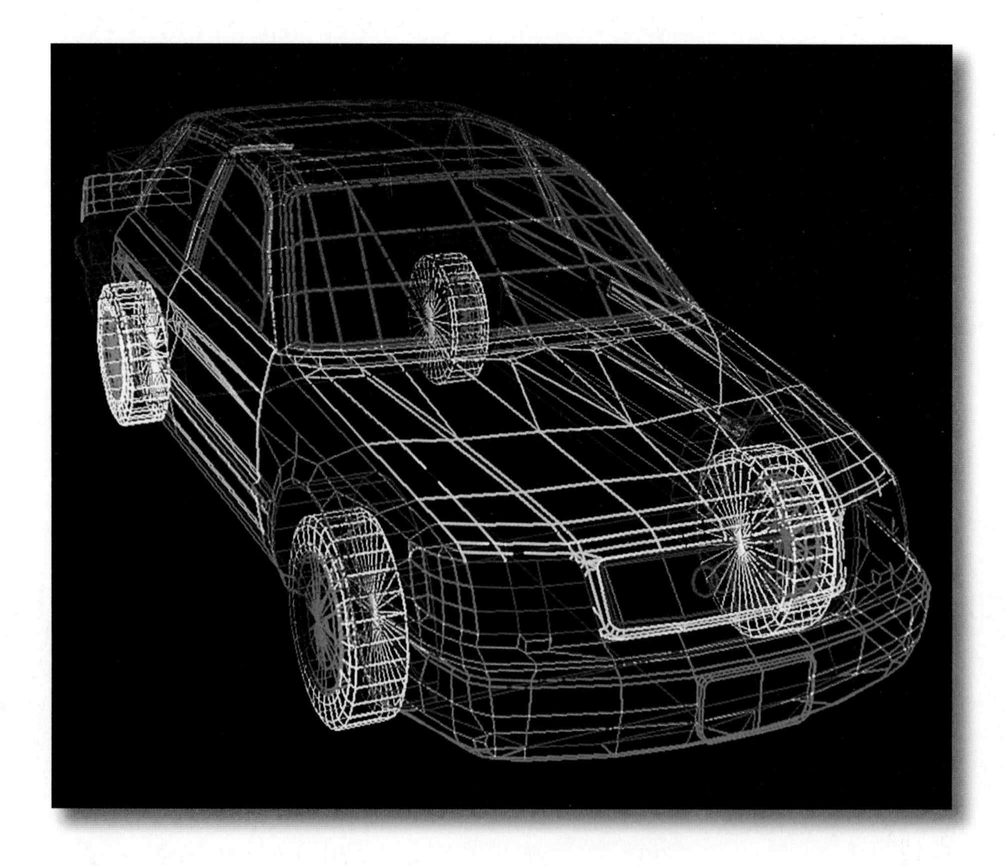

Wired Model After a wireframe object is drawn, a drafter can change it in many ways. *What advantage might a wireframe structure have over a rendered object if testing shows structural weaknesses?*

Automated Animation Physical movement in an animation could include thousands of individual drawing cells. *What advantage would an animator using a computer have over an animator who is drawing individual cells on paper?*

Solid Models

The most realistic kind of modeling uses solid models. They show the shape, area, and volume of an object, or how much an object can hold of something.

Rendering

Rendering give a drawing depth. It produces the reflections that are created when light hits an object. A powerful computer and rendering software can help make 3D models look lifelike.

Animation

To make a drawing move, the drafter adds animation. By simulating moving parts, engineers can use special software to test a design before it is ever built. By combining 3D drawing, rendering, and animation, drafters can create images for architectural presentations, scientific visualization, electronic games, and motion pictures.

section 5.1 assessment

After You Read Self-Check

1. Give three reasons why people make freehand sketches.
2. Name the three types of pictorial drawings.
3. Identify an object that is represented by an isometric drawing. An oblique drawing. A perspective drawing. A section drawing.

Think

4. What scale would you use to sketch a multiview drawing of a simple shape, such as a box of tissues? On a separate sheet of paper, freehand sketch this shape. Be sure to use accurate dimensions.

Practice Academic Skills

English Language Arts/Writing

5. Explain in one paragraph what is meant by the saying: "A picture is worth a thousand words."

STEM Mathematics

6. Jonas designed a new tool by using CAD software. In the specifications for the tool, its measurements are in inches. He would like to sell the tool in Europe where people use the metric system. So he needs to convert the measurements. He knows that 1 inch equals 25.4 millimeters. If the tool is 18 inches long, what is its length in millimeters?

Math Concept **Measurement Conversion** Either multiply or divide to convert measurements.

1. You know that 1 inch = 25.4 millimeters.
2. To find the value for 18 inches, multiply both sides of the equation by 18.

For help, go to **glencoe.com** to this book's OLC and find the Math Handbook.

section
5.2
Drafting Applications

Reading Guide

Before You Read **Connect** How is drafting used in the real world?

Content Vocabulary
- working drawing
- assembly drawing
- schematic diagram

Academic Vocabulary
- specific
- consider

Graphic Organizer
Draw the section diagram. Use it to organize and write down information as you read.

Criteria and Constraints When Designing

1. ______________________
2. ______________________
3. Size of rooms
4. ______________________

Go to **glencoe.com** to this book's OLC for a downloadable graphic organizer and more.

TECHNOLOGY STANDARDS

STL 3 Relationships & Connections

STL 8 Attributes of Design

STL 17 Information & Communication Technologies

STL 19 Manufacturing Technologies

ACADEMIC STANDARDS

Science

NSES Content Standard E Abilities of technological design

Mathematics

NCTM Geometry Analyze characteristics and properties of two- and three-dimensional geometric shapes and develop mathematical arguments about geometric relationships.

STL *National Standards for Technological Literacy*

NCTM *National Council of Teachers of Mathematics*

NCTE *National Council of Teachers of English*

NSES *National Science Education Standards*

NCSS *National Council for the Social Studies*

The Importance of Drafting

Why do designers and builders need drafting?

Architects, engineers, technicians, scientists, and tradespeople need technical drawings that will help them complete jobs. For example, drawings that provide all the information needed to build a house look very different from drawings used to build a computer. Drafters become specialists in fields such as architecture, map making, engineering, manufacturing, and electronics.

As You Read

Identify What are some different types of drawings?

Drafting for Manufacturing

How does industry use drawings?

Without technical drawing, modern industry would never have developed. The machines that are used by manufacturers and all the products made by these machines started out on the drawing boards of drafters.

Manufactured objects begin as **working drawings** that include all the information necessary to build the specific product or structure without any further instructions. The complete working drawing package consists of:

- Multiview drawings
- Pictorial drawings
- Assembly drawings

The **assembly drawings** show workers or consumers how to put parts together to make different objects, from a house to a piece of furniture. See **Figure 5.6**.

Define What information is included in working drawings?

Drafting for Electricity and Electronics

Why do electrical drawings look different than other drawings?

Electrical items have circuits, electric components, and electronic components. Just as you have your own way of talking to your friends, drafters who work in the electric/electronics field have their own way of expressing things.

Imagine This...

Like Star Trek

On the TV show *Star Trek,* replicators made real things, not just models. Researchers at the Sandia National Laboratories have been working to turn this sci-fi idea into real technology. They call it "Laser Engineered Net Shaping" (LENS®). Their model uses CAD and a laser-guidance system to make an almost-finished product from powdered metals. The final machine will produce finished parts and products. *What other use might LENS technology have?*

Go to **glencoe.com** to this book's OLC for answers and to learn more about new rapid prototyping technology.

Figure 5.6 **Assembly Drawing**

Following Instructions An assembly drawing shows how an object should be put together. *Why do assembly drawings usually show parts in unassembled but aligned locations?*

Figure 5.7 Symbols and Lines

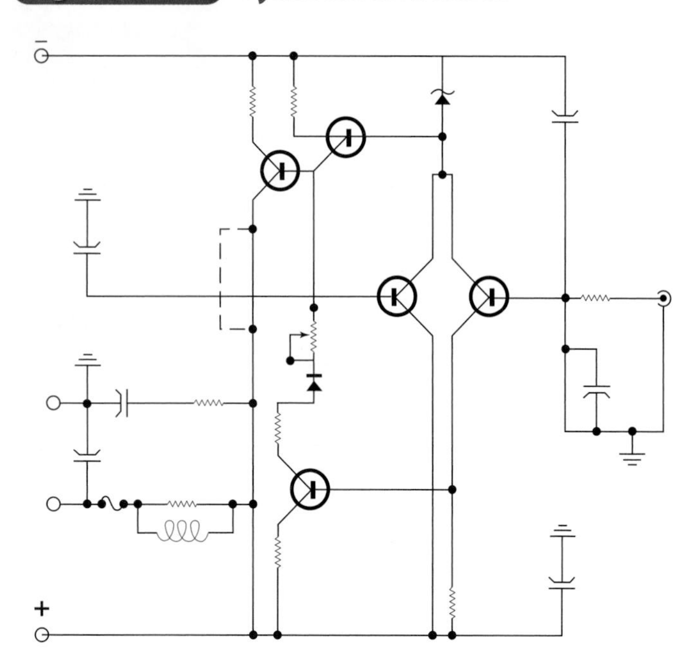

Schematic Symbols A schematic drawing uses lines and symbols to show electrical circuits and components. *Why are symbols and lines used rather than small pictures?*

Drawings that show how electric and electronic components are connected together are called **schematic diagrams**. See **Figure 5.7**. Drafters have created symbols to represent all electrical parts, such as a battery or plug, switch, light bulb, socket, transistors, and integrated circuits. These symbols help make the drawings immediately understood by the people who will manufacture a product from a design.

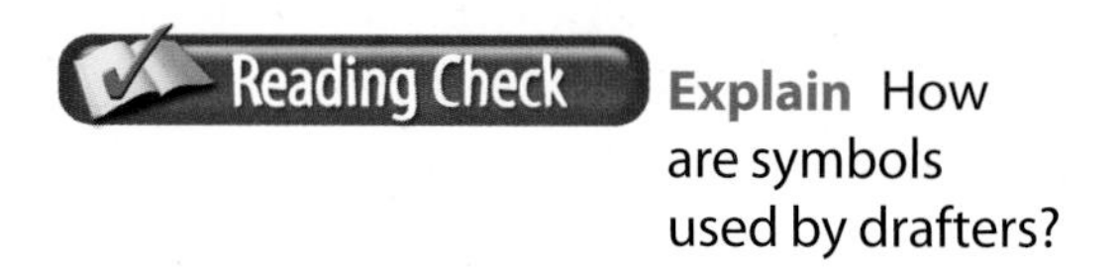

Explain How are symbols used by drafters?

Drafting for Engineering and Construction

What information is included in drawings for a construction project?

Architects and engineers use design and construction knowledge that has been developed over centuries. The first designers probably outlined plans for their structures in the dirt or on cave walls so that helpers could work together to build or create.

A set of construction drawings includes a complete description of the shape and size of the structure's foundation, the inside and outside of the structure, the electrical system, and the plumbing. See **Figure 5.8**. Pictures are included that show how the structure should look when finished. The drawings might also include driveways, parking lots, landscaping, and roads.

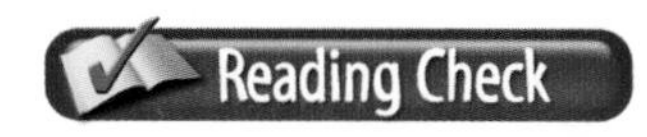

Explain Is all design and construction knowledge new?

Drafting with Criteria and Constraints

Certain locations are susceptible to earthquakes, termites, floods, and ground slides. When building in these areas, people must plan and build a structure that can survive the elements. The architect's goal is to design a structure big enough to meet the client's needs and small enough to fit the client's pocketbook. Designers must consider all criteria and constraints when developing a design.

Figure 5.8 Architectural Drawing

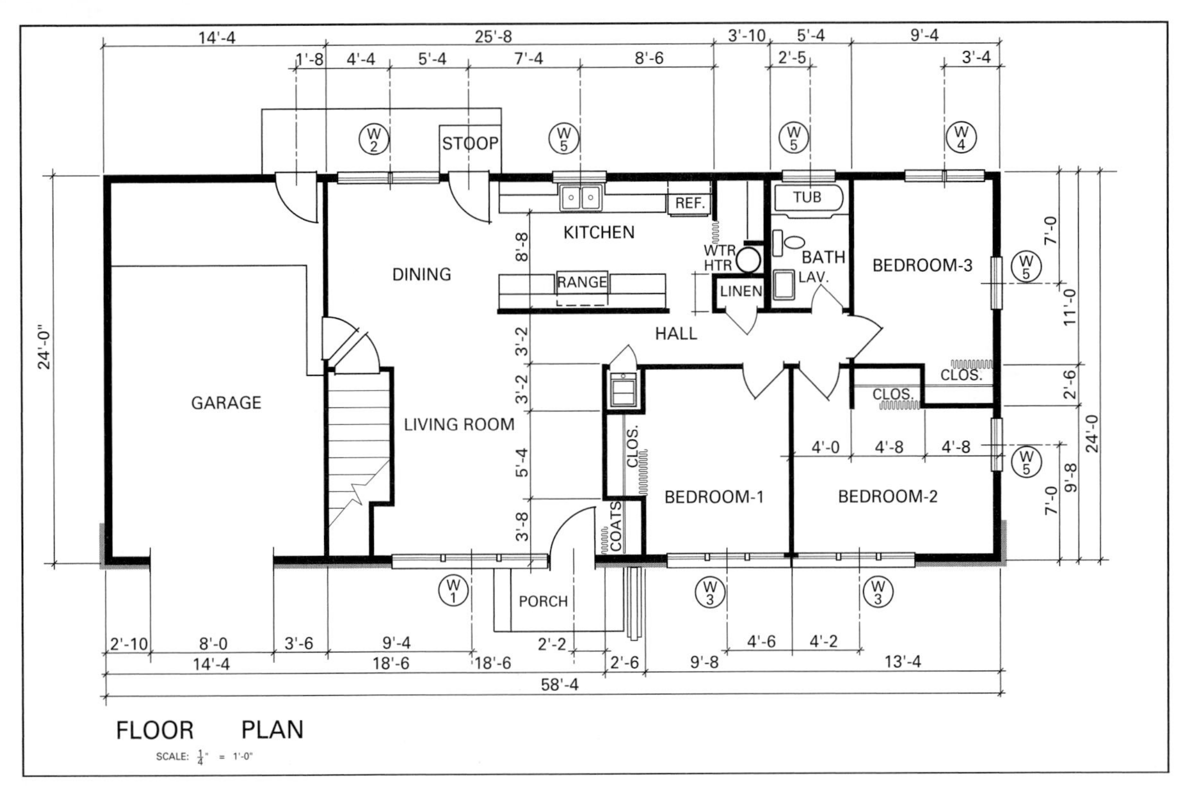

Building from a Drawing An architectural drawing gives the size and shape of the structure. Accuracy is very important. *If the plans are not accurate, what will happen when they are used to build the structure?*

section 5.2 assessment

After You Read Self-Check

1. List the different types of drawings included in a set of working drawings.
2. Describe schematic diagrams.
3. Identify the architect's goal when designing a structure.

Think

4. Explain why you think the builders of the Pyramids in Egypt used or did not use construction plans.

Practice Academic Skills

English Language Arts/Writing

5. Create a list of instructions for a simple task, such as brushing your teeth or making a peanut butter sandwich. Then make sketches that describe the steps. Your goal is to make the process clear to someone who is learning the English language.

STEM Mathematics

6. Marcus is designing a cabinet for his brother's bathroom. He needs to figure out the volume so his brother will know how much it will hold. The cabinet is 13 inches wide, 20 inches long, and 9 inches deep. What is the volume?

Math Concept **Determining Volume** When determining volume, be sure to express your answer in cubic units.

1. Multiply the length times the width times the height (or depth).
2. Write the answer in cubic inches.

For help, go to **glencoe.com** to this book's OLC and find the Math Handbook.

section

5.3 Building Models and Prototypes

Reading Guide

Before You Read **Preview** What are models and how are they used?

Content Vocabulary

- model
- prototype
- rapid prototyping
- visualization software

Academic Vocabulary

- structure
- accurate

Graphic Organizer

Draw the section diagram. Use it to organize and write down information as you read.

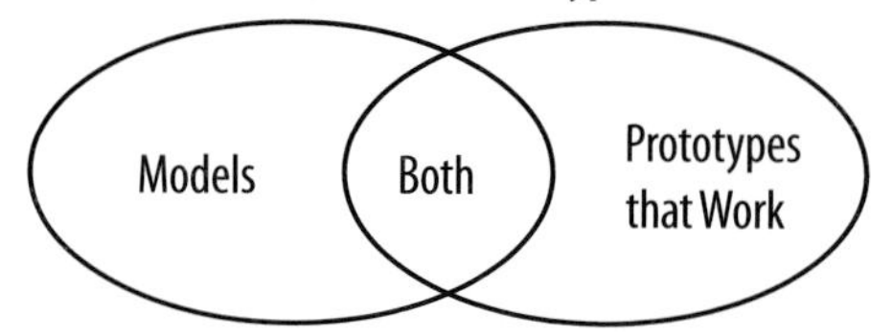

Go to **glencoe.com** to this book's OLC for a downloadable graphic organizer and more.

TECHNOLOGY STANDARDS

STL 3 Relationships & Connections

STL 9 Engineering Design

STL 11 Design Process

STL 12 Use & Maintenances

ACADEMIC STANDARDS

Science

NSES Content Standard E Abilities of technological design

English Language Arts

NCTE 10 Use first language to develop competency in English language arts and content across the curriculum.

STL	*National Standards for Technological Literacy*
NCTM	*National Council of Teachers of Mathematics*
NCTE	*National Council of Teachers of English*
NSES	*National Science Education Standards*
NCSS	*National Council for the Social Studies*

Using Models and Prototypes

What is the difference between a model and a prototype?

Many projects require construction of models and prototypes. A **model** does not actually work; it just shows how the product, building, or system will look when it is built. But a **prototype** is a *working* model of a proposed project. Models and prototypes can be crafted by hand, by machine, or with a computer system.

As You Read

Predict How are models and prototypes made?

A prototype can be used to test product reliability and safety before it is mass-produced. The drawings might be changed many times.

Ethics in Action

Can Designs Be Stolen?

Designers and inventors often build on the work of others. When Steve Jobs and Steve Wozniak built the Apple II, they were improving on the personal computer, not inventing it. What is the difference between building on someone else's work and stealing it?

Give Credit When you steal an idea, you use it without giving the designer credit and/or payment. When you build on someone's idea, you might improve on it or add to it in some way.

English Language Arts/Writing

Getting a Patent Patents issued by the government can protect inventors. Find out how to apply for a patent and what kind of protection is offered.

1. Write a short report on the patent process.
2. Present the report to your class.

At one time, airplane manufacturers had no choice but to build wood models of their airplane designs. They would fly these models in wind tunnels. If the model had good flying characteristics, the next step would be to build a prototype that actually worked.

Powerful computers allow designers and engineers to create virtual 3D models and prototypes. For example, NASA uses the oldest and the newest technologies to create designs for future flight.

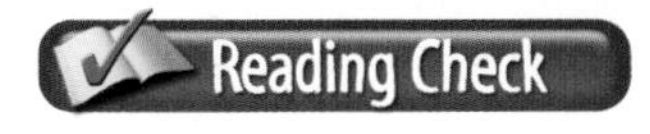

Connect What products that you use might have prototypes?

Rapid Prototyping

What is rapid prototyping?

Rapid prototyping takes place in a special machine that can turn CAD drawings into 3D models. **Rapid prototyping** uses a laser and a vat of special light-sensitive plastic to create 3D models. Thin layers of plastic build the object.

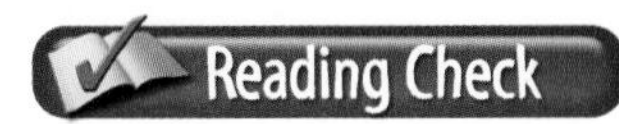

Define What is rapid prototyping?

Space Test Many people think that the Orion Crew Exploration Vehicle (CEV) is a design throwback to the Apollo program that took us to the moon about a half century ago. *Do you think a vehicle that travels in space requires a model or a prototype or both? Why or why not?*

Instant Prototype A rapid prototyping machine allows a designer to quickly create a model or prototype of a product. *Before rapid prototyping, how did craftspeople create models and prototypes?*

Computer Visualization

How do computer models help scientists and engineers understand data?

In the past, to understand tiny structures, scientists turned to LEGO® blocks and Tinker Toys® to construct models. CAD **visualization software** is now used to create virtual models of these structures.

For example, scientists used CAD to create a model of a heart. Medical equipment gathered the data on the size and shape of the heart's muscles, veins, arteries, valves, and chambers. This information was used to create a 3D drawing. The drawing was then rendered and animated. Scientists are using this accurate heart model to test new medicines and explore possible cures.

Engineers use the same tools to develop models, gather data, and perform data analysis. Huge structures like dams can be tested for strength during simulated earthquakes. Models of skyscrapers can be tested for wind resistance.

section 5.3 assessment

After You Read Self-Check

1. Define rapid prototyping.
2. Explain the purpose of a model. A prototype.
3. Identify the type of software used for computer visualization.

Think

4. Explain why scientists or designers might use LEGO® blocks or Tinker Toys® to build models. List some advantages and some disadvantages of using these materials.

Practice Academic Skills

English Language Arts/Writing

5. Do some Internet research to find designs of model airplanes. Choose one and build it, and then test it in a wind tunnel. Write a technical, step-by-step, one-page instruction sheet of how to build it, and then write a report on the test results.

STEM Science

6. Design software has changed the design industry in many ways. The way things are designed and put into production relies more and more on computers. Everything from pop cans to airliners has part or all of its design created on computers using computer design programs. Research several different design applications. Write a paragraph describing different design applications and their uses.

Exploring Careers in Technology

Nathan Gross

GRAPHIC DESIGNER

Q: *What do you do?*

A: I design graphics that are displayed on products, such as bicycles. Before I start the design, I visit retail stores and browse through catalogs, magazines, and the Internet for trends and ideas. I may sketch some ideas for graphics first. If I already have an idea in mind, I will start designing graphics on the computer. Once the design is complete, it is printed on a special film and applied directly to the bicycle. The bicycle is then approved and shown to buyers for stores. If the buyers decide to buy the bicycle, the design files are shipped out for production.

Q: *What kind of training and education did you need to get this job?*

A: As a child, I was always drawing and painting. I decided to take some computer-design and art classes in college. When I discovered the world of graphic design, I realized what I was born to do. I learned the fundamentals of graphic design and how to use current software that enables graphic designers to do their jobs.

Q: *What do you like most about your job?*

A: The coolest thing about my job is being able to go to a retail store and see a product on the shelf that I helped create. Equally satisfying is that no two work days are alike. Almost every day I get to start a new project that has the potential to be seen by a worldwide audience.

English Language Arts/Writing

Design and Promote Think of and create some colorful graphic designs that might appear on a bicycle, a skateboard, or another product of your choice.

1. Choose a product for which you will create designs.
2. Freehand sketch or use a computer graphics program to illustrate the designs.
3. Write a one-page report, describing the designs, how they will be used, and why they might help the sale of the product.

Go to **glencoe.com** to this book's OLC to learn more about this career.

Real-World Skills

Listening, observation, creativity

Academics and Education

English language arts, technology, computer design, graphic arts, marketing

Career Outlook

Growth as fast as average for the next ten years

Source: *Occupational Outlook Handbook*

chapter 5 Review and Assessment

Chapter Summary

Section 5.1 Drawings are used by scientists, architects, engineers, designers, and illustrators to show what objects look like. A drafter makes a freehand sketch to show the shape and size of an object. To draw an accurate picture, drafters use rulers, compasses, and other mechanical drawing tools. A multiview drawing shows the front, top, and right views. A section drawing shows internal structure. Pictorial drawings are also done.

Section 5.2 Drafters represent an object by using different kinds of lines. Thickness and length are determined by the rules of drafting. Dimensions give the size and location of the object's features. In CAD, the designer creates and tests his or her designs on the computer. Most made objects are first drawn. Drawings larger or smaller than the real object are drawn to scale; the parts are in the correct proportions.

Section 5.3 A model is a replica of the product; however, it does not work. A prototype does work and may be tested for reliability or safety. Rapid prototyping makes replicas from plastic. The replicas are first drawn with CAD software, which is also used to make models for scientific and engineering visualization for research and testing.

Review Content Vocabulary and Academic Vocabulary

1. On a sheet of paper, use each of these terms and words in a written sentence.

Content Vocabulary

- technical drawing
- drafting
- scale drawing
- multiview drawing
- dimension
- pictorial drawing
- isometric drawing
- oblique drawing
- perspective drawing
- section drawing
- CAD
- working drawing
- assembly drawing
- schematic diagram
- model
- prototype
- rapid prototyping
- visualization software

Academic Vocabulary

- individual
- relationship
- specific
- consider
- structure
- accurate

Review Key Concepts

2. **Explain** the difference between freehand sketching and technical drawing.
3. **Describe** the alphabet of lines and scale drawing.
4. **Identify** the purposes of technical drawing.
5. **Give** examples of drafting applications.
6. **Identify** the purpose of working drawings.
7. **Identify** the purpose of models and prototypes.
8. **Describe** rapid prototyping.
9. **Explain** the purpose of scientific engineering.
10. **Explain** the purpose of electrical drawings.

Real-World Skills

11. **Graphic Communication** Consider the different aspects of an object needed to produce it. Sketch of an item you use every day. Include information a manufacturer would need to make it.

Technology Skill

12. **Writing Directions** Directions are often given in a simple step-by-step manner.
    a. Use the Internet to find a map of your community.
    b. Write step-by-step directions to get from your school to another location, but do not indicate the location. Give your directions to another student to test them.

WINNING EVENTS

Automotive Engineer

Situation Your team will design and build a dragster model. It must meet certain criteria and constraints for safety, weight, and size. The model will be powered by a CO_2 cartridge.

Activity Make a series of sketches of the model and produce a technical drawing of it. Then make a wooden model and conduct wind tunnel tests. Make changes as needed. Use these specifications: minimum weight = 50g; dragster body dimensions with wheels: length: minimum = 200mm, maximum = 305 mm, height: maximum = 75 mm; width at axles: minimum = 35mm, maximum = 42mm.

Evaluation The sketches and technical drawings will be evaluated by these criteria:
- Clearly communicates your design ideas
- Technically correct

The dragster:
- Safety (Stays on the track, does not break)
- Speed
- Appearance

Go to **glencoe.com** to this book's OLC for information about TSA events.

Academic Skills

English Language Arts

13. Choose a product that could benefit from using computers in its design. Write a few paragraphs describing the item and how CAD and manufacturing could improve it.

STEM Mathematics

14. The design software for a manufacturing company costs \$5,400; 75 cents of the selling price of each item will pay for the software. How many items must be sold to pay for the software?

Math Concept **Expense** Determining expense involves dividing the cost of one item by the number of items sold.

Standardized Test Practice

Directions Choose the letter of the best answer. Write the letter on a separate piece of paper.

1. What is the area of a 12 feet by 48 feet piece of sheet metal?
   A 144 square feet
   B 1,248 square feet
   C 4 square feet
   D 576 square feet

2. Rapid prototyping can turn drawings into three-dimensional objects.
   T
   F

Test-Taking Tip When taking a test, do not worry about how quickly other test-takers finish; take your time and concentrate.

chapter 5

TECHNOLOGY LAB

Make a Back-Massage "Vehicle"

The wooden van in the drawing has hexagonal wheels. It is sold in novelty stores as a back massager. When you roll it on someone's back, it can help the person relax and feel better. A technical drawing "breaks down" a three-dimensional object, and shows how it looks from different angles. Multiview drawings help a builder create the same object by using only these drawings as a guide.

Tools and Materials

- ✓ Drawing board
- ✓ Drawing tools
- ✓ Computer
- ✓ Computer printer
- ✓ Drawing software
- ✓ Toy factory software
- ✓ Modeling clay
- ✓ 2 × 4 lumber
- ✓ 3/4-inch plywood
- ✓ Woodworking tools
- ✓ Thumbtacks
- ✓ Sandpaper
- ✓ Drill set
- ✓ Drill press
- ✓ Wood glue
- ✓ Paper
- ✓ Pencils
- ✓ Dowels
- ✓ Scroll saw
- ✓ Wood vises

Set Your Goal

Your goal for this lab is to design your own back-massaging car or truck. You will then build it by using your own technical drawings as guides.

Know the Criteria and Constraints

In this lab, you will:

1. Make a multiview drawing that provides the size and shape of your vehicle and any other information needed to build it.
2. Make a pictorial drawing of your vehicle.
   - Your drawings may be done by hand or with a CAD program.
   - Your vehicle can have small or large wheels of any shape *except* round shape.
3. Prepare a list of materials needed to construct the vehicle.

Design Your Project

Follow these steps to design your project and complete this lab.

1. Draw sketches of different car or truck shapes using drawing software or by hand, or design these shapes using modeling clay.
2. Select the design that best meets the requirements.
3. Prepare full-size drawings or models of front, top, and right side views of your final design. These views should give a complete description of your vehicle's shape.
4. Measure all the dimensions and add these numbers to the drawings.
5. Make a pictorial drawing of your vehicle.
6. Prepare a list of all materials needed to construct your vehicle.

! SAFETY

Reminder

In this lab, you will be using tools and materials. Be sure to follow appropriate safety procedures and rules so you and your classmates do not get hurt.

7. Build your vehicle:
   - Make the body out of 2 × 4 lumber.
   - Make the wheels out of plywood.
   - Cut the axles for the wheels from dowels.
   - Locate and drill the holes for the axles. The holes must be large enough for the axles to turn freely.
   - Locate and drill the wheel axle joints.
   - Sand all parts.
   - Slip the axles through the body, and slip the wheels onto the ends of the axles.
   - Glue the axle ends into the wheels. Be sure that no glue gets into the body of your vehicle or your wheels will not be able to turn.
8. Try out your massager on a classroom volunteer.

SIDE VIEW

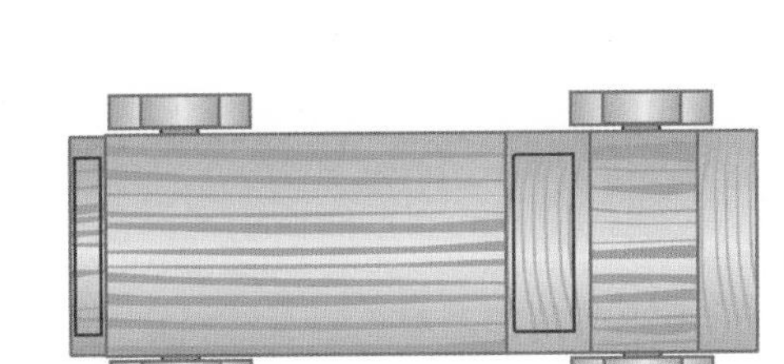

TOP VIEW

Evaluate Your Results

After you complete the lab, answer these questions on a separate piece of paper.

1. Could you build your vehicle using only the size and shape descriptions on your drawing?
2. How could you have improved your multiview drawing?
3. If other students built your vehicle from your plans, would they end up with exactly the same finished product? Explain why or why not.

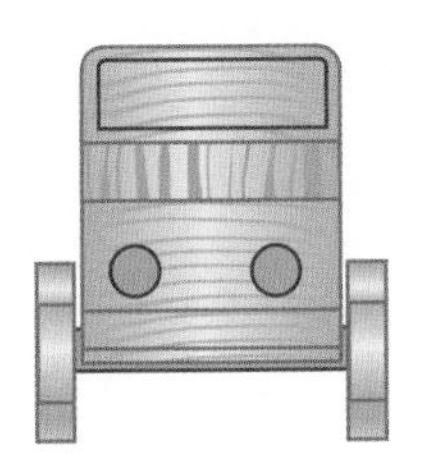

FRONT VIEW

Academic Skills Required to Complete Lab				
Tasks	**English Language Arts**	**Math**	**Science**	**Social Studies**
Research different vehicle and wheel shapes and design several of them.	✓		✓	
Draw with computer software or drawing tools, or build clay models, of three views of the vehicle design, measuring and labeling all dimensions.		✓	✓	
Build final vehicle.	✓	✓	✓	
Create presentation.	✓		✓	
Present to the class.	✓		✓	

chapter 6

Technology Connections

Sections

What You'll Learn

- **List** ways in which science and technology are connected.
- **Identify** ways in which mathematics and technology are connected.
- **Discuss** ways in which language arts and technology are connected.
- **Describe** ways in which social studies and technology are connected.

Explore the Photo

Technology Today Working with and developing technology require knowledge in many areas. Learning science, mathematics, as well as language arts and social studies can be the key to technology of the future. *Why might social studies be important for technology?*

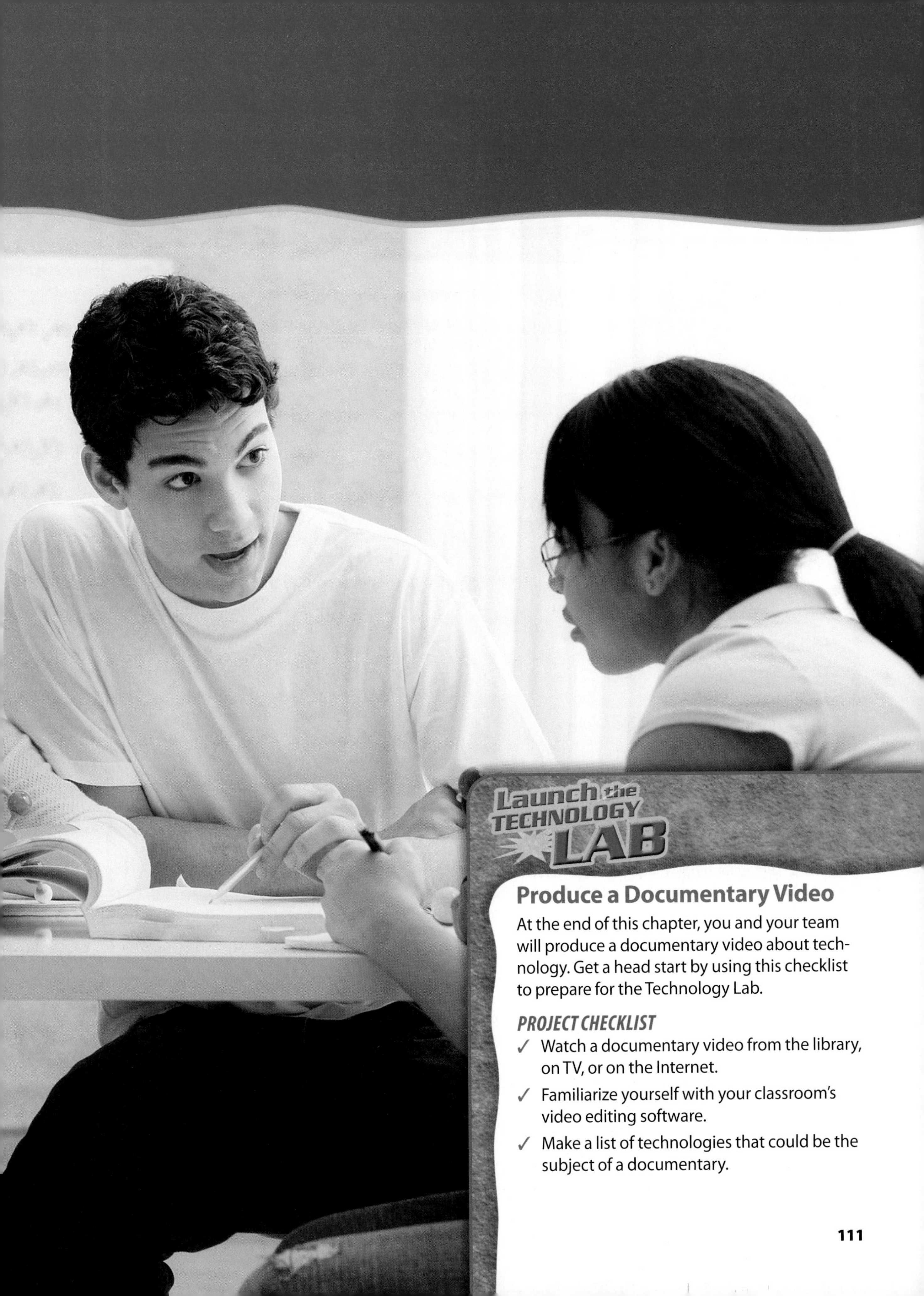

Produce a Documentary Video

At the end of this chapter, you and your team will produce a documentary video about technology. Get a head start by using this checklist to prepare for the Technology Lab.

PROJECT CHECKLIST

✓ Watch a documentary video from the library, on TV, or on the Internet.

✓ Familiarize yourself with your classroom's video editing software.

✓ Make a list of technologies that could be the subject of a documentary.

section 6.1

Science and Mathematics

Reading Guide

 **Before You Read** **Connect** Why are science and mathematics necessary for technology?

Content Vocabulary

- hypothesis
- scientific theory
- scientific law

Academic Vocabulary

You will see these words in your reading and on your tests. Find their meanings at the back of this book.

- volume
- area

Graphic Organizer

Draw the section diagram. Use it to organize and write down information as you read.

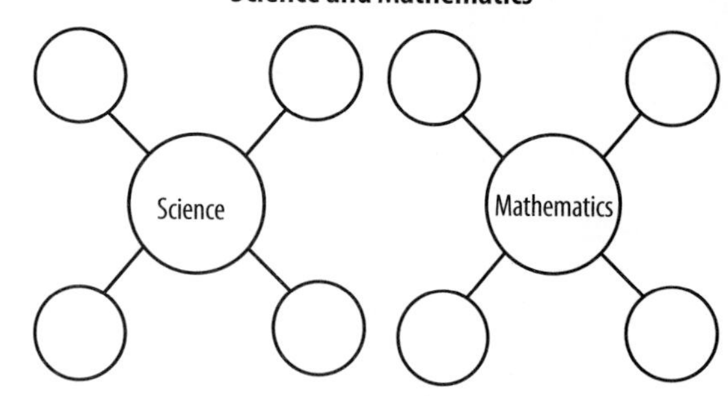

 Go to **glencoe.com** to this book's OLC for a downloadable graphic organizer and more.

TECHNOLOGY STANDARDS

STL 3 Relationships & Connections

STL 7 Influence on History

STL 10 Troubleshooting & Problem Solving

ACADEMIC STANDARDS

Mathematics

NCTM Algebra Understand patterns, relations, and functions

Science

NSES E Understandings about science and technology

STL *National Standards for Technological Literacy*

NCTM *National Council of Teachers of Mathematics*

NCTE *National Council of Teachers of English*

NSES *National Science Education Standards*

NCSS *National Council for the Social Studies*

Science

What is the scientific method?

Early humans created very simple technologies using crude tools and natural materials. They learned what worked by trying different combinations. Modern technology is quite different. It depends heavily on science and mathematics.

As You Read

Predict How does mathematics help in the development of technology?

Scientific Method

All scientists apply the same process to solve problems. This process is called the "scientific method."

1. Scientists make observations and form questions.
2. Next, they gather information about what was observed.
3. They use this information to form a **hypothesis**.
4. Scientists then develop an experiment to test it.
5. They carefully analyze the results of the experiment.
6. They then repeat the experiment to see whether they get the same results. Finally, they present their conclusions.

Scientists use technology as a tool for scientific discovery. Their discoveries often lead to the development of new technologies.

Scientific Theories and Laws

To develop new technology, engineers must have a clear understanding of the principles, theories, and laws of science. Scientific conclusions that have been carefully developed through experimentation are called **scientific theories**. Over time, these theories are tested again and again. Eventually, they may be accepted as **scientific laws**, which are theories proven true and accepted as fact. In time, scientific law can be challenged by new knowledge.

For example, Newton's third law of motion states that the greater an object's mass, the greater its attraction force. Newton's law explains why apples fall to the earth and why the planets in our solar system revolve around the sun. However, telescopes show that light, which has no mass, bends in a strong gravitational field. Albert Einstein developed his theory of relativity to account for this warping of light by the massive objects in our universe.

Most scientists think that Newton's law and Einstein's theory cannot be reversed. But NASA researchers are working on an anti-gravity machine, which would actually revolutionize air and space travel.

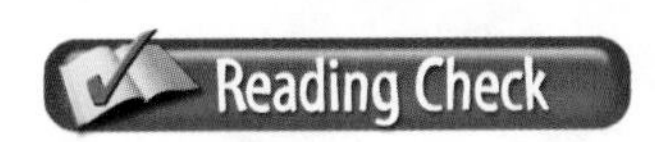

Summarize What is the relationship between science and technology?

Academic Connections
Math

Looking for Profit Mathematics is needed to calculate the profit from selling a product. Profit is the money earned after all expenses are paid.

Apply You and a friend decide to build dog houses to sell during summer vacation. One load of lumber and other materials to build 12 dog houses costs $648. If you sell them for $80, how much total profit will you earn?

Mathematics

How can mathematics help us design technology products?

Few subjects are more important to technology than mathematics. Engineers use mathematics to determine things such as the flight characteristics of airplanes and the shape of cell phones. Mathematics is often the reason products look the way they do.

A narrow river might require only a simple bridge. A larger river might require a complex bridge with suspension cables. Mathematical calculations tell engineers which design is better.

Gravity at Work According to legend, Newton's third law of motion was inspired by an apple falling from a tree. *What is the difference between scientific theory and scientific law?*

EcoTech

Sea Robot

As Earth's environment changes, so does the ocean. The oceans affect global weather and sea life. So, technologists at Rensselaer Polytechnic Institute devised a solar robot that "swims" underwater, calculating sea levels, algae blooms, and the melt and drift of ice.

Try This Almost anything you do to help the environment can help our oceans. Many cleaning products are toxic and get washed into the sea. Read the labels of cleaning products at the market. Make a list of environmentally friendly products and share it in class.

Working with Robots

Industrial robots are important machine tools in manufacturing. Engineers make mathematical calculations to determine where robots should be used and for what tasks.

For example, hydraulic robots use pressurized oil, and pneumatic robots use pressurized air. Pneumatic robots operate at an air pressure of about 100 psi (pounds per square inch). Hydraulic robots operate at about 3,000 psi. A hydraulic robot can lift about 30 times as much weight because the pressure is 30 times higher. If a pneumatic robot can lift 20 pounds, a hydraulic robot could lift up to 600 pounds ($30 \times 20 = 600$ pounds).

A robot having an arm extension speed of 40 inches per second is not unusual. How long would it take such a robot to move its arm five feet? The robot could do it in 1.5 seconds! (Convert feet to inches: 5 feet $\times$ 12 = 60 inches. Divide this result by 40.)

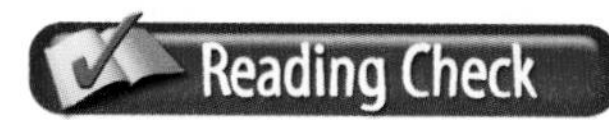

Describe What are some of the ways that engineers use mathematics?

Building Ships

Isambard Kingdom Brunel was England's greatest 19th century technologist. He designed and built the world's first all-metal steamship, the *SS Great Britain*. Many people scoffed at him because they knew that metal sinks easily. However, Brunel knew that it is easy to make metal float if you give it the proper shape. As long as the density of a solid is less than that of water, the solid will float. Density is weight divided by **volume**. The density of water is 0.0361 pounds per cubic inch (pci). Any solid that is less dense than 0.0361 pci will float.

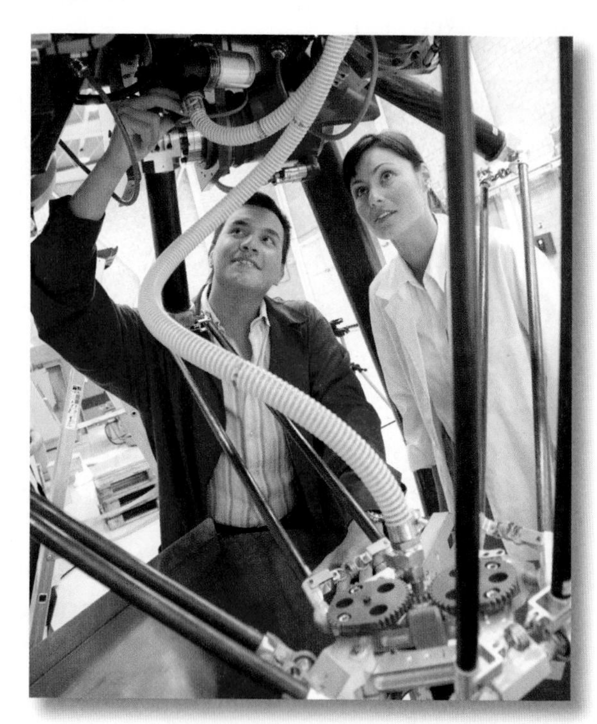

Real Robots Many mathematical calculations determine how much weight this robot can lift. *This robot is hydraulic and can lift 360 pounds. How many pounds could it lift if it were a pneumatic robot?*

Applying Math

Suppose you wanted to make the flat piece of copper shown in **Figure 6.1** into a boat. The piece of copper weighs 1.3 pounds, and its area is 64 square inches. Find its volume by multiplying its area by its height, or thickness, which is $\frac{1}{16}$ inch. The volume of the flat copper piece is 4 cubic inches ($64 \times \frac{1}{16} = 4$). In order to make the copper float, you must increase its volume. You can do this by folding it into a box.

- The box's volume is length times width times height:
  Volume = 6 inches × 4 inches × 2 inches
  Volume = 48 cubic inches
- The density of the boxy copper boat is its weight divided by its volume:
  Density = 1.3 pounds (divided by) 48 cubic inches
  Density = 0.0271 pci

The density of the copper boat (0.0271 pci) is less than the density of water (0.0361 pci), which means it will float. The *SS Great Britain* weighed 6.89 million pounds—but its volume was large enough, and it could also float.

Figure 6.1 Making Metal Float

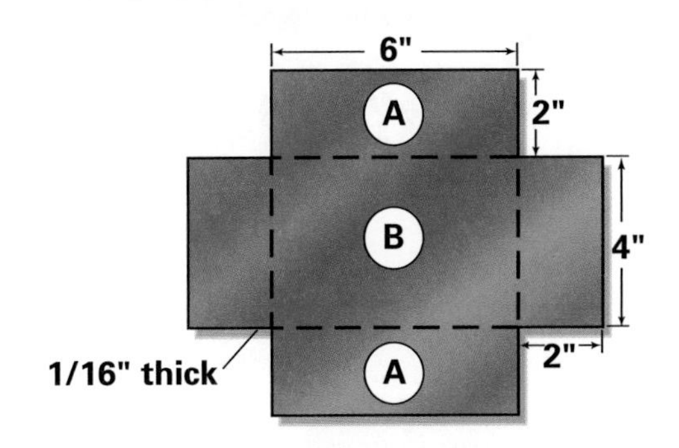

Density As long as the density of the copper boat is less than the density of water, it will float. *In addition to density, what other factor would be important for keeping this box afloat?*

After You Read **Self-Check**

1. Tell the difference between a hypothesis and a scientific law.
2. Explain the importance of calculating a robot's ability to extend its arm.
3. Describe how it is possible for a very large, heavy ship to float.

Think

4. Suppose you observed that certain metals could be melted at lower temperatures than previously thought. Explain how you would establish your observation as a scientific law.

Practice Academic Skills

English Language Arts/Writing

5. Divide a sheet of paper into two columns. Label one column "Science" and the other "Mathematics." Think of ways in which science and mathematics are connected to technology in your school and at home and write them in the appropriate column.

STEM **Mathematics**

6. Brianna wants to be a race car driver. She is able to practice at a course that is 2.25 miles long. If it takes her 1 minute and 15 seconds to complete one lap, what is her average speed in miles per hour?

Math Concept **Average Speed** Average speed is total distance divided by time.

1. When calculating average speed, make sure to solve using the correct units.
2. To convert something from miles per minute to miles per hour, multiply by 60.

For help, go to **glencoe.com** to this book's OLC and find the Math Handbook.

section 6.2 Language Arts and Social Studies

Reading Guide

Before You Read **Preview** How are language arts and social studies linked to technology?

Content Vocabulary

- humanities
- Information Age
- shadowing program

Academic Vocabulary

- element
- region

Graphic Organizer

Draw the section diagram. Use it to organize and write down information as you read.

Technology and the Working World

Occupation		Related Technologies
Health	→	X-ray machines, microscopes

Go to **glencoe.com** to this book's OLC for a downloadable graphic organizer and more.

TECHNOLOGY STANDARDS

STL 17 Information & Communication Technologies

ACADEMIC STANDARDS

English Language Arts

NCTE 7 Conduct research and gather, evaluate, and synthesize data to communicate discoveries

NCTE 11 Participate as members of literacy communities

Social Studies

NCSS 8 Science, technology, and society

STL *National Standards for Technological Literacy*

NCTM *National Council of Teachers of Mathematics*

NCTE *National Council of Teachers of English*

NSES *National Science Education Standards*

NCSS *National Council for the Social Studies*

Technology and the Humanities

Which subjects are called the "humanities"?

Some subjects that you study in school are called the **humanities**. They include language arts, social studies, and art. In this section, we will look at how technology is connected to language arts and social studies.

As You Read

Connect How will studying technology help you in your future career?

Language Arts

What is the relationship between language arts and technology?

Spoken and written communication helps develop technology and teach it. Technology provides for better communication.

Language and Technology in the Past

Early humans probably used hand gestures to communicate. Eventually, they developed a written language to pass along information.

The invention of the plow helped change our ancestors from hunters and gatherers to farmers. This led to property ownership and the need for records, making written and spoken language even more important.

In the 15th century, Johannes Gutenberg developed movable metal type. That led to the manufacture of inexpensive books, which paved the way for universal education. Before Gutenberg, few people learned reading and writing. As more people became educated and communication increased, inventions multiplied. The printed word helped spread the use of technology.

Language and Technology Today

How is language connected to technology today? If you invented a new kind of computer, you would need language to program it and to create plans for producing it. You would also need language to create a company, hire employees, keep business records, and communicate with customers. People who work with companies overseas need world language skills.

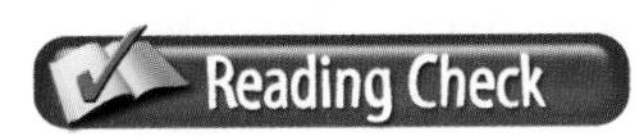

Recall What led to universal education?

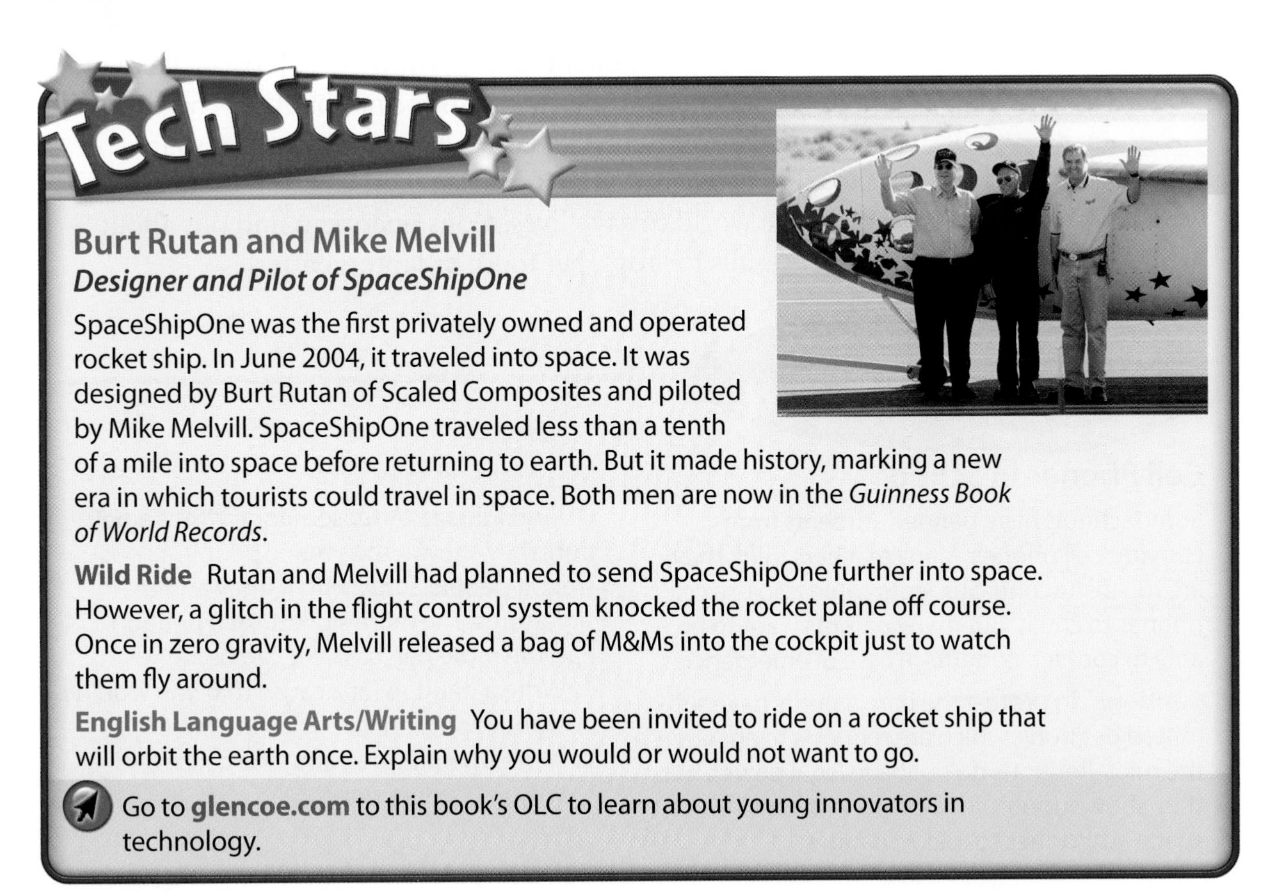

Burt Rutan and Mike Melvill

Designer and Pilot of SpaceShipOne

SpaceShipOne was the first privately owned and operated rocket ship. In June 2004, it traveled into space. It was designed by Burt Rutan of Scaled Composites and piloted by Mike Melvill. SpaceShipOne traveled less than a tenth of a mile into space before returning to earth. But it made history, marking a new era in which tourists could travel in space. Both men are now in the *Guinness Book of World Records.*

Wild Ride Rutan and Melvill had planned to send SpaceShipOne further into space. However, a glitch in the flight control system knocked the rocket plane off course. Once in zero gravity, Melvill released a bag of M&Ms into the cockpit just to watch them fly around.

English Language Arts/Writing You have been invited to ride on a rocket ship that will orbit the earth once. Explain why you would or would not want to go.

Go to **glencoe.com** to this book's OLC to learn about young innovators in technology.

BusinessWeek Tech NEWS

Amazon's E-Book

Gadget makers have been trying to sell us on e-books—digital devices that display text. Amazon [came] up with the first e-book to function independently of a computer. When you purchase a "Kindle," you take the reader out of the box, go to Amazon's Kindle store, make your book selection, and it downloads in seconds.

Critical Thinking *How might using an e-book affect studying English language arts?*

Go to **glencoe.com** to this book's OLC to read more about this news.

Social Studies

How do technology and social studies work together?

The study of technology deals with more than things. It deals with how people use technology and how it changes their lives. You learn about these interactions in social studies.

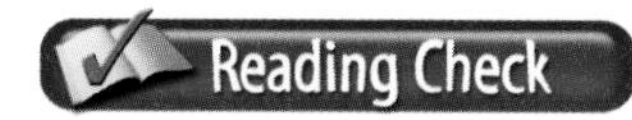

Predict What is social studies?

The Industrial Revolution

The Industrial Revolution began about 1750 and resulted in worldwide social changes. Before the Industrial Revolution, most countries' economies were based on farming. Then they became based on factory production.

Today we are experiencing another revolution as we move into the **Information Age**. We still need factories, but more people make their living by creating, processing, and distributing information.

The Information Age

It is difficult to say when the Information Age began. It may have started in 1844 when Samuel Morse tapped out the first long-distance telegraph message. Or perhaps it started in 1944 when Howard Aiken operated the world's first large-scale digital computer. Whenever it began, the Information Age is here to stay.

With the world linked by computers, everyone can share specialized knowledge. Doctors can send information immediately around the world to save lives. Consumers can find out about product recalls for toys, pet food, or laptop batteries.

Ethics in Action

Cell Phones in School

Some schools have banned students from carrying cell phones. Many teachers think they are too distracting. Students could also use cell phones to cheat. But many parents want to be able to contact students in case of emergencies.

Petitions To protest the ban, parents have submitted petitions, which are requests to change the rules. Petitions do not have legal power, but they show support for a cause, and can persuade authorities to make changes.

English Language Arts/Writing

Online Causes After someone writes a petition, they show it to as many people as possible. If people agree with it, they add their signatures. The more signatures a petition has, the more persuasive it can be.

1. Write a petition for a cause that you would want to post online.
2. Write a paragraph about the cause and why you would sign it.

Visualizing the World

The rapid exchange of information has changed the way we visualize the world. Television, computers, cell phones, and other Information Age products are elements that have changed the way schools educate students. Today's young people have immediate access to large amounts of information.

Global positioning system (GPS) satellites help map the world with great accuracy. A small receiver picks up satellite data to tell you exactly where you are located. GPS is useful for hikers, travelers, and for ship captains planning long ocean voyages. It also helps engineers construct roads in undeveloped regions.

Figure 6.2 **The U. S. Department of Education Career Clusters**

Career Cluster	Job Examples
Agriculture, Food, and Natural Resources	farmer, ecologist, veterinarian, biochemist
Architecture and Construction	contractor, architect, plumber, building inspector
Art, Audio/Video Technology, and Communications	graphic designer, musician, actor, journalist, filmmaker
Business, Management and Administration	executive assistant, receptionist, bookkeeper, business owner
Education and Training	teacher, trainer, principal, counselor, financial planner
Finance	bank teller, tax preparer, stockbroker, financial planner
Government and Public Administration	soldier, postal worker, city manager, nonprofit director
Health Science	pediatrician, registered nurse, dentist, physical therapist
Hospitality and Tourism	chef, hotel manager, translator, tour guide
Human Services	social worker, psychologist, child care worker
Information Technology	Web designer, software engineer, technical writer
Law, Public Safety, Corrections, and Security	attorney, police officer, firefighter, paralegal
Manufacturing	production supervisor, manufacturing engineer, welding technician, quality technician
Marketing, Sales, and Service	sales associate, retail buyer, customer service representative
Science, Technology, Engineering, and Mathematics	lab technician, marine biologist electrical engineer, cryptanalyst
Transportation, Distribution, and Logistics	pilot, railroad conductor, truck driver, automotive mechanic

Career Clusters The United States Department of Education groups careers into 16 career clusters that are based on similar job characteristics. *How might career clusters help you explore careers?*

GPS on Your Cell Phone

Imagine knowing exactly where your friends are when they call you. Some cell phones have a built-in GPS feature. This is special software that allows users to share their location with their friends. If the friends do not have a GPS-capable model, they will receive a text message with a Web link to a customized map. *Would you like people to know exactly where you are all the time?*

Go to **glencoe.com** to this book's OLC for answers and to learn more about GPS and cell phone technology.

Business in the Information Age

We also conduct business differently in the Information Age. In retail stores, universal product code (UPC) readers rapidly sense the price of food and clothing. People can get through checkout lines faster or even check out their own purchases.

With a credit card and an Internet Service Provider, you can pay bills, make banking transactions, and buy products online.

Technology and the Working World

Economics involves the goods and services we produce. It also relates to how people earn a living. Today's jobs require workers with technology-based abilities. In our Information Age, many jobs are available for people with education and training.

Technology and Careers

Try this activity. With your teacher's guidance, divide the class into groups of students with similar occupational interests. See **Figure 6.2** on page 119. Each group should research its chosen field to determine what technological devices are used and how workers in each field might need to use:

- English language arts
- Science
- Mathematics
- Social studies
- World languages

With your teacher's help, establish a **shadowing program** in which you and your classmates spend a day in the work environment of your chosen field. Share your experiences with the class.

section  6.2 assessment

After You Read **Self-Check**

1. Name at least three subjects included in the humanities.
2. Describe how Johannes Gutenberg influenced both language arts and technology.
3. Explain how the Information Age has changed education.

Think

4. Explain how the global positioning system could be used to help find lost children.

Practice Academic Skills

English Language Arts/Writing

5. Do some research to find the difference between the Internet and the World Wide Web. Write a paragraph explaining your findings.

STEM Science

6. The Information Age has linked the world through computers and other technologies. One such technology is the Global Positioning System (GPS). GPS allows for accurate positioning around the world. Research how GPS was developed and the different applications that utilize GPS. Write a few paragraphs describing what you find.

Exploring Careers in Technology

Beth Heller

SCIENCE TEACHER

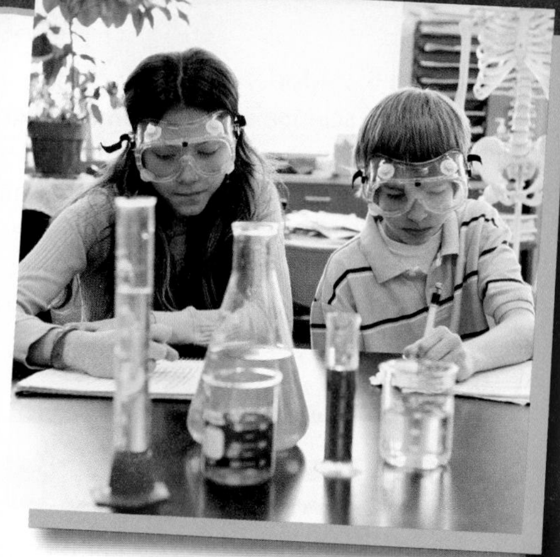

Q: *How did you get interested in being a science teacher?*

A: The teachers I had in school made learning science fun! We did all kinds of neat, hands-on activities, and I was inspired to do the same.

Q: *What is a typical day like for you on this job?*

A: Before the students arrive, I answer parent e-mails, grade papers, go over my lesson plans, or set up labs. I also spend my early mornings attending faculty meetings and meeting with students who have questions. Each day, I teach six classes and supervise a study hall period.

Q: *What do you like most about your job?*

A: I really enjoy interacting with so many different students and teaching a variety of science subjects. I enjoy being able to watch the students grow up and mature. It amazes me how uniquely every person changes, along with their interests.

Q: *What can you tell aspiring students about preparing for a career as a science teacher?*

A: Take as many different science classes as you can, especially physics and chemistry. A psychology course will help you understand how the brain works and how to present new topics to students. See if your school will allow you to become a teacher's assistant, tutor, or mentor. These activities can give you an idea of what teaching is like. We need more good teachers!

English Language Arts/Writing

Write an E-mail Letters and e-mail have the same basic elements but different purposes. Send an e-mail message to a classmate comparing e-mail to letter writing. List the advantages and disadvantages of each form.

1. Copy your teacher on the e-mail. To do this, add your teacher's e-mail address to the CC line.
2. Ask for a response from your classmate. Request that they "Reply to All", so that your teacher will receive the response as well.
3. Be clear and concise. Make sure you proofread your message before you send it.

Go to **glencoe.com** to this book's OLC to learn more about this career.

Real-World Skills

Speaking and listening, problem-solving, organization

Academics and Education

Science, English language arts, mathematics

Career Outlook

Growth as fast as average for the next ten years

Source: *Occupational Outlook Handbook*

chapter 6 Review and Assessment

Chapter Summary

Section 6.1 Modern technology depends heavily on science and mathematics. All scientists apply the same scientific method to solve problems. Scientific conclusions that have been carefully developed through experimentation are called "scientific theories." Over time they may be accepted as scientific laws. Few subjects are more important to technology than mathematics.

Section 6.2 We need verbal and written communication skills in order to develop technology and teach it to others. Social studies teaches us that past inventions have improved people's lives, changed economies, and altered governments. Our society has moved into the Information Age. Today's jobs require workers with technology-based abilities.

Review Content Vocabulary and Academic Vocabulary

1. On a sheet of paper, use each of these terms and words in a written sentence.

Content Vocabulary
- hypothesis
- scientific theory
- scientific law
- humanities
- Information Age
- shadowing program

Academic Vocabulary
- volume
- area
- element
- region

Review Key Concepts

2. **Describe** the connections between science and technology.
3. **Discuss** the connections between mathematics and technology.
4. **Explain** the scientific method.
5. **Identify** what mathematical calculations tell us about robots.
6. **Discuss** ways in which language arts and technology are connected.
7. **Describe** how social studies is connected with technology.
8. **Define** the Information Age.
9. **Discuss** why the Information Age and technology are important to a career.
10. **Explain** the difference between technology and the humanities.

Real-World Skills

11. **Written Communication** Write instructions for doing a simple task, such as putting on a jacket or tying shoes. Give your instructions to a classmate to use and ask for feedback. Describe how well the instructions were written.

Technology Skill

12. **Global Communications** In the past, the only way to communicate with people in other countries was through mail or on the phone.
    - **a.** Use the Internet to find a Web site that allows you to e-mail someone in another country.
    - **b.** Write an email to someone in another country. Say why you are writing and ask for a reply.

Academic Skills

Social Studies

13. Research the Information Age on the Internet and write a few paragraphs on how it has affected the way goods and services are produced.

STEM **Mathematics**

14. Sunjana works for a delivery service. She is paid $5.15 an hour, plus half of her cost for gasoline. Her car is old and gets only 21 miles per gallon of gas. If she drives a total of 86 miles one day, and gas costs $2.59 a gallon, how much should she be reimbursed for gas?

Math Concept **Multi-Step Problems** Multi-step problems require extra attention. Note the information given and determine what to solve for so you can eliminate unnecessary information.

WINNING EVENTS

Summer Work

Situation You want to get a summer job by showing employers you are the best candidate.

Activity Conduct job research, write a cover letter and a résumé, and practice being interviewed.

- Letters and résumés should be one page.
- With classmates, develop interview questions. Take turns interviewing each other.

Evaluation You will be evaluated on how well you meet these performance indicators:

- A list of five jobs
- Cover letter—well written, featuring your strengths related to the position
- Résumé—showing work history, related activities, grades, or awards
- Practice interviews—answers, poise, voice

Go to **glencoe.com** to this book's OLC for information about TSA events.

Standardized Test Practice

Directions Choose the letter of the best answer. Write the letter on a separate piece of paper.

1. What is the circumference of a circle with a radius of 4.24 inches?
   Pi = 3.14
   - **A** 13.3136 inches
   - **B** 26.6272 inches
   - **C** 13.6272 inches
   - **D** 26.3136 inches
2. A scientific conclusion that is developed through careful experimentation is a scientific law.
   - **T**
   - **F**

Test-Taking Tip Always focus on the question at hand; do not let your mind wander.

chapter 6

TECHNOLOGY LAB

Produce a Documentary Video

Have you ever seen a documentary? A documentary gives facts and information about real events and people rather than telling a fictitious story. However, some documentaries unfold like a story. The best keep viewers interested by showing action rather than just talking about something.

Tools and Materials

- ✓ Camcorder
- ✓ Tripod
- ✓ Computer
- ✓ Presentation and video editing software
- ✓ Art materials
- ✓ Props for your video

Set Your Goal

You and your team will produce a documentary video. The subject will be the ways in which a particular technology of your choice (or an event involving that technology) is connected to at least three subjects taught in school. Subjects may include history, mathematics, science, English language arts, economics, and art.

Know the Criteria and Constraints

In this lab, you will:

1. Your documentary must examine the connections between the technology you choose and at least three different school subjects. The connections may involve how the technology was developed or how it is used. They must be described in your documentary by a narrator or with written or printed signs.
2. Your documentary must be at least ten minutes long but no longer than 20 minutes.
3. You must produce a script for either the narrator or the actors.
4. Your teacher must approve your choices before you begin.

Design Your Project

Follow these steps to complete this lab.

1. Brainstorm with your teammates to select a technology that interests you. Think of at least three subjects you can connect with your technology.
2. Be creative and have fun. For example, you could make a documentary about the sinking of the *Titanic* ship by making and filming a model.
3. Create a plan and a script for your documentary.
   - What will you show and how will you show it?
   - Keep in mind the audience, purpose, and nature of your message.
   - How much time will each part of your film take?
4. Rehearse before filming.
   - Gather or build props. If you are doing an experiment, make sure it works before you film it.
   - Create signs or other graphics.
5. Shoot your documentary.

Evaluate Your Results

After you complete the lab, answer these questions on a separate piece of paper.

1. Was it easy or difficult to find connections to other subjects? Explain.
2. How long did it take to plan, write, and produce your documentary?
3. Was editing necessary? Did you use video editing software or just the camera's editing function? Explain.

Academic Skills Required to Complete Lab

Tasks	English Language Arts	Math	Science	Social Studies
Research the subject of your documentary.	✓		✓	✓
Explore how the topic connects to subjects of study in school.	✓			✓
Write a plan for the film.	✓	✓	✓	
Shoot the film, write script, or act in it, and edit film.	✓	✓		
Present to the class.	✓		✓	

Technology Time Machine

On Technology's Trail

Play the Game This time machine will travel to the past to show you technology since its beginnings more than 2.5 million years ago. People use technology every day to extend their abilities and satisfy their wants and needs. To operate the time machine, you must know the secret code word. To discover the code, read the clues, and then answer the questions.

Clue 1

Prehistoric Times Human beings have been using technology since they first invented primitive tools from materials such as wood, stone, or bone. Prehistoric people also learned how to make fire, which gave them power.

Clue 2

8000 B.C.E. People developed skills to raise crops and animals for food. The development of agriculture promoted civilization as people no longer had to move around searching for food. They settled down in small groups to grow crops and tend their land.

Clue 3

4000 B.C.E. The first real civilizations were based on agriculture. But when the plow was developed, civilization took a giant leap forward.

Clue 4

3500 B.C.E. Although no one knows for sure, many experts believe the wheel was invented around the same time as the plow and the wagon. Until the first automobile was developed, the wagon remained the most basic form of transportation.

Clue 5

1700 B.C.E. Up until this time, construction methods had been slow to develop, but construction technology began to rapidly change. Simple rectangular structures were transformed with columns and beams, as seen in Greek architecture.

Clue 6

27 B.C.E. (start of the Roman Empire) The ancient Romans were groundbreaking engineers. They developed roads, bridges, tunnels, and aqueducts, some of which are still used today. Similar achievements were taking place in China and Central America.

Clue 7

1881 People dreamed of digging a canal across the Isthmus of Panama in Central America. This would shorten the route for ships traveling between the Atlantic and Pacific Oceans. Construction faced many problems, including workers dying from yellow fever. The Panama Canal opened for passage in 1914.

Clue 8

2000s Today computer programs can draw floor plans, figure out electrical and plumbing systems, and design doors and windows. Some can build 3D models of your design and landscaping ideas.

Crack the Code

On a piece of paper, write the answers to these questions:

1. What opened for passage in 1914?
2. What did the Romans build over land during the Roman Empire?
3. The Atlantic and the Pacific are two of these.
4. Primitive tools were made of wood, stone and what else?
5. Civilization developed when people could grow crops and tend to this.
6. The Romans were skilled ___________.
7. Prehistoric people found power when they learned how to do this.
8. What word describes the design of early buildings?

Now write down the first letter of each answer. Put them together to discover the secret code word!

Hint Solving these often leads to technological developments.

unit 1 Thematic Project

Redesigning an Invention

In Unit 1, you learned about how products are designed. Creating a new invention or innovation takes time and energy to plan. Inventors must build prototypes, try out, and problem-solve a design. They need to know about physics, power sources, CAD, and other elements of engineering.

Filling a Need Once a need is determined, the engineers get to work inventing. Developing a new product can be expensive. Engineers take many steps before finalizing an invention. 3D software shows how parts fit and work together. A prototype proves parts fit. When the invention works, parts are made.

Form and Function Inventors have made flops by forgetting about form or function. A beautiful chair may be uncomfortable, or an economical hybrid car may be ugly. Designers need to keep beauty and usefulness in mind.

This Project In this project, you will redesign an everyday invention.

Your Project

- Choose any everyday object to redesign, such as a stapler, a chair, or spatula. Explain why it needs to be redesigned.
- Choose and complete one of these tasks:
  1. Create a two-dimensional model of your redesign, by freehand or CAD. Label parts.
  2. Fill out a patent application for your design.
  3. Create a working model of your invention.
  4. Create a 1-minute commercial.
- Write a report.
- Create a presentation with posters, video, or presentation software.
- Present your findings to the class.

Tools and Materials

- ✓ Computer
- ✓ Internet access
- ✓ Trade magazines
- ✓ Word-processing software
- ✓ Presentation software
- ✓ Posterboard
- ✓ Colored markers

The Academic Skills You'll Use

- Communicate effectively.
- Speak clearly and concisely.
- Use correct spelling and grammar in a written report.
- Conduct research using a variety of resources.
- Incorporate reading, writing, and speaking with viewing, representing, and listening.

English Language Arts

NCTE 8 Use information resources to gather information and create and communicate knowledge.

Science

NSES Content Standard F Science and technology in society

Step 1 Choose Your Topic

You can choose any product to redesign for your project. Examples include:

- Ballpoint pen
- Pillow
- Soda-pop can
- Clothes hanger
- Coffee table
- Night light
- Hand-soap pump
- Messenger bag

 *If you have thought there must be a better _____, design one!*

Step 2 Do Your Research

Research your project. Your fact finding may include a combination of these ideas:

- Ask for feedback from someone who has used the product.
- Search for articles about your topic—try libraries and online magazine archives.
- Look at how the item was made in the past.
- Study blueprints or specifications for similar products.
- Ask for tips from someone who does drafting.

 *Keep your improvements simple!*

Step 3 Explore Your Community

Find someone in your community who knows something about designing products—a designer or a teacher. Ask how the job connects to engineering design. Or visit a factory.

 *Try visiting a small factory or workshop.*

Step 4 Create Your Project

Your project should include:

- 1 research project (design, blueprint, patent application, video, or model)
- 1 report
- 1 presentation

Project Checklist

	Objectives for Your Project
Visual	✓ Make a poster, blueprint, model, video, or slide presentation to illustrate your project.
Presentation	✓ Make a presentation to your class and discuss what you have learned. ✓ Turn in research and names from your interview to your teacher.

Step 5 Evaluate Your Presentation

In your report and presentation, did you remember to:

- Demonstrate your research and preparation?
- Engage your audience?
- Back up statements with facts and evidence?
- Use visuals effectively?
- Speak slowly and enunciate clearly?

 **Rubrics** Go to **glencoe.com** to the book's OLC for a printable evaluation form and your academic assessment form.

GLOBAL TECHNOLOGY

Musical Innovations

Innovations are improvements to existing inventions—and they all rely on technology. Robotics technology by a German company called *Tronical GmbH* allows a guitar to tune itself in two seconds. The Powertune System uses self-tuning technology. Its processor directs motors on six tuning pegs of a guitar to tighten or loosen the strings to find the right pitch.

Critical Thinking *What innovations might improve other musical instruments?*

German

hello	hallo
goodbye	Auf Wiedersehen
How are you?	Wie geht es dir?
thank you	danke
You're welcome	Keine Ursache

 Go to **glencoe.com** to the book's OLC to learn more and to find resources from **The Discovery Channel.**

unit 2 Energy and Power Technologies

Chapters In This Unit:

Unit Thematic Project Preview

Designing Alternative Power Plants

As part of this unit, you will learn about different forms of energy. Electric companies can convert energy from many sources—water, the sun, wind, nuclear energy, or coal.

As you read this unit, use this checklist to prepare for the project at the end of this unit:

PROJECT CHECKLIST

✓ Make a list of alternative energy sources.

✓ Think about which one you would like to explore further.

✓ Do online research and find an example of this energy source being used today.

WebQuest Internet Project

Go to **glencoe.com** to this book's Online Learning Center (OLC) to find the WebQuest activity for Unit 2. Begin by reading the Task. This WebQuest activity will help you find out how alternative fuels can reduce air pollution and provide other benefits.

Explore the Photo

Power for Our World Since the beginning of the 20th century, people have depended on the power of electricity to operate everything from lighting to music. Recently alternative forms of energy to create electricity, such as solar power, have emerged. *Look around the classroom and make a list of everything that relies on electricity to function.*

chapter 7

Energy and Power for Technology

Sections

What You'll Learn

- **List** the different forms of energy.
- **Explain** renewable, nonrenewable, and unlimited energy resources.
- **Identify** the most common forms of power.
- **Describe** uses for each form of power.
- **Explain** how energy and power technologies are selected.
- **Identify** forms of pollution resulting from energy and power technologies.
- **Discuss** methods to slow depletion of energy resources.

Explore the Photo

The Power of Wind Large wind farms made up of wind turbines are a source of energy for many communities. *Where might be the best places for these wind farms?*

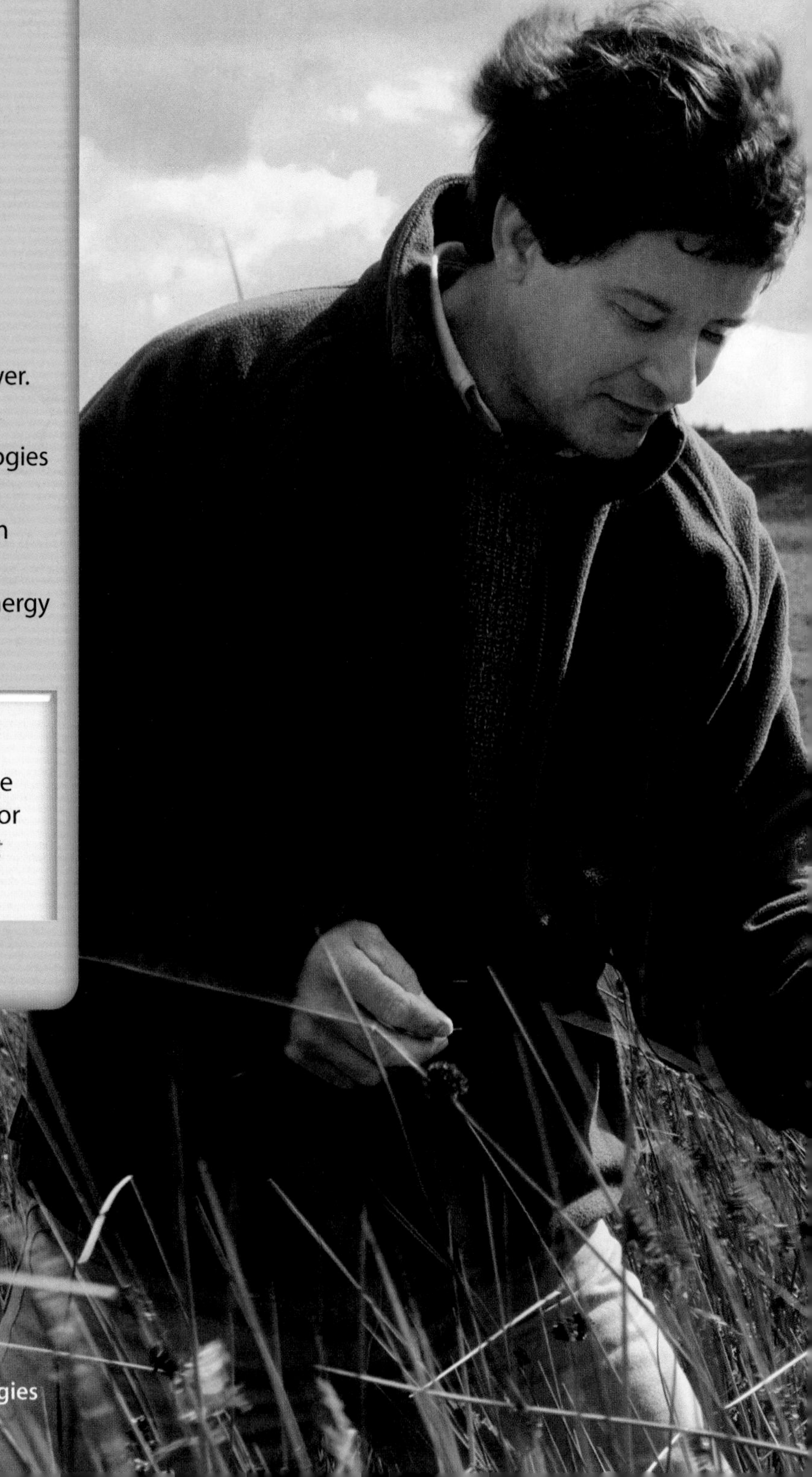

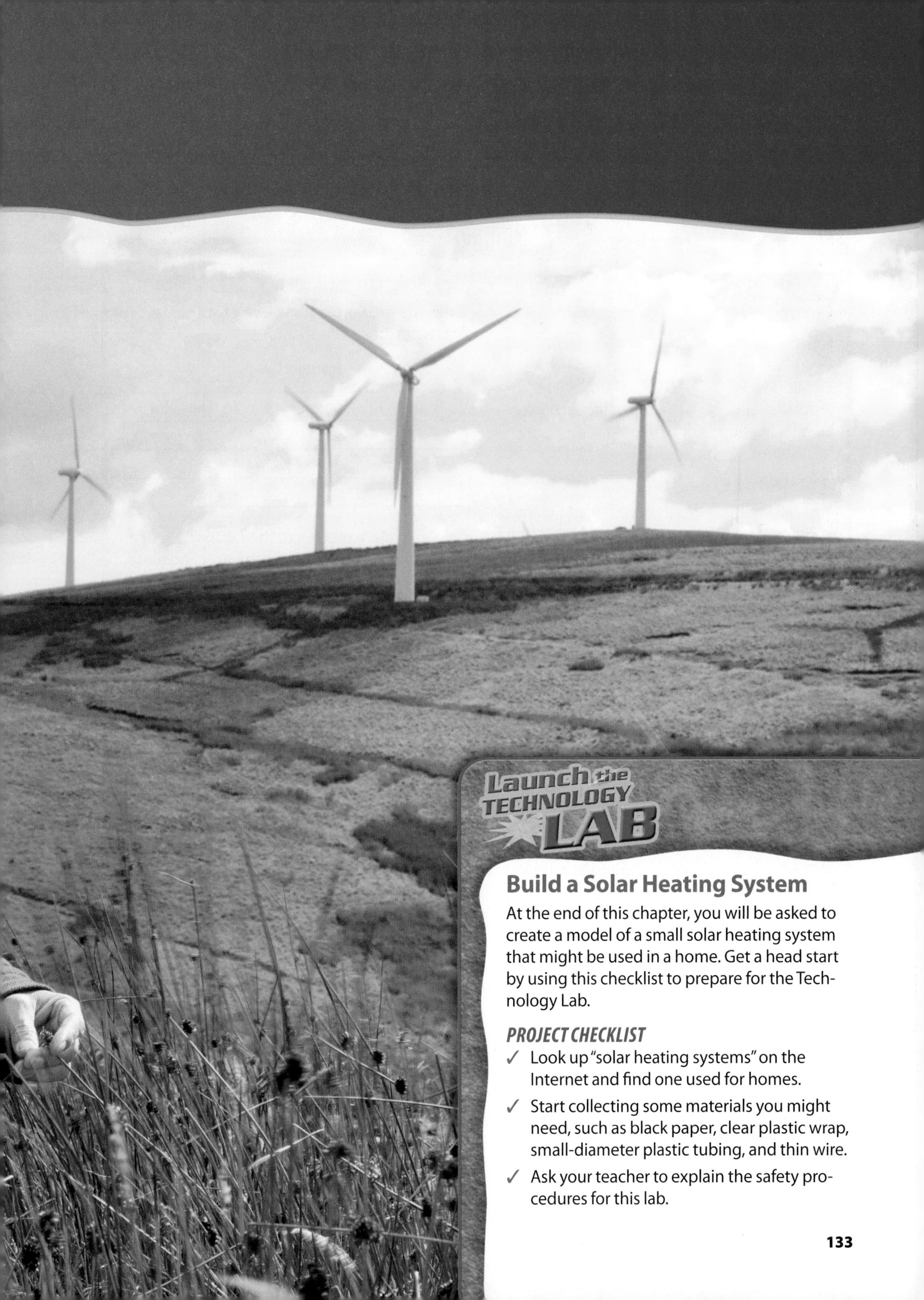

Build a Solar Heating System

At the end of this chapter, you will be asked to create a model of a small solar heating system that might be used in a home. Get a head start by using this checklist to prepare for the Technology Lab.

PROJECT CHECKLIST

- ✓ Look up "solar heating systems" on the Internet and find one used for homes.
- ✓ Start collecting some materials you might need, such as black paper, clear plastic wrap, small-diameter plastic tubing, and thin wire.
- ✓ Ask your teacher to explain the safety procedures for this lab.

section

7.1 Energy Basics

Reading Guide

Before You Read **Preview** What is energy?

Content Vocabulary

- energy
- calorie
- fossil fuel
- solar heating system
- solar cell
- wind farm
- hydroelectric power
- geothermal energy

Academic Vocabulary

You will see these words in your reading and on your tests. Find their meanings at the back of this book.

- vehicle
- constant

Graphic Organizer

Draw the section diagram. Use it to organize and write down information as you read.

Sources of Energy

Renewable	Nonrenewable	Unlimited
Food	Oil	Solar energy

Go to **glencoe.com** to this book's OLC for a downloadable graphic organizer and more.

TECHNOLOGY STANDARDS

STL 3 Relationships & Connections

STL 5 Environmental Effects

STL 11 Design Process

STL 16 Energy & Power Technologies

ACADEMIC STANDARDS

Mathematics

NCTM Problem Solving Apply and adapt a variety of appropriate strategies to solve problems.

English Language Arts

NCTE 6 Apply knowledge of language structure and conventions to discuss texts.

STL *National Standards for Technological Literacy*

NCTM *National Council of Teachers of Mathematics*

NCTE *National Council of Teachers of English*

NSES *National Science Education Standards*

NCSS *National Council for the Social Studies*

Understanding Energy

Can we use energy directly from nature?

What is energy? Is it strength, speed, or motion? Actually, **energy** is the ability to do work. Nature does not usually give us energy in a form we can use directly. It is as if the energy were hiding—waiting for someone to find it. For example, coal is a source of energy, but it is just a black rock. Technology makes it possible to find and release the energy hidden in the rock, and then put that energy to work. Technologists are always looking for new and better ways to use nature's energy.

As You Read

Identify Think of some examples of the different forms of energy.

Forms of Energy

How many forms of energy are there?

As you go through an average day, you use many different forms of energy. Electrical energy powers light bulbs. Thermal, or heat, energy from your home's furnace keeps you warm. Mechanical energy, used by buses or cars, might get you to school.

All energy in nature can be grouped into these six forms:

1. Mechanical energy, or energy of motion—pedaling a bicycle
2. Thermal, or heat, energy—a campfire for cooking marshmallows
3. Electrical energy—a bolt of lightning
4. Chemical energy—a battery in a cell phone
5. Nuclear energy—a submarine engine
6. Light, or radiant, energy—solar cells for a calculator or a house

Energy cannot be created or destroyed, but it can be changed from one form to another. For example, to use coal to produce useful electricity, the form of the energy must change many times before it becomes electricity. **Figure 7.1** Changing Forms of Energy shows the path from the black rock to electricity in these steps: Burning coal produces heat; heat changes water to steam; steam powers a turbine that spins a generator; and the generator produces electricity.

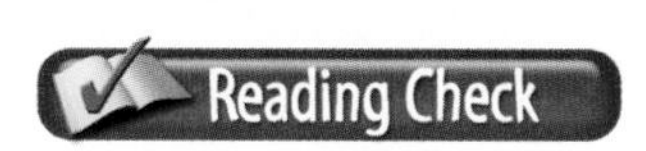

Recall What are the six forms of energy in nature?

Figure 7.1 **Changing Forms of Energy**

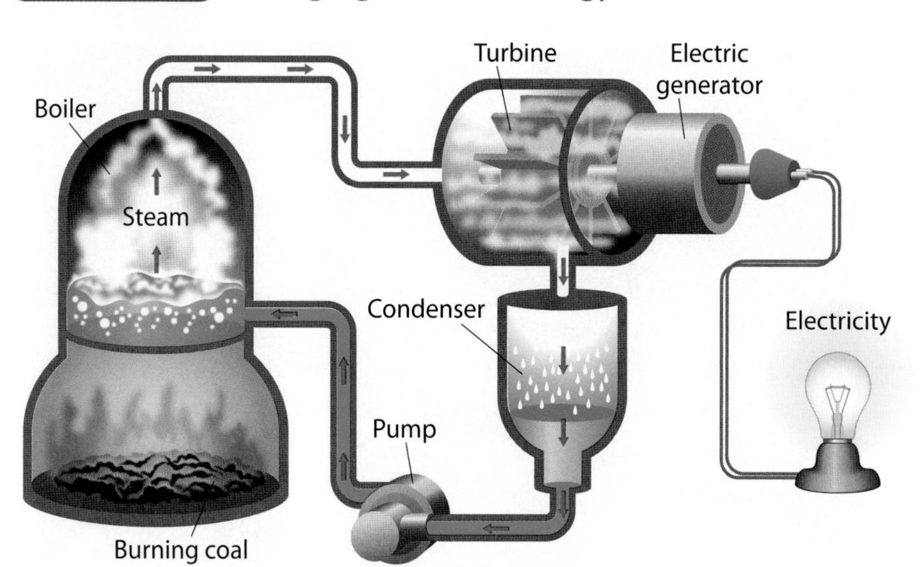

Coal Power Raw energy sources can be converted into useful forms of energy. *What energy is created from coal?*

Energy Resources

Today's energy sources fall into what three groups?

Let's take a closer look at today's sources of energy. They can be divided into three groups: renewable, nonrenewable, and unlimited.

Renewable Sources

Renewable energy sources come from plants and animals. They can be replaced or renewed when we need more. Two examples include food and alcohol.

Food

Your body requires food for the energy you use in walking, blinking your eyes, thinking, and all your other activities. Food energy is measured in **calories**. Some foods contain more energy than others; more food energy means more calories.

Alcohol

Alcohol is a liquid made from crops such as corn and sugar cane. It can be used as fuel in special automobile and truck engines. These engines can operate with regular gasoline, an alcohol-gasoline mixture, or pure alcohol. They are called "flexible-fuel" vehicles.

Automobiles that run on alcohol fuel produce less air pollution than those that use gasoline. When added to gasoline, an alcohol called "ethanol" can extend fuel supplies. A mixture of 10 percent ethanol and 90 percent gasoline is sold at some service stations. All ordinary car engines can run on this fuel.

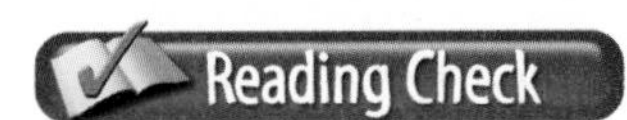

Recall What might be some examples of renewable, nonrenewable, and unlimited energy sources used by your family every day?

Burning Fuel Swimming uses a lot of energy. *From what do we get that energy? What units are used to measure that kind of energy?*

Figure 7.2 Forming Fossil Fuels

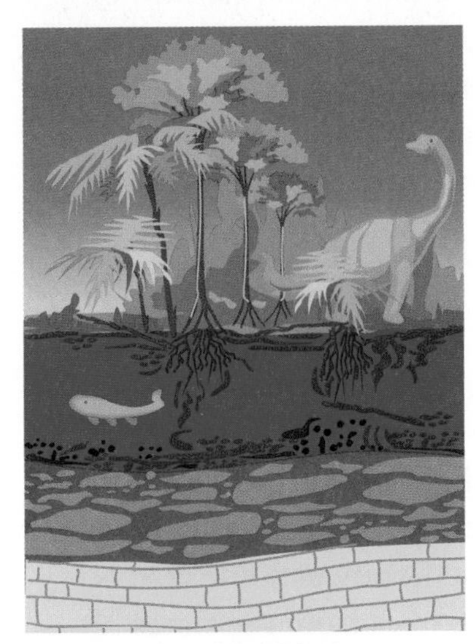

The formation of fossil fuels began when plants and animals died, and their remains formed thick layers on the bottom of swamps.

Eventually, the remains were covered by thick layers of earth.

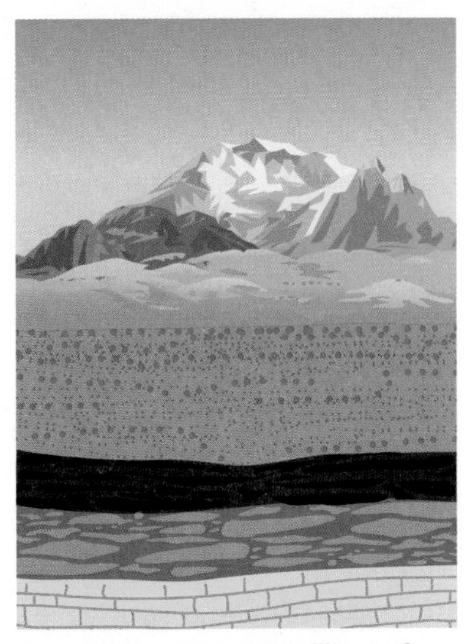

Pressure and heat over millions of years slowly changed the remains into fossil fuels.

Ancient Energy Sources Coal, oil, and natural gas are fossil fuels that have formed over millions of years. *What type of source of energy are fossil fuels?*

Nonrenewable Sources

Nonrenewable sources of energy cannot be replaced once they are gone. Coal, oil, natural gas, and uranium are examples of these sources.

Coal, oil, and natural gas are **fossil fuels**, or fuels made from fossils. A fossil is what remains from a plant or animal that lived long ago. Coal, oil, and natural gas are formed from these once-living plants and animals. (See **Figure 7.2.**)

Coal

Millions of years ago, as plants died, their remains fell to the ground. Over time and under pressure, these plants formed thick layers, which became coal seams. A seam is a strip of coal between other rock layers. It can be close to the surface or deep underground. Many seams are only two or three feet thick. Power companies use coal to generate electricity.

Oil

You can thank animals that lived millions of years ago for the gasoline used in today's cars. When these animals died, their remains combined with the remains from plants to form crude oil. The fuels for cars, trucks, locomotives, airplanes, and ships come from oil. More of our energy comes from oil than from any other source. Oil is also turned into products, such as plastics, paint, and asphalt.

Academic Connections
Science

Your Carbon Footprint A carbon footprint is a measure of how much carbon dioxide enters the atmosphere as a result of your daily activities. Carbon dioxide is a major greenhouse gas that comes from burning fossil fuels. What you do has an effect on your carbon footprint. That includes turning on a light switch, riding in a car, and buying food. The average U.S. carbon footprint is about 3,500 pounds per person each month.

Apply Determine your own carbon footprint. Use a search engine and the keywords *carbon footprint calculator.*

Natural Gas

The slow transformation of plant and animal matter into oil also produces natural gas. As a result, natural gas is often near oil deposits. It is a flammable gas, which means that it burns easily.

The United States produces more natural gas than any other country. Many industrial processes use natural gas. It is mostly used as a fuel for home heating and cooking. What kind of energy heats your home?

Uranium

Uranium is not a fossil fuel, but it is also a nonrenewable source of energy. Unlike fossil fuels, uranium does not come from plants and animals. It is a radioactive, rocklike mineral that comes from the ground. Uranium is used for fuel in nuclear power plants, which produce electricity. It is also used to power some U.S. Navy ships. Most of our supply of uranium comes from the states of New Mexico and Wyoming.

Uranium develops a large amount of heat during a controlled nuclear reaction. For example, in power plants, the heat changes water into steam, and the steam operates generators that produce electricity.

The amount of energy in uranium is amazingly high. One pound of uranium can produce as much electricity as 3 million pounds of coal could produce. There are more than 400 nuclear reactors in the world. The United States has more than 100, which is more than in any other country. These nuclear reactors produce about 17 percent of all electricity in the United States.

On Land and Sea One picture is a drilling rig for extracting crude oil from the ground. The other is an offshore oil rig, which resembles a community where workers can stay for weeks. *Where do the offshore oil rigs find oil?*

Energy Conversion Nuclear power plants convert nuclear energy into electrical energy. *What is the white cloud coming from one of the cooling towers?*

Unlimited Sources

Although we are using up our nonrenewable sources of energy, there are several other sources of energy that will never run out. They are known as *unlimited sources*. We will never use up all the solar energy, the energy from wind, the energy from flowing water, or the geothermal energy on our planet.

Solar Energy

Solar energy comes from the sun. Unlike coal, oil, wood, or many other sources of energy, solar energy is available all over the world. We use the sunlight, or sun's rays, for light, electricity, and heat. Many homes today are practicing energy conservation by using **solar heating systems**.

One type of system uses large flat panels called "solar collectors," which are mounted on a roof. Water flows through tubes in the solar collectors and is warmed by the sun. The heated water continues moving into the building and heats the home's interior. Another type of system uses only special windows and walls to take advantage of the sun's warming rays. Sunlight streams in and warms the interior. Some of the warmth is absorbed by the walls. Then the walls radiate the heat back into the room at night.

Imagine This...

Driving on Air

Imagine driving a car that runs on air! A French company is experimenting with a three-passenger car that runs on compressed air. The *Air Car* travels 125 miles on one charge of compressed air at a top speed of 68 mph. It would take a few minutes to refill the air tanks at special stations. The company plans to make the car in India and 12 other countries. *What might be the advantages and disadvantages of this air car?*

Go to **glencoe.com** to this book's OLC for answers and to learn more about several alternative engines.

Solar energy can also produce electricity. This happens when sunlight strikes wafer-thin **solar cells**. These cells are also known as *photovoltaic* (foh-toh-vohl-TAY-ik) *cells,* or photocells. Orbiting satellites above Earth get their electricity from solar cells built on the satellites. Solar cells are not used as much because their electricity costs much more to produce than electricity from a power plant.

Wind

The motion of air across the earth has filled the sails of ships and turned windmills for centuries. More recently wind has been used to turn propellers connected to generators that produce electricity. The wind spins the bladed rotors of large wind turbines at about 15–17 revolutions per minute. The blades capture the energy of the wind, which is channeled to a gearbox in the "nacelle" (the turbine housing). From there, the energy flows to a generator, where it is converted into electricity. The wind turbine is controlled by advanced computers.

Wind farms consist of a group of many wind turbines. They exist in many states and countries. Wind farms are located in areas known to have fairly **constant** winds. Certain regions of the country provide enough wind for wind turbines to operate regularly. Wind turbines and other necessary equipment (like large batteries for storing electricity) can be expensive. It is not usually practical for one house to have its own wind turbine, although in earlier days, people used simple windmills for pumping water.

Energy in Space Solar cells on the International Space Station produce electricity from sunlight. They generate the same amount of electricity that 55 U.S. homes might use. *Why might this space station use solar cells?*

Flowing Water

We use flowing water to generate **hydroelectric power**. *Hydro* means "water." A controlled amount of water flows through pipes in a dam and into a turbine. A turbine resembles a pinwheel. A spinning turbine, connected to a generator, creates electricity. About 7 percent of our electricity comes from hydroelectric dams. The Grand Coulee Dam in Spokane, WA, is the largest source of U.S. water power. See **Figure 7.3**.

Figure 7.3 Water Power

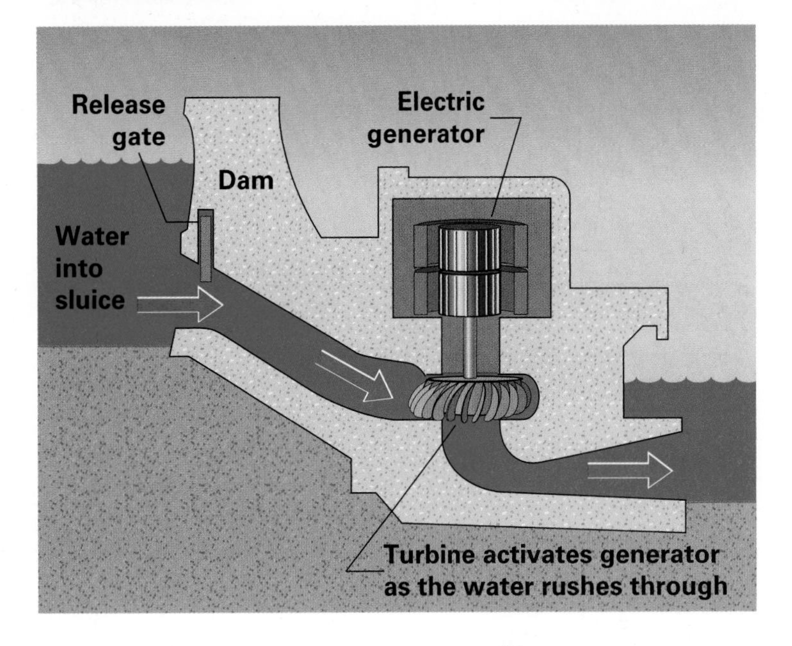

Going with the Flow Water flowing through the dam spins a turbine connected to a generator. *What is another term for water power?*

Geothermal Energy

Molten rock lies far beneath the earth's crust. This is where we get **geothermal energy**—heat produced under the earth. Hot water or steam is created when underground water comes in contact with hot materials and surfaces as a spring. The steam can be used to produce electricity. Some geothermal electric power plants are located along the west coast of the United States.

section 7.1 assessment

After You Read Self-Check

1. List the six forms of energy. Give an example of each form.
2. Explain how fossil fuels are formed.
3. Tell where a house's solar collectors would be located.

Think

4. Some solar houses have no windows on one side. Explain why you think this is so.

Practice Academic Skills

English Language Arts

5. Build a model of an offshore oil-drilling rig. Put it on display. Write a one-page description of how the rig operates.

Mathematics

6. The local electric company has a unique way of billing. They charge a monthly minimum of $4.75, which includes the first 50 kilowatt hours (kwh). They charge $0.0822 for the next 450 kwh, $0.0536 for the next 1,500 kwh, and above that they charge $0.0514 for every kwh. How much would a bill be for 2,650 kwh?

Math Concept **Multi-Step Problems** Most multi-step problems require extra attention to solve.

1. Make notes to help you organize the steps that you need to take.
2. Eliminate any information that is given but not needed.

For help, go to **glencoe.com** to this book's OLC and find the Math Handbook.

section

7.2

Converting Energy to Power

Reading Guide

 **Before You Read** **Preview** What is the difference between energy and power?

Content Vocabulary

- power
- horsepower
- pneumatic power
- hydraulic power
- load
- efficiency

Academic Vocabulary

- goal
- technical

Graphic Organizer

Draw the section diagram. Use it to organize and write down information as you read.

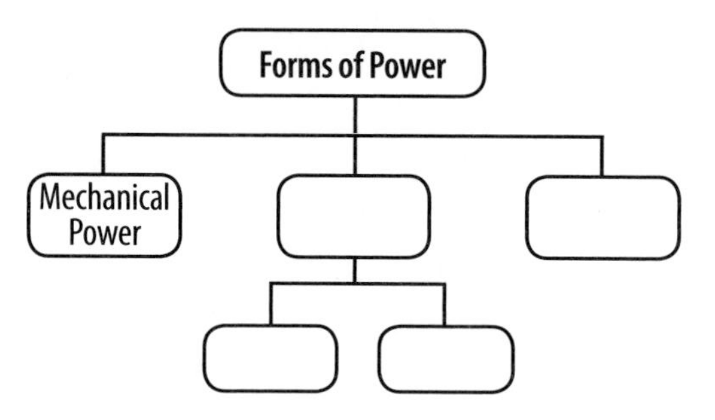

 Go to **glencoe.com** to this book's OLC for a downloadable graphic organizer and more.

TECHNOLOGY STANDARDS

STL 4 Cultural, Social, Economic & Political Effects

STL 5 Environmental Effects

STL 16 Energy & Power Technologies

ACADEMIC STANDARDS

Science

NSES Content Standard E Abilities of technological design

English Language Arts

NCTE 7 Conduct research and gather, evaluate, and synthesize data to communicate discoveries.

STL *National Standards for Technological Literacy*

NCTM *National Council of Teachers of Mathematics*

NCTE *National Council of Teachers of English*

NSES *National Science Education Standards*

NCSS *National Council for the Social Studies*

Power and Work

Is energy different from power?

Although the words *energy* and *power* are related to each other, the two words have different meanings. Power is a measure of the work done over a certain period of time. It is a way of rating how quickly the work is done. You can cut grass more quickly by using a riding lawn mower than you can using a push-type lawn mower. In both cases, the goal of the work is the same: cutting the same area of grass. However, the engine in the riding mower allows you to do it more quickly, so it is more powerful.

 **As You Read**

Question What are different types of power?

One common measure of power is **horsepower**. A three-horsepower lawn mower engine produces as much power as three horses. If you are in very good physical condition, you might be able to develop about 0.2 horsepower for several minutes.

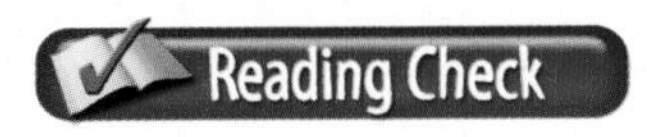

Recall What is the definition of power?

Forms of Power

What are the most common forms of power?

We commonly use three forms of power: mechanical, electrical, and fluid. The photo below illustrates these forms of power as they are used every day.

Mechanical Power

Mechanical energy is the energy that is involved in motion. When you pedal a bicycle, you convert mechanical energy to power. You use the up-and-down motion of your leg muscles to provide the mechanical power that moves the bicycle.

Electrical Power

We convert fossil fuel energy into electrical power in power plants. That power is then used for things such as lighting our homes, operating air conditioners, and running electric motors.

Three Forms of Power The three most commonly used forms of power are mechanical, electrical, and fluid power. *What form of power is used for a refrigerator?*

Fluid Power

When gases or fluids, such as air or water, are put under pressure, they can control and transmit fluid power. An electric motor is often used to pressurize the fluid. The greater the pressure they are under, the greater the force they exert.

Fluid power that is produced by using pressurized gases is called **pneumatic power**. Pneumatic (noo-MAT-ik) power operates tools in some automobile repair shops and factories. Compressors are used to pressurize air for power tools and paint sprayers, and to inflate tires. Putting liquids under pressure results in **hydraulic power**. Hydraulic power is popular for heavy construction equipment and factories because it provides more force than pneumatic power can produce.

Power Systems

Power systems drive and provide propulsion (motion or force) to other technological products and systems. Power systems must have a source of energy, a process, and a load. The energy source is part of the input in a power system. The process converts the energy into a form that can do work. The **load** is the output force. For example, your uncle might use a small pickup truck to carry 500 pounds of mulch for a flower garden. The load is the force that the truck must exert to carry that much mulch.

Power systems have been important to the development and growth of our culture. For example, 100 years ago, it took months to travel around the world. Ideas, customs, and products moved slowly to reach people. Today, thanks to machines (airplanes) that fly very fast, almost every city on our planet is only hours away from every other city. As a result, ideas, customs, and products spread quickly.

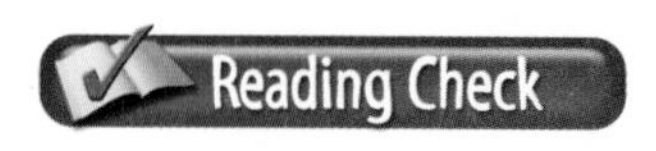

Analyze What kind of power has affected cultures around the world? Why?

Selecting Energy and Power Technologies

How would you decide which energy type is best for a specific purpose?

You are faced with many choices and decisions every day. Some are technical decisions. Selecting a wristwatch is an example. You can buy a digital watch that displays numbers. However, you may want an analog model that has hands. Your decision is based on each item's advantages and disadvantages. Engineers and designers select specific energy and power technologies by considering their advantages and disadvantages.

Lewis Latimer

Inventor of Carbon Filaments for Light Bulbs

Lewis Latimer was born in Massachusetts in 1848. He was the son of escaped slaves from Virginia. Latimer studied Thomas Edison's work on the light bulb and improved its design. In 1881, he invented the Latimer electric lamp, which used inexpensive carbon filaments instead of paper filaments. This bulb could burn longer. A few years later, Latimer became the only black member of Thomas Edison's 24-member engineering division at the Edison Company.

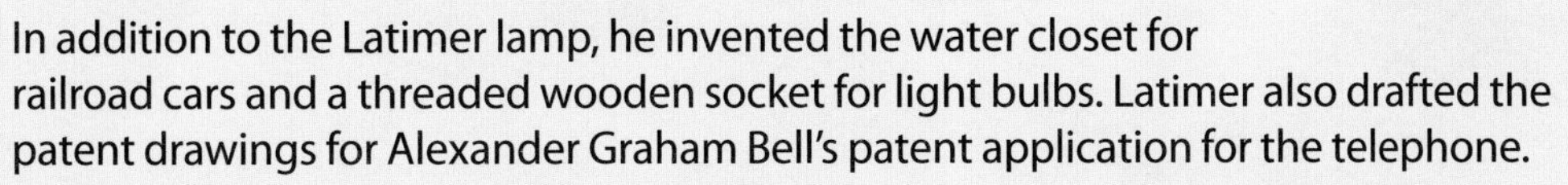

In addition to the Latimer lamp, he invented the water closet for railroad cars and a threaded wooden socket for light bulbs. Latimer also drafted the patent drawings for Alexander Graham Bell's patent application for the telephone.

Jack-of-All-Trades Lewis Latimer had many interests. He was an inventor, draftsman, engineer, author, poet, musician, a devoted family man—and a philanthropist.

English Language Arts/Writing Research and write a brief history of the light bulb from Thomas Edison's original design to today's energy-saving, compact fluorescent light bulbs.

Go to **glencoe.com** to this book's OLC to learn about young innovators in technology.

Factors to Consider

What kinds of things must be considered when selecting energy and power technologies? Engineers consider efficiency when designing products and systems. **Efficiency** is the ability to achieve a desired result with as little effort and waste as possible. An efficient machine, for example, does a lot of work compared to the amount of energy it uses. A bicycle is a very efficient machine. Pedaling a bike at 12 miles per hour requires only 0.10 horsepower.

However, much of the energy we use is not used efficiently. Have you ever felt the hood of a car that has been running? It probably felt hot. All engines that perform work release heat energy into the environment. Overall, only about 20 percent of the energy content of gasoline is actually converted into mechanical power in an ordinary car engine. The rest is wasted, mostly in the form of heat.

Other factors that must be considered include cost, availability, ease of use, environmental and social impacts, and time required. Many trade-offs may be involved.

Consider these common situations:

- Face-to-face speech works very well when you are near another person. A telephone is better when you are farther apart. At what distance do we change from face-to-face speech to using a phone?

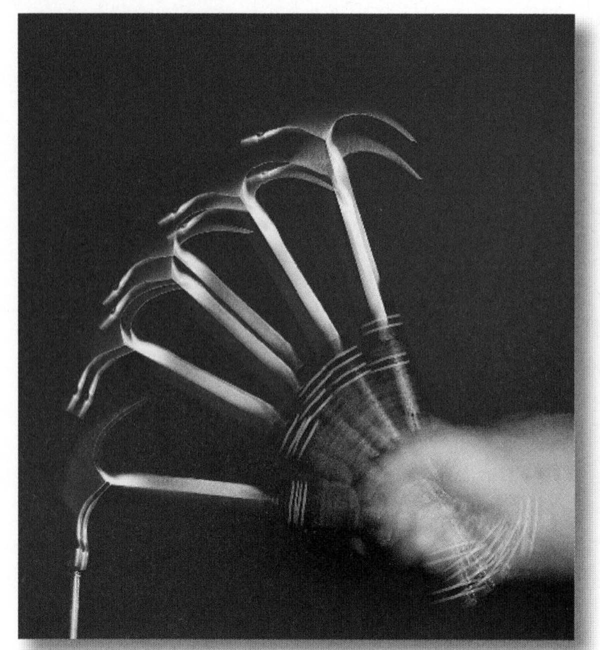

The Best Tool? A traditional hammer does not require electricity. Most carpenters prefer to use a pneumatic nailer for many nails. *If a carpenter used a hammer instead of a nailer to build a house, how might the cost of the project be affected?*

- Most cars use gasoline engines. They do not use jet engines. Why not?
- Some carpenters use ordinary hammers and nails. Some use pneumatic nailers. What would be some advantages and disadvantages of each method in terms of power?

section 7.2 assessment

After You Read Self-Check

1. Explain the difference between power and energy.
2. Name the three forms of power.
3. Define the word *efficiency*.

Think

4. Identify the energy source, the process, and the load for a bicyclist riding down the street. For a power saw cutting wood.

Practice Academic Skills

English Language Arts

5. Write a script for a 60-second radio commercial advertising and promoting one form of power. Talk about its advantages and the ways it is used.

STEM Mathematics

6. Erich built an internal combustion engine. His calculations show that it is capable of producing 65 horsepower. If the engine only produces 70 percent of its total horsepower, how much would it produce?

Math Concept **Percents** Percents can be thought of as parts of the whole.

1. Divide the percent by 100 to convert it to a decimal.
2. Find the value by multiplying by the decimal.

For help, go to **glencoe.com** to this book's OLC and find the Math Handbook.

section 7.3

Impacts of Energy and Power Technology

Reading Guide

Connect What could be a positive and a negative effect of using energy?

Content Vocabulary

- acid rain
- greenhouse effect
- energy conservation
- recycle

Academic Vocabulary

- percent
- aware

Graphic Organizer

Draw the section diagram. Use it to organize and write down information as you read.

Ways to Protect the Environment

1. Reduce energy use. ______
2. ______
3. ______
4. ______
5. ______
6. ______

Go to **glencoe.com** to this book's OLC for a downloadable graphic organizer and more.

TECHNOLOGY STANDARDS

STL 3 Relationships & Connections

STL 4 Cultural, Social, Economic & Political Effects

STL 5 Environmental Effects

STL 6 Role of Society

ACADEMIC STANDARDS

Science

NSES Content Standard F Science and technology in society

English Language Arts

NCTE 12 Use language to accomplish individual purposes.

STL *National Standards for Technological Literacy*

NCTM *National Council of Teachers of Mathematics*

NCTE *National Council of Teachers of English*

NSES *National Science Education Standards*

NCSS *National Council for the Social Studies*

The Effects of Using Energy

Besides air pollution, what other negative effects are caused by misusing technology?

The use of technology can have unintended consequences. For example, Americans have good transportation systems, countless electrical devices, good housing, and food. This way of life and a successful economy have been the result in part of advancements in technology. The use of energy has made technology possible. Unfortunately, not all results have been good. Our energy consumption has created some serious problems. People are striving to find ways to balance financial needs with protecting the earth's environment.

List Think of things people can do to offset the negative effects of technology.

BusinessWeek Tech NEWS

Green Fuel for Jet Liners

Aviation giant Boeing teamed up with a handful of airlines to find an efficient and environmentally friendly way to power jets. Among the resources being considered is algae, the green gunk found in stagnant ponds and unkempt aquariums. Algae-based fuels hold up better in extreme conditions and cost less to harvest than crop-based fuels.

Critical Thinking *What are some extreme conditions at which a jet operates?*

Go to **glencoe.com** to this book's OLC to read more about this news.

Pollution

What are some different causes of pollution?

Pollution results when contaminants—unwanted elements—get into our environment, whether in our air, our water, or our land.

Fossil Fuels

More than 90 percent of our energy comes from burning fossil fuels. You may already know that burning produces a great many pollutants. Burning fills the air with haze and sometimes makes your eyes burn. It can also be a serious health threat. Each fuel produces many pollutants, but there is usually one pollutant that is particularly serious in each case.

Acid Rain

All coal has a small amount of sulfur. When coal burns, it creates sulfur dioxide (SO_2). The sulfur dioxide combines with water vapor and oxygen in the air to form a weak sulfuric acid. This acid mixes with nitric acid (NO_2), another pollutant, and falls to the earth as **acid rain**. See **Figure 7.4**. It can kill fish, crops, and trees. Acid rain also damages monuments and statues.

Carbon Monoxide

Gasoline forms carbon monoxide when it burns. Carbon monoxide is an odorless, colorless, and poisonous gas. When you breathe it into your lungs, it reduces the ability of your blood to carry oxygen. If a person has a lung problem, too much carbon monoxide can make the problem worse.

Effects of a Warmer Climate Global warming increases the temperature of air and water, causing glaciers to melt. *What are some causes of global warming?*

Reducing Acid Rain

There is no complete solution to the problem. However, acid rain can be reduced if power companies and industries use coal containing less sulfur. They can also install special equipment to remove up to 90 percent of the sulfur dioxide from their smoke.

Figure 7.4 Effects of Fossil Fuels

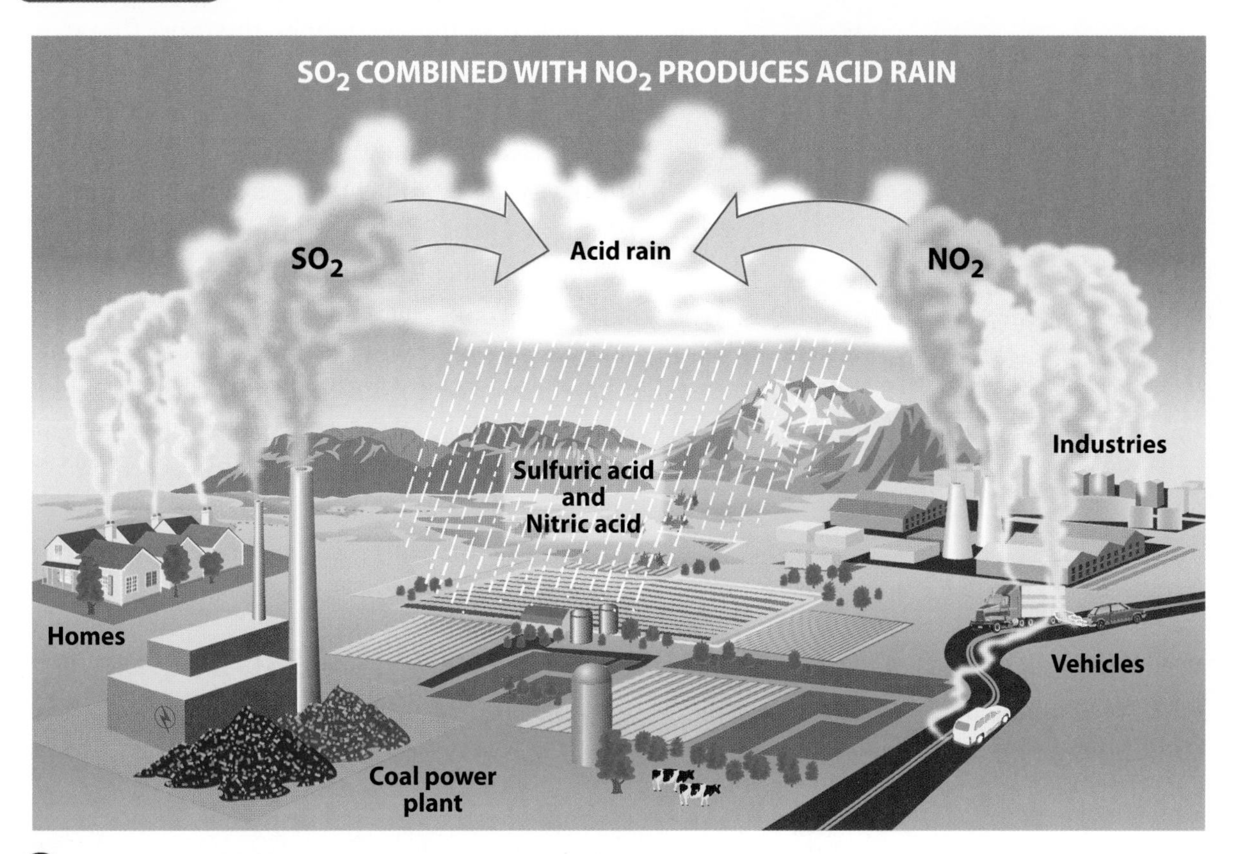

Acid Rain Acid rain is created when sulfur dioxide and nitric acid in the air mix with water vapor and fall to earth. *What is the greenhouse effect?*

The Greenhouse Effect

Burning fossil fuels also contributes to the **greenhouse effect**—the heating of the earth's atmosphere. This occurs when too much carbon dioxide builds up in the air. The carbon dioxide prevents heat from escaping. If we produce too much heat, and it cannot escape, the temperature of the earth's atmosphere may increase, creating the greenhouse effect.

Warmer temperatures might mean longer growing seasons for crops. More food could possibly be produced for the world's increasing population.

However, the polar ice caps would melt significantly. That melted ice would raise the ocean level and cause flooding of some seacoast cities. The changing climate also affects ecosystems and plant and animal life. These effects of "global warming" have already occurred in some areas of the world.

Waste Heat

Waste heat is also a water pollutant. Heat produced by power plants is discharged into lakes, rivers, and oceans. Water plants and animals can be harmed.

A Different Kind of Waste Nuclear waste is transferred to sites like this one under very strict safety guidelines. *What problems can be caused by nuclear waste?*

Nuclear Waste

Nuclear pollution may be the most threatening pollution of all. Nuclear, or radioactive, waste is a solid left over after nuclear fuel is used up. The waste remains dangerous for many years and can cause serious health problems. Proper disposal of this material is difficult and is an important social issue.

The waste is placed in special concrete containers. The area where they are stored is constantly checked for radioactive leakage. Some people think that nuclear waste should be sent into space with rockets. What do you think? Future generations may be able to figure out how to use this waste in a positive way.

Depletion of Resources

What can we do to maintain a supply of energy for our future needs?

Our earth has a limited supply of some energy sources like fossil fuels. However, people need more and more energy as the world's population grows. Also, each person uses more energy now than in the past, partly because there are so many technological conveniences available for us to use.

We have to be sure that there is enough energy to provide for human needs. One way to do that is to develop alternate sources, such as solar energy. Another way is through **energy conservation**. Energy conservation is the management and efficient use of energy sources.

Ethics in Action

Emissions Trading

To reduce pollution, governments in the European Union have developed a system called "emissions trading." They give businesses credits in exchange for reducing pollution. One credit allows a business to release a certain amount of pollutants. If the business pollutes too much, it must buy more credits.

Fair Trade Critics say this policy is hard to enforce. They say it is too easy to cheat. Also, richer businesses can pay more to pollute more.

English Language Arts/Writing

Go Local Emissions trading focuses on individual responsibility. Critics say we need to reduce pollution as a community.

1. In a small group, brainstorm ways you can reduce pollution at your school.
2. Compare your list with other groups. As a class, pick the top three things you would do to reduce pollution and write a class statement.

Energy can be conserved if we recycle the materials and products we use. To **recycle** means to use again. Metal, glass, paper, and some plastics can be recycled. Reusing aluminum cans is one of the best ways to save energy. Recycled cans require 80 percent less energy to produce than cans made from raw materials. All glass can be recycled. Paper is usually recycled into bags, paper towels, and packaging materials. Materials that can be recycled have a special triangular symbol on them.

Recycling reduces our use of natural resources. It also reduces the amount of solid waste sent to landfills. Most cities and towns have a recycling center or a civic organization that arranges routine collections. Check to see how recycling is handled in your own community.

What You Can Do

How can you help?

Energy is used by our hospitals, fire fighters, schools, and in our homes. We cannot completely stop using energy even though it causes pollution and some sources are limited. It is also true that we cannot ignore these things just because we want energy. We have to find a proper balance between the two.

We can reduce our energy use. We can pass laws and obtain publicity in the media. This can make everyone aware of the problems and what is being done by our political leaders. We can control how much pollution enters the environment from each energy source. We can reduce the amount of pollution. Less pollution is created when we use solar, wind, water, or geothermal energy. We can develop more and better ways to use these unlimited energy sources.

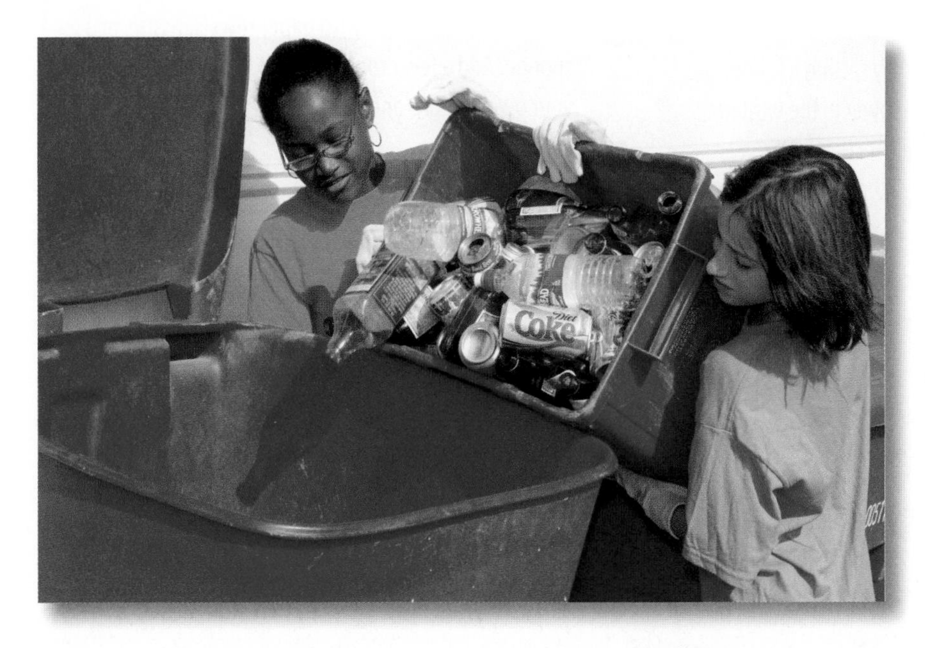

Recycling Counts Everyone can help our environment by recycling certain kinds of waste materials. *What types of materials can be easily recycled?*

EcoTech

Downsize Your Computer

Desktop computers use twice as much energy as laptop computers use. Unlike a desktop computer, a laptop can be unplugged after its battery is charged. You can use it for a few hours. A laptop computer does not need a separate monitor.

Try This Do research at stores or online to compare the estimated costs of laptops with desktop computers. Consider the costs of purchasing the computer, monitor, and any other equipment, as well as the cost of energy usage. What is the difference?

What can you do? You might think that since you cannot vote, your opinion and ideas do not count. That is not true. You can influence the adults around you. They may listen to you and even follow your advice if they think that you are sincere and you know what you are talking about.

Here are some things to do to conserve energy and resources:

1. Keep informed. Know what is happening in your community, state, country, and world. Read newspapers, Web sites, and magazines; watch the news; and communicate with others in your community. Talk to knowledgeable adults. Have discussions with your friends.
2. Set your home thermostat at or below 65 degrees (Fahrenheit) in the winter. You will use less energy because the heating system will operate for less time.
3. If your home is air-conditioned, set the thermostat at or above 78 degrees in the summer. You will use less energy because the air conditioner will operate for less time.
4. Use less hot water by spending less time in the shower.
5. Replace light bulbs with low-energy fluorescent ones and turn off all unnecessary lights.
6. Walk or ride a bike. Use buses, trains, or subways instead of automobiles.
7. Practice recycling metal, glass, paper, and plastics at school and at home. Start with aluminum cans and paper.
8. Use renewable or unlimited energy sources whenever possible.

section 7.3 assessment

After You Read Self-Check

1. Explain how energy is saved by setting a home thermostat at 65 degrees or lower in the winter.
2. Define the term *greenhouse effect*.
3. Identify the type of pollution that may be the most threatening.

Think

4. Discuss some possible ways to save energy that are not included in this section.

Practice Academic Skills

English Language Arts

5. See how difficult it can be to clean up an oil spill. Place a small amount of cooking oil in a large bowl of water. Try and soak up the oil with paper towels, tissue, or a sponge. Add a few drops of liquid detergent. Present your observations and conclusions to the class in a written report.

STEM Science

6. Research and write a few paragraphs on the dangers of nuclear waste and the problems involved with its disposal. Discuss alternate ways to deal with nuclear waste. Generate your own ideas about what to do with the waste.

Exploring Careers in Technology

Jason Lewis

MECHANICAL ENGINEER

Q: *What do you do?*
A: I work in the technology department of Vestas, the world's leading supplier of wind power technology. My job involves improving wind turbine reliability and performance by resolving mechanical issues.

Q: *What do you do on a typical day?*
A: I travel to wind farms to conduct inspections and gather data to performing analysis of technical solutions. I also communicate the progress of an investigation to a customer.

Q: *What kind of training and education did you need?*
A: Getting my job started with a basic education in mechanical engineering. The job also requires an ability to solve problems, experience related to wind energy, and strong communication skills.

Q: *What do you like most about your job?*
A: I really enjoy being part of the development of a clean and renewable energy source. Wind power can help meet modern energy demands while reducing dependence on fossil fuels.

Q: *How did you get interested in your job?*
A: I knew I wanted to be an engineer in an industry where my work made a difference. A year after college, I attended the Global Wind Energy Conference. I returned home later that week committed to finding a job in wind energy.

English Language Arts/Writing

The Evolution of Wind Power Wind power has been used for centuries. Find out how people have harnessed the power of the wind from the past to the present.

1. Research the history of wind power. Find out how it was first used and where.
2. Find some images or photographs of machines that have used wind power to operate.

Go to **glencoe.com** to this book's OLC to learn more about this career.

Real-World Skills

Analytical thinking, communication

Academics and Education

Physics, mathematics, writing, mechanical engineering

Career Outlook

Growth as fast as average for the next ten years

Source: *Occupational Outlook Handbook*

chapter 7 Review and Assessment

Chapter Summary

Section 7.1 Energy is the ability to do work, and power is the measurement of work done. Renewable energy sources come from plants and animals. Nonrenewable sources of energy cannot be replaced. Unlimited sources of energy include solar, wind, hydro, and geothermal energy.

Section 7.2 Power is measured when energy is converted from one form to another. All power systems include a source of energy, a process, and an output load.

Section 7.3 Burning fuels produces serious threats to our health and environment, including the greenhouse effect, global warming, acid rain, and carbon monoxide. We can help control pollution by making everyone aware of the problems. We can reduce the amount of pollution that enters our environment, and we can develop more and better ways of conserving energy.

Review Content Vocabulary and Academic Vocabulary

1. On a sheet of paper, use each of these terms and words in a written sentence.

Content Vocabulary

- energy
- calorie
- fossil fuel
- solar heating system
- solar cell
- wind farm
- hydroelectric power
- geothermal energy
- power
- horsepower
- pneumatic power
- hydraulic power
- load
- efficiency
- acid rain
- greenhouse effect
- energy conservation
- recycle

Academic Vocabulary

- vehicle
- constant
- goal
- technical
- percent
- aware

Review Key Concepts

2. **List** the different forms of energy.
3. **Identify** renewable, nonrenewable, and unlimited energy resources.
4. **Explain** how fossil fuels are formed.
5. **Identify** the most common forms of power.
6. **Describe** uses for each form of power.
7. **Explain** how to select energy and power technologies.
8. **Identify** forms of pollution.
9. **Describe** ways to slow depletion of energy resources.
10. **Discuss** the disposal of nuclear waste.

Real-World Skills

11. Understanding Energy Costs Look at the monthly bills your family has received over the past year for electricity. Write a paragraph or two describing the differences in electricity usage and the rates you were charged from season to season.

Technology Skill

12. Generate Your Own Power There are several ways to make a small power generator you can use for simple tasks such as lighting a light bulb.

a. Research the topic of creating a power generator using the Internet. If possible, build a power generator.

b. Write an explanation of how and why your power generator works.

Academic Skills

Social Studies

13. Some communities charge for recycling, while others include it with trash disposal. Write about local recycling services and their prices.

STEM **Mathematics**

14. Amy is a waitress. She made \$33.50 in tips. Amber only made \$12.25 in tips. Julia, a coworker, made 1.5 times as much as Amy made, minus what Amber made. How much did Julia make?

Math Concept **Algebra** Some problems are easier to solve if you write an equation. Think of an equation as a sentence that explains what you are trying to solve for. The unknowns are represented by letters such as x or y.

TSA WINNING EVENTS

TECHNOLOGY STUDENT ASSOCIATION

Inventor

Situation Your team is to design a device to reduce energy consumption, increase recycling, or both. It may be as simple as a container to deposit aluminum cans or as complex as an electronic switch for home lighting.

Activity Working as a team, design a device that meets the criteria below. You are required to complete the following:

- Brainstorm and choose the best idea.
- Develop a series of sketches.
- Build a model or working prototype.
- Present your invention to the class.

Evaluation You will be evaluated on how well you meet the following criteria:

- Proposal is well thought-out.
- Sketches communicate the team's ideas.
- Model clearly demonstrates your invention.
- Presentation is well organized and clear.

Go to **glencoe.com** to this book's OLC for information about TSA events.

Standardized Test Practice

Directions Choose the letter of the best answer. Write the letter on a separate piece of paper.

1. What is 38 percent of \$212.34, rounded to the nearest cent?
   A \$80.70
   B \$80.69
   C \$81.68
   D \$81.69

2. Hydraulic power is fluid power produced by putting a gas under pressure.
   T
   F

Test-Taking Tip When taking a multiple choice test, read the question before you look at the answer choices. If you come up with the answer in your head before looking at the possible answers, the choices given on the test will not throw you off or trick you.

chapter 7

TECHNOLOGY LAB

Build a Solar Heating System

Most solar collectors are black because dark colors absorb the sun's rays. The simplest kind of solar collector has a black heating plate and a glass or plastic cover. Sunlight strikes the plate, and the plate becomes hot. The heat is trapped by the glass or plastic cover. The trapped heat is transferred to water flowing through tubing. The warmed water is sent where heat can be used.

Tools and Materials

- ✓ Two 8 × 12-inch sheets of polystyrene foam (Styrofoam®)
- ✓ 6 × 10-inch sheet of corrugated cardboard
- ✓ 6 × 10-inch sheet of black paper
- ✓ Clear plastic wrap
- ✓ 5 feet of small-diameter flexible plastic tubing
- ✓ Small plastic bottle and bowl
- ✓ Thin wire & cutters
- ✓ Drill
- ✓ Razor knife
- ✓ Rubber bands
- ✓ Tape
- ✓ Red food dye
- ✓ Thermometer
- ✓ Clothespin
- ✓ Wristwatch
- ✓ Gloves

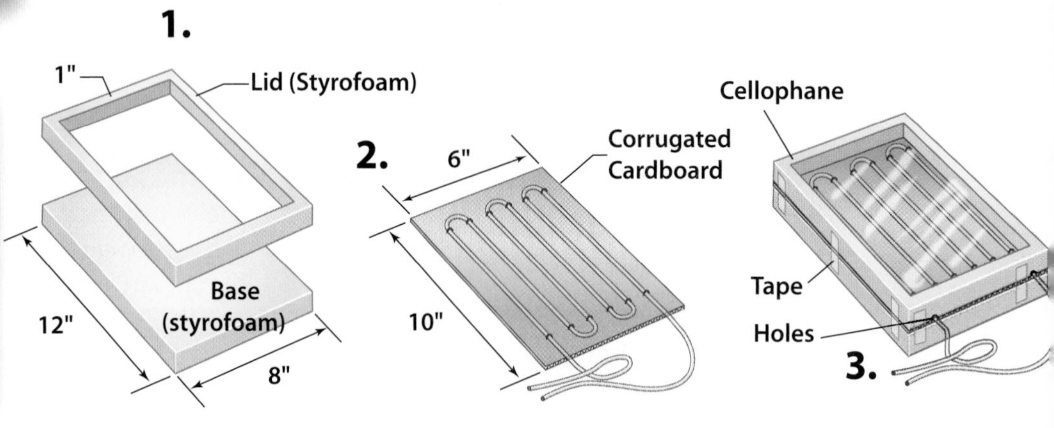

SAFETY

Heat Caution!

Be sure to follow all safety precautions for preventing burns. Wear gloves when handling hot items.

Set Your Goal

Your goal is to make a small solar heating system similar to those used in houses. Then you will test it to see how well it works.

Know the Criteria and Constraints

In this lab, you will:

1. Make your solar collector from Styrofoam, plastic tubing, and other materials.
2. Keep your water supply in a plastic bottle.
3. Make your system as watertight as possible.
4. Compare the efficiency of your system with others.

Design Your Project

Follow these steps to complete this lab.

1. Look at the illustrations to become familiar with the procedure.
2. Use the razor knife to cut the Styrofoam to the sizes shown. Save the center cut.
3. Make the solar heating plate:
   - Place the black paper on top of the corrugated cardboard.
   - Arrange plastic tubing in S-shaped curves on black paper.

1. 1" Lid (Styrofoam) Base (styrofoam) 12" 8"

2. 6" 10" Corrugated Cardboard

3. Cellophane Tape Holes

- Make as many S curves as will fit. Leave about 18 inches of extra tubing at each end.
- Fasten the tubing to the black paper and cardboard using wire.
- Poke wire through the cardboard. Twist ends together underneath.
- Tape the clear plastic wrap over the opening in the Styrofoam lid.
- Place the heating plate on the Styrofoam base. Notch holes in the lid for the tubing and place the lid over the base. Tape together.

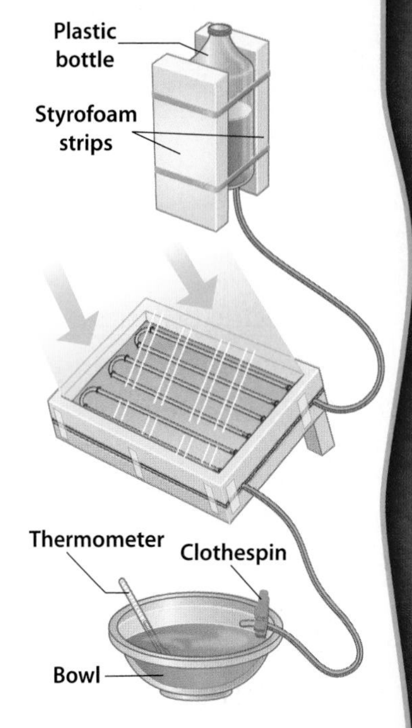

4. Now complete the solar heating system as depicted:
   - Drill a hole in the bottom of the plastic bottle so the tubing fits.
   - Force one of the free ends of the tubing into the hole.
   - If necessary, seal the connection with waterproof glue.
   - Make a bottle stand with Styrofoam.
   - Place the solar collector where the sun's rays will hit the heating plate. Place it at a 45° angle.
   - Put the end of the tubing in bowl. Clamp it with a clothespin.
   - Add red food dye to the bottle of water.
   - Place the bottle at least six inches above the collector.
5. Your system is complete. Now test it.
   - Open the clothespin to allow the water to flow through the tubing.
   - Measure the water's temperature. Close the clothespin.
   - Allow the water to remain in the collector for 2 to 5 minutes.
   - Open the clothespin and measure the water's temperature again.

Evaluate Your Results

After you complete this lab, answer these questions on a separate piece of paper.

1. What was the highest water temperature you measured?
2. Cover up half the collector. That is like creating 50 percent cloud cover. Does the water reach only half the maximum temperature?
3. What could you do to improve your solar collector?

Academic Skills Required to Complete Lab

Tasks	English Language Arts	Math	Science	Social Studies
Build heating plate.	✓	✓	✓	
Build the solar heating system.	✓	✓	✓	
Heat water in the sun.		✓	✓	
Measure water temperatures before and after heating.	✓	✓	✓	
Compare your results with the results of other students.	✓	✓	✓	

chapter 8

Electricity to Electronics

Sections

What You'll Learn

- **Define** electricity.
- **Identify** the three types of electricity and the differences between them.
- **List** uses for direct and alternating current.
- **Discuss** voltage, amperage, and resistance.
- **Use** Ohm's law to determine measurements in a circuit.
- **Tell** the purpose of an electric meter.
- **Explain** the difference between a conductor and an insulator.
- **Name** some uses for semiconductors and superconductors.
- **Describe** the two basic types of electrical circuits.
- **Summarize** the different types of electronic signal transmissions.
- **Tell** how fiber optic cables work.
- **Compare** analog and digital signals.

Explore the Photo 

Strike One! A bolt of lightning carries a tremendous amount of electricity. *Why is it dangerous to walk through an open field during a thunderstorm?*

Launch the TECHNOLOGY LAB

Build a Simple Alarm System

At the end of this chapter, you will be asked to build a simple alarm system. Get a head start by using this checklist to prepare for the Technology Lab.

PROJECT CHECKLIST

- ✓ Review your classroom's safety rules and procedures.
- ✓ Ask your teacher how to properly use a wire stripper, metal punch, and drill.
- ✓ Brainstorm some possible uses for your alarm system.

section

8.1 Electricity Basics

Reading Guide

Before You Read **Connect** How are static electricity and lightning related?

Content Vocabulary

- atom
- electricity
- static electricity
- DC
- circuit
- AC
- frequency
- voltage
- transformer
- amperage
- resistance
- Ohm's law
- wattage

Academic Vocabulary

You will see these words in your reading and on your tests. Find their meanings at the back of this book.

- transmit
- vary

Graphic Organizer

Draw the section diagram. Use it to organize and write down information as you read.

Measuring Electricity

Term	Definition	Unit of Measure
Voltage		
Amperage		
Resistance		

Go to **glencoe.com** to this book's OLC for a downloadable graphic organizer and more.

TECHNOLOGY STANDARDS

STL 3 Relationships & Connections

STL 11 Design Process

STL 16 Energy & Power Technologies

ACADEMIC STANDARDS

Science

NSES Content Standard B Motion and forces

Mathematics

NCTM Algebra Represent and analyze mathematical situations and structures using algebraic symbols.

STL *National Standards for Technological Literacy*

NCTM *National Council of Teachers of Mathematics*

NCTE *National Council of Teachers of English*

NSES *National Science Education Standards*

NCSS *National Council for the Social Studies*

What Is Electricity?

Can static electricity be dangerous?

Have you ever touched a doorknob, and received a shock? In this chapter, you will learn why these shocks are only annoying and why lightning, which is another form of static electricity, is extremely dangerous. You will also learn why some devices require very little electricity while others require a lot. Every electric circuit contains voltage, amperage, and resistance. They determine what electricity can do and how dangerous it can be.

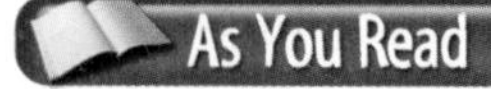

Predict What is the purpose of an electric meter?

The Atom

To understand what electricity is and how it functions, you need to understand more about **atoms**. Atoms are the building blocks from which all things are made. They are made of several atomic particles:

- Protons
- Electrons
- Neutrons

Atoms contain an equal number of positively charged protons and negatively charged electrons. Most atoms also contain the atomic particles called neutrons, which have no charge at all. See **Figure 8.1**, The Atom.

Figure 8.1 The Atom

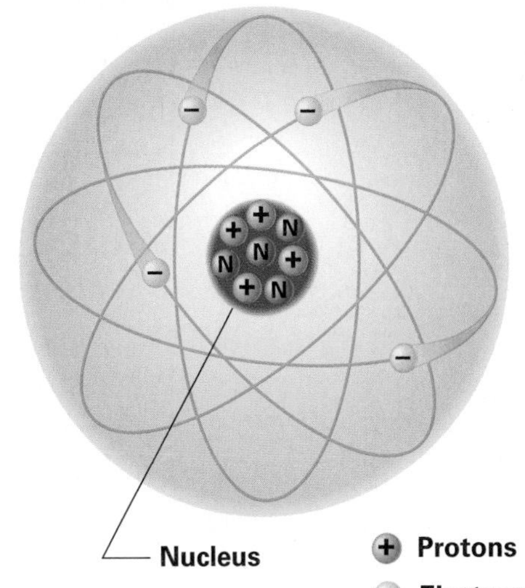

Inside the Atom Negatively charged electrons travel around the nucleus. The nucleus contains positively charged protons and neutral neutrons. *What happens when an atom has too many electrons?*

Electrons and Protons

An atom is in balance when it contains the same number of electrons and protons. Atoms obey the same rules of repulsion (pushing away) and attraction (moving together) as magnets. When an atom has too many or too few electrons, it pushes or pulls the extra electrons off to its neighboring atoms. This flow of electrons from one atom to another is called **electricity**. See **Figure 8.2**.

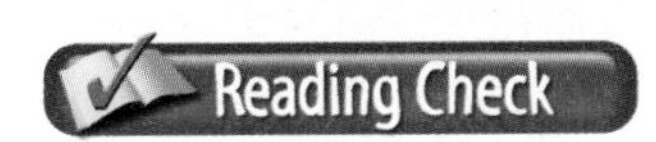

Define What is electricity and how does it work?

Figure 8.2 The Flow of Electrons

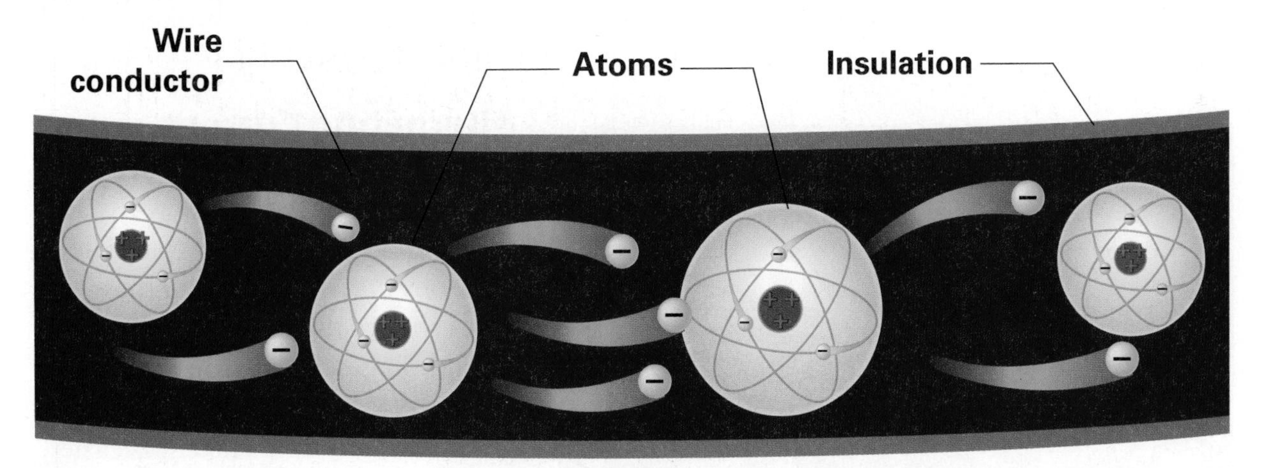

Understanding Electricity Electrons in a circuit move from atom to atom. When an atom loses electrons, it becomes positively charged. When an atom gains electrons, it becomes negatively charged. *Why do scientists relate the behavior of magnets to the behavior of atoms?*

Types of Electricity

What are some different types of electricity?

Is the electricity that comes out of the wall outlets in your home the same as the electricity in batteries? Are static electricity and lightning different from the electricity that powers your home?

Static Electricity

Static electricity occurs when atoms have built up "extra" electrons. These extra electrons are ready to jump to new atoms. When super-charged atoms come close to an object that can conduct electricity, a spark occurs. When you receive a shock from ordinary static electricity, you are not hurt because the static charge lacks power. Static electricity can, however, destroy sensitive electronic circuits and ignite flammable liquids and gases.

Lightning is a very strong discharge of static electricity. If it hits a power line, the electricity surge can destroy electrical equipment in the area and cause blackouts.

Direct Current

DC, or direct current, is the flow of electrons in only one direction. At one time, all of the electricity used in the United States was direct current. Your local phone company still supplies DC current. Many people keep their original local phone service because the phones almost always work during a power failure.

Devices powered by batteries also use direct current. When plugged in, these devices convert the wall outlet electricity to direct current. Batteries contain stored electricity. When batteries are in use, electrons flow out from the negative terminal of the battery, and then back through the battery's positive terminal. A **circuit** is the pathway electricity takes. If a circuit is broken, electricity stops flowing, and the system stops working.

A String of Lights A series circuit provides a single path for current. *What happens if one of the bulbs burns out?*

Alternating Current

The wires that come into your home can be traced all the way back to a power generating station. The electrons that flow through these wires change direction 120 times per second. This is the reason that this electricity is called **AC**, or alternating current. Alternating current is easier to transmit and control than direct current.

Each back-and-forth motion is called a *cycle*. The electricity that powers your home is measured at 60 cycles per second. The number of cycles per second is the **frequency** of the alternating current. (In the metric system, cycles are called *hertz* or Hz.)

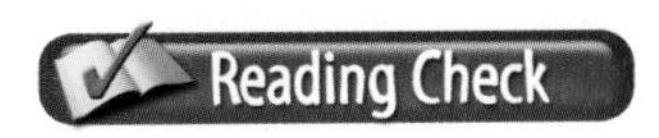

Summarize What are types of electricity?

Electrical Flow

What are voltage, amperage, and resistance?

You cannot see electricity flowing through a circuit. However, its movement through a circuit is similar to that of water through a pipe. You need water pressure to push water through a pipe, and you need electrical pressure to push electricity through a wire.

The pressure and strength of the electric current as well as the opposition to the current are known as:

- Voltage
- Amperage
- Resistance

Voltage

Voltage is the pressure that pushes electricity through an electric circuit. Voltage pressure is measured in units called *volts*. A **transformer** is a device that can change alternating current from one voltage to another. Some equipment and appliances require less voltage than others require. High voltages can be sent over distances inexpensively.

Amperage

Amperage is the strength of the electrical current. It is measured in units called *amperes*. Amperage is measured in electrons per second. In cases of electric shock, it is the amount of current (amperage) that makes electricity dangerous enough to kill.

Resistance

Resistance is anything that opposes or slows the flow of electricity. It is measured in units called *ohms*. Resistance in a water pipe is determined by the diameter, length, bends, and kinks in the pipe. The resistance in an electric circuit is determined by the electric wire's diameter, length, and temperature. For example, the resistance of a wire *decreases* as the wire gets fatter. The resistance *increases* as the wire gets longer. If you add a component to the circuit, like an electrical appliance, resistance increases.

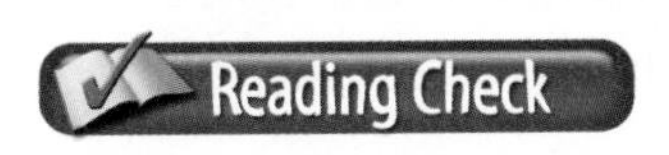

Identify What do voltage, amperage, and resistance mean?

EcoTech

Hit the Lights

Traditional light switches can be forgotten and left on. Motion detectors and timed switches will automatically turn off unnecessary lighting. Controlled light switches allow for lower-level lighting, which uses less energy.

Try This Instead of lighting an entire room, use a small lamp at your desk or chair.

Figure 8.3 Ohm's Law

What Is Needed To use this Ohm's law diagram, put your finger over the value you want to find. This will give you the correct equation to use. *If the voltage is 120 and the resistance is 10, what is the amperage?*

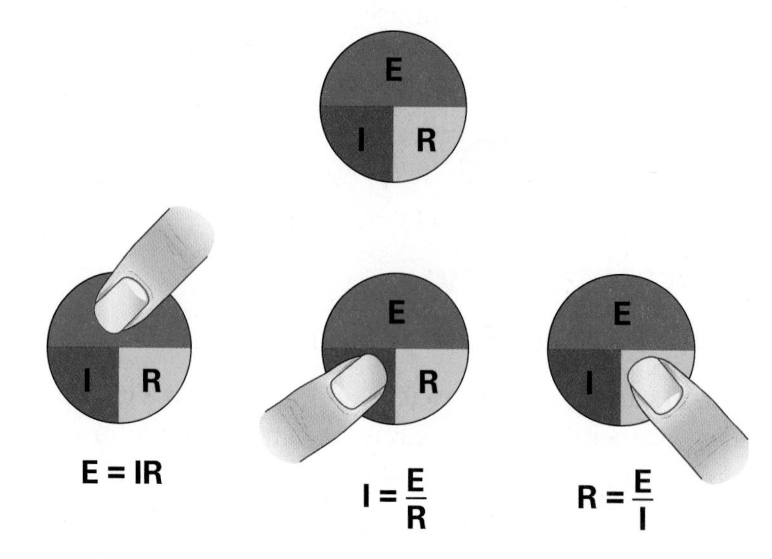

Ohm's Law

What can you determine by using Ohm's law?

Building an electric device often means joining many different electrical systems together. These separate systems are all needed to fulfill the requirements of the device. The components must get the proper amount of electricity or the device will burn out, not function properly, or refuse to work. When you design an electric circuit, you must know the power needs of the components.

Ohm's law is a mathematical formula that is used to determine the voltage (E), amperage (I), or resistance (R) of an electric circuit. Ohm's law states:

$$\text{Voltage} = \text{Amperage} \times \text{Resistance or } E = I \times R$$

The law gives the electrical designer a way of determining exactly what the circuit will need to make it work efficiently. (See **Figure 8.3**.)

Academic Connections — Social Studies

Neglected Metrics Many electric measurements are in metrics. Most of the world uses the metric system of measurement.

Apply Find out why the metric system has been slow to catch on in the United States. What social changes might make it more accepted?

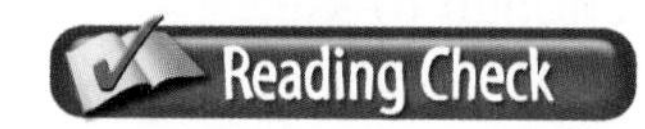

State How can Ohm's law help you design electrical systems?

The Electric Meter

What does an electric meter do?

Your local electric company keeps track of the electricity that your family uses. When you turn on an electric appliance in your home, the wheels of your electric meter turn. The moving wheels are measuring the electricity that is being consumed. The more appliances that you run, the faster the wheels turn, and the bigger your electric bill becomes.

Measuring Electricity

The electricity that your appliances use is measured in watts. **Wattage** (W) is a measurement of the electrical power an appliance will need to run. It is the voltage of the electric circuit multiplied by the amount of amperage needed to run the appliance ($W = E \times I$).

Your electric meter measures kilowatt hours. Your lights and other appliances consume the equivalent of one kilowatt hour of electricity when they consume 1,000 watts of electricity.

Measuring Usage Electric companies use meters to keep track of the electricity that your family uses. When you turn on an electric appliance, the wheels of the meter turn. *What happens to the spin of the wheels on this meter if you turn off your appliances?*

Energy Efficiency

You can compare wattage efficiency of an appliance to gasoline efficiency in a car. The more efficient the car or electrical appliance, the less you will pay to keep it running.

Most electrical appliances, including light bulbs, indicate how much electricity (wattage) they will consume. You want to purchase the most energy efficient appliances possible. If you buy less efficient appliances, you will spend more on electricty over time. Look for Energy Star ratings on packages or labels and purchase the appliance or light bulb that is the most energy efficient.

Electric utility companies encourage people to use less electricity during high-demand periods. When the utility company cannot keep up with the demand, an electrical blackout is likely to occur.

section 8.1 assessment

After You Read Self-Check

1. Explain what causes atoms to become positively or negatively charged.
2. Define voltage and amperage.
3. Identify which unit is used for measuring voltage and amperage.

Think

4. Based on your understanding of atoms, explain what causes electricity to flow through a circuit.

Practice Academic Skills

STEM Science

5. Build a small DC electric motor. (Note: Your teacher must approve all experimental plans before you hook up any device to an electric current.)

STEM Mathematics

6. An electrical engineer is analyzing the power needs of a computer circuit she is designing. She knows that the voltage is 110 and the resistance is 11. What is the amperage of the computer circuit?

Math Concept **Using Formulas** Ohm's law states that Voltage (E) equals amperage (I) times resistance (R), or $E = I \times R$.

1. In a formula, quantities are represented by letters.
2. Plug known quantities into the formula, and then solve for the unknown.

For help, go to **glencoe.com** to this book's OLC and find the Math Handbook.

section 8.2

Controlling Electrical Flow

Reading Guide

Preview How is electrical flow controlled?

Content Vocabulary

- conductor
- insulator
- semiconductor
- superconductor
- series circuit
- parallel circuit

Academic Vocabulary

- generate
- series

Graphic Organizer

Draw the section diagram. Use it to organize and write down information as you read.

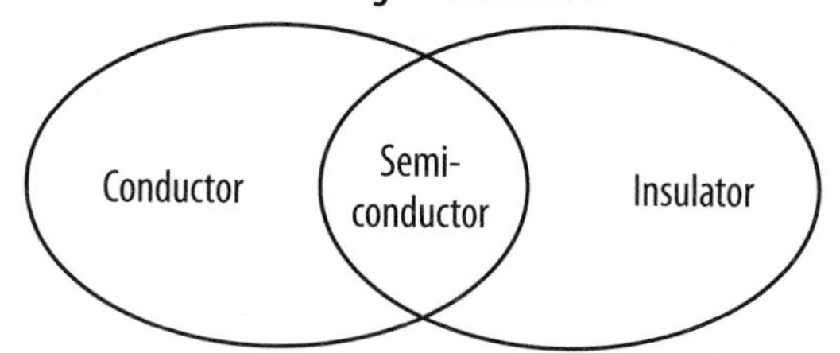

Go to **glencoe.com** to this book's OLC for a downloadable graphic organizer and more.

TECHNOLOGY STANDARDS

STL 2 Core Concepts of Technology

STL 3 Relationships & Connections

STL 12 Use & Maintenance

STL 16 Energy & Power Technologies

ACADEMIC STANDARDS

Science

NSES Content Standard B Structure and properties of matter

Mathematics

NCTM Geometry Use visualization, spatial reasoning, and geometric modeling to solve problems.

STL *National Standards for Technological Literacy*

NCTM *National Council of Teachers of Mathematics*

NCTE *National Council of Teachers of English*

NSES *National Science Education Standards*

NCSS *National Council for the Social Studies*

Conductors and Insulators

How is the flow of electricity controlled?

Conductors are materials that contain atoms that have a very weak hold on their electrons. Copper, aluminum, gold, and silver are excellent conductors of electricity. Think of electrical conductors as highways on which electricity travels.

Insulators are materials that contain atoms that have a very tight grip on their electrons. Rubber, plastic, and ceramics make good insulators. Insulation prevents the electrons from leaving their intended path.

Connect How do circuit breakers protect your home and school?

Semiconductors and Superconductors

What are semiconductors and superconductors?

Semiconductors are materials that can act as either conductors or insulators. Most semiconductors are made using the silicon found in ordinary sand. Manufacturers add different ingredients to the silicon to control their electrical conducting properties.

Semiconductors are found in most electrical and electronic devices. Without them, your TV, computer, MP3 player, and digital camera would not exist.

Superconductors have no measurable resistance to electricity. They make it possible to generate more productive and efficient electrical power. Superconductor technology is built into new electric generators and transformers. They also play a critical role in magnetic resonance imaging (MRI) machines, new super-efficient industrial motors, and magnetic levitation trains.

Recall What are some uses for semiconductors and superconductors?

Charles F. Brush

Wind Energy Pioneer

In 1888, Charles F. Brush introduced the world to the wind turbine, an automatic windmill used to generate electricity. As one of the founders of the American electrical industry, Brush invented the electrical DC system used in the public electric grid, as well as the first commercial electrical arc light, and an efficient method for manufacturing lead-acid batteries. His wind turbine was huge with a rotor diameter of 50 feet. Its 144 rotor blades were made of cedar wood. The Brush windmill ran for 20 years and charged the batteries in the cellar of his mansion.

The principles of Brush's windmill did not change much over the years, until about 1980 when computers took control of the turbines. But his insightful use of a natural, renewable resource has proved successful over the long term. In Denmark, wind-powered generators produce about 20 percent of the country's electricity use.

A Legacy Brush's company, Brush Electric, was based in Cleveland, Ohio. He sold it in 1889. By 1892, the company merged with Edison General Electric Company under the name of General Electric Company (GE). More than 100 years have passed, and GE is one of the world's leading wind turbine suppliers.

English Language Arts/Writing Describe in a few paragraphs how you feel wind power can benefit the earth.

Go to **glencoe.com** to this book's OLC to learn about young innovators in technology.

Ethics in Action

The Three Gorges Dam

The Three Gorges Dam in Hubei province, China, is the largest hydroelectric river dam in the world. It provides electricity for nine provinces and two cities. It also controls flooding, which is a major problem on the Yangtze River.

Progress at a Price However, the dam has caused more than a million people to leave their homes because of rising waters. In addition, many ancient sites and artifacts have been lost. The dam has permanently altered the ecosystem in Hubei province.

English Language Arts/Writing

Trading Places Imagine that you live in a small village in China. The government is building a huge new dam. It will bring electricity to the region, but your village will be destroyed. You will have to leave your home.

1. Write a letter to an American student.
2. Describe how the situation makes you feel.
3. Explain what you would want the government to do.

Electrical Circuits

What makes up an electrical circuit?

An electrical circuit begins at a power source and ends back at that same power source. The circuit must have at least one device that consumes electricity, such as a buzzer or light bulb.

An electric switch is used to open and close the circuit to turn things on and off. There are two types of electrical circuits:

- Series circuit
- Parallel circuit

Series Circuits

In a **series circuit**, the electricity takes one path through multiple electrical devices. A major disadvantage of series wiring is that if one device burns out, all the devices in the circuit stop working. Devices cannot operate independently.

When batteries are connected in a series, each one increases the voltage (pressure) in the circuit. When you place three 1.5-volt batteries together in a series, the circuit is powered by 4.5 volts. Do you understand why?

Light bulbs and other electrical devices can also be wired together in a series. In this type of circuit, the electricity must pass through each device on its way to the next. Again, if any item in the circuit breaks down, power is lost to the entire circuit. Most of the circuits in your home are not series circuits.

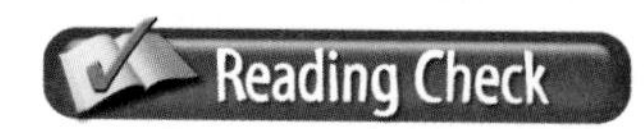

Identify What are the two basic types of electrical circuits?

Figure 8.4 Parallel Circuit

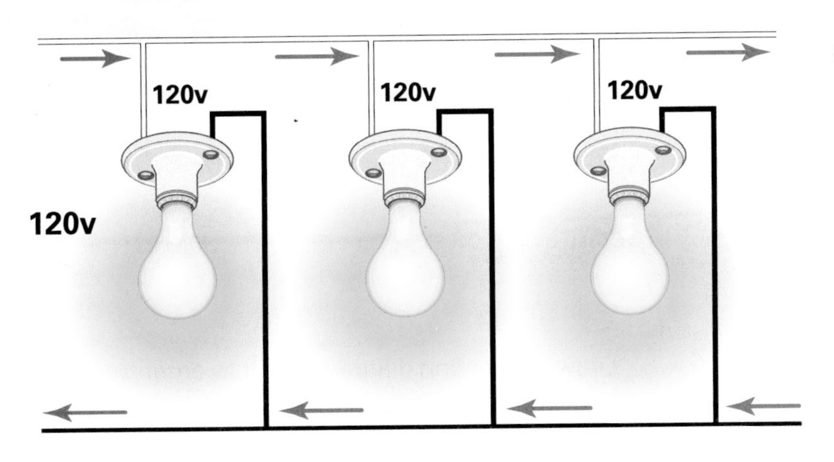

Multiple Pathways A parallel circuit provides individual electrical pathways for the current to flow to each energy-consuming device. *How can you tell if a string of lights is wired into a parallel or a series circuit?*

Parallel Circuits

In a parallel circuit, electricity flows along separate paths to each individual energy-using device in the circuit. See **Figure 8.4**. If one device burns out, the electricity does not stop flowing to the other devices in the circuit.

Most of the circuits in your home are parallel circuits. All the electric outlets in your home are connected to a fuse box or circuit-breaker panel. These safety devices shut off the electricity in case of a power overload. A power overload can superheat electrical wires and start a fire.

section 8.2 assessment

After You Read Self-Check

1. Compare and contrast electrical insulators and electrical conductors.
2. Describe a semiconductor.
3. Define superconductors and how they are used.

Think

4. Name some possible causes for a blown fuse.

Practice Academic Skills

STEM Science

5. With your teacher's help, wire low-voltage, DC light bulbs together in a series circuit and a parallel circuit. Remove a bulb from each circuit. What happens?

STEM Mathematics

6. Jorge wanted to find out how many 9-volt batteries connected in a series are required to create a 120-volt circuit. So, he divided 120 by 9 and concluded that he needed $13\frac{1}{3}$. Write a sentence explaining Jorge's mistake, and tell the correct answer.

Math Concept **Discrete Quantities** Sometimes a fraction or decimal solution to a problem does not make sense.

1. Think about the question asked in the problem.
2. Be sure your answer makes sense given the situation described in the problem.

For help, go to **glencoe.com** to this book's OLC and find the Math Handbook.

section
8.3 Electronics

Reading Guide

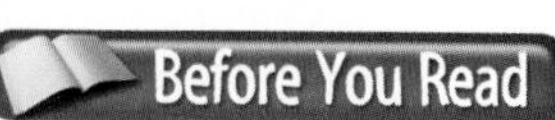

Before You Read **Connect** What is an electronic device?

Content Vocabulary

- electronic device
- electromagnetic wave
- transmitter
- fiber optic cable
- laser
- analog signal
- digital signal
- transistor
- integrated circuit

Academic Vocabulary

- network
- compare

Graphic Organizer

Draw the section diagram. Use it to organize and write down information as you read.

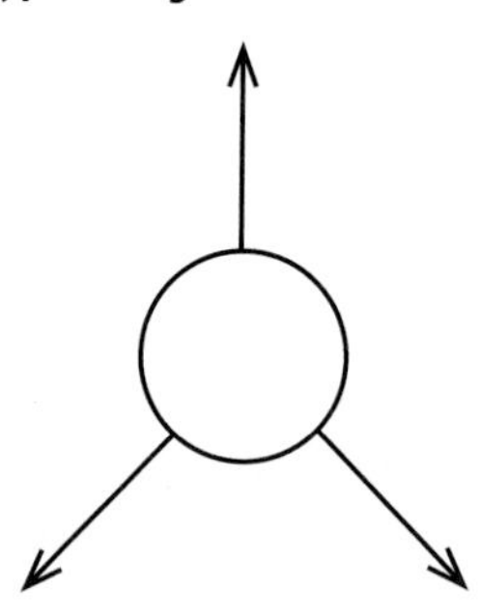

Go to **glencoe.com** to this book's OLC for a downloadable graphic organizer and more.

TECHNOLOGY STANDARDS

STL 2 Core Concepts of Technology
STL 3 Relationships & Connections
STL 12 Use & Maintenance
STL 16 Energy & Power Technologies

ACADEMIC STANDARDS

Science
NSES Content Standard E Abilities of Technological Design
English Language Arts
NCTE 12 Use language to accomplish individual purposes.

STL *National Standards for Technological Literacy*
NCTM *National Council of Teachers of Mathematics*
NCTE *National Council of Teachers of English*
NSES *National Science Education Standards*
NCSS *National Council for the Social Studies*

Electronic Devices

What is an electronic device?

The words *electric* and *electronic* are often used interchangeably. To keep things simple, electric devices generate and utilize electricity as a source of power. The electricity lights electric bulbs and rings door bells. An **electronic device** can change one form of energy, such as sound or light, into an electrical signal to be used in the device or transmitted.

As You Read

Predict What is a fiber optic cable?

Electronic devices allow for information to be processed and transmitted. This processing takes place in self-contained electronic games and in complex robotic systems. Your telephone, TV, radio, and computer are all electronic devices.

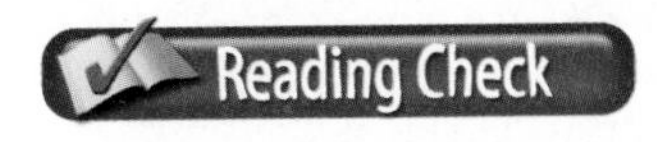

Describe What is the difference between an electric device and an electronic device?

Signal Transmission

How can electronic signals be sent?

Electronic signals can be sent along a wire, through the air on electromagnetic waves, or as pulses of laser light through fiber optic cables.

Wire Transmission

Wire transmission includes all communication that takes place over wires. Telephones, cable TV, computer-to-printers, and even computer-to-monitors all use wire transmission.

To transmit a signal along a wire, the signal must be converted into a series of electric pulses. Alexander Graham Bell's 1876 telephone changed the sound of the speaker's voice into a varying electric current. Wires carried these electric pulses to the receiving phone. The electric pulses were then converted back into varying vibrations that replicated the sound of the sender's speech.

Today's land-line telephones have been modified and improved many times, but they still use the transmission principle developed by Bell more than 100 years ago. Your call is transmitted through a cable. Your phone signal reaches a computer exchange network, where it is directed to the proper receiver.

Atmospheric Transmission

The magnetic field created when electricity flows through a wire can also carry a message through the air. This very weak magnetic field is converted into **electromagnetic waves** by passing through a strong magnetic field. The electromagnetic wave that is produced carries the original signal off into the atmosphere.

Remote Control These students are participating in a robotics competition that involves many schools. They use electronic communication systems to communicate with robots they have built. *Are the robots the senders or the receivers?*

Whose Line Is It? Thousands of phone conversations can go through each of these fiber optic phone lines at the same time. *If you send multiple messages through the same cable at the same time, what keeps them from scrambling together?*

The device used to send the electromagnetic wave that carries the signal into the atmosphere is called a **transmitter**. The transmitters in cell phones and personal remote controls are very small, so their signals must be picked up reasonably close to the location where they are sent.

Fiber Optic Transmission

Fiber optic cables are made from very pure glass. The term *optic* is used to describe something related to light. Each glass strand in a cable can be thinner than half the thickness of a human hair.

The glass strands have an outer glass coating with different reflective characteristics. This causes the rays of light traveling at the core to stay inside in spite of twists and turns. These cables can carry a signal for very long distances. If necessary, the sender can easily boost the power so messages will not be lost.

The signals that are sent along fiber optic cables are powered by lasers. **Lasers** are very powerful, very narrow beams of highly focused light. All of these light rays have exactly the same wavelength. The beam that is made by a laser can be visible light or invisible infrared light.

By transmitting signal beams that are on different wavelengths, engineers can send many messages along the same fiber optic strand at the same time. When they reach their destination, the different wavelengths are separated.

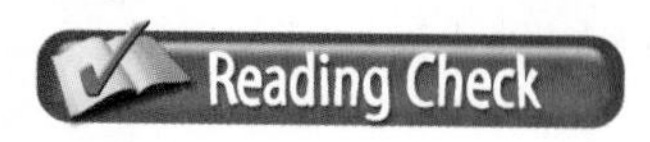

List What are the basic types of electronic signal transmissions?

Analog and Digital Signals

Which signal is faster—analog or digital?

Electric **analog signals** change continuously, and they are used in all kinds of systems. Alexander Graham Bell learned how to turn electricity into analog signals. The current in his telephone circuit varied with the intensity of the person's speech. If your watch has hands that rotate smoothly, it is an analog watch. The second hand on an analog watch never stops moving.

A **digital signal** is analog information that has been converted into digital information. The changing analog information is sampled frequently and turned into distinct, separate values. Your digital watch represents the seconds as exact numbers. It starts and stops continuously. The time is never "between" one second and the next.

Because digital signals can be compressed to take up less space in storage, sent faster, and stored longer than analog signals, they are preferred for electronics. For example, when you speak into a phone, your voice, which is analog information, is changed into a digital signal. However, this does not mean that digital signals are more accurate than analog signals.

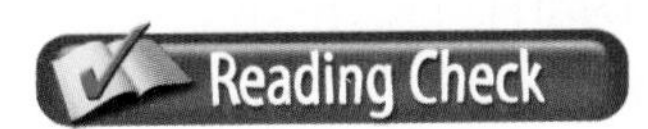

Contrast How are analog and digital signals different?

Imagine This...

Smaller Is Better

Imagine if electronic devices were embedded in roadways and could automatically send out traffic reports. Engineers are testing new wireless sensors no bigger than a coin for a variety of uses. They are also monitoring energy use in supermarkets and the movement of enemy troops on battlefields. *How could small, wireless sensors be used in your school or home?*

Go to **glencoe.com** to this book's OLC for answers and to learn more about different wireless sensors.

Size and Speed

Will electronic devices continue to shrink in size?

Each new improvement in electronics is built on past knowledge. In the 1940s, vacuum tubes became the first electric "switch" in electronic equipment. They resembled light bulbs, used a lot of electricity, and gave off a lot of heat.

Smaller and Smaller Microchips may be incredibly small, but they can process huge amounts of information. *Do you think this trend of shrinking electronics will continue for the foreseeable future? Why or why not?*

Compact Storage Microchips process a lot of information. *What material is used to make chips?*

1950s

Starting in the 1950s, vacuum tubes were replaced by the **transistor**, a small, cool running device that could also act as an electric switch. The transistor was tiny **compared** to a vacuum tube. However, thousands of hand-soldered connections were needed to join transistors and other components into completed circuits.

1960s

The next size reduction took place in the 1960s. The first **integrated circuit** contained one transistor and two other electric components on a single chip of silicon. Very tiny passageways in the silicon connected the tiny components together.

1970s to the Present

By the 1970s, technologists were placing thousands of integrated circuit components on to a single chip. Today these tiny microchips contain millions of components.

section 8.3 assessment

Self-Check

1. Describe the ways in which electronic signals can be transmitted.
2. Explain how Alexander Graham Bell's original telephone worked.
3. Compare and contrast an integrated circuit and a vacuum tube.

Think

4. Mika is singing a song. State whether her song is analog or digital information and explain why.

Practice Academic Skills

English Language Arts/Writing

5. Look up the meaning of the prefixes *trans-* and *tele-*. Write two sentences about the origin and meaning of each prefix and give one examples of words that use them.

STEM Science

6. With your teacher's help, use a low-power laser to send sound signals through a fiber optic cable. Describe and summarize the experiment in a one-paragraph essay.

Exploring Careers in Technology

Stephanie Martin

ELECTRICAL ENGINEER

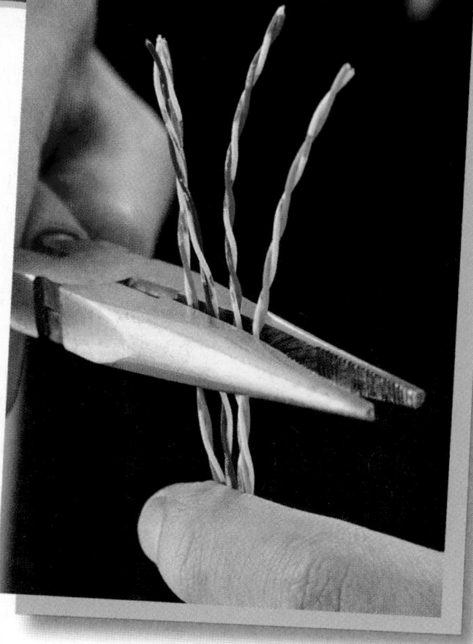

Q: *What do you do?*
A: I work in the field of electronic warfare, which is preserving the electromagnetic spectrum for friendly use while denying its use to the enemy. By jamming an enemy radar or communication system, you can prevent them from attacking friendly forces. On a typical day, I might research a target system, model a scenario to estimate effectiveness, conduct a field test, brief a group that will be using jammers, communicate with war fighters, or write a report. I also travel extensively for field testing and conferences.

Q: *What kind of training and education did you need to get this job?*
A: I studied electrical engineering with an emphasis on communications systems. I also worked as an intern during college. Most of my training has been on the job. I have attended classes on electronic warfare, and I am currently working on my master's degree in electrical engineering by taking evening classes through Johns Hopkins University.

Q: *What do you like most about your job?*
A: I like the people I work with and traveling to new places. By far the greatest thing about my job is being able to see something I have worked on help protect people and save lives.

English Language Arts/Writing

Specialization There are many kinds of engineers. Most engineers specialize in one particular field, such as aerospace, biomedical, chemical, nuclear, or civil engineering.

1. Research a particular field of engineering and write about it in a one-page report.
2. Remember your audience: Write a description that will make your classmates interested.
3. Make it exciting. Include details about what people do in the field, what a typical day is like, and why this field of engineering is important.

Go to **glencoe.com** to this book's OLC to learn more about this career.

Real-World Skills

Listening and speaking, problem solving, troubleshooting

Academics and Education

Mathematics, science, engineering, English language arts

Career Outlook

Growth as fast as average for the next ten years

Source: *Occupational Outlook Handbook*

chapter 8 Review and Assessment

Chapter Summary

Section 8.1 Electricity is the movement of electrons from atom to atom. Electricity is forced through wires by its voltage. A wire's resistance is determined by its thickness, length, and its ability to conduct electricity.

Section 8.2 Insulators cover conductors because they resist the flow of electricity. Superconductors have no electrical resistance. In a series circuit, electricity must pass through each device. If one device burns out, the electricity stops flowing. Parallel circuits have separate pathways.

Section 8.3 We use electromagnetic waves and fiber optic cables to carry messages. By transmitting signal beams on different wavelengths, we can send many messages on the same fiber optic strand at the same time.

Review Content Vocabulary and Academic Vocabulary

1. On a sheet of paper, use each of these terms and words in a written sentence.

Content Vocabulary

- atom
- electricity
- static electricity
- DC
- circuit
- AC
- frequency
- voltage
- transformer
- amperage
- resistance
- Ohm's law
- wattage
- conductor
- insulator
- semiconductor
- superconductor
- series circuit
- parallel circuit
- electronic device
- electromagnetic wave
- transmitter
- fiber optic cable
- laser
- analog signal
- digital signal
- transistor
- integrated circuit

Academic Vocabulary

- transmit
- vary
- generate
- series
- network
- compare

Review Key Concepts

2. **Compare** the three types of electricity.
3. **Define** voltage, amperage, and resistance.
4. **Describe** the purpose of an electric meter.
5. **Tell** the difference between an insulator and a conductor.
6. **List** some uses for semiconductors and superconductors.
7. **Identify** the basic types of electrical circuits.
8. **Discuss** the different types of electronic signal transmissions.
9. **Explain** how fiber optic cables work.
10. **Describe** the difference between analog and digital signals.

Real-World Skills

11. **Safety and Electricity** Communicate electrical safety concepts to young children. Research childhood accidents related to electricity, where most of them occur, and why. What are the most important concepts that should be explained to young children? Design a poster or public service ad for TV.

Technology Skill

12. **Test for Conductivity** Create a simple circuit using a battery, wire, and a light bulb. Use it to test the conductivity of materials you find at home or at school.
    a. Test items such as paper clips, cloth, metal objects, and water.
    b. Chart your results by listing the name of the object, what it is made from, and whether it conducts.

WINNING EVENTS

Electrical Engineer

Situation Design an electrical device that provides light, sounds an alarm, or serves some other purpose.

Activity Brainstorm ideas with a teammate. Sketch your ideas; make a schematic diagram of your best idea. Then build your device and present it to your classmates.

Evaluation You will be evaluated on how well you meet the following criteria:

- Proposal—good idea
- Schematic—clear and technically correct
- Device—does what it is supposed to do
- Complexity—evident

Go to **glencoe.com** to this book's OLC for information about TSA events.

Academic Skills

Social Studies

13. Write a one-page report about the history of electricity. You might research an inventor, such as Benjamin Franklin, Thomas Edison, or Nicola Tesla, and his or her role in developing the technology of electricity.

STEM **Mathematics**

14. What is the amperage to run a 120-volt hair dryer that draws 2,400 watts?

    Math Concept **Using Formulas** The two formulas that relate to how electricity works are: $E = I \times R$ and $W = E \times I$. E is voltage, or the pressure that pushes electricity through a circuit; I is amperage, or the flow of current; R is resistance to that flow; and W is wattage, or the power an appliance uses.

Standardized Test Practice

Directions Choose the letter of the best answer. Write the letter on a separate piece of paper.

1. Which statement is true about lightning?
   - **A** Lightning is a form of static electricity.
   - **B** Lightning occurs when atoms lose most of their electrons.
   - **C** Lightning is a form of alternating current.
   - **D** Lightning only hits metal objects that are good conductors.
2. When an atom loses electrons, it becomes positively charged.

   T **F**

Test-Taking Tip When studying for a test, write important ideas, definitions, and formulas on flash cards. Use them for quick reviews.

chapter 8

TECHNOLOGY LAB

Build a Simple Alarm System

All alarm systems include a circuit, a switch, and some kind of ringer. In an alarm clock, the switch is connected to a clock, and the ringer might be a radio that wakes you up. A modern security alarm system may include sensors that react to physical movement or changes in temperature.

Tools and Materials

- ✓ Clothespin
- ✓ 6 inches of #20 bell wire
- ✓ Wire stripper
- ✓ Metal punch
- ✓ Drill
- ✓ 9/64 drill bit
- ✓ 3½ × 4½ × ¾ -inch wood base
- ✓ 9-volt battery
- ✓ 9-volt battery snap
- ✓ Piezoelectric buzzer
- ✓ Two 6/32-inch machine bolts
- ✓ Four 6/32-inch nuts
- ✓ Six small washers
- ✓ Four ½-inch #6 wood screws
- ✓ One 1-inch #7 wood screw
- ✓ Electrical tape
- ✓ Wooden tongue depressor
- ✓ String
- ✓ 3½ × ¾ inch tin plate strip

⚠ SAFETY

Electrical Safety

When using tools, machines, and wiring, be sure to always follow appropriate safety procedures and rules.

Set Your Goal

For this activity, you will build the alarm shown in the illustration. When your alarm is completed, try to find a serious or funny way to use it.

Know the Criteria and Constraints

In this lab, you will:

1. Build and test your alarm to be sure it works.
2. Find a creative use for the alarm.
3. Demonstrate your alarm to the class. You can use visual aids.

Design Your Project

Follow these steps to build your alarm and complete this lab.

1. Create the base. Cut and bend the metal strip that holds the battery.
2. Cut holes in the battery holder with the metal punch.

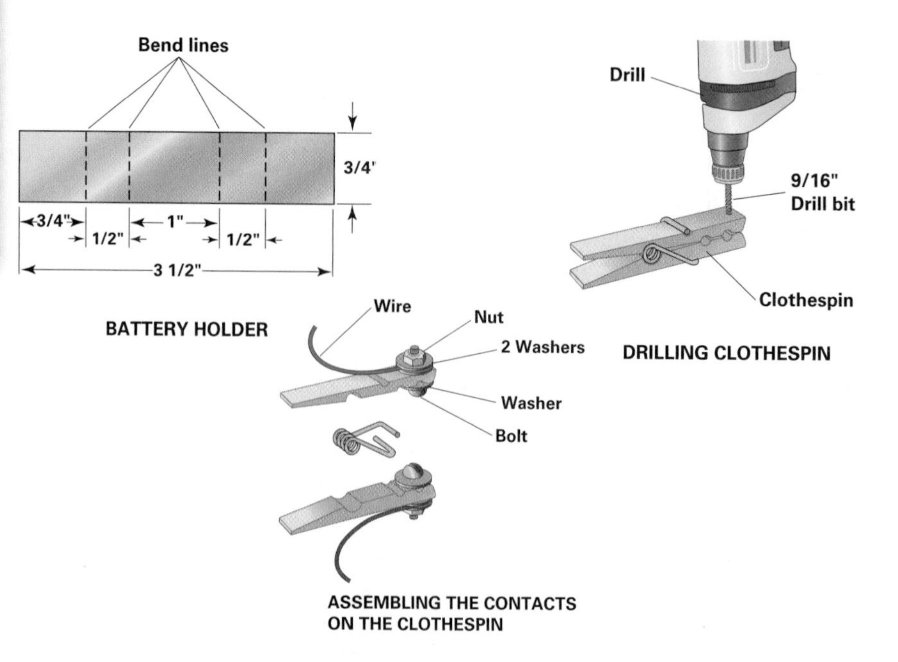

3. Use #6 wood screws to attach the holder and buzzer to the base.
4. Build the switch using the clothespin.
   - Drill a 9/64-inch hole through the two closed ends of the clothespin.
   - Slip a bolt and washer through each hole. Secure each bolt with a nut.
   - Strip ½ inch of insulation from wire ends and slip two washers onto the bolts' ends. Hook one end of the bell wire between washers and secure in place with a nut.
   - Repeat procedure to hook the buzzer's black wire to the other side.
5. Secure the clothespin by screwing the #7 wood screw through the spring.
6. Splice the buzzer's red wire to the battery snap's red wire. Splice the free end of the bell wire to the black wire of the battery snap. Tape your connections with electrical tape.
7. Place a tongue depressor between the clothespin contacts. Connect the battery snap.
8. Test the alarm by pulling out the tongue depressor.
9. Adapt your alarm to work where you plan to install it.
10. Create a video or other visual aid of your demonstration.

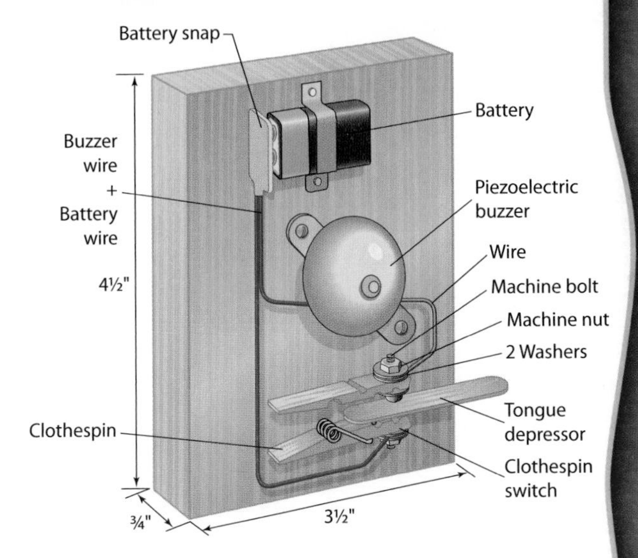

THE SIMPLE ALARM SYSTEM

Evaluate Your Results

After you complete this lab, answer these questions on a separate piece of paper.

1. Why does the buzzer ring after you pull the tongue depressor?
2. Can a larger model of this alarm protect your home? Why or why not?
3. What is the relationship of the battery's voltage and the buzzer size?

Academic Skills Required to Complete Lab

Tasks	English Language Arts	Math	Science	Social Studies
Research ways to use an alarm.	✓			✓
Build the alarm.		✓	✓	
Write presentation content.	✓		✓	
Create presentation of how student will use the alarm.	✓			✓
Present to the class.	✓		✓	

Technology Time Machine

The Essence of Energy

Play the Game This time machine will travel to the past to show how experiments in energy and electricity led to today's technological achievements. To operate the time machine, you must know the secret code word. To discover the code, read the clues, and then answer the questions.

Clue 1

200 B.C.E. Greek inventor Hero of Alexandria invented the pump originally to bring water up from wells more than 2,000 years ago. He could never have imagined its many uses today. The automobile might never have been possible without the pump.

Clue 2

600s Wind power for ships goes back thousands of years. In the Middle Ages, windmills were designed to manufacture, grind grain, and operate water pumps by converting kinetic energy. Windmills may have originated in Persia, now Iran.

Clue 3

1698 For many years pumps were used to drain water from flooded mines, but they were not very powerful. Thomas Savery of England invented the steam engine to add power to the pumps and drain the mines.

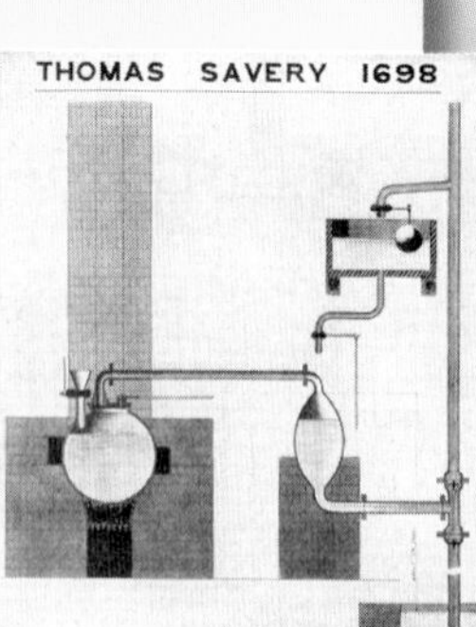

Clue 4

1700s The development of steam-powered engines progressed quickly in Europe. One steam engine could supply the power needed to run all the machines in a factory. Steam locomotives hauled heavy loads to far-off places. Steamships provided safe, fast, dependable transportation. It was the very beginning of the Industrial Age.

Clue 5

1769 Nicolas-Joseph Cugnot, a French military engineer, built the first self-propelled road vehicle by attaching a steam engine to a "horseless" carriage. The automobile was born.

Clue 6

1878 Inventor Thomas Edison began his research on electric lighting. He learned that a carbon arc light produced an extremely bright light by sending electricity across a gap between two carbon terminals. With his associates, Edison developed an incandescent lamp that produced a glow by passing electricity through a filament or wire.

Clue 7

1879 Edison and his associates discovered a carbon filament made from burned sewing thread. This produced the first incandescent light bulb for use in people's homes. A year later, Edison realized that bamboo filaments increased the life of the bulbs.

Clue 8

1904 Electricity was a hot new product. Demand for streetlights soared. Some materials were in short supply. An inventor named Leo Baekeland developed a synthetic material to insulate electrical coils so the technology became more available.

Crack the Code

On a piece of paper, write the answers to these questions:

1. What did Cugnot attach a steam engine to when he built a self-propelled vehicle?
2. What is essential to technology?
3. Where did Greek inventor Hero come from?
4. What inventor was most responsible for bringing electric light into our homes?

Now write down the first letter of each answer. Put them together to discover the secret code word!

Solar collectors can trap this and save money on bills for its use.

unit 2 Thematic Project

Designing Alternative Power Plants

Energy and Power In Unit 2, you learned about different forms of energy. The power you receive by flipping a switch may seem automatic. But your electric company can convert energy from many sources—hydro, solar, wind, nuclear, or coal—to create the power to run our homes and businesses.

Voltage Inventory Most homes in the United States are wired with 120 volts of electricity. This means a standard plug can supply that amount of power. Look around your house. Count sockets, wall plugs, extension cords, and power strips. Multiply that number by 120 to find the potential voltage.

One Step Forward We can conserve energy to help our planet. If we all do just one thing each day, the task will not seem so overwhelming. Turn off a light in an empty room or use compact fluorescent bulbs.

This Project In this project, you will design an alternative power plant.

Your Project

- Choose an alternative energy source.
- Choose and complete one task:
  1. Write a proposal. State why your source is possible, efficient, cost-effective, and Earth-friendly.
  2. Create a two-dimensional model of a power plant. Use freehand drawing or CAD. Label your model.
  3. Create a 1-minute commercial.
- Write a report.
- Create a presentation with posters, video, or presentation software.
- Present your findings to the class.

Tools and Materials

- ✓ Computer
- ✓ Internet access
- ✓ Trade magazines
- ✓ Word-processing software
- ✓ Presentation software
- ✓ Posterboard
- ✓ Colored markers

The Academic Skills You'll Use

- Communicate effectively.
- Speak clearly and concisely.
- Use correct spelling and grammar in a written report.
- Conduct research using a variety of resources.
- Incorporate reading, writing, and speaking with viewing, representing, and listening.

English Language Arts

NCTE 4 Use written language to communicate effectively.

NCTE 8 Use information resources to gather information and create and communicate knowledge.

Science

NSES Content Standard F Science and technology in society

Step 1 Choose Your Topic

You can choose any energy source for your project. Examples include:

- Hydroelectric
- Wind
- Solar
- Clean coal
- Atomic
- Geothermal
- Natural gas
- Biodiesel
- Biomass
- Biofuels

Tip! *Explore all energy sources to see which is most interesting!*

Step 2 Do Your Research

Research your project. Your fact finding may include a combination of sources. Write answers to these questions:

- How was the energy source used in the past?
- What do old and new articles in libraries and online sites say about your topic?
- What are the blueprints or specifications of similar systems?
- How would someone who does CAD or drafting do a sketch?

Tip! *Keep your design as simple as possible!*

Step 3 Explore Your Community

Tour the local electric company. Ask about the future of power in your community. Ask for feedback on your design.

Tip! *Research the company before you go.*

Step 4 Create Your Project

Your project should include:

- 1 research project (proposal, design, blueprint, model, or video)
- 1 report
- 1 presentation

Project Checklist

	Objectives for Your Project
Visual	✓ Make a poster, blueprint, model, video, or slide presentation to illustrate your project.
Presentation	✓ Make a presentation to your class. ✓ Discuss what you have learned. ✓ Turn in research and names from your interview to your teacher.

Step 5 Evaluate Your Presentation

In your report and/or presentation, did you remember to:

- Demonstrate your research and preparation?
- Engage your audience?
- Back up statements with facts, details, and evidence?
- Use visuals effectively?
- Speak slowly and enunciate clearly?

Rubrics Go to **glencoe.com** to the book's OLC for a printable evaluation form and your academic assessment form.

GLOBAL TECHNOLOGY

A Sunny Source of Power

In the past Portugal depended on fossil fuels for energy. But the country is one of Europe's sunniest areas, and so a company called Catavento built a solar power plant in Serpa, Portugal. It is the first large solar installation there with 52,000 photovoltaic modules (solar panels). Located among olive trees on hillside pastures, the plant produces electricity for 8,000 homes. It also reduces greenhouse gas emissions, required by the European Union.

Critical Thinking *What other countries might be able produce solar energy? Why?*

Portuguese

hello	hallo
goodbye	adeus
How are you?	Como é vai?
thank you	obrigado
You're welcome	Nãoha de qué

Go to **glencoe.com** to the book's OLC to learn more and to find resources from **The Discovery Channel.**

unit 3

Communication Technologies

Chapters In This Unit:

Unit Thematic Project Preview

Shopping Online or In-Store

As part of this unit, you will learn about different communication technologies, including graphic, photographic, multimedia, and computer technologies. All of these technologies contributed to the growth of e-commerce and shopping over the Internet.

As you read this unit, use this checklist to prepare for the project at the end of this unit:

PROJECT CHECKLIST

✓ Think about the types of businesses that have Web sites.

✓ Explore different stores' Web sites.

✓ Consider if the Web sites have online purchasing features.

WebQuest Internet Project

 Go to **glencoe.com** to this book's Online Learning Center (OLC) to find the WebQuest activity for Unit 3. Begin by reading the Task. This WebQuest activity will help you discover the different products and systems that depend on communications satellites to work.

Explore the Photo

Satellite Communication Satellite communication technology is used for cell phones, television, radio, GPS systems, and more. *Why do you think some communications systems moved into outer space?*

chapter 9

Communication Systems

Sections

What You'll Learn

- **Apply** the systems model to communication.
- **Identify** communication subsystems.
- **Discuss** different forms of communication.
- **Describe** the different modes of communication.
- **Explain** how communication technology has affected modes of communication.
- **Identify** the positive impacts of communication technology.
- **Identify** the negative impacts of communication technology.

Explore the Photo 

Convergence Cell phones like the iPhone combine an iPod, camera, e-mail, IM, and Internet access. This combining is called "convergence." *What technology tool do you use the most? Why?*

Create an Ad

At the end of this chapter, you will be asked to create an advertisement for a brand new communication device, such as a cell phone, plasma TV, or video game. Get a head start by using this checklist to prepare for the Technology Lab.

PROJECT CHECKLIST

✓ Begin Internet research to learn about inventions.

✓ Find images of interesting communication inventions.

✓ Gather ads for current devices in magazines and newspapers.

section 9.1

Introducing Communication Technology

Reading Guide

 **Before You Read** **Connect** How do you use technology for communicating?

Content Vocabulary

- communication technology
- graphic communication
- sound waves
- electromagnetic carrier wave
- telecommunication

Academic Vocabulary

You will see these words in your reading and on your tests. Find their meanings at the back of this book.

- technology
- factor

Graphic Organizer

Draw the section diagram. Use it to organize and write down information as you read.

Forms of Communication

Biological Communication			
Speaking			

 Go to **glencoe.com** to this book's OLC for a downloadable graphic organizer and more.

TECHNOLOGY STANDARDS

STL 2 Core Concepts of Technology

STL 17 Information & Communication Technologies

ACADEMIC STANDARDS

English Language Arts

NCTE 4 Use written language to communicate effectively.

NCTE 12 Use language to accomplish individual purposes.

STL *National Standards for Technological Literacy*

NCTM *National Council of Teachers of Mathematics*

NCTE *National Council of Teachers of English*

NSES *National Science Education Standards*

NCSS *National Council for the Social Studies*

Communication and Technology

What are some examples of communication technology?

Imagine you are standing in the hallway at school talking to your friends. Are you using communication? Yes. Communication is sending, receiving, and responding to messages. Are you using communication technology? No. When you are talking face-to-face, you are not using communication technology.

However, if you communicate by using a written note, Instant Messaging (IM), or a cell phone, then you are using communication technology. **Communication technology** is the transfer of messages (information) among people and/or machines through the use of technology. This processing of information can help people make decisions, solve problems, and control machines.

As You Read

Predict Think of different ways to send and receive messages.

The knowledge, skills, and tools that were the foundation of past and current communication technology are also the foundation for new technologies and improved ones.

The Systems Model

How does communication fit the systems model?

As you know, systems can be charted on a diagram. **Figure 9.1** breaks systems into input, process, output, and feedback. This is the systems model.

Communication systems include all the inputs, processes, outputs, and feedback associated with sending and receiving messages (information). The message is the input. How the message is moved is the process. The reception of the message at the other end is the output. Feedback may relate to static or clarity.

Real-World Systems

Suppose you write an article for the school newspaper about the computer lab. Your words, pictures, time you spend, and the computer you use are inputs. Putting the newspaper together and printing it are parts of the process. The primary output is the newspaper.

Figure 9.1 **Communication System: A Newspaper Process**

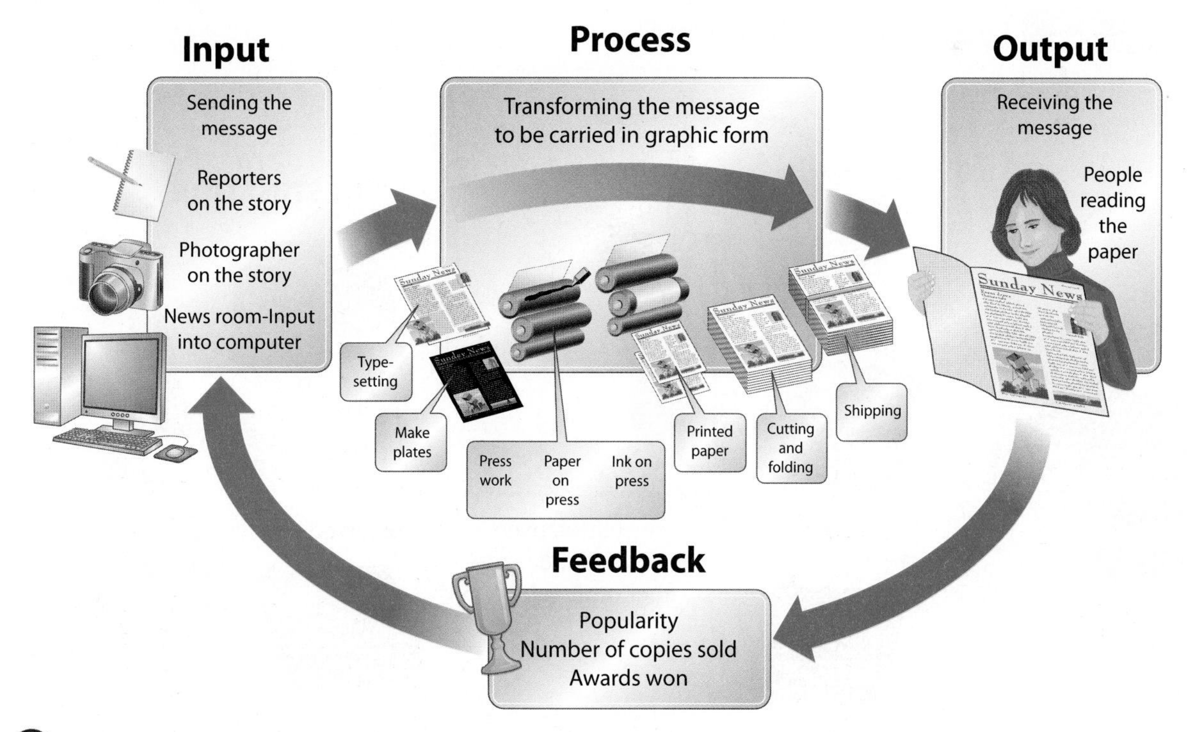

A System The process of making a traditional newspaper is an example of a communication system. *Give some examples of inputs.*

When you read a book or listen to an MP3 player, you are on the output end of the communication system, receiving the message. When you use a telephone, a computer, or a video camera, you are controlling both the input and output of the system. What parts of a system are you involved with when you play a video game, watch television, or type an e-mail or IM? Computers, iPods, and video recorders are communication systems that may contain the input, process, and output devices all in one unit.

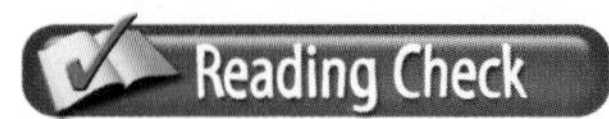

Define What is an input and an output?

Communication Subsystems

How would you use a decoder?

Communication systems usually include several subsystems that help transmit information. See **Figure 9.2**. The subsystems are made up of these elements: a source, an encoder, a transmitter, a receiver, a decoder, and a destination.

The source is the sender, which could be a person or a machine with a message to send. The encoder changes the message so that it can be transmitted. When you write a note, type on a computer keyboard, or talk into a telephone, you are encoding your message. Your message could be written on paper or be sent as an electronic signal.

Figure 9.2 **Systems and Subsystems**

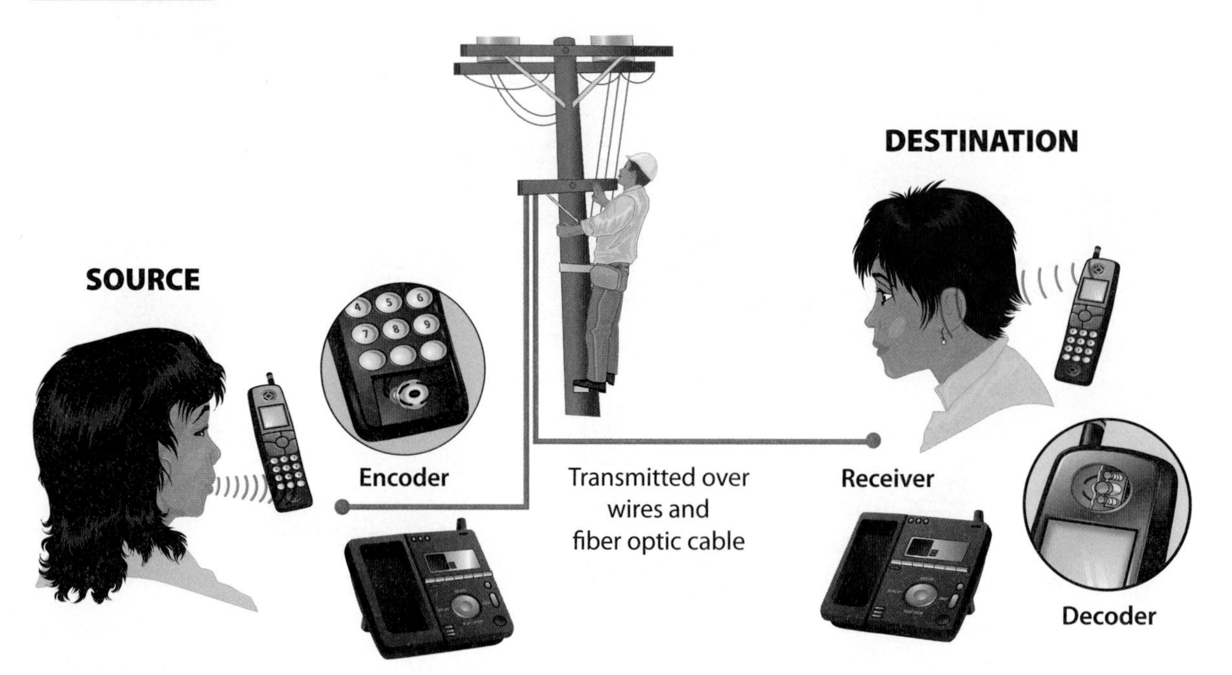

Systems within a System Most communication systems have several subsystems. *What are some examples of a destination?*

Tim Berners-Lee

Inventor of the World Wide Web

When Tim Berners-Lee was a teenager in England, he built a computer using a soldering iron and an old television. Later, he worked as a computer programmer at CERN, a physics laboratory in Geneva, Switzerland. There, he wrote a program with Robert Cailliau for storing information by using "random associations." This idea led to "hyperlinking," which is how the World Wide Web works.

Tim said the World Wide Web should be like a "global hypertext project." It should help people work together through a "web" of documents. The documents could be opened from anywhere on the Web.

The WC3 In 1994, Tim set up the World Wide Web Consortium (WC3) at the Massachusetts Institute of Technology (MIT). The group manages Web development around the world. Of course, the WC3 has its own Web site at w3.org.

English Language Arts/Writing Write an article for your school newspaper about spam and how it affects using the Web.

Go to **glencoe.com** to this book's OLC to learn about young innovators in technology.

The receiver of the message at the final destination could be a person or a machine. The message is then decoded, which means symbols on paper must be read or electronic impulses must be turned into information that a person or machine can understand.

Multiple Subsystems

Some communication systems have two subsystems designed to store and retrieve information. A telephone answering machine, an iPod, and a DVD burner are all parts of communication systems that can store and retrieve information for later use.

Video Games and Subsystems

When you play a video game, the controller encodes and transmits your moves to the machine's central processing unit. The machine determines its own moves and encodes this information so it can be displayed on the screen. You and the machine take turns being the source of information and the final destination for information until one of you wins the game. As you play, you are using the tools of communication to process information, solve problems, and make informed decisions.

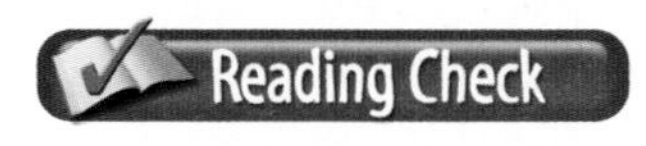

Identify Name the six elements of communication subsystems.

Computers You Can Wear

Imagine wearing a computer that is a light-weight, voice-activated box with a headset and eyepiece. When you look into the eyepiece, you see a computer screen that appears as if it is a few feet away. Researchers at NASA are working on a model called a "Wearable Augmented Reality Prototype (WARP)" for astronauts to wear so their hands are free for other tasks. *How might wearing your computer help you in everyday communication?*

Go to **glencoe.com** to this book's OLC for answers and to learn more about NASA and technology.

Message Design

What factors are important when designing a message?

If you were to create a Web page for a social networking Web site, you would probably think about the people who might view your site. You would want to share things that would interest them. The design of any message is influenced by the intended audience, or receiver of the message.

Other Factors of Message Design

The medium used—in this case, a social networking Web site—is another factor. Your message must be designed to match the selected communication medium. Other media might include:

- Print newspapers
- Billboards
- TV commercials

The nature of the message would also affect the design of the message. A story about the school computer lab would be set up differently than instructions for building the computer lab.

Also, the language used to express the ideas is important when designing a message. The language you use for writing or speaking, such as English, is one example. Symbols, measurements, and drawings are other kinds of languages people can use to communicate clearly. Do you and your friends have your own language that you use when you write instant messages or e-mails?

For example, instructions to workers for building a computer lab would include drawings (graphic images), measurements, symbols, and special vocabularies (construction terms) that the workers would understand. Ideas would be expressed in a common language of the construction industry for clear communication.

The Purpose of a Message

Every message has a purpose. That purpose may be to inform, persuade, entertain, control, manage, or educate. For example, a set of building plans is a type of message that is absolutely necessary for a building project. Can you read a floor plan? The purposes of a floor plan might be to control, manage, and educate.

An article about the school computer lab would probably inform and educate the people who read your school newspaper. Perhaps the writer thinks more students should use the computer lab after school. In that case, the article would be designed to persuade. If the article warns students not to play computer games, the article's purpose would be to control or manage. If the article included a funny story about a homework-eating computer, it could be meant to entertain. All the ways to communicate, including graphic and electronic means, can be used for these purposes.

Forms of Communication

What is one way to identify or group communication systems?

Although the different forms of communication may overlap, communication systems can be grouped by the way they carry messages. Let's look at some different ways that you can transmit a message, including biological communication, graphic communication, wave communication, and telecommunication.

Biological Communication

Ordinary biological communication is not part of technology. You will study it in depth in science courses. However, it is important for you to know that communication can take place without technology and that most living things communicate.

Biological communication includes all forms of communication that use natural methods, such as the voice, ears, arms, and hands, to transmit and receive messages. Examples include speaking (language), facial expressions, and hand signals.

Graphic Communication

Graphic communication includes all forms of communication that send and receive messages visually through the use of drawn or printed pictures and symbols. Printing is the most common example. Magazines, newspapers, messages on clothing, billboards, road signs, and computer images are all forms of graphic communication. In this form of communication, people send and receive information through reading, writing, drawing, and painting.

Wave Communication

Wave communication refers to all forms of communication that move through air, water, outer space, or some other medium in waves. They use the newest and oldest technological inventions.

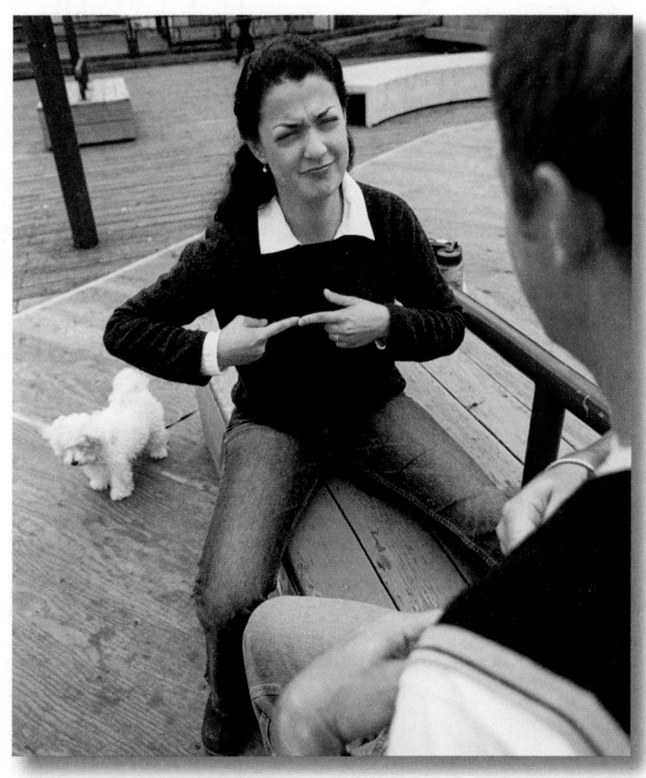

Biological Conversation One type of biological communication is "signing." American Sign Language (ASL) helps people with hearing disabilities communicate. *What other forms of communication might aid people with disabilities?*

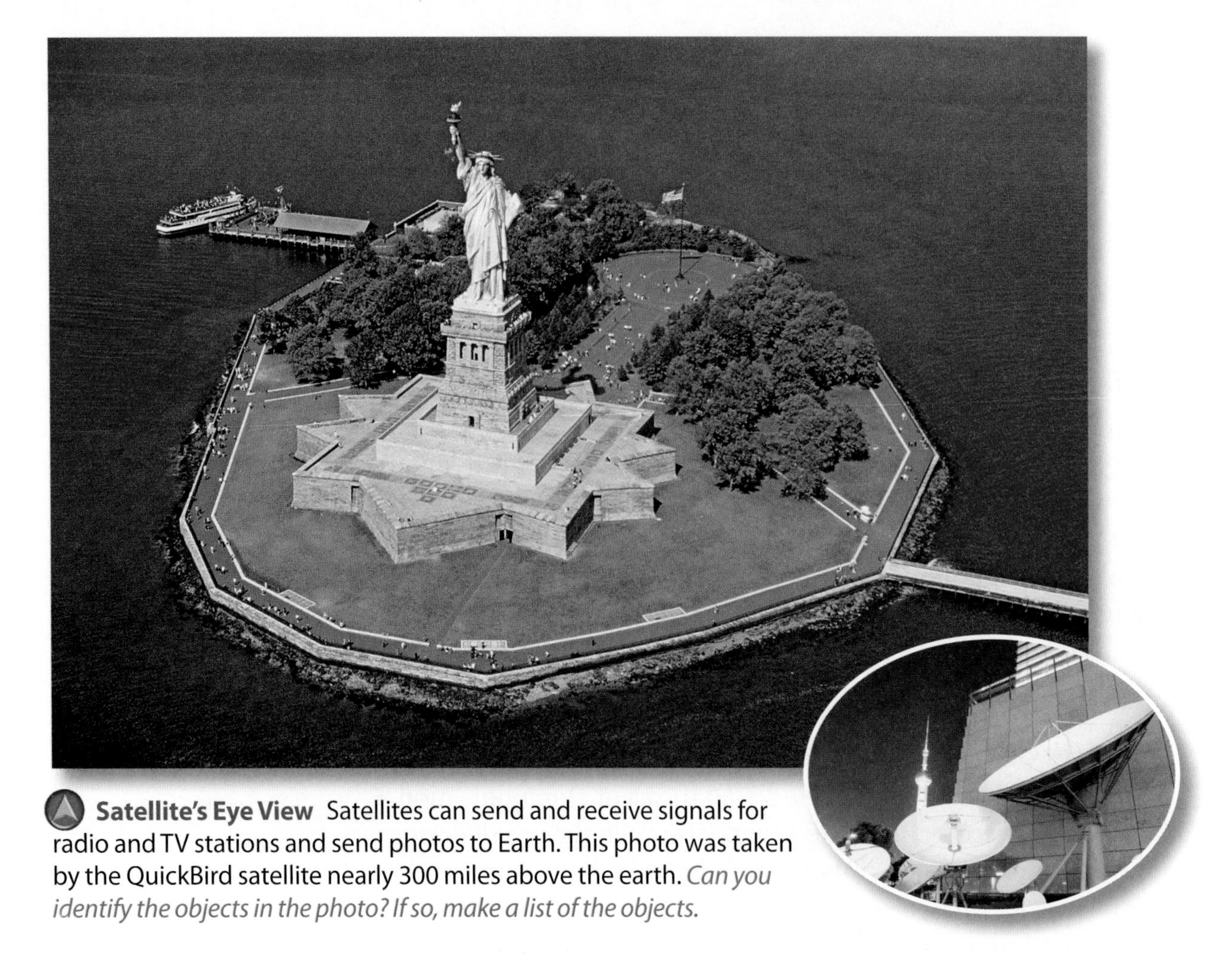

Satellite's Eye View Satellites can send and receive signals for radio and TV stations and send photos to Earth. This photo was taken by the QuickBird satellite nearly 300 miles above the earth. *Can you identify the objects in the photo? If so, make a list of the objects.*

Sound Waves

Sound waves are vibrations traveling through air, water, or some other medium that can be perceived by the human ear. Our early ancestors used hollow logs as drums to send coded messages. Banging on the logs caused the air to vibrate with sound waves that reached people far away. Musical instruments and face-to-face communication depend on sound waves.

Electromagnetic Carrier Waves

Our radio and TV programs are converted into electrical signals. These signals are carried by **electromagnetic carrier waves**, which are waves of electromagnetic energy that carry signals.

Large antennas and dishes receive the signals. The audio part of a TV signal and the full broadcast of a radio transmission are converted into sound waves, which you hear through speakers.

Light Waves

If your telephone company uses fiber optic cables, the signal is transmitted through the cable as waves of light. When you use a camera to take a picture, light waves from your subject bring the image to your eyes and to the camera lens.

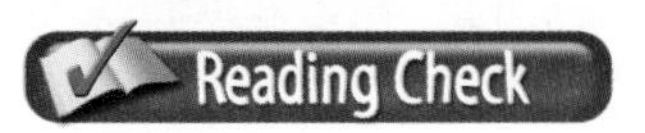

List What are four ways to transmit a message?

Telecommunication

Communication over a distance is **telecommunication**. Today most telecommunication systems use electronic or optoelectronic devices. Have you ever used a telecommunication machine or device? You have if you have used a phone, television, or radio.

Satellites and Telecommunication

Satellites are also telecommunication devices. Satellites placed 22,300 miles above the earth and traveling at the same speed that Earth spins are in a "geosynchronous orbit." This means that the satellite always stays above the same part of the earth. Its lack of movement in relation to the ground could give the impression that the satellite was attached to Earth with a very long pole. When a satellite is in a geosynchronous orbit, it seems to move with the objects on the ground.

Uses of Satellites

Satellites can help produce maps, provide climate information, track weather patterns, and even observe what people are building in other parts of the world. The United States has many spy satellites looking down on other countries as the satellites circle the earth. Some of these spy satellites can take detailed photos by using equipment similar to that used on the Hubble Space Telescope. It is possible for some satellites to see the details of an object that is smaller than a golf cart.

Academic Connections
Science

Now Hear This! You have your own built-in sound receivers—your ears. Sound waves striking your eardrums make them vibrate.

Apply Find an illustration of the ear in an encyclopedia at school. Make a drawing showing the main parts of the ear. Label the parts and functions.

section 9.1 assessment

After You Read Self-Check

1. Name the factors are involved in message design.
2. List the six purposes of a message.
3. Define telecommunication.

Think

4. Explain why smoke signals would be classified as a form of telecommunication.

Practice Academic Skills

English Language Arts/Writing

5. Until the late 1950s, many people shared telephone lines, or party lines. Only one party could use the phone at a time, because you might hear another person talking. Write a paragraph comparing today's social networking technologies to ones your parents used at your age.

STEM Mathematics

6. Shanda wants to create an Internet ad for China and Thailand and has to calculate the combined population of the two countries. If China has 1.306 billion people and Thailand has 65 million, what is the total population?

Math Concept **Representing Large Numbers** When you add larger numbers, make sure that they are expressed in the same form.

1. Express the number for China's population, 1.306 billion, as 1,306 million.
2. Add 1,306 million to 65 million to find the total combined population.

For help, go to **glencoe.com** to this book's OLC and find the Math Handbook.

section

9.2 Modes of Communication

Reading Guide

Before You Read **Preview** Why might new modes of communication be necessary?

Content Vocabulary

- mode
- machine-to-machine communication

Academic Vocabulary

- response
- process

Graphic Organizer

Draw the section diagram. Use it to organize and write down information as you read.

Modes of Communication

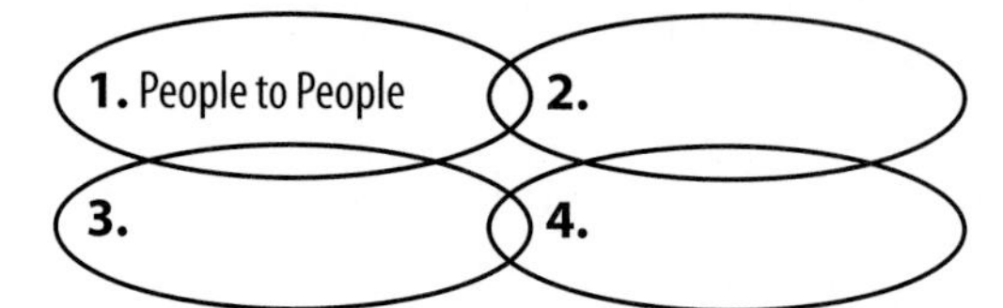

Go to **glencoe.com** to this book's OLC for a downloadable graphic organizer and more.

TECHNOLOGY STANDARDS

STL 1 Characteristics & Scope of Technology

STL 2 Core Concepts of Technology

STL 11 Design Process

STL 17 Information & Communication Technologies

ACADEMIC STANDARDS

English Language Arts

NCTE 4 Use written language to communicate effectively.

NCTE 12 Use language to accomplish individual purposes.

STL *National Standards for Technological Literacy*

NCTM *National Council of Teachers of Mathematics*

NCTE *National Council of Teachers of English*

NSES *National Science Education Standards*

NCSS *National Council for the Social Studies*

Evolving Modes of Communication

What progress have people made in their ability to communicate with each other?

Technology has given us new modes of communication. A **mode** is a way of doing something. Originally, "people-to-people" communication was the only mode. It is still the most basic mode of communication. Over time people have learned to create new and more powerful modes of communication. They gained the knowledge and skills needed to build complex communication devices and used the mode called "people-to-machines" communication. People also created graphic communication systems to transmit their messages using the printed word, which uses the mode

As You Read

Connect Do you communicate with machines? Which ones?

called "machines-to-people" communication. Finally, people developed communication based on electrical signals that are sometimes used for the mode called "machines-to-machines" communication.

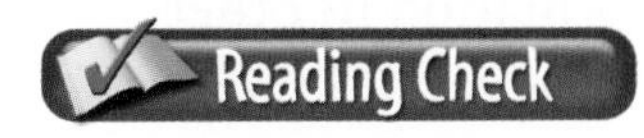

Identify What are the four modes of communication?

People to People

People-to-people communication is one mode, but it was not always the way it is today. At one time people had to communicate with each other within the limits of their own physical makeup (biological communication).

For example, a baby points, cries, grunts, stamps its feet, grabs what it wants, or speaks "baby talk" to be understood. Our earliest ancestors probably communicated in a similar way. They used simple sounds and gestures to get a response.

People to Machines

Until the development of electronic communication devices, people were talking only to other people. The machines they built could only carry the message, and this message could only be understood by other people. Today, however, people are communicating with the machines that they have created. Some examples of people communicating with machines include someone setting a digital video recorder (DVR or TiVo) to record a favorite television show; a computer programmer typing a program into a computer; and a person using a keyboard and a controller while playing a computer game.

Machines to People

Machines also send messages to people. A whistling teapot is a kind of machine that tells you the water is boiling. An alarm system tells you someone has entered a protected store or home. Also, a smoke/carbon monoxide detector can sense fire or carbon monoxide and warn you of the danger. Can you think of other examples of machines communicating with people?

EcoTech

Treeless Paper

Making paper from trees can use a lot of energy and toxic chemicals. But paper can also be made from hemp, bamboo, or kenaf—a plant that grows quickly and uses eco-friendly chemicals for production.

Try This To save paper, write on both sides of your paper, use scrap paper—and recycle it.

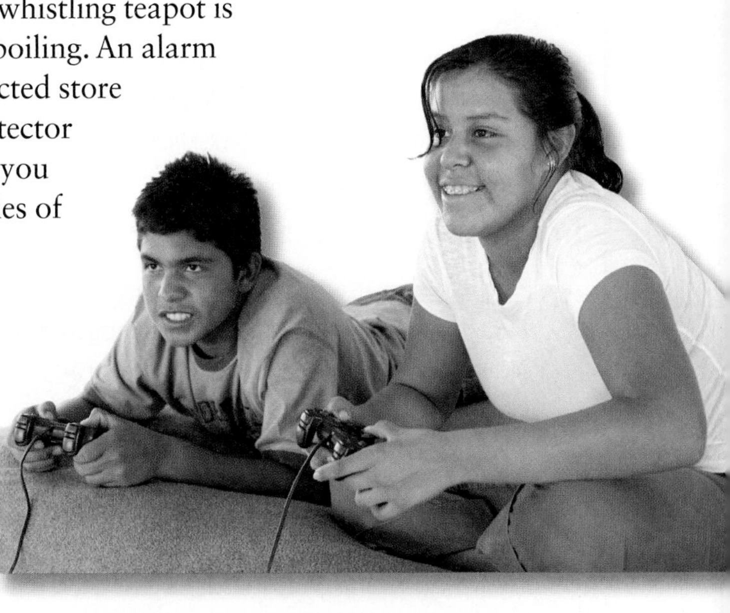

Recreational Communication Communication happens each time you play a computer game. *Why might this be considered people-to-machine communication?*

Machines to Machines

Machine-to-machine communication is the transfer of messages from one machine, such as a computer, to another machine. Machine-to-machine communication is also known as M2M. All types of machines can be adapted to send information to other machines. Many machines use wireless technology to do this. Machine-to-machine communication is quite common today, and there are many examples of this particular mode of communication. Besides transferring information, it is used by individuals and manufacturers for a variety of other processes.

Your computer gives instructions to your printer, telling it to print your report. In an automated factory, computers attached to sensors control the flow of raw materials and the operation of the machines. The assembly and finishing processes used to make a product are also controlled by machines communicating with other machines. Even the packaging of the finished product may be handled by machines, under the direction of still other machines. Many businesses and manufacturers depend on machine-to-machine communication to produce the products people buy and use around the world.

section 9.2 assessment

Self-Check

1. Explain how a baby who cannot talk communicates.
2. Discuss how electronic devices have changed modes of communication.
3. Describe a mode.

Think

4. Identify the mode of communication in these examples: A computer program tells a robot how to paint a car; a computer spell-checker finds errors in a report; a friend passes a note.

Practice Academic Skills

English Language Arts/Writing

5. Use a computer graphics program, scanner, magazine pictures, and your words to create a Web page or a large poster. The theme of the project is: "Communication happens between people and machines."

STEM Mathematics

6. There are different terms used to describe the amount of memory storage on a computer hard drive—byte, kilobyte, megabyte, and gigabyte. The smallest unit is the byte, the basis for the kilobyte, equaling 1,000 bytes. Most personal computers come with memory storage measured in gigabytes: 1,000 kilobytes = 1 megabyte. Using the same ratio of bytes to kilobytes, how many megabytes make up one gigabyte? One terabyte?

Math Concept **Numbers and Operations: Ratios** A ratio compares two numbers.

1. Notice that 1 equals 1,000 in each ratio.
2. Substitute 1 megabyte for one kilobyte to find how many megabytes make up a gigabyte.
3. Substitute 1 megabyte for one gigabyte to find how many gigabytes make up a terabyte.

For help, go to **glencoe.com** to this book's OLC and find the Math Handbook.

section 9.3

Impacts of Communication Technology

Reading Guide

Connect How has communication technology changed the way people communicate?

Content Vocabulary
- personal privacy
- biometrics
- tolerance

Academic Vocabulary
- method
- invest

Graphic Organizer
Draw the section diagram. Use it to organize and write down information as you read.

Impacts of Communication Technology

	Technology #1 ______	Technology #2 ______
Positive		
Negative		

Go to **glencoe.com** to this book's OLC for a downloadable graphic organizer and more.

TECHNOLOGY STANDARDS

STL 4 Cultural, Social, Economic & Political Effects

STL 5 Environmental Effects

ACADEMIC STANDARDS

English Language Arts

NCTE 1 Read texts to acquire new information.

Science

NSES Abilities necessary to do scientific inquiry.

STL *National Standards for Technological Literacy*

NCTM *National Council of Teachers of Mathematics*

NCTE *National Council of Teachers of English*

NSES *National Science Education Standards*

NCSS *National Council for the Social Studies*

The Impacts of Technology

How has communication technology affected the world?

When people say the world is getting smaller, they mean technology allows us to communicate instantly with almost anyone anywhere. Communication technology is neither good nor bad, but the use of its products and systems can have good and bad consequences. Political, social, cultural, economic, and environmental issues are influenced by communication technology.

Political Impacts

What is one way that communication technology affects politics?

Political decisions and world news are brought to you daily via 24/7 broadcasts, newspapers, magazines, and the Internet.

As You Read

Predict How might communication systems help people of different cultures get along with each other?

BusinessWeek Tech NEWS

Building a Super Cell Phone

The newest wave of tech entrepreneurs is transforming our mobile phones into personal computers. "The most common digital device in the world is the cell phone," says Motricity's chief technology officer. Young entrepreneurs are looking to social networking, Internet video, and online photo-sharing. They extend those applications by putting them on cell phones.

Critical Thinking *What modes of communication would apply to a super cell phone? Why?*

Go to **glencoe.com** to this book's OLC and read more about this news.

For the presidential nomination and the 2008 election, the candidates raised and spent more than $1 billion to communicate their messages. With all this information, it is difficult to separate the best candidate from the one who has the most money.

Social and Cultural Impacts

How important is communication technology to you?

Experts sometimes refer to you and your friends as *Generation @*, or the *wired generation*. It is possible that your social network depends on communication technology. You may use different methods of communication, such as cell phones, e-mail, instant messaging, social networking, interactive gaming, blogging, and podcasting.

In addition, communication technology has affected other aspects of society and culture. Education, personal privacy, and tolerance for other cultures have all been influenced by communication technology.

Education

Knowledge was at one time passed on from one person to another by word of mouth. The key to the advancement of education and our technology was the development of communication technology. It helped us to learn from the achievements of people who lived far away or long ago.

Before the invention of printing, only churches, royalty, and the very wealthy owned books. Printing made it possible for more people to own books. The growth of the printing industry was very important to education and the spread of technology.

As printing developed and literacy increased, people began to learn from other people. Today so much information is available that our era is called "The Information Age." Knowledge can promote equality across cultures and provide new opportunities.

Personal Privacy

Personal privacy is the right of individuals to keep certain information from public view. Invasion of personal privacy can be a form of harassment, which is repeated hostile annoyance. Banking, stock market transactions, and credit card purchases are monitored by communication systems. Personal information about individuals is on the Internet.

Biometrics is the science of measuring a person's unique features, such as fingerprints, facial features, voice, and retina of the eye. A biometric reader converts your fingerprints or facial

Ethics in Action

Misinformation on the Net

Mistakes, deliberate falsehoods, and personal opinion appear in newspapers, books, the Internet, and magazines everyday. This can lead to harassment. Both ethical and unethical people use communication media.

Just the Facts How can you protect yourself from misinformation? Stay informed using many sources. For example, during an election, consider what each side says.

English Language Arts/Writing

Comparing News Select a news event and compare the coverage that it receives in a newspaper, a TV news broadcast, and a news Web site.

1. Use a spreadsheet to chart and compare how each covers the story.
2. Write a paragraph summarizing your conclusions.

other images on file to identify you. Retina scans and voiceprints may become as common as fingerprints for identifying people. They may even replace keys and locks. Biometric security systems make it harder for thieves to take money from your bank account or use your credit card, but they may reduce privacy.

Tolerance

Our communication systems have made us more aware of the different cultures that share our planet. Through television and the Internet, you have been invited into the homes of families of different cultures, religions, and nationalities. They, in turn, have learned about us. The awareness of the customs and traditions of others can promote **tolerance**, which is an acceptance of others. Tolerance can reduce discrimination, which is most often negative selective treatment due to intolerant attitudes.

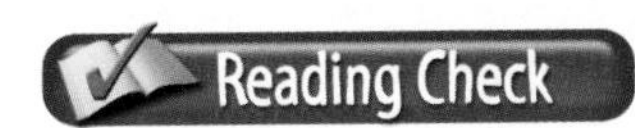

Examine What might be a positive effect of communication technology? Why?

Economic Impacts

How has communication technology affected the world economy?

Because communication technology is so fast today, news travels around the world at lightning speed. What happens to our stock market today affects global stock markets within hours instead of days. People may invest money in certain stocks and cause those stock prices to increase or decrease. Communication technology has changed how products are advertised or sold via Web sites.

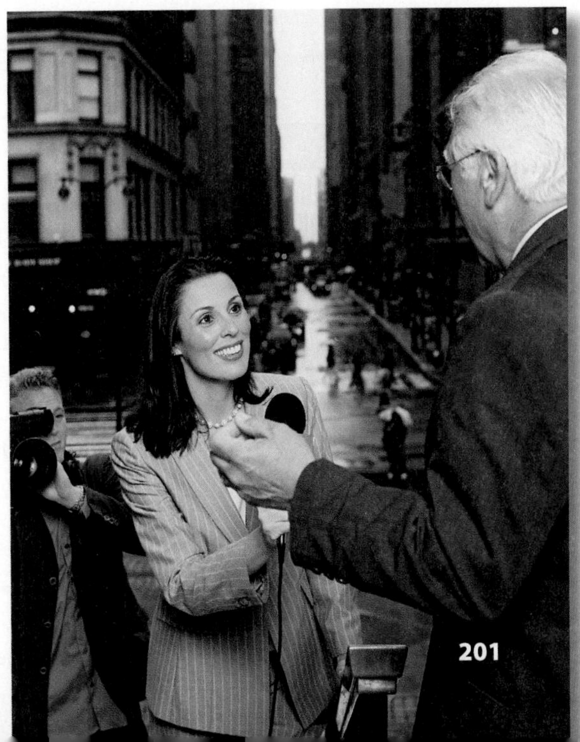

Global News Television news reporters report and broadcast live from almost anywhere in the world. *Do you think technology has improved understanding of other cultures? Why or why not?*

Blending In The tree on the right is a cell tower with fake branches to make it look real. *Why might a cell phone company design this?*

Environmental Impacts

How has the environment changed due to communication technology?

You have already learned that systems consist of input, process, output, and feedback. Unfortunately, many processes produce what we want, but they may also harm our environment.

Many communication systems use paper, which is made out of trees. At one time trees were cut down to feed paper mills without concern for future needs. The computer, copy machine, and fax machine increased our demand for paper. Today paper manufacturers have replanting programs to replace trees.

Papermakers also must not pollute nearby rivers and streams with waste. Many chemicals, metals, and plastics are used to manufacture communication equipment. Our environment is harmed when equipment is not recycled or disposed of properly.

Power lines and transmitters that carry our communication signals affect the appearance of communities. The acronym NIMBY stands for Not In My Back Yard. This slogan is identified with local residents who wish to prevent construction of things they find unsightly or dangerous, such as cell towers and power lines.

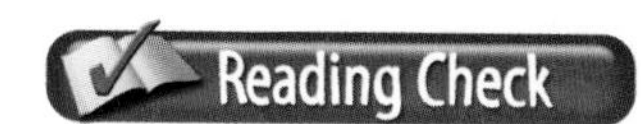

Describe What is one disadvantage of communication technology?

Economics and the Environment

Cost and profit are important to businesses. Environmental and economic concerns may compete. People may not be aware of the dangers of a new system until it is used. Government agencies and consumer groups try to protect us from these dangers.

section 9.3 assessment

Self-Check

1. Describe who owned books before printing.
2. Explain how communication technology has made our world smaller.
3. List two negative and two positive effects of communication technology.

Think

4. World events appear on TV as they happen. Discuss how this instant communication affects our world.

Practice Academic Skills

English Language Arts/Writing

5. In your words, write some definitions for the word *biometric*, its prefix, *bio*, and its base word, *metrics*.

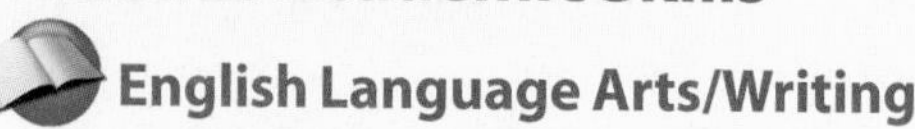

Science

6. If something is biodegradable, living organisms can break it down into materials that do not harm the environment. Most packaging material is plastic. Some plastics can be recycled but are not biodegradable. Design a biodegradable package for a product.

Exploring Careers in Technology

Michael Eisenberg

SOUND ENGINEER

Q: *What do you do?*

A: I am a freelance sound designer and engineer for theatrical and performing arts events. My typical day includes changing the sound engineering design to ensure that all elements of a production run smoothly and to make sure that the crew understands and executes my design.

Q: *What kind of training and education did you need to get this job?*

A: In high school and later in college, I took several courses in physics, calculus, and electrical engineering. I use math almost everyday to make decisions, like where to locate the sound equipment to achieve the most efficient sound design, while using the least number of microphones and preserving the sound quality.

Q: *What do you like most about your job?*

A: I enjoy being able to help enhance the performance experience of an audience. I am always curious about the audience's likes and dislikes, and I want them to hear something different.

Q: *How did you get interested in your job?*

A: I became extremely interested in sound design and mixing in high school. The theater teachers at my school sparked my interest and gave me opportunities and an outlet to do something that I felt I could call my own.

English Language Arts/Writing

Career Search Create a list of ten careers that require speaking and listening skills.

1. Using a spreadsheet program, make a chart of the different careers, including their characteristics.
2. Select the career that interests you the most.
3. Research the type of formal education you would need to pursue the career you chose.

Go to **glencoe.com** to this book's OLC to learn more about this career.

Real-World Skills

Speaking, listening, problem-solving

Academics and Education

Physics, mathematics, English language arts, music, electrical engineering

Career Outlook

Growth as fast as average for the next ten years

Source: *Occupational Outlook Handbook*

chapter 9 Review and Assessment

Chapter Summary

Section 9.1 The communication system includes input, which is the message; the process, or how the message moves; output, or receiving the message; and feedback, which includes information about clarity. Communication subsystems have a source, an encoder, a transmitter, a receiver, a decoder, and a destination. Biological, graphic, wave, and telecommunication are all forms of communication.

Section 9.2 The modes of communication include people to people, people to machines, machines to people, and machines to machines. Communication technology has made new modes of communication possible through graphic communication systems, print, and modes based on electrical signals.

Section 9.3 Positive impacts of communication technology are rapid mass communication systems that report news and politics. More information is available for making choices. People gain a better understanding of cultures. Technology may improve security systems. Negative impacts might be invasion of personal privacy. Using more paper for printouts, chemicals for manufacturing, and using electrical systems may be harmful.

Review Content Vocabulary and Academic Vocabulary

1. On a sheet of paper, use each of these terms and words in a written sentence.

Content Vocabulary
- communication technology
- graphic communication
- sound waves
- electromagnetic carrier wave
- telecommunication
- mode
- machine-to-machine communication
- personal privacy
- biometrics
- tolerance

Academic Vocabulary
- technology
- factor
- response
- process
- method
- invest

Review Key Concepts

2. **Identify** the elements of a communication system.
3. **Define** personal privacy.
4. **Give** two examples of biometric scanning.
5. **Discuss** how biometric scanning leads to less personal privacy.
6. **Explain** sound waves.
7. **Explain** social networking Web sites.
8. **Describe** NIMBY.
9. **Explain** why is this era is sometimes called "The Information Age."
10. **List** uses for geosynchronous satellites.

Real-World Skills

11. Listen Actively Practice active listening during a conversation with a classmate, friend, or family member. Pay close attention to body language, tone of voice, speed, and volume, and try to understand the message. Respond to the speaker with comments or questions. After the conversation, write a paragraph describing the experience.

STEM Technology Skill

12. Buying a Bicycle Many buyers use the Internet to find new and used products in certain price ranges.

a. Use the Internet to find Web sites that offer new and used bikes for sale. Research information about the cost of a dirt bike, racing bike, or another kind of bike.

b. How much is the same model that is three years old and one that is five years old?

Academic Skills

Social Studies

13. Write a report about what your community is doing to recycle paper. Interview school, town, and civic leaders. Write a plan for a school-wide recycling program with their assistance.

STEM Mathematics

14. Create a pie chart with the following data for a presentation on your company's spending. 65 percent employee salaries, 15 percent rent, 10 percent utilities expenses, 8 percent maintenance, and 2 percent miscellaneous expenses.

Math Concept **Pie Charts** A pie chart illustrates data as pieces of the whole. The size of a pie piece corresponds to the percentage of the whole that one piece of data represents. Hint: Draw a circle and divide it into the number of percentages listed.

TSA WINNING EVENTS

TECHNOLOGY STUDENT ASSOCIATION

Graphic Designer

Situation You are preparing for a competition that applies communications technologies. Assume the role of a graphic designer and work creatively under constraints to design a solution to a problem.

Activity Design and produce several types of graphic communications that promote a TSA chapter's activities.

Evaluation You will be evaluated on how well you meet these performance indicators:

- Design and produce a newsletter.
- Design and produce an effective sponsor support request on chapter letterhead.
- Design and produce a business card for your chapter.

Go to **glencoe.com** to this book's OLC for information about TSA events.

Standardized Test Practice

Directions Choose the letter of the best answer. Write the letter on a separate piece of paper.

1. What element is not part of a communication subsystem?

A decoder **C** transmitter

B encoder **D** retriever

2. Electrical signals are the basis of a mode of communication.

T

F

Test-Taking Tip In a multiple-choice test, answers should be specific. Read the questions first, and then read all answer choices. Eliminate incorrect answers.

chapter 9

TECHNOLOGY LAB

Create an Ad

For any invention to become important, people must know about it. They must also be convinced that they need it. If you wanted to tell everyone about a new invention, what mass communication system would you use? In this lab, you will get the chance to design an ad using the communication system of your choice.

Tools and Materials

- ✓ Markers, pens, pencils
- ✓ Posterboard
- ✓ Scissors
- ✓ Glue
- ✓ Magazines, books
- ✓ Computer and printer
- ✓ Presentation software
- ✓ Graphics software
- ✓ Scanner
- ✓ Copy machine
- ✓ Video camera

Set Your Goal

Your goal for this lab is to create an advertisement for a brand new communication device. Some examples might include an iPhone, plasma TV, iPod, or the latest video game. You can also advertise your own futuristic invention. Use the keywords *future* or *new communication* to search the Internet for more ideas.

Know the Criteria and Constraints

In this lab, you will:

1. Work in a group of two or three, but no more than four people.
2. Create an interesting print ad, a Web page, a PowerPoint presentation, or a short video advertisement.
3. Design an ad that is serious or funny!
4. Include this information in your ad:
   - The inventor of the device
   - When it was invented
   - What it does
   - Why it is important to the development of technology

Design Your Project

Follow these steps to design your project and complete this lab.

1. Choose the invention that your group wants to advertise.
2. Choose the communication method that you will use—print, Internet, presentation software, or video.
3. Brainstorm the theme of your ad. You can study advertisements that you have seen. Be creative and have fun!
4. Say in your ad who invented the device, when the invention was invented, what it does, and why it is important to the development of technology.
5. Produce your ad using the communication method you chose. You can:
   - Paste up your artwork and text for your graphic ad
   - Use a computer to make a Web page
   - Use PowerPoint presentation software, or
   - Use a video camera to shoot your commercial.
6. Share your advertisement with the class.

Evaluate Your Results

After you complete this lab, answer these questions on a separate piece of paper.

1. What do you think makes a new technological development successful?
2. What are the input, process, and output phases of the communication system that you used for your ad?
3. If you were to redo your ad, what would you do differently? Why?

Academic Skills Required to Complete Lab

Tasks	English Language Arts	Math	Science	Social Studies
Research new and future communication devices.	✓		✓	✓
Gather sample advertisements.	✓		✓	✓
Write ad copy that communicates effectively.	✓			
Create presentation.	✓	✓		
Present ad to the class.	✓			

chapter 10

Computer Technologies

Sections

What You'll Learn

- **Identify** the main parts of a computer system.
- **Explain** why binary code is important to computer function.
- **Name** several computer input and output devices.
- **Describe** artificial intelligence and identify ways in which it can be used.
- **Discuss** wi-fi, WiMAX, and distributed computing.

Explore the Photo 

Anywhere You Want It The first computers were the size of large rooms. They performed simple operations and consumed massive amounts of energy. Today's personal computers are powerful devices that fit on your lap and can connect to the Internet wirelessly. *What part of this computer is the input device?*

Program a Computer to Control a Machine

At the end of this chapter, you will be asked to build a motorized robotic machine. You will then write a program that will control your robot. Get a head start by using this checklist to prepare for the Technology Lab.

PROJECT CHECKLIST

- ✓ Do some Internet research on robots that other students have created.
- ✓ Read the LEGO Mindstorms™ Education NXT system user's guide.
- ✓ Do the control experiment that comes with the system.

section

10.1

Computer Systems

Reading Guide

Before You Read **Preview** How much do you already know about computer systems?

Content Vocabulary

- CPU
- ROM
- RAM
- program
- operating system
- binary code
- computer virus

Academic Vocabulary

You will see these words in your reading and on your tests. Find their meanings at the back of this book.

- component
- function

Graphic Organizer

Draw the section diagram. Use it to organize and write down information as you read.

Acronym	Complete Phrase
CPU	
RAM	
ROM	

Go to **glencoe.com** to this book's OLC for a downloadable graphic organizer and more.

TECHNOLOGY STANDARDS

STL 2 Core Concepts of Technology
STL 3 Relationships & Connections
STL 9 Engineering Design
STL 11 Design Process

ACADEMIC STANDARDS

Science
NSES Content Standard B Motion and forces

Mathematics
NCTM Number and Operations Understand numbers, ways of representing numbers, relationships among numbers, and number systems.

STL *National Standards for Technological Literacy*
NCTM *National Council of Teachers of Mathematics*
NCTE *National Council of Teachers of English*
NSES *National Science Education Standards*
NCSS *National Council for the Social Studies*

The Central Processing Unit (CPU)

What are the three main parts of the CPU?

Computers are part of our daily lives. Computers control automobile engines, DVD players, and cell phones. In the future, you will use computers to make informed personal, business, and governmental decisions. In order to do well in our society, it is important that you become computer literate.

A computer is an electronic device that can calculate, store, and process data. A computer system is made of different components. (See **Figure 10.1.**) Each component has a special job to do.

As You Read

Predict How does a computer know how to run a program?

Basic Operations

Inside your computer are tiny pieces of silicon called "integrated circuits," or "microchips." Integrated circuits have many electrical circuits burned into them. The circuits act like switches. Sometimes they let electricity flow. Other times they shut it off.

The **CPU** (central processing unit) is the largest and most important integrated circuit on the computer's motherboard. It performs all basic operations. A CPU is like a highway system. Information travels the circuits of the CPU to get processed.

The CPU's control unit guides the flow of information. The arithmetic/logic unit performs mathematical calculations with data sent by the control unit. The memory unit stores that information before and after processing. These three parts of the CPU work together.

ROM and RAM

The CPU has two types of memory. **ROM**, or read-only memory, contains the basic information that the computer needs to perform any operation. It is permanent and cannot be changed, deleted, or erased. The other type of CPU memory is **RAM**, or random access memory. All data that you feed to your computer is put into its RAM. The RAM temporarily stores data and software instructions. When you turn your computer off, this information is lost.

Compare What is the difference between ROM and RAM?

Figure 10.1 **A Basic Computer System**

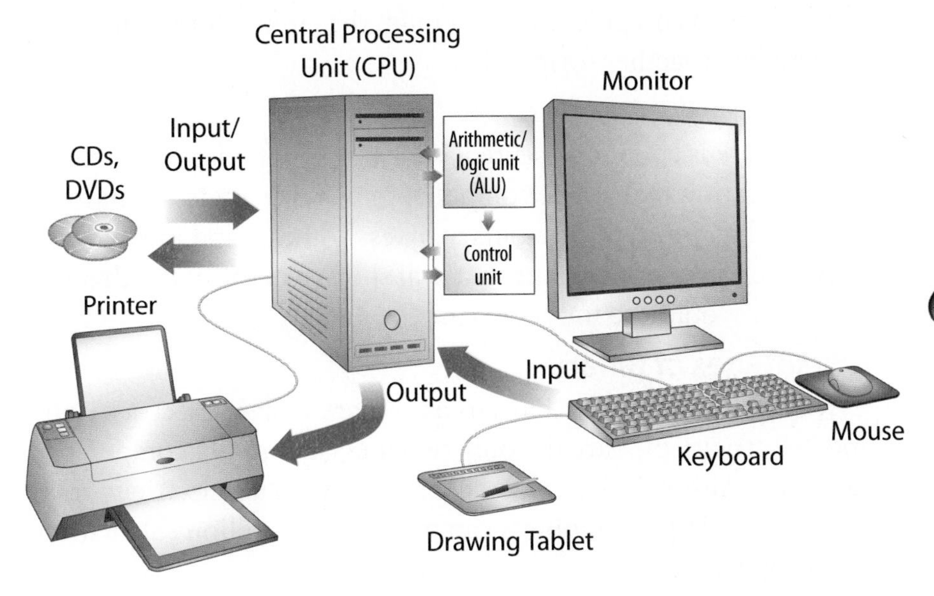

Plugging In A basic computer system includes the CPU, monitor, printer, keyboard, and other useful input and output devices. *Why are CDs and DVDs listed as input and also output devices?*

The Motherboard The CPU is located on the computer's motherboard. Memory chips, expansion slots, and controllers are also on the motherboard. *Why would a computer manufacturer include expansion slots?*

Computer Programs

What is the function of a computer program and how does a computer understand it?

A computer **program**, or software, is a set of instructions that the computer follows to do its work. The program controls the computer. It tells the CPU exactly how to handle all the data that is entered into the machine. The program turns the computer into a game machine or tells it to perform other functions. Computer programs are often stored on the computer's hard disk drive.

Operating Systems

Many components inside the computer have **operating system** programs that are permanently burned into their microchip circuits. When the computer is turned on, these programs first tell the computer how to run its hardware. In a sense, each time you turn your computer on, it reads an entire instruction book on how to operate.

Binary Code

The computer program and all the information that the computer will use must be converted into **binary code**, which is code that the computer can understand or read. Binary code uses only the numbers 1 (one) and 0 (zero). Each 1 (electricity on) or 0 (electricity off) is a bit.

Bits

A bit is the smallest piece of information that a computer can use. A computer sends or receives these 1s and 0s in the form of small electrical pulses. A 1 means a pulse flows through the circuit; a 0 means no pulse flows through the circuit. Stringing eight of these bits together forms a byte.

Bytes

Each byte is code represented as 1s and 0s for a letter, number, or punctuation mark. The computer converts every letter and number that you type into binary bytes. You can find free binary conversion programs online that will show you the codes that a computer would use for your name.

Advantages of Binary

Why does a computer have to use binary code? "Electricity on" and "electricity off" are the only two messages that a computer can sense. Although binary code seems slow to us, electricity travels very fast. Supercomputers can make over a trillion calculations per second.

Computer Viruses

A **computer virus** is a set of destructive instructions that someone has written and hidden inside a Web page, an e-mail attachment, or even a computer image. Viruses can take different forms. Some are merely annoying. Others do serious damage.

Most viruses are passed from computer to computer by users whose computers do not have proper virus protection. If your computer's virus protection is not up to date, your computer can catch viruses. You could unknowingly pass them on to friends and family through e-mails, file sharing, or by trading homemade CDs and DVDs.

Spyware and Worms

Spyware is a kind of virus that teaches your computer to spy on you, then secretly sends this information back to its creator. If the virus quickly duplicates itself and worms its way into every memory location in your computer, it is called a "worm."

Trojan Horses

A "Trojan Horse" is not a virus. It often takes the form of a fake e-mail that looks like it came from a bank, a major company, an official from a foreign country, or from a Web site you like to visit. If you click to respond, it will take you to a fake log-in location or provide questions designed to steal information about your personal identity and accounts.

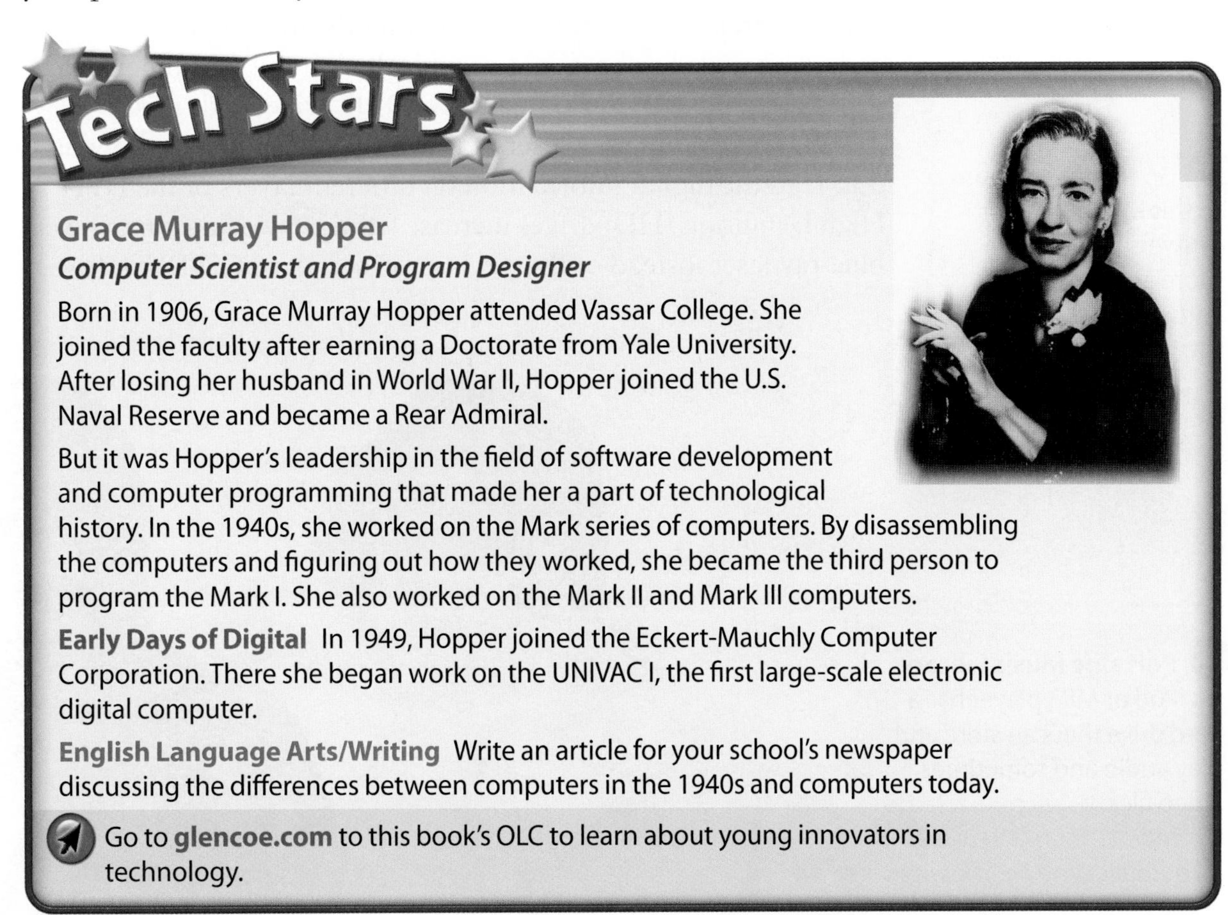

Tech Stars

Grace Murray Hopper

Computer Scientist and Program Designer

Born in 1906, Grace Murray Hopper attended Vassar College. She joined the faculty after earning a Doctorate from Yale University. After losing her husband in World War II, Hopper joined the U.S. Naval Reserve and became a Rear Admiral.

But it was Hopper's leadership in the field of software development and computer programming that made her a part of technological history. In the 1940s, she worked on the Mark series of computers. By disassembling the computers and figuring out how they worked, she became the third person to program the Mark I. She also worked on the Mark II and Mark III computers.

Early Days of Digital In 1949, Hopper joined the Eckert-Mauchly Computer Corporation. There she began work on the UNIVAC I, the first large-scale electronic digital computer.

English Language Arts/Writing Write an article for your school's newspaper discussing the differences between computers in the 1940s and computers today.

Go to **glencoe.com** to this book's OLC to learn about young innovators in technology.

Computers without Computers

Imagine life without personal computers. All you need is a small handheld device to access online programs and store all of your files on the Internet. Companies like Google already offer photo storage, word processing, and spreadsheet applications as free Web-based services. In the future, you may be able to do everything you now do on a personal computer with very little hardware. *What might be the advantages and disadvantages of using online software and online storage space?*

Go to **glencoe.com** to this book's OLC for answers and to learn more about online software and storage.

Anti-Virus Software

The FBI reports that viruses and spyware crimes, and the software to protect against them, cost billions of dollars each year. Anti-virus program packages usually include a firewall program to block known dangers and a virus program to find and destroy viruses that do get into your computer. These programs must be updated frequently to protect against new attacks.

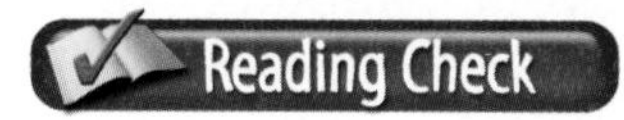

Recall Why is binary code important to the functioning of computers?

Disk Drives

What is the purpose of a computer disk drive?

The computer's disk drive allows data to be written to storage (recorded memory) or read from storage. If you were to look inside a hard-disk drive case, you would see a stack of round metal-oxide platters. Hard drives use electromagnetism to write messages onto these disks. When the playback head passes along the disk, it picks up this magnetic coded message. The message is then converted back into an electronic signal.

CDs and DVDs

Commercial CDs and DVDs contain information in binary code. This code has been microscopically burned into the plastic disk as tiny pits. A laser on your CD/DVD drive reads this.

DVD drives contain a laser that can change its focus, allowing it to read the digital information on different layers of the DVD. High Definition (HD) drives increase storage by using a narrower, blue-ray laser instead of the red laser in standard CD/DVD drives.

Portable Music Library An iPod or MP3 player has a hard drive that can store and play audio and sometimes video files. *When connected to a computer, is an MP3 player an output device or an input device?*

Input Devices

What kinds of devices send data to computers?

Any device that can send information to a computer is an input device. The list includes disk drives, keyboards (for computers and musical instruments), mouses, joysticks, scanners, drawing tablets, touch screens, digital cameras, and video cameras. Input devices also include specialized equipment used in scientific research, engineering, medicine, industry, and music.

Converting Input

Each input device must have a way of converting its data into the binary code that a computer can process. When you press down on a keyboard letter, you cause contacts that are under the key to send the binary code for that letter. This coded message is sent into a memory location by the CPU and onto your computer screen. The CPU controls and uses this binary information according to the instructions provided by the program.

A Electronic Music This synthesizer is both an input and an output device. MIDI (Musical Instrument Digital Interface) technology allows computers and electronic musical instruments to communicate. *How do you think the musical information changes as it is sent on to the computer's CPU?*

Input and Computer Screens

The surface of a computer screen is divided into horizontal and vertical coordinates. They are similar to the horizontal and vertical coordinates on a world map that we call latitude and longitude. The computer uses these coordinates to locate things on the screen.

When you move a mouse, the row and column location of the pointer changes. The input device converts that movement into a binary electronic signal that the computer can understand.

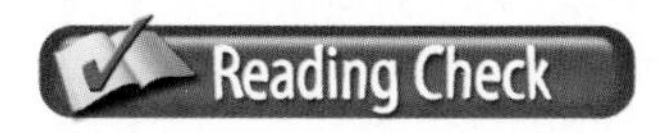

List What are some examples of input devices?

Output Devices

What devices are used for output?

Any device that can receive information from a computer is an output device. This includes the computer's disk drives, monitor, printers, speakers, sound card, video card, headphones, and music synthesizers. Many other specialized output devices have been created for use in science, engineering, medicine, and industry.

Input/Output Devices

Did you notice that disk drives and music synthesizers are also included on the input device list? Any device that can send and also receive data from a computer is considered to be an input/output device.

Figure 10.2 Color Ink-Jet Printers

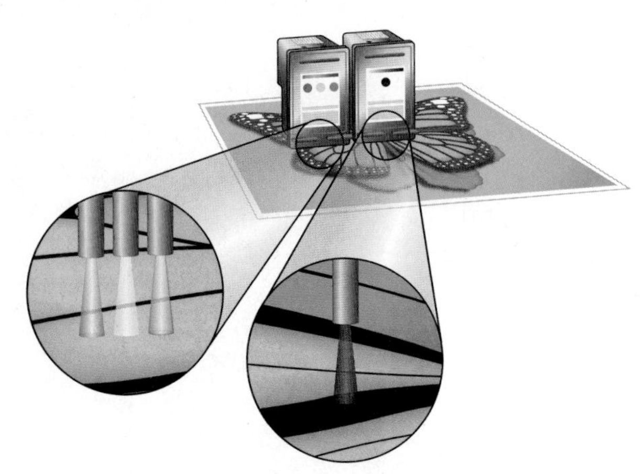

Primary Colors Color ink-jet printers have three primary color ink jets, plus a black jet. *What four color inks produce a color image?*

Printers

The image on your computer screen is called a "soft copy" because it is temporary. To make permanent copies of computer-generated material, you can attach a printer to your computer. The permanent copies are called "hard copies."

Printers usually form letters as a series of dots. The CPU tells the printer the exact pattern and how many dots to print. The dots produced by a laser printer are spaced so close together that it is impossible to see them individually. The signal that determines the letter is the trigger for a laser beam.

Color Printing

Ink-jet printers print by squirting small dots of ink onto the paper. (See **Figure 10.2.**) Many color ink-jet printers use multiple ink cartridges. Only the black cartridge is used when you print in black. One or more additional cartridges contain the three primary ink colors. They work in combination with the black cartridge to print every color of the rainbow.

section 10.1 assessment

After You Read Self-Check

1. Describe the purpose of a computer program.
2. Identify the two numbers used in binary code.
3. Explain the difference between ROM and RAM.

Think

4. List some effects of computers on our society.

Practice Academic Skills

English Language Arts/Writing

5. Design the ideal computer. List its features and create a mock-up of the computer. Create an advertising poster. Prepare a presentation for the class.

STEM Mathematics

6. Here is a place value chart for the decimal number system. It is based on 10s. It uses these counting numbers: 0, 1, 2, 3, 4, 5, 6, 7, 8, and 9. What does the 1 in 1,000 stand for?

thousands	hundreds	tens	ones
1000	100	10	1
10^3	10^2	10^1	10^0

Create a similar place value chart for the binary number system based on 2s. A binary system uses only these counting numbers: 0 and 1. What does the 1 in 1000 stand for in the binary system?

eights	fours	twos	ones
1000	100	10	1
2^3	2^2	2^1	2^0

Math Concept **Number Systems** The binary system works just like the decimal system except that each place value is a power of 2 instead of a power of 10.

For help, go to **glencoe.com** to this book's OLC and find the Math Handbook.

section 10.2

Computers on the Cutting Edge

Reading Guide

Before You Read **Connect** What will computers be like in the future?

Content Vocabulary

- AI
- expert system
- wi-fi
- WiMAX
- distributed computing

Academic Vocabulary

- error
- analyze

Graphic Organizer

Draw the section diagram. Use it to organize and write down information as you read.

Distributed Computing Projects

1. ______
2. ______
3. ______

Go to **glencoe.com** to this book's OLC for a downloadable graphic organizer and more.

TECHNOLOGY STANDARDS

STL 1 Characteristics & Scope of Technology
STL 7 Influence on History
STL 9 Engineering Design
STL 11 Design Process
STL 17 Information & Communication Technologies

ACADEMIC STANDARDS

English Language Arts
NCTE 11 Participate as members of literacy communities

Science
NSES G Historical Perspectives

STL *National Standards for Technological Literacy*
NCTM *National Council of Teachers of Mathematics*
NCTE *National Council of Teachers of English*
NSES *National Science Education Standards*
NCSS *National Council for the Social Studies*

Artificial Intelligence

What is artificial intelligence?

How do you think the computer might evolve during your lifetime? If computers continue to develop at their current rate, will they surpass humans at most tasks?

People sometimes talk about the intelligence of computers. They are not really intelligent at all. A computer gets no satisfaction when it solves a problem. It can only run programs and process data. Section 10.1 explains how computers follow instructions, so that computer error is often really human error. "Thinking" computers are, at this time, just "science fiction."

As You Read

Infer Do you think computers make decisions on their own?

Academic Connections
Language Arts

Language Challenged U.S. military troops sent to other countries are equipped with hand-held computers that use speech recognition software. When a word in English is spoken into the *Phraselator*, it shows the word on a small screen, then translates it and broadcasts it in the chosen language.

Apply Do you think a common language spoken everywhere in the world would be a good thing? Write a paragraph giving reasons for your opinion.

AI Programs

AI (artificial intelligence) programs, however, give the impression that a computer can think. The programmer has provided the computer with a number of answers that will be triggered by certain requests.

Video and Computer Games

In video and computer games, AI programs control the characters who are not controlled by the player. These characters seem to make their own decisions. If the programmer has given the computer a wide range of responses, the computer-controlled characters will seem more real and complex.

Expert Systems

Some artificial intelligence programs are called **expert systems**. In these systems, information from experts in a particular field is stored in a computer's memory. When the computer is asked a question, it uses this information to answer correctly. A medical expert system, for example, might diagnose diseases. If a doctor or nurse provides it with a list of symptoms, it will match that list against all known diseases.

Deep Fritz

In 2006, Vladimir Kramnik, the world chess champion, played Deep Fritz, which is now considered the world's top chess computer program. Deep Fritz is capable of "thinking" millions of moves per second. How many moves a human can examine per second is unknown. Deep Fritz won this competition, which raises the question: Will there come a time where human intelligence cannot match the AI programs it has created?

Immobots

Another new development in AI is immobots, or "immobile robots." Developed by researchers at the Massachusetts Institute of Technology, most immobots do not move around, but they control a machine that probably does.

Human vs. Machine World chess champion Vladimir Kramnik tries to prove the superiority of the human brain over the computer brain. Deep Fritz, the computer, won the match. *If Deep Fritz is an expert system, what kind of experts contributed to its memory?*

Ethics in Action

Wi-Fi: Hijacking or Sharing?

Many users leave their wireless networks unsecured. Anyone can access these networks and surf the Internet for free. Some people think there is nothing wrong with this. The laws regarding wireless network access are unclear.

No Free Lunch If you do use other people's wireless networks, they could be charged for using more bandwidth. Also, you might cause their connections to run slowly or prevent them from getting online.

English Language Arts/Writing

Analogies We use analogies to compare words and ideas. We can also use analogies to help us decide if something is ethical or not.

1. Michael's neighbor leaves his wireless network unlocked, so Michael uses it. Cathy's neighbor leaves her front door unlocked, and so Cathy goes inside her house.
2. Compare these two situations. Are Michael and Cathy being ethical? Write an answer in two paragraphs.

Most AI software programs follow a long list of complex rules to solve problems. Unfortunately, it is hard to think of all the possible difficulties that could occur with complex machinery and write a rule for them.

Immobot software is different. The word *immobot* means "immobile robot." It includes a model of the machine's system. When a problem occurs, the immobot studies the system, finds the source of the problem, and determines a way around it. The goal is for the immobot to respond to unexpected situations on its own and to learn from its experience.

Speech Recognition

AI is also being used for speech recognition. The average computer does not have the processing power or the noise suppression tools needed for speech recognition software to work perfectly. However, researchers are combining vision inputs with sound inputs so that a computer can read your lips as it listens to what you say. The gradual improvements to computers and speech programs combined with computer vision may eventually free you from your computer keyboard.

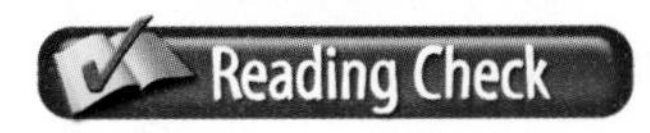

Identify What are some uses for artificial intelligence?

Wireless Computing

What is the difference between wi-fi and WiMAX wireless computing?

Wi-fi and WiMAX can both provide a wireless connection to printers, media readers, external hard drives, and almost any other piece of hardware that once required a cable.

EcoTech

High-Tech Waste

The rapid advancement of technology means that electronic devices become obsolete faster than ever. Most monitors contain lead; CPUs contain mercury; and batteries contain cadmium—all toxic materials. Monitors and other "e-waste" must be disposed of in a specialized manner.

Try This Look up the location of a local disposal center for electronic goods. Take your old and unused e-waste to the center.

Wi-Fi

Wi-fi is a short range wireless connection to the Internet. Based on wireless radio, wi-fi allows you to access your network and the Internet from anywhere within a range of a few hundred feet from a base station. With the right equipment and software, computers with wi-fi can be online out of doors and in public areas.

WiMAX

WiMAX is a long-range wireless connection to the Internet. WiMAX is also based on wireless radio. However, your computer can be 30 miles from the radio tower that is providing the signal. Individual tower transmitter signals mesh together, so you and your computer can be on the move without losing a signal.

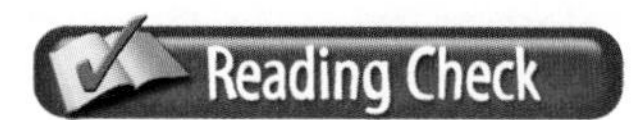

Compare What is the difference between wi-fi and WiMAX?

Distributed Computing

Can your computer help search for extraterrestrial intelligence?

Millions of computer users donate their computer downtime for science, mathematics, and technological research. **Distributed computing** allows networks of computers worldwide to analyze research data to help solve problems. Millions of computers have more processing power than the most powerful supercomputers.

section 10.2 assessment

After You Read Self-Check

1. Explain how immobots are different from AI systems.
2. Define an expert system.
3. Describe wi-fi and WiMAX.

Think

4. Compare and contrast artificial intelligence and human intelligence.

Practice Academic Skills

STEM Science

5. Working in groups, research different distributed-computing projects. Select the topics that interest the majority of students and set up a distributed-computing screen-saver program in your classroom.

STEM Mathematics

6. Did you know that 512 IBM server computers have the same processing power as 8,192 Apple Power PC computers? Using mental math, estimate how many Apple Power PC computers are equal in processing power to one IBM server computer. Write a sentence or two telling how you estimated the number.

Math Concept **Rounding to Estimate** Rounding numbers to a convenient place value can help you compute mentally. When rounding numbers, look at the digit to the right of the place to which you are rounding. If the digit is 5 or greater, round up. If it is less than 5, round down.

For help, go to **glencoe.com** to this book's OLC and find the Math Handbook.

Exploring Careers *in* Technology

Nate Couture

COMPUTER FORENSICS SPECIALIST

Q: *What is a typical day like at your job?*

A: I don't have a typical day. When I'm wearing my network administrator hat, some of my duties include server maintenance and installation, network infrastructure administration, and storage administration. As an IT security officer, I develop security policies, evaluate security concerns, monitor network weaknesses and attacks, and apply new solutions to help protect the college.

Q: *What kind of training and education did you need to get your job?*

A: As a network administrator, I lean heavily on my associate's degree in computer networking. My bachelor's degree in digital forensics improved my skills and allowed me to step into the role of IT security officer. I also did an internship with Vermont State Police. As an intern, I maintained the computer forensic equipment, modified the evidence-collection database, and conducted identity-theft research.

Q: *Why is your job important?*

A: Digital forensics and security are particularly important in higher education where there is more openness and personal freedom than might be found in the corporate world. My goal is to balance that openness with the goal of keeping the community safe from attacks on servers, worm outbreaks, Trojan infections, and attempts to steal sensitive data.

English Language Arts/Writing

Citing Sources The Internet makes it easy to find articles and other information for your research papers. It is very important, however, to cite the source when you quote someone else's material.

1. Use the Internet to research current trends in online use, such as social networking, blogging, or online classes.
2. Write a one-page report, using at least one quote from a specific source.
3. Make sure to attribute the quote to the person who said or wrote it.

Go to **glencoe.com** to this book's OLC to learn more about this career.

Real-World Skills

Problem solving, observation, speaking, and listening

Academics and Education

Mathematics, computer science, English language arts

Career Outlook

Growth faster than average for the next ten years

Source: *Occupational Outlook Handbook*

chapter 10 # Review and Assessment

Chapter Summary

Section 10.1 A computer is an electronic device that calculates, stores, and processes information. A computer program provides instructions that tell the computer what to do. Computer systems include input devices, such as keyboards, and output devices, such as monitors. All computers have two types of memory: ROM and RAM.

Section 10.2 Artificial intelligence programs can solve problems and make decisions ordinarily handled by humans. Expert systems gather information from experts in a particular field and store it in the computer's memory. Wi-fi and WiMAX are both wireless connections to the Internet, but WiMAX is newer and more long range. Distributed computing networks use many computers' downtime for a variety of research projects.

Review Content Vocabulary and Academic Vocabulary

1. On a sheet of paper, use each of these terms and words in a written sentence.

Content Vocabulary

- CPU
- ROM
- RAM
- program
- operating system
- binary code
- computer virus
- AI
- expert system
- wi-fi
- WiMAX
- distributed computing

Academic Vocabulary

- component
- function
- error
- analyze

Review Key Concepts

2. **Describe** a computer system's main parts.
3. **Explain** the purpose of integrated circuits.
4. **Identify** ROM and RAM and the difference between them.
5. **Tell** the difference between an operating system and a computer program.
6. **Summarize** the importance of binary code to computers.
7. **Describe** how viruses can damage a computer.
8. **List** some examples of computer input and output devices.
9. **Discuss** artificial intelligence and some ways it can be used.
10. **Explain** wi-fi, WiMAX, and distributed computing.

Real-World Skills

11. **Understanding Viruses** Go to your library and/or use the Internet to read about computer viruses and the live viruses that cause illness in your body. How are they alike and different? Write a few paragraphs explaining why they both share the name *virus*.

Technology Skill

12. **Lasers** Lasers are used to read the memory disks in computers. What other uses do they have? Research the various uses for lasers.
    - a. List uses for lasers in today's world.
    - b. Categorize the uses into groups such as Lasers in Computers, Lasers in Medicine, and Lasers in Space Science, etc.

Academic Skills

Social Studies

13. Research the history of the personal computer. Focus on one aspect, such as miniaturization or computer viruses. Create and display a timeline showing your research.

STEM **Mathematics**

14. Franco receives a joke e-mail that he forwards to his friends. The e-mail is infected with a computer virus. Of the 40 people in his address book, 30% forward the joke to 15 friends each. 25% of those friends forward the joke to 10 people each. How many computers will be infected with the virus? Explain.

Math Concept **Identifying Operations** In a word problem, find key words to determine what operations to use.

WINNING EVENTS

Computer Traffic Control Programmer

Situation Design a computer-controlled model of a traffic control system for school.

Activity Working as a team, brainstorm the design of your system. Develop rough sketches of roads and sidewalks surrounding your school and the sensors, lights, and other mechanisms you will use. Write the computer program which will control these devices. Build a scaled working model of your design.

Evaluation The model will be evaluated using the following criteria:
- Safety—reliable
- Intelligent—appropriate solution
- Bonus points—includes a subsystem that digitally photographs speeding cars

Go to **glencoe.com** to this book's OLC for information about TSA events.

Standardized Test Practice

Directions Choose the letter of the best answer. Write the letter on a separate piece of paper.

1. Which is NOT true about integrated circuits?
   - **A** They are made from silicon.
   - **B** They act like switches.
   - **C** Each chip has a single circuit burned into it.
   - **D** They are called microchips.
2. Data fed into your computer as you use it is stored in RAM.

   T **F**

Test-Taking Tip If each item on a test is worth the same number of points, do not spend too much time on questions that are confusing.

chapter 10

TECHNOLOGY LAB

Program a Computer to Control a Machine

It took less computer power than there is in today's automobiles to land the first people on the moon. Today computers control the fuel system, engine, and many other parts of a car. Have you ever programmed a computer to control a motor-powered machine?

Tools and Materials

- ✓ LEGO Mindstorms™ Education NXT Invention System
- ✓ Computer system

Set Your Goal

Your goal for this activity is to build a motorized robot machine using a LEGO Mindstorms™ Education NXT Invention System. The motors and sensors that are part of your machine will be controlled by a computer program that you will write.

Know the Criteria and Constraints

In this lab, you will:

1. Use the problem-solving process to create your robot machine.
2. Test your computer program thoroughly before downloading it.
3. Download your program into your machine by using the RCX transmitter.

Design Your Project

Follow these steps to complete this lab.

1. Read the user's guide that comes with the LEGO Mindstorms™ Education NXT system. Complete the control experiment before building a machine of your own.
2. Pick one machine and follow the directions for assembly. Use the problem-solving process as you work.
3. Test motors and sensors, following the user's guide.
4. Use the manual to learn the programming language for the RCX controller.
   - You must use different commands to control your machine's motors and sensors.
   - The programming language lets you "talk" to the computer by using phrases that the machine (through the software) can understand.
5. Plan out the sequence of commands that will tell the computer how to control your machine.
6. Test each command, one at a time, before downloading to your machine's microprocessor. Troubleshoot solutions for any malfunctions.
7. Use the infrared transmitter to transfer your program to your robot machine.
8. Demonstrate the operation of your machine to the class.

Evaluate Your Results

After you complete this lab, answer these questions on a separate piece of paper.

1. In computer control systems, the machine under control is often equipped with optic sensors, touch sensors, and motors. How do these subsystems play a part in controlling the machine?
2. If you were to design and program another machine, what would you do differently?

Academic Skills Required to Complete Lab

Tasks	English Language Arts	Math	Science	Social Studies
Read user's guide and build robot.	✓		✓	
Learn programming language.	✓	✓	✓	
Write the computer program.	✓	✓	✓	
Test commands on robot.			✓	
Present to the class.	✓		✓	

chapter 11

Graphic Communications

Sections

What You'll Learn

- **Identify** common printing processes.
- **Explain** the difference between dynamic digital printing and traditional printing.
- **Describe** the process of producing a graphic message.
- **Compare** photographs and line art.

Explore the Photo

Printing Technology Ink dots make up the different color graphics of a printed picture, such as a magazine advertisement. *Do you think a home computer printer operates like a regular printing press? Why or why not?*

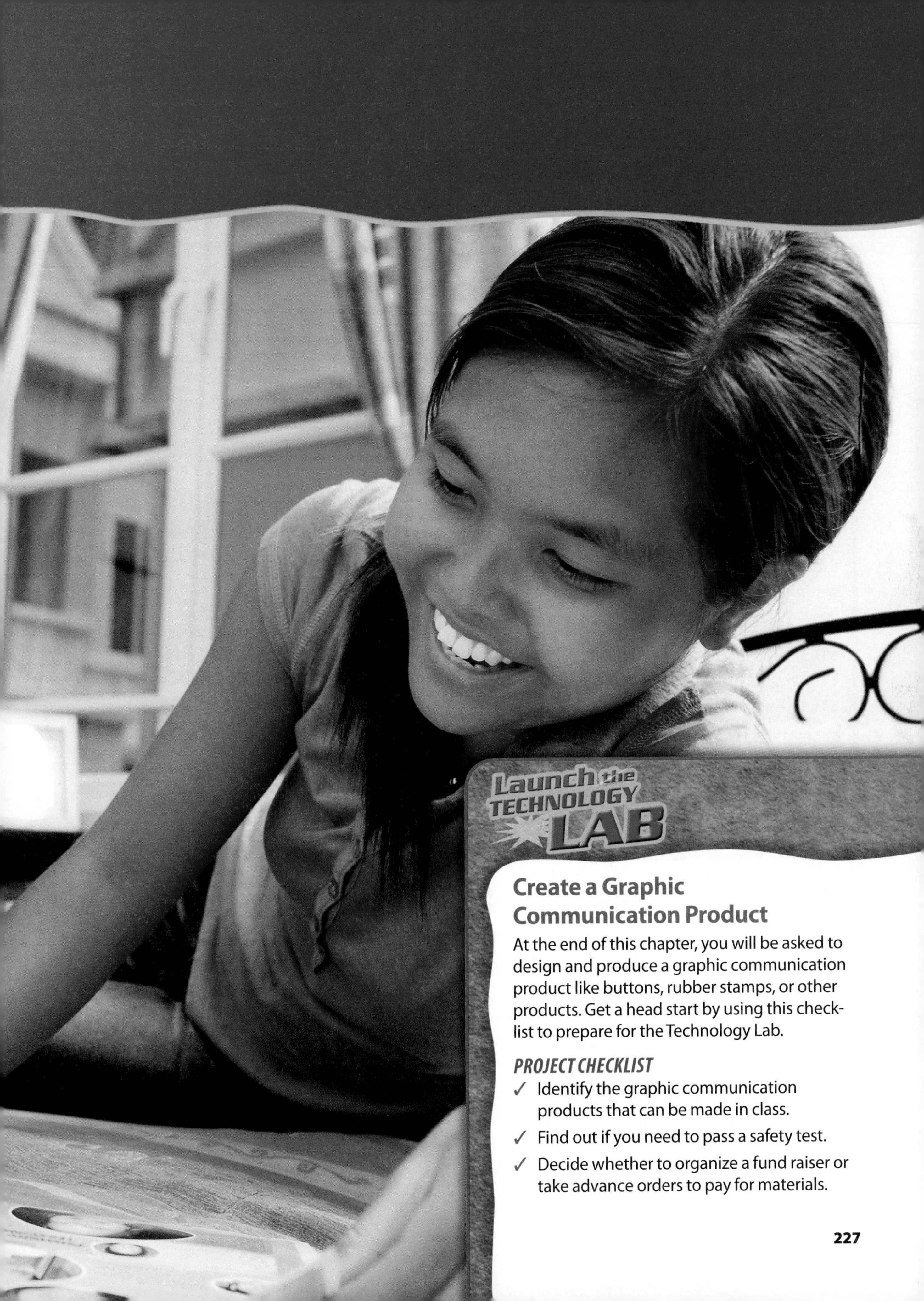

Create a Graphic Communication Product

At the end of this chapter, you will be asked to design and produce a graphic communication product like buttons, rubber stamps, or other products. Get a head start by using this checklist to prepare for the Technology Lab.

PROJECT CHECKLIST

✓ Identify the graphic communication products that can be made in class.

✓ Find out if you need to pass a safety test.

✓ Decide whether to organize a fund raiser or take advance orders to pay for materials.

section

11.1 Printing Processes

Reading Guide

Connect Do you think a copy machine is a type of printer? Why?

Content Vocabulary

- letterpress printing
- flexography
- gravure printing
- lithography
- serigraphy
- dynamic digital printing
- xerography
- ink-jet printing

Academic Vocabulary

You will see these words in your reading and on your tests. Find their meanings at the back of this book.

- journal
- purpose

Graphic Organizer

Draw the section diagram. Use it to organize and write down information as you read.

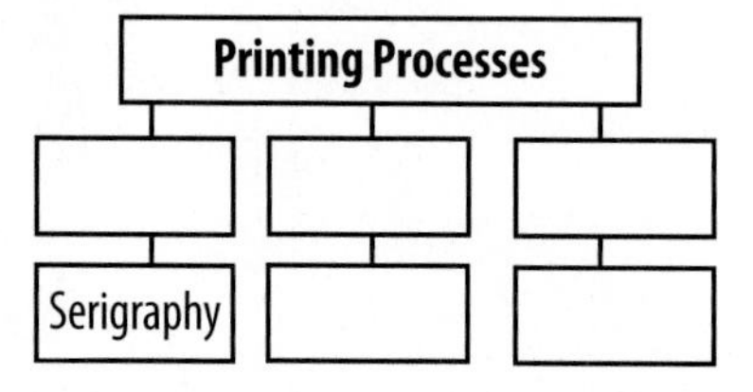

Go to **glencoe.com** to this book's OLC for a downloadable graphic organizer and more.

TECHNOLOGY STANDARDS

STL 5 Environmental Effects

STL 12 Use & Maintenance

STL 17 Information & Communication Technologies

ACADEMIC STANDARDS

Social Studies

NCSS 2 Time, continuity, and change

English Language Arts

NCTE 11 Participate as members of literacy communities.

STL *National Standards for Technological Literacy*

NCTM *National Council of Teachers of Mathematics*

NCTE *National Council of Teachers of English*

NSES *National Science Education Standards*

NCSS *National Council for the Social Studies*

Importance of Printing

Why is printing so important to society?

If all forms of printing suddenly disappeared, there would be no books, magazines, journals, or newspapers to read. A supermarket would have thousands of items without labels. There would be no paper money and no circuit boards for computers and televisions. Electronic formats would also disappear since graphic communication processes are used to create CDs, DVDs, e-books, and Internet pages.

There are many types of printing. This chapter discusses these types of printing: letterpress, gravure, lithography, photographic, serigraphy, and dynamic digital printing processes.

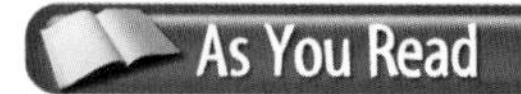

Identify What types of printing are used for what purposes?

Letterpress Printing

What part of a letterpress printing plate does the printing?

Letterpress printing uses metal type that is very similar to the type created by Johannes Gutenberg in the 15th century. In **letterpress printing**, the printing surface is a mirror image that is raised above the rest of the printing plate. (See **Figure 11.1**.) Another term for letterpress printing is *relief printing*. Letterpress printing plates can also be made of plastic and rubber, which is becoming more common.

Increased demand for printed materials continues to spur the development of newer, faster specialized equipment. Letterpress improvements did not match the improvement in speed and function of other printing technologies, and so its use has diminished. Letterpress still holds on to a small market share in areas where other printing technologies cannot match the printing characteristics of its metal, rubber, and plastic plates.

Flexography

When the printing plates are made of rubber, the process is called **flexography**. This form of letterpress printing is still popular. For example, rubber stamps can print on corrugated boxes, wallpaper, food packaging, paper bags, and several other flexible materials.

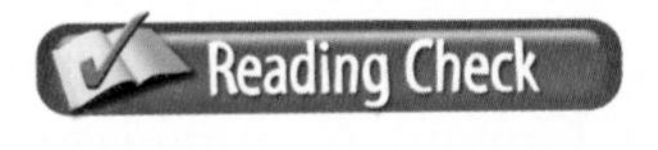

Identify What is one type of letterpress printing?

Figure 11.1 **Letterpress Printing Process**

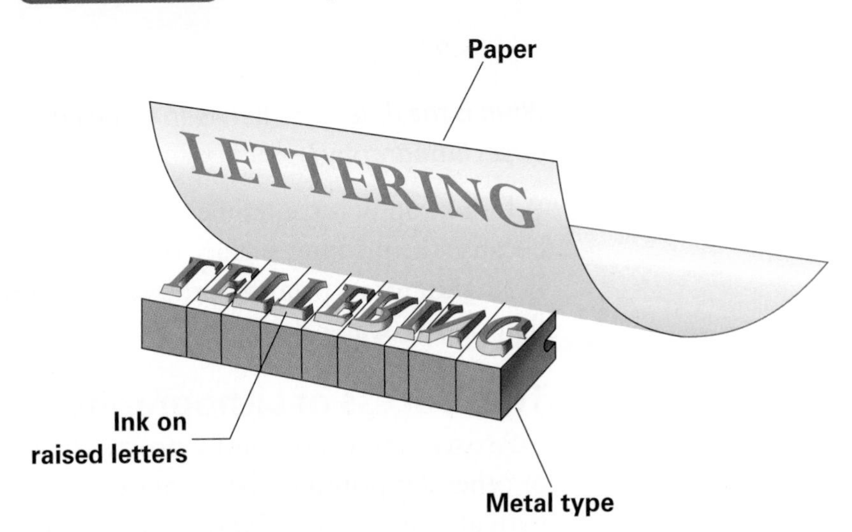

Letterpress Style In letterpress printing, the printing surface is raised above the rest of the printing plate. *What is a common form of letterpress printing that is used in crime investigations?*

Figure 11.2 Gravure Printing Process

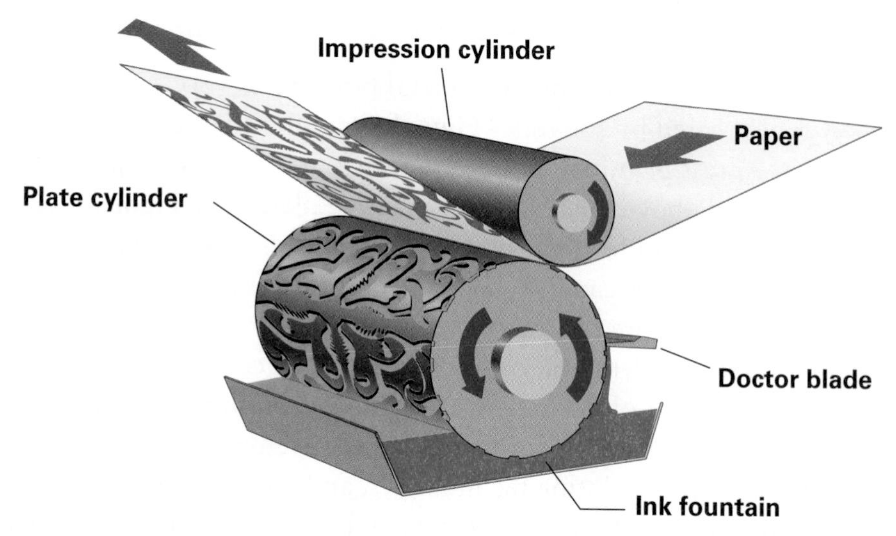

Printing with Etched Plates A gravure printing plate has inkwells etched or carved into its surface. *How might the number of ink wells and the depth of these wells affect the quality of printing?*

Gravure Printing

What part of a gravure printing plate does the printing?

In **gravure printing**, the printing plate is usually made of metal and has depressions, or pits, in its surface. (See **Figure 11.2.**) The pits vary in depth and thickness and function like miniature inkwells. The deeper the inkwell, the more ink it will hold, and the darker it will print. Gravure printing is also called "intaglio printing." It is used for printing magazines, paper money, and postage stamps.

Lithography

What is the difference between stone lithography and offset lithography?

Lithography is sometimes called "planographic printing" because the printing surface of the printing plate is perfectly flat. **Lithography** is a printing process based on the scientific principle that oil and water do not mix.

The Process of Lithography

Artists perform the entire process by hand with a limestone slab or other flat printing plate. They create an image on their plate with an oily medium. Water is applied to the plate, and then ink is applied. Ink sticks only to the oily image and is repelled by the water from the rest of the plate. Paper is pressed against the plate, and the ink transfers to the paper.

Offset Lithography

In offset lithography, the image is usually placed on a thin metal plate through a photographic process. The plate is mounted on an offset press, which prints the image to a rubber blanket. Then the blanket transfers the image to the paper. (See **Figure 11.3**.)

Offset printing is currently the fastest, most popular, and most economical method of printing. A Web offset press prints from a roll of paper. The fastest Web-feed press can print more than 100,000 copies per hour. This textbook was printed on an offset lithography press.

This kind of press produces a variety of print products. Its main purpose is to print:

- Business forms
- Magazines
- Newspapers
- Books

Photographic Printing

What is the plate used in photographic printing?

Photography is considered to be a printing technique. Basically, when you make a photographic print of a picture you have taken, light is projected through a plate that is called a "negative." The light is projected onto a light-sensitive material, such as special photographic paper. After processing, the image appears. Chapter 12 discusses more about photography.

Imagine This...

E-Textbooks

An e-book can be downloaded to your computer or a hand-held reading device. Instead of reading a paper book, you read text on a screen. E-books have been available, but consumers have not been buying them. In the future, instead of carrying heavy schoolbooks, students might prefer to keep their "textbooks" in a small device like an Apple iPhone®. *Would you switch to e-textbooks if the reading device allowed you to text-message, surf the Internet, and make phone calls? Why or why not?*

Go to **glencoe.com** to this book's OLC for answers and to learn more about e-books and technology.

Figure 11.3 **Lithography Printing Process**

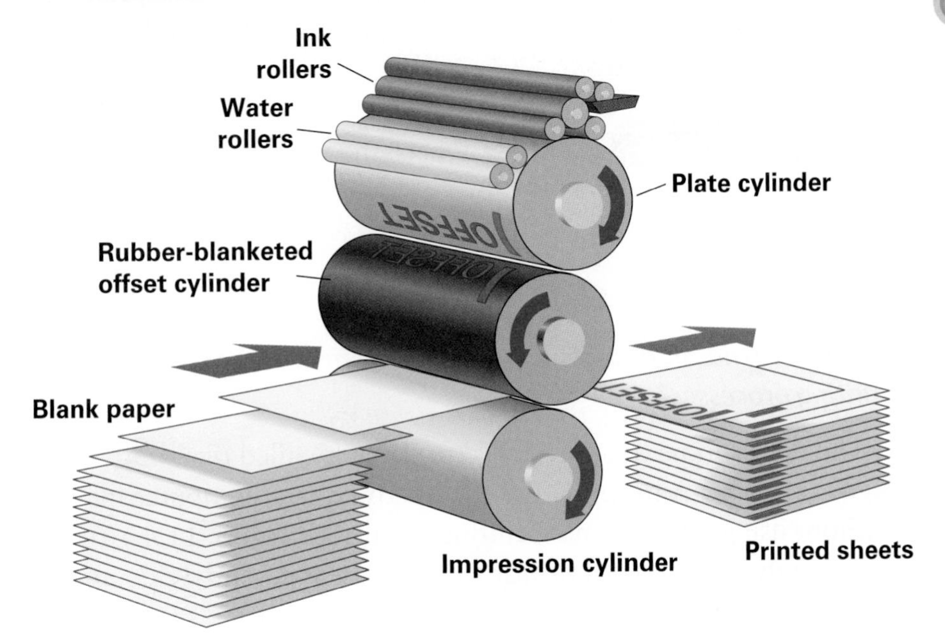

Offset Printing In offset lithography, the flat printing plate transfers the image to a rubber blanket. The rubber blanket then prints the image onto paper. *Why is the image on an offset-lithography printing plate not reversed like a mirror image?*

T-Shirt Art In this production shop, workers are screen printing logos and art on many T-shirts. *How is ink transferred from a serigraphy printing plate to the printed object?*

Serigraphy

Why are designs and sayings printed on shirts usually by serigraphy?

In **serigraphy**, or screen process printing, the printing plate is an open screen of silk, nylon, or metal mesh. The openings in the screen are sealed in the non-printing areas. Ink is then forced through the open areas of the screen by the movement of a tool called a squeegee. Serigraphy is used to print designs on shirts, patterns on fabrics, labels on glass and plastic bottles, and details on plastic and wood toys.

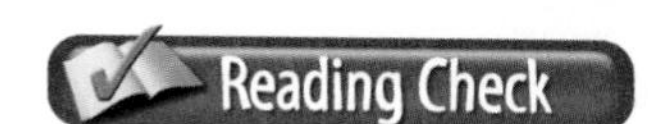

Contrast What is the difference between serigraphy and lithography?

Dynamic Digital Printing

How does dynamic digital printing differ from the other printing processes?

The newest type of commercial printing is called **dynamic digital printing**. A machine prints directly from a computer file without using a traditional printing plate. (See **Figure 11.4**.) You perform dynamic digital printing every time you send a document from your computer to your printer.

All digital printing machines need to refresh the image before each page is printed. This means that every page in a continuous printing run can be different. This is a perfect printing technology for people who want to place personalized information in each document that they print and mail in bulk. This technology also allows an entire book to be printed one copy at a time. Your local library might soon have an Espresso Book Machine to print books on demand. Type "Espresso Book Machine" into your favorite search engine to find out more.

As dynamic digital printing gains in speed and efficiency, it will replace many of the more traditional methods of printing. These changes may cause some workers to retrain or lose their jobs. They will need to learn new skills. Dynamic digital printing processes include:

- Xerography
- Electron beam imaging
- Electrophotography
- Magnetography
- Ink-jet printing

Xerography

You use xerography, or electrostatic printing, whenever you use a laser printer or a basic copy machine at school, at a library, or at a copy shop. This process uses a special metal plate in the shape of a cylindrical drum, static electric charges, and a powder that is called a "toner." **Xerography** is the printing process that transfers negatively charged toner to positively charged paper.

Figure 11.4 **Dynamic Digital Printing**

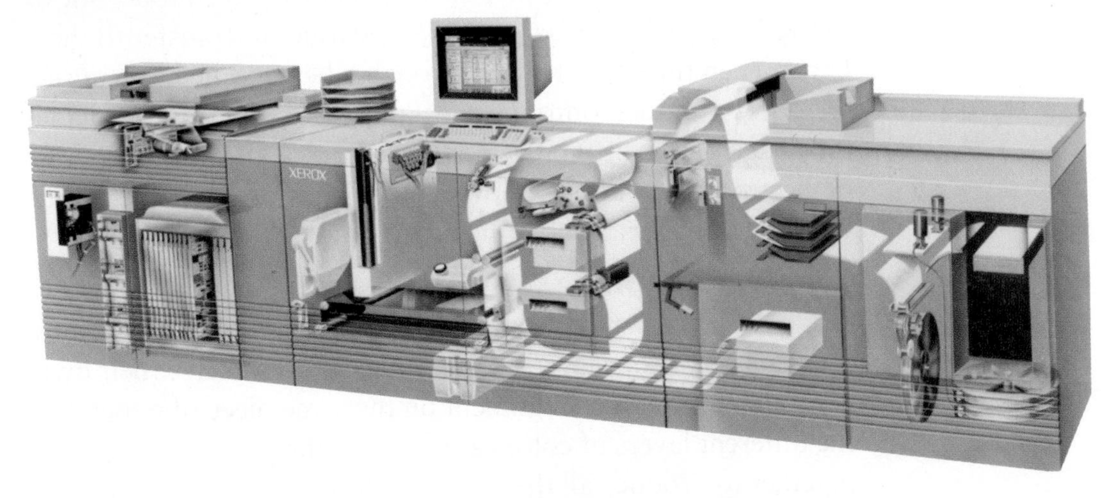

All in One This dynamic digital printing press prints, collates, and binds a book in one operation. *How may this technology affect employment?*

Tech Stars

Chester Carlson

Inventor of Xerographic Process

Chester Carlson (1906–1968) invented the photocopier. As a young man, he worked as a patent analyzer for an electrical component maker. This job required reviewing and preparing many drawings and documents, all of which he had to copy by hand. Being near-sighted and arthritic, this proved difficult for him. Carlson decided to come up with a better way.

He worked at his home on an idea for a reproduction technique based on photoconductivity. Carlson developed a process in which some materials changed their electric properties when exposed to light. In 1942, he patented this process and called it "electric photography."

Patent Patience It took years of trying to sell his new patented technology. In 1960, the Haloid Company, later known as *Xerox*, bought Carlson's idea. The Xerox copy machine was born.

English Language Arts/Writing Write a few paragraphs on the importance of the photocopier in your everyday life.

Go to **glencoe.com** to this book's OLC to learn about young innovators in technology.

Parts of Xerographic Printers

The drum, or printing plate, conducts electricity when exposed to light, but it acts as an insulator when the light is off. The machine projects an image onto the statically charged drum. The lit area of the drum conducts electricity and loses its static charge. The image area remains dark and is, therefore, still charged. The image becomes visible when the negatively charged toner sticks to the drum. A static charge causes the toner to transfer to the paper. Then the paper travels through a heater. The heat causes the powder to melt onto the page and become permanent. Finally, the image on the drum is refreshed for the next print. Xerography, or electrostatic printing, allows you to print or duplicate a large quantity of material very quickly.

Color Xerography

A color electrostatic printer separates the original image into four images or four layers. These images each receive their own color toner in perfect alignment on the same piece of paper. When the different layers of colors are fused to the paper, they work together to produce all the colors of the original as one image. Electrostatic printing allows for quick printing of newsletters, advertisements, and business forms.

Electron Beam Imaging (EBI)

Electron beam imaging uses an electron beam instead of a laser to create the image on the drum. Streams of electrons carry toner to the drum, instantly forming the image. The rest of the process is very similar to laser printing. Currently electron beam imaging is the fastest dynamic digital printing process, with press runs capable of producing a 400-page book every ten seconds.

Electrophotography and Magnetography

Electrophotography machines use tiny light-emitting devices to expose a light-sensitive, rotating electronic film. The entire page is transferred to the printing drum in a single flash of light.

Magnetography machines have special drums that allow individual points on the drum's surface to be instantly magnetized. In a thousandth of a second, a complete magnetized image is transferred to the drum for printing. Magnetography uses a magnetized toner. The fusing of this toner to paper is very similar to laser printing.

Ink-Jet Printing

In **ink-jet printing**, there are no printing plates or drums. The printer actually has very tiny spray guns that shoot their ink to the printing surface. Since the spray guns do not touch the material being printed, this method can be used to print on uneven or fragile surfaces.

In multicolor ink-jet printing, a separate ink gun is used to apply each color ink. With this process, many colors can be printed at the same time.

In Living Color With the development of graphics programs and color ink-jet printers, many people can afford to print color images for newsletters, reports, photos, greeting cards, and other materials. *How does an ink-jet printer operate?*

EcoTech

Recycling Billboards

Billboards are a form of advertisement. However, some people see them as eyesores. Some can even hurt the environment. After being used, the toxic PVC vinyl "posters" that are pasted on billboards are dumped in landfills. In response, one Los Angeles art gallery is reusing them as canvases for artists.

Try This Think of one new and creative way to reuse a non-recyclable item that gets thrown away. Share this idea with your class and try it at home if possible.

Do you think ink-jet printers are slow? Then do an Internet search for "Kodak Stream" and learn about a color ink-jet printer that spits ink like a machine gun on paper that is moving 24 miles an hour.

Impacts of Graphic Communication

What are some positive and negative impacts of printing technologies?

For more than 500 years, printing has helped distribute knowledge to every corner of our world. Every aspect of our society, every business, every workplace, and every government agency needs graphic communication to function. These are positive impacts of graphic communication.

However, some changes due to graphic communication technologies have not been positive. Economic factors and environmental concerns are sometimes opposed. Waste products from paper mills and printing facilities have polluted rivers and streams. Many of the substances once used in these industries have been linked to cancer and other debilitating diseases.

The undesired effects of past technology can motivate inventors to develop new technologies that can solve past problems. For example, new technologies have been developed to break down wastes from graphic communication industries in order to protect local communities and waterways.

section 11.1 assessment

After You Read Self-Check

1. Explain letterpress printing.
2. Explain gravure printing.
3. Discuss characteristics of lithography.

Think

4. Identify the difference between dynamic digital printing and traditional methods.

Practice Academic Skills

English Language Arts/Writing

5. Using the Internet and the library, do research about the positive and negative impacts of graphic communication on society or on the environment. Write a one-page report and present it to the class.

STEM Mathematics

6. An offset printing press is set up with four plates, one for each of the four colors used in a brochure. If the press can produce 42,000 brochures in one hour, how many can it produce in one minute?

Math Concept **Problem Solving** Sometimes a word problem contains information that is not needed to solve it.

1. Think about the question asked in the problem.
2. Be sure to eliminate any information that is given but not needed.

For help, go to **glencoe.com** to this book's OLC and find the Math Handbook.

section

11.2 Producing a Graphic Message

Reading Guide

Before You Read **Preview** Which colors do you think are used in four-color process printing? Why?

Content Vocabulary

- copy
- print
- layout
- line art
- halftone
- four-color process printing
- desktop publishing

Academic Vocabulary

- identify
- publication

Graphic Organizer

Draw the section diagram. Use it to organize and write down information as you read.

Producing a Graphic Message

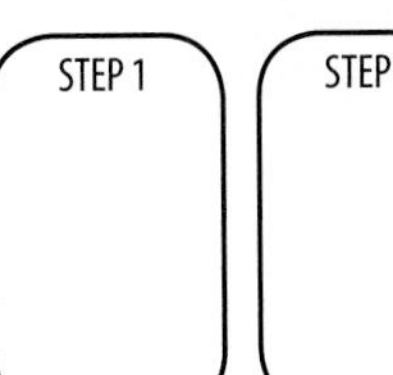

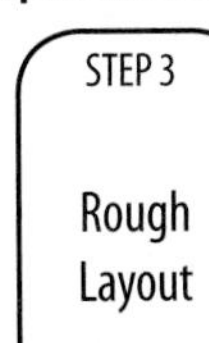

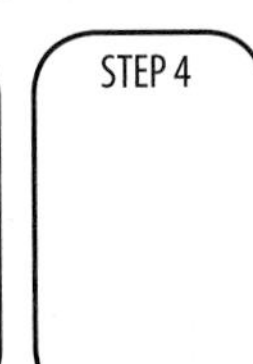

Go to **glencoe.com** to this book's OLC for a downloadable graphic organizer and more.

TECHNOLOGY STANDARDS

STL 4 Cultural, Social, Economic & Political Effects

STL 8 Attributes of Design

STL 17 Information & Communication Technologies

ACADEMIC STANDARDS

Science

NSES Content Standard E Abilities of Technological Design

English Language Arts

NCTE 9 Develop an understanding of diversity in language use across cultures.

STL	*National Standards for Technological Literacy*
NCTM	*National Council of Teachers of Mathematics*
NCTE	*National Council of Teachers of English*
NSES	*National Science Education Standards*
NCSS	*National Council for the Social Studies*

Producing a Graphic Message

In graphic communication, what is the difference between copy and prints?

Before being printed, every graphic message, such as an advertisement, poster, or announcement, must be designed by a designer. After the message is designed, there are steps to take for preparing it for printing. These steps are called "prepress operations." At this stage before printing, the message is called **copy**. After printing, the copies of the message are called **prints**. This section will discuss how pages are designed and illustrations are prepared for printing.

Compare What is more important in a graphic message—words or pictures?

Academic Connections
Math

Making Things Fit
Pictures often have to be reduced or enlarged to fit the space in a layout. When one dimension (measurement) changes for fitting, the other dimension changes proportionately. If you change the first dimension of a 3 × 5 photo to 6 inches, the second dimension must then become 10 inches.

Apply Calculate the missing number in these photo sizes: 4 × 5 = 6 × ?; 5 × 7 = 7.5 × ?; 11 × 14 = 1.57 × ?

STEM

Message Design

How do graphic designers develop their ideas?

The first step of developing a message design is to identify the message and decide how to convey it to your audience. The arrangement of type, artwork, and photographs on a page is called a **layout**. Designers often create a number of sketches of possible layouts. (See **Figure 11.5**.) In the second step, the designers select the best layout or brainstorm improvements. In the third step, they create a comprehensive layout of all the details. The fourth and last step is the creation of camera-ready copy, which is a replica of how the item will appear. Today almost all prepress operations are done on computers.

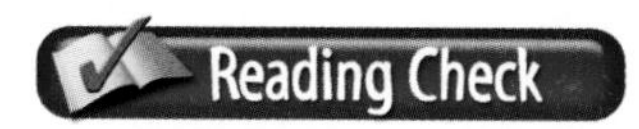

List What are the typical steps of the message design process?

Illustrations

What is the difference between photographs and line art?

When drawings are made of solid lines and shapes, they are called **line art**. However, a photograph has shades of color which are referred to as *gradation of tones*, which include black areas, white areas, and other areas that are light to dark shades of gray.

To print a photograph, you need to create a **halftone**, which converts the tones into a series of dots. The spacing and size of the dots determine if you see black, white, or gray. You can create a halftone by scanning the photograph into a computer file or by re-photographing it through a halftone screen. Digital cameras shoot all photographs as halftones.

Figure 11.5 **Message Design**

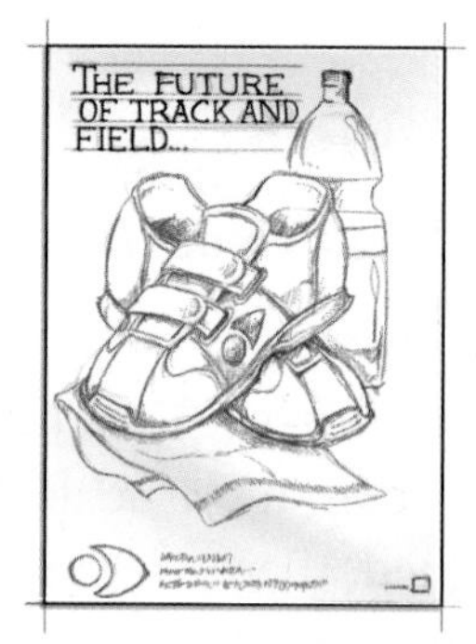

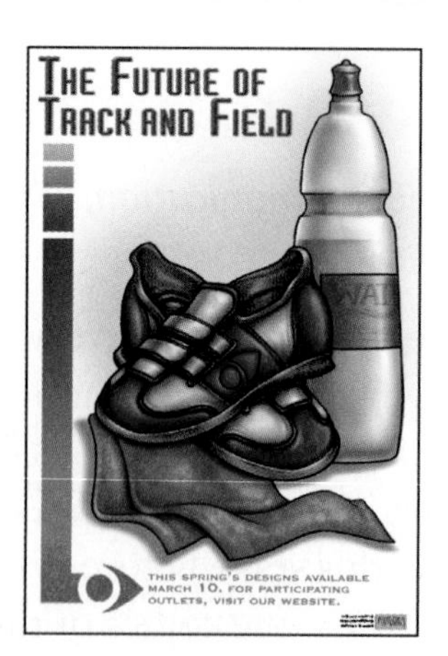

Planning the Design Many designers use hand-drawn sketching for preliminary drawings, while others use computers to create artwork. *Why would a designer prepare many drafts or sketches?*

Black

Magenta

Full-color image

Cyan

Yellow

Four Colors of the Spectrum All the colors of the rainbow can be reproduced on a printing press by using ink in four colors—magenta, cyan (blue), yellow, and black. *How does reflected light change these colors into every color of the rainbow?*

Reproducing Colors

How are colors produced?

All colors in nature can be reproduced through the printing of three primary ink colors plus black. **Four-color process printing** combines black, magenta (similar to red), cyan (blue), and yellow to print the colors of a rainbow.

When you see a color-printed object, you see the image in full color because light is reflected from the inks. White light contains the wavelengths of all colors. The inks absorb (subtract) some of the wavelengths of white light and reflect the color wavelengths. So, the inks are called "subtractive" colors.

To create color images, a process called "color separation" is used. In the past, halftones were printed on top of one another to produce the illusion of all colors. Today most color separations are created by using computer programs and scanners.

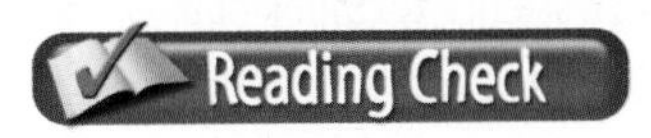

Reading Check **Explain** What is four-color process printing?

Desktop Publishing

What is the purpose of desktop publishing software?

When you use a computer to design and produce a publication, you are doing **desktop publishing**. Desktop publishing software is also used to convert electronic images into the formats for publishing CDs, DVDs, e-books, and Internet pages.

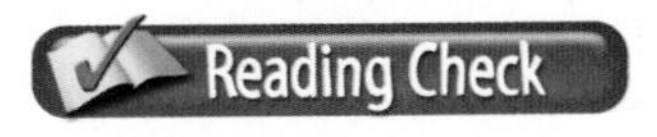

Reading Check **Define** What is desktop publishing?

Ethics in Action

The Right to Copy

Have you ever seen a funny story or joke on an Internet Web site and wanted to copy it? The Internet makes it easy to copy and transmit material. However, when material is copyrighted, it should not be reproduced without permission from the owner of the material.

Copyright Protection You must assume that everything you see on the Internet is copyrighted. Using copyrighted material without permission is considered stealing. Lawmakers are working on new laws to regulate copying material found on the Internet.

English Language Arts/Writing

Changing Laws Research the laws that protect any material found on the Internet. Find out if there are any exceptions to the laws.

1. Use the library or the Internet to research current and future copyright laws related to writing, music, and films or television shows.
2. Write about a writer or company using material you have created and why you would or would not be protected. The material can be a story, a song, or a film.

Teamwork and Desktop Publishing

Students can use desktop publishing software to set up school newspaper pages exactly the way they will look when the pages are printed. Time is limited, so a team must work efficiently. The member of the team who is the best typist could do the typing, and the best artist might create the layout. The student who is most computer literate might take care of routine hardware and software problems, system management, and file backup. Most school newspapers are team efforts with students from different grades working together, enhancing each other's skills.

section 11.2 assessment

Self-Check

1. Define a layout.
2. Explain line art.
3. Describe a halftone.

Think

4. Explain why film photographs and digital photographs are handled differently for printing.

Practice Academic Skills

English Language Arts

5. Use the Internet and the library to do research on the history or evolution of the basic printing processes. Make a poster or presentation slides to illustrate your research. Write a one-page report on how people's needs and wants changed with that evolution.

STEM Science

6. Magenta, yellow, cyan, and black inks are used as primary colors in the printing process to create all other colors. Find a color wheel in a science textbook or a book about painting. Use it to help figure out which of the four primary colors are used to make each of these colors:
   - Green
   - Orange
   - Dark blue
   - Purple
   - Olive green

Exploring Careers in Technology

Audrey Yamada

SCREEN PRINTER

Q: *What do you do?*
A: I own a screen-printing business. We're not limited to printing on T-shirts and sweatshirts. We also provide printed promotional items such as totes, caps, mugs, and other items. Whatever our customers would like to embellish, we try our best to fulfill.

Q: *How did you become interested in your field?*
A: My brother started the company, and I inherited the position of owner/operator. He had a good crew already, which made the transition and learning process quite easy for me. Once I started, I liked the business immediately.

Q: *What do you enjoy most about your job?*
A: The most rewarding part of my job is when customers are happy and excited about a job that we did for them. We tend to work with creative people, which is just great! Bands sell apparel on tour; artists make shirts to commemorate exhibitions; and magazines make them for readers. We can help them. Keeping them happy is important since most of my new business comes from word of mouth.

Q: *What do you do to keep up with the industry?*
A: We go to trade shows where we can talk to our peers. Knowing the latest screen-printing processes, software, and trends is essential. This allows us to offer the best quality, speed, and products.

English Language Arts/Writing

Write a Proposal Write a proposal to a possible client explaining how you can help his or her business.

1. Using a word-processing document, describe your services as well as the technology you offer.
2. On paper or with an illustration program, depict possible promotional items that you can make for the client.
3. Using a spreadsheet program, create a table listing your services and their prices.

Go to **glencoe.com** to this book's OLC to learn more about this career.

Real-World Skills

Speaking, listening, problem-solving

Academics and Education

Mathematics, English language arts, marketing

Career Outlook

Growth slower than average for the next ten years

Source: *Occupational Outlook Handbook*

Review and Assessment

Chapter Summary

Section 11.1 Graphic communication is also called "printing." It includes printing with ink on paper and publishing CDs, DVDs, e-books, and Web pages. The type of printing process used is determined by the part of the printing plate that prints. Letterpress prints from the raised part of the plate. Gravure prints from inkwells etched into the plate. Lithography prints from a plate that is flat. Photographic printing uses light. Serigraphy prints through little holes in the plate. Dynamic digital printing prints directly from a computer file. Electron beam imaging uses an electron beam to create an image. Electrostatic printing prints from a plate charged with static electricity. Ink-jet printing sprays ink directly on the material being printed.

Section 11.2 Desktop publishing is used to print newspapers, stationery, booklets, and other items. Computers and electrostatic printers have made this type of printing possible.

Review Content Vocabulary and Academic Vocabulary

1. On a sheet of paper, use each of these terms and words in a written sentence.

Content Vocabulary

- letterpress printing
- flexography
- gravure printing
- lithography
- serigraphy
- dynamic digital printing
- xerography
- ink-jet printing
- copy
- print
- layout
- line art
- halftone
- four-color process printing
- desktop publishing

Academic Vocabulary

- journal
- purpose
- identify
- publication

Review Key Concepts

2. **Explain** the importance of printing to society.
3. **Contrast** letterpress, gravure, offset lithography, and serigraphy processes.
4. **Discuss** the impact of dynamic digital printing.
5. **Identify** the positive and negative impacts of printing technologies.
6. **Describe** the steps in prepress operations.
7. **Explain** how to produce a graphic message.
8. **Identify** the difference between photographs and line art.
9. **Define** color separation.
10. **Discuss** a situation in which desktop publishing is useful.

Real-World Skills

11. Understanding Printing Costs Produce a one-page flyer advertising a charity event such as a fun run or golf outing. Contact local graphic designers, paper retailers, and copy stores to obtain bids from each. Create a budget for producing 1,000, 5,000, and 10,000 flyers. Include in your budget fixed costs and costs that vary.

 **Technology Skill**

12. Create a Layout Design a half-page public service ad to promote a club or organization. Include a headline, some copy, and an illustration or photograph.

- **a.** Create a layout for your ad on paper.
- **b.** Use art or drawing software to create a layout for the same ad. Compare the approaches in a paragraph.

Academic Skills

 **Social Studies**

13. Write a one or two-page report about the history of printed money and the steps printers take to discourage counterfeiting. Discuss the problems and solutions that have improved digital technology. Examples include changing the look of paper money.

STEM Mathematics

14. April found that she could produce 1,000 copies of an advertising brochure for $550. If she produces 2,500 copies, the cost per copy will be 10 percent less. What is the cost per copy if she prints 1,000 copies? What is the cost per copy if she prints 2,500?

Math Concept **Percents and Decimals** To change a percent to a decimal, drop the percent sign and divide by 100 by moving the decimal point two places left.

WINNING EVENTS

Graphic Designer

Situation Your principal has asked you to design a new letterhead for the school, business cards for the teachers, and a new layout and design for the school's newsletter.

Activity With a team, develop several different sketches. Use computer graphic design tools to develop possible designs. Select a design and have the new business cards and letterhead printed. Incorporate the design in your newsletter.

Evaluation The project will be evaluated by the following criteria:

- Originality
- Appearance
- Communication

 Go to **glencoe.com** to this book's OLC for information about TSA events.

Standardized Test Practice

Directions Choose the letter of the best answer. Write the letter on a separate piece of paper.

1. Which phrase best describes a halftone?
   - **A** camera-ready copy
   - **B** a layout of lines and shapes
   - **C** a pattern of dots
   - **D** electronic imaging
2. Ink-jet printing technology uses a system of revolving drums with plates attached.
   T **F**

Test-Taking Tip If each item on a test is worth the same number of points, do not spend too much time on questions that are confusing.

chapter 11

TECHNOLOGY LAB

Create a Graphic Communication Product

Tools and Materials
- ✓ Button machine
- ✓ Button parts
- ✓ Button print cutter
- ✓ Computer/printer
- ✓ Shirt transfers
- ✓ Heat transfer press or electric iron
- ✓ Shirts
- ✓ Linoleum blocks
- ✓ Linoleum block cutters
- ✓ Proof press
- ✓ Printer's ink
- ✓ Bench hook
- ✓ Paper
- ✓ Padding cement
- ✓ Paper cutter
- ✓ Spiral binding machine
- ✓ Spirals
- ✓ Engraver

You can use graphic communication processes to develop many different products to keep, sell, or share. Companies and organizations use these kinds of products for advertising, promotions, and good public relations. Can you think of some products that have logos or product names printed on them?

Set Your Goal

Your goal is to create a product by using graphic communication processes. Buttons, logos on shirts, and notebook covers are only a few of the possibilities. Buttons can be designed with funny sayings or images on them. Notebooks with covers in school colors are always popular.

Know the Criteria and Constraints

In this lab, you will:

1. Choose a product to manufacture. Your team will make enough of your product to share.
2. Brainstorm ways to organize the tasks for maximum efficiency.
3. Establish a method for checking product quality.

Design Your Project

Choose one of the following projects and follow the steps for that project.

Buttons

1. Draw or use a graphics program to create designs; for example, a school emblem.
2. Collect photos from friends or family, and transform the photos into buttons. Take orders for photographic buttons.
3. Let people design their own buttons. Make a form as a circle the size of the button.
4. Cut the design out using the button print cutter.
5. Place the design and button materials into the machine and press the button.
6. Remove the button.

Notebooks or Pads of Paper

1. Draw a linoleum block design for the cover of a small assignment pad or spiral notebook.
2. Transfer the design to the block. Carve it out using linoleum block cutters and a bench hook.
3. Print the design on cover stock at the proof press.
4. Cut paper at the paper cutter.
5. Attach cover and paper together to make a pad or a spiral notebook.

Print Logo on Shirts

1. Create your design on a computer.
2. Print your design using transfer instructions.
3. Design your transfer for printing as a mirror image.
4. Apply the transfer to the shirt following instructions.

Reminder

You must pass a safety test before you use the tools and machines needed for this activity. Be sure to follow appropriate safety procedures and rules so you and your classmates do not get hurt.

Evaluate Your Results

After you complete the lab, answer these questions on a separate piece of paper.

1. What safety rules did you follow?
2. What project did you enjoy the most? Why?
3. What would you do differently?

Academic Skills Required to Complete Lab

Tasks	English Language Arts	Math	Science	Social Studies
Brainstorm efficiency procedures.	✓			✓
Design image to be printed.			✓	
Learn to use graphics equipment.	✓		✓	
Print image.			✓	

chapter 12

Photographic Technologies

Sections

What You'll Learn

- **Name** the main parts of a camera.
- **Compare** film and digital cameras.
- **Describe** how images are captured on film.
- **Explain** how images are captured with a digital camera.
- **Summarize** how film negatives are processed.
- **Discuss** how digital images are processed.

Explore the Photo 

Making a Splash You have to open this camera's waterproof battery compartment to find its rechargeable battery and SD flash memory card. *Do you think this is a film or digital camera? Why?*

Launch the TECHNOLOGY LAB

Process Photographs

At the end of this chapter, you will be asked to process and edit film and/or digital pictures. Get a head start by using this checklist to prepare for the Technology Lab.

PROJECT CHECKLIST

- ✓ Pass your classroom's safety test for film processing.
- ✓ Select an old photo you want to edit or repair.
- ✓ Practice repairing photos with your classroom's photo-editing software.

section 12.1

Cameras

Reading Guide

Before You Read **Predict** What are the main parts of a camera?

Content Vocabulary

- lens
- shutter
- focus
- aperture
- CCD
- CMOS

Academic Vocabulary

You will see these words in your reading and on your tests. Find their meanings at the back of this book.

- compensate
- demonstrate

Graphic Organizer

Draw the section diagram. Use it to organize and write down information as you read.

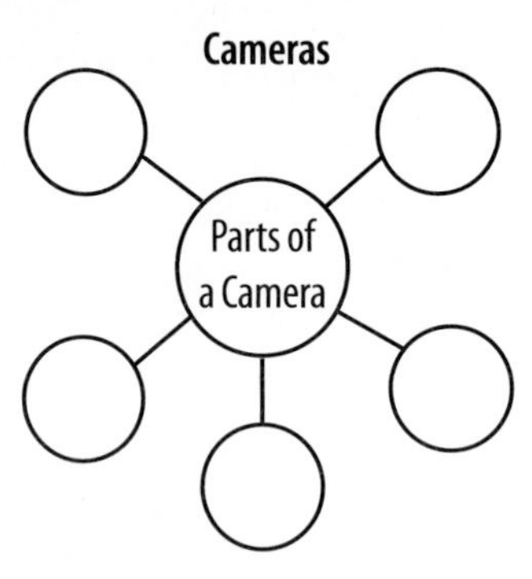

Go to **glencoe.com** to this book's OLC for a downloadable graphic organizer and more.

TECHNOLOGY STANDARDS

STL 5 Environmental Effects
STL 7 Influence on History
STL 11 Design Process
STL 13 Assessment

ACADEMIC STANDARDS

Science
NSES Content Standard B Interactions of energy and matter

Social Studies
NCSS 8 Science, technology, and society

STL *National Standards for Technological Literacy*
NCTM *National Council of Teachers of Mathematics*
NCTE *National Council of Teachers of English*
NSES *National Science Education Standards*
NCSS *National Council for the Social Studies*

The Evolution of Photography

When was the first digital camera sold?

The first consumer film camera went on sale in 1888. The first consumer digital camera, the Apple QuickTake 100, went on sale in 1994. Once introduced, each of these photographic technologies went through a slow, methodical process to improve their performance, one product cycle at a time. Today photography is changing from a chemical technology into a digital technology.

As You Read

Contrast What makes film and digital cameras different?

Parts of a Camera

What is the purpose of a camera lens?

Digital cameras and film cameras have the same basic components: a lens, shutter, aperture, and viewfinder. See **Figure 12.1**. Film cameras record images on light-sensitive film. Digital cameras record images electronically. Otherwise, they are basically the same.

Figure 12.1 Parts of a Camera

Film
Aperture
Lens
Shutter

Similarities Film cameras like this one and digital cameras have the same components. *How do digital cameras record images?*

Lens

Photography uses reflected light from a scene to record images. The **lens** on your camera focuses this reflected light and magnifies the image. A moderately priced camera might have a zoom lens that can be adjusted from wide angle to telephoto. A wide-angle lens makes close objects appear further away, and a telephoto lens makes far away objects appear closer.

To learn about new lenses, type "Octopus Camera Lens" or "Ceramic Lens" into your favorite Internet search engine.

Shutter

The **shutter** lets the reflected light from the scene enter the camera. It blinks open for a very short time. On more expensive cameras, shutter speed is adjustable. A slow shutter speed can help compensate for low-light conditions. The photo on this page demonstrates how a fast shutter setting keeps moving objects in **focus**, which is the sharpness of the image.

Aperture

The **aperture** is an adjustable opening that controls how much light will enter the camera when the shutter opens. Most inexpensive cameras have an aperture that cannot be adjusted. The aperture setting effects how much of a picture is in focus.

Shutter Speed Shutter speed helps determine what is in focus. For the photo on the left, the shutter speed was slow. For the photo on the right, the shutter speed was fast. *How did the photographer keep the cyclists in focus in the second picture?*

Range of Focus The photo on the left was taken using a large aperture. Only one car is in sharp focus. The photo on the right was taken using a small aperture. *What aperture setting would you use to shoot in low light?*

The size of the aperture is indicated by units called "f-stops." A larger f-stop number indicates a smaller aperture opening. A smaller aperture opening allows a larger area of focus. (See **Figure 12.2.**)

Viewfinder

The viewfinder on a camera allows you to view your picture before you take the shot. Film and digital single lens reflex cameras (SLRs) let you view your subject directly through the camera lens. Most digital cameras, including SLRs, have a small video screen that displays the picture and functions as a viewfinder and stored-picture viewer.

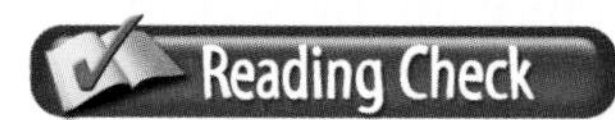

Recall What is a camera's aperture?

Figure 12.2 **Aperture Settings**

Field of Focus The relationship between the f-stop and the size of the opening is shown here. *What aperture setting would keep a viewer's attention on a single element of a photograph? Why?*

Film and Camera Memory

What is used to record an image?

Do you and your friends take photographs using film or digital cameras? Do you have a digital camera or one in a cell phone?

Film

In a film camera, photographic film records the image. This film is made with different levels of sensitivity to light. This light sensitivity is known as *film speed*. The International Standards Organization (ISO) uses a number system to rate the speed of film. If the number is higher, the film speed is "faster." If you know you will be taking photos where the light levels are low, you should use faster (higher number) films.

Digital

A digital camera does not use film. Light from the image falls on the CCD or CMOS microchips that convert light into an electrical signal. At one time, CCD sensors produced higher-quality images than CMOS sensors, but CMOS sensors consumed less energy than the more expensive CCDs. Today both sensors are equal in picture-taking quality and energy consumption. Digital cameras store their images in the camera's memory or on separate flash memory cards. Unlike film, a digital camera's memory can be erased and reused many times.

Academic Connections
Science

Looking at Lenses
Lenses are curved on one or both sides. The curvature determines the lens's effect on light waves. Concave lenses spread the light waves. Convex lenses bring them together.

Apply Obtain a series of photo lenses. Use them to observe the same object to determine their effect. Write two paragraphs about the strengths and weaknesses of different types of lenses.

section 12.1 assessment

Self-Check

1. Name at least three camera components.
2. Describe the purpose of a camera aperture and shutter.
3. Explain what creates the electrical signal in a digital camera.

Think

4. Forecast how trends in photography, such as combining cameras and telephones will evolve.

Practice Academic Skills

English Language Arts/Writing

5. The first camera was the *camera obscura*. Research the *camera obscura* and answer the following questions in a short report: How did they work? What does *camera obscura* mean? Who used them, and how are they similar to modern cameras?

Mathematics

6. The shutter speed on Anna's digital SLR camera is set at $\frac{1}{125}$. She wants to photograph a friend running in a race at a shutter speed that is four times faster. What shutter speed should Anna choose?

Math Concept **Understanding Operations** When thinking through this problem, do not be confused by the word *faster*. In this case, the *smaller* the fraction of a second, the *faster* the shutter speed.

1. Think about the question asked in the problem.
2. Divide the starting shutter speed by four.

For help, go to **glencoe.com** to this book's OLC and find the Math Handbook.

section 12.2

Recording the Image

Reading Guide

 **Before You Read** **Connect** How do cameras capture images?

Content Vocabulary
- latent image
- photosite

Academic Vocabulary
- undergo
- equip

Graphic Organizer
Draw the section diagram. Use it to organize and write down information as you read.

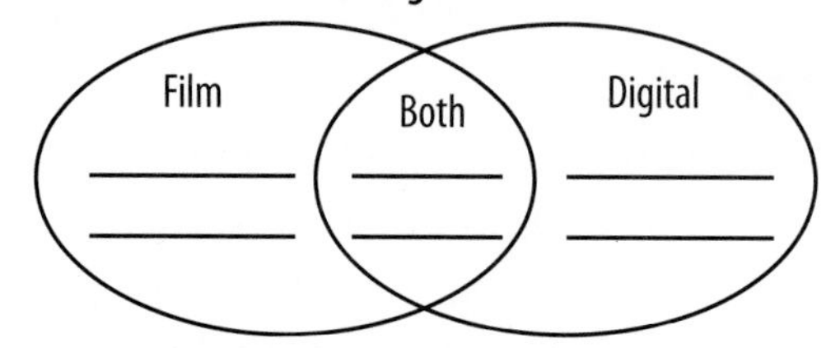

 Go to **glencoe.com** to this book's OLC for a downloadable graphic organizer and more.

TECHNOLOGY STANDARDS

STL 1 Characteristics & Scope of Technology

STL 2 Core Concepts of Technology

STL 17 Information & Communication Technologies

ACADEMIC STANDARDS

Mathematics

NCTM Number and Operations Understand numbers, ways of representing numbers, relationships among numbers, and number systems.

Science

NSES Content Standard G Science as a human endeavor

STL *National Standards for Technological Literacy*

NCTM *National Council of Teachers of Mathematics*

NCTE *National Council of Teachers of English*

NSES *National Science Education Standards*

NCSS *National Council for the Social Studies*

Images on Film

How does photographic film record a color image?

Photographic film consists of a sheet of thin plastic that is coated with chemicals that are very sensitive to light. Color film has three layers of chemicals. Each layer is sensitive to one of the three primary colors of light—red, green, or blue. When color film is exposed, the chemicals in each of these layers undergo changes that record the image.

The image stored on the film is invisible until after the film is developed. This invisible image is called a **latent image**.

 **As You Read**

Predict How do movie cameras create the illusion of motion?

Figure 12.3 Digital Camera

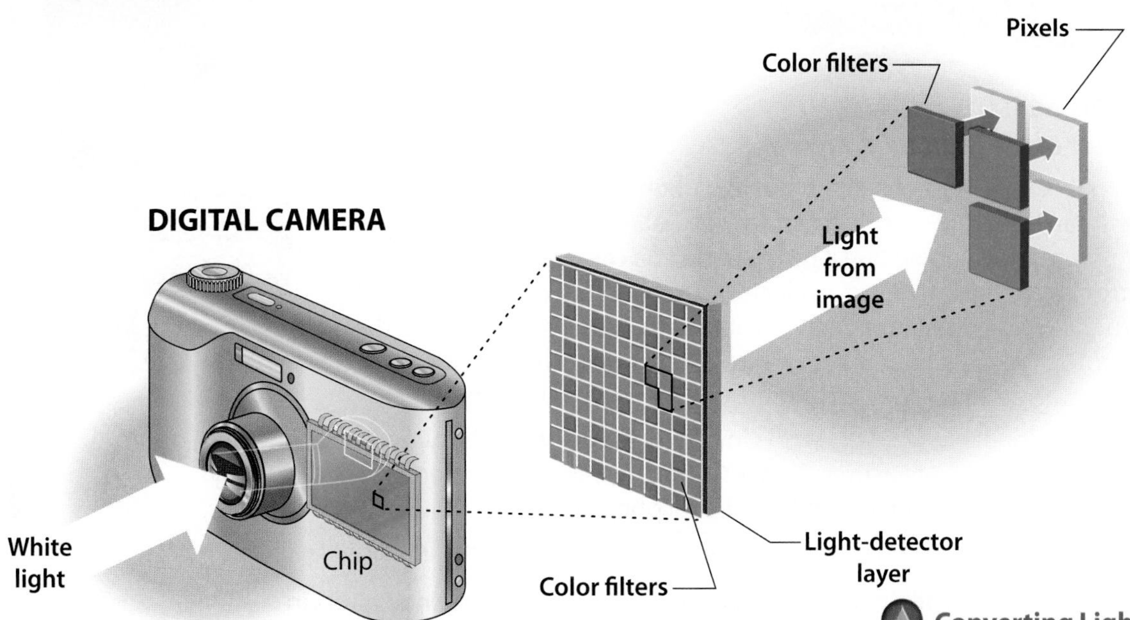

Converting Light In a digital camera, the light goes through filters and light sensors to create the tones and colors on the final photograph. *In what ways are color filters on a digital camera's microchip similar to four-color process printing? How are they different?*

Digital Images

How does a digital camera record a picture?

In digital photography, a CCD or CMOS microchip records pictures. See **Figure 12.3**. The microchip contains millions of **photosites**, which are tiny, light-sensitive cells that convert light into an electrical charge.

The brighter the light that strikes a single photosite sensor, the greater the electrical charge. These charges become a photo file stored in the camera's memory or removable flash memory card.

Ethics in Action

Protecting an Image

Digital technology is so advanced that almost anything is possible. For example, the famous actor Fred Astaire died in 1987. Nine years later, he appeared in a new television commercial, dancing with a vacuum cleaner.

Happily Ever After? Fred Astaire's widow authorized use of his image for the commercial. But the relatives of other deceased celebrities want to stop people from cashing in on their fame. They feel that a person's likeness and image should be protected.

English Language Arts/Writing

Now Showing Advertisers have already used deceased actors to promote soft drinks, clothing lines, and personal computers. Thanks to digital technology, you could see them starring in new movies.

1. Henry David Thoreau, the great American writer, once said, "Men have become the tools of their tools." What did he mean by this?
2. Do you agree or disagree? Write a one-paragraph response.

Steven Sasson

Inventor of the Digital Camera

Fresh out of college, Steven Sasson was working as an electrical engineer at Eastman Kodak Company. His supervisor, Gareth Lloyd, gave him an assignment. Lloyd asked him if he could build a camera with solid-state electronics and imagers, and an electronic sensor that gathers optical information. Using components available at the time, he assembled the first prototype of a digital camera. It was 1975. About the size of a toaster, the camera's first images were blurred. But Sasson fixed that problem.

Digital World By 1978, Sasson and Lloyd's invention was finely tuned, and they were issued a U.S. patent. At Eastman Kodak Company, Sasson has been involved in many other innovative ideas in the world of digital photography.

English Language Arts/Writing Write a short essay about the first digital camera.

 Go to **glencoe.com** to this book's OLC to learn about young innovators in technology.

Color Filters

To record the colors in the scene, color filters are placed in front of the photosites. The filters separate the light into red, green, or blue—the three primary colors of light. Each photosite records only one color, and each color produces a different electrical signal. These varying signals create the range of tones and colors that appear in the final photograph. (See again Figure 12.3.)

Pixels

The image from each photosite appears as a tiny dot called a "pixel." Pictures shot with a digital camera are collections of pixels. Cameras equipped with a CMOS use filters to record colors.

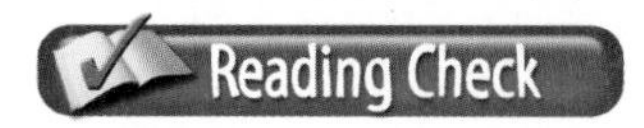

Describe How are digital images recorded?

Motion Pictures

How are motion pictures recorded?

The invention of the film camera led to the invention of the movie camera. Film and digital movie cameras produce movies using the same components in a still camera. They are designed to rapidly photograph many still pictures per second.

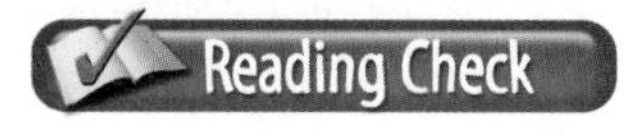

Compare In what way are movie and still cameras similar?

Figure 12.4 Film Playback

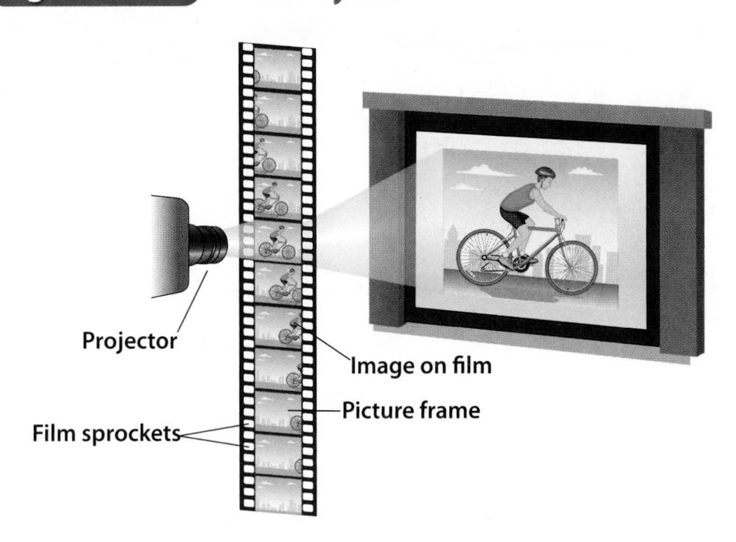

Action! A film movie camera shoots 24 still pictures per second. When the movie is played back, your brain thinks you are seeing motion. *Why does video shoot at 30 frames per second instead of 24 frames per second?*

Multi-Use Cameras

Camera makers realized they could combine the still camera and the movie camera. At first these dual digital cameras were superior in one type of photo shooting and inferior in the other. Newer digital cameras are true multi-use machines.

Frame Speed

Film cameras capture 24 frames per second, and video cameras capture 30 frames per second. The faster recording speed of video relates to the way video is refreshed on video screens. When played back, the images create an illusion of motion. Our eyes cannot see the quick change from image to image. See **Figure 12.4**.

section 12.2 assessment

After You Read Self-Check

1. Describe the effect that exposure to light has on film.
2. Summarize how a digital camera records a picture.
3. Compare a CCD to a personal computer's CPU.

Think

4. Discuss the meaning of the word *photography*.

Practice Academic Skills

English Language Arts/Writing

5. Cameras can show us places and things we cannot usually see, such as very small things, things that are very far away, and things that are under the ocean. Write a short report about this.

STEM Mathematics

6. The fastest camera can shoot 20,000,000 frames per second. How many frames can it shoot in one minute?

Math Concept **Large Numbers** Working with large numbers is often easier if you eliminate as many zeroes as possible.

1. Think of 20,000,000 as 20 million. Multiply 20 by 60 seconds. $20 \times 60 = 1{,}200$.
2. Think of a simpler way to express 1,200 million.

For help, go to **glencoe.com** to this book's OLC and find the Math Handbook.

section 12.3

Processing the Image

Reading Guide

Before You Read **Preview** Do all photographs require processing?

Content Vocabulary
- developer
- negative

Academic Vocabulary
- determine
- feature

Graphic Organizer
Draw the section diagram. Use it to organize and write down information as you read.

Processing the Image

1. ______________________
↓
2. ______________________
↓
3. ______________________

Go to **glencoe.com** to this book's OLC for a downloadable graphic organizer and more.

TECHNOLOGY STANDARDS

STL 2 Core Concepts of Technology

STL 4 Cultural, Social, Economic & Political Effects

STL 11 Design Process

ACADEMIC STANDARDS

Science

NSES Content Standard E Abilities of Technological Design

English Language Arts

NCTE 12 Use language to accomplish individual purposes.

STL *National Standards for Technological Literacy*

NCTM *National Council of Teachers of Mathematics*

NCTE *National Council of Teachers of English*

NSES *National Science Education Standards*

NCSS *National Council for the Social Studies*

Film Processing

How is film developed?

Commercial film processing centers develop your film by using a wet chemical process. This process is performed in a machine that looks very similar to a very large photocopier.

Negatives

Film processing must take place in total darkness since film is sensitive to all light. Your film is first bathed in a **developer** solution that turns the latent images into negatives. A **negative** is exposed film in which normally light-colored areas appear dark, and dark-colored areas appear light. Other chemical solutions in a "stop bath" are then used to end the film's sensitivity to light.

As You Read

Predict Why is digital photography less damaging to the environment?

Making Prints

To print your photographs, light is projected through your negatives onto photographic paper. The distance between the negative and the paper determines the print size. Most automated equipment is pre-set to print all pictures to a single size.

Color Prints

Color print photographic paper has three layers that are sensitive to the three primary colors of light. Light passing through the negative causes a chemical reaction in the paper. The proper colors are produced after the paper travels through a number of chemical baths.

Darkrooms

Professional photographers and hobbyists develop and print their own film by hand in a photo darkroom. Film must still be developed into negatives in total darkness, but this process can be performed in a light-tight developing tank. Special color safelights can be used during the printing process.

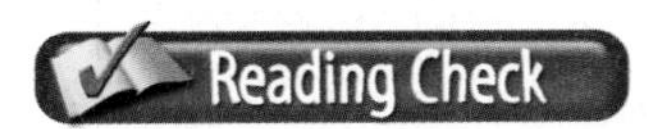

Summarize How are film negatives processed?

Digital Processing

What do you need to process digital pictures?

Chemicals are not needed to process digital pictures. This makes digital photography much friendlier to the environment. Computer software is used to transform digital photographs into prints. You need to download the pictures from the digital camera into your computer or take or upload your memory card's information for processing.

Imagine This...

Snapshots of the Future

You have probably seen holograms on credit cards or in science fiction movies. Holograms are 3D images created using mirrors and lasers. Until recently, the images did not look very real. But French holographer Yves Gentet developed a film coating that produces holographic images so true to life you have to touch them to know the difference. In the future, your HDTV and computer monitor will have images that seem to float off the screen. *How would holograms change TV, movies, and the Internet?*

Go to **glencoe.com** to this book's OLC for answers and to learn more about holograms.

Darkroom In a darkroom, an enlarger is used to make the photo the desired size. *Does a darkroom have to be totally dark all of the time? Why?*

EcoTech

Digital and Green

Digital photography eliminates the use of highly toxic darkroom chemicals. However, there are more ways to lessen the effect on the environment. You can also use solvent inks and recycled paper for prints.

Try This Make a list of ten everyday activities in which you can choose to use nontoxic alternatives.

Downloading Digital Images

The download process usually involves placing the camera or a removable flash memory card in a cradle attached to a USB port on your computer. The software copies the digital files into computer memory.

You can then use the software that comes with your camera. Photoshop®, Google's Picasa®, and Hyperstudio® are programs used to crop, enlarge, improve color, and improve contrast. If the feature is available with your software, you can even change your photo into a work of art.

Explain How can you process digital images?

Printing Digital Images

New home and commercial printers have also been designed to print your pictures without using a computer. To use this type of printer, you plug in the memory card, view the pictures on a screen, and pick the pictures you want to print on special paper. Digital pictures can be printed at home, at a local store, or uploaded to an online photo printing company. These companies make it easy to electronically share your pictures with friends and family.

section 12.3 assessment

After You Read Self-Check

1. Identify how the image on a negative differs from the original scene.
2. Explain why digital processing is easier on the environment than film processing.
3. Recall what causes a chemical reaction in photographic paper when you develop a color print.

Think

4. Describe how you think digital cameras and processing have affected film-processing businesses.

Apply

English Language Arts/Writing

5. Using either a film or a digital camera, create a photo essay about technology and the future. As you work, keep in mind your audience, medium, purpose, and the nature of your message. Write an introduction to your photo essay and a caption for each photo.

STEM Science

6. Draw a diagram showing how the eye works. Then draw another diagram showing how a camera lens works. Write a paragraph or two discussing the similarities and differences between the two.

Exploring Careers *in* Technology

Ben Clark

FREELANCE PHOTOGRAPHER

Q: *What do you do?*

A: I am a freelance photographer. Assignments or projects come from clients such as lifestyle magazines, clothing companies, skateboard gear manufacturers, event coordinators, and stock film agencies. I also shoot photos for artistic purposes.

Q: *What kind of training and education did you need to get this job?*

A: I started taking photos when I was in high school and concentrated on fine art photography when I attended college. After graduating, I had to go out, gain experience, and learn the business side of my work. I now run my own business and create my own jobs.

Q: *What do you enjoy most about your job?*

A: It isn't repetitive, and I don't have to be at a desk from 9 to 5. Every assignment is different. When I arrive at a shoot, I have to figure out what kind of photos I can get out of my subject. I consider the person's appearance and mood, the environment, and the client's expectations. When I'm finished with a job, I move on to the next one.

Q: *How did you become interested in your field?*

A: I've always had an interest in cameras, and I really enjoy the process of photography. I still feel excitement when I snap a picture and anticipate how the image will look. Using digital photography, I can create the image that I need.

English Language Arts/Writing

Make a Brochure Write a description of photographic services offered by a photographer. It can be portraiture, event photography, product shots, or other jobs.

1. Using a word-processing document, describe your services as well as the technology you offer.
2. Using a spreadsheet program, create a table listing your services and their prices.
3. Combine the pieces to a make a brochure. It can be factual or artistic, but should include information such as services, prices, and contact information.

Go to **glencoe.com** to this book's OLC to learn more about this career.

Real-World Skills

Speaking, listening, interpersonal, problem solving

Academics and Education

Mathematics, English language arts, marketing

Career Outlook

Growth as fast as average for the next ten years

Source: *Occupational Outlook Handbook*

Review and Assessment

Chapter Summary

Section 12.1 Photography is an important tool of communication technology. It is changing from a chemical to a digital technology. A camera has several main parts: the lens, shutter, aperture, and viewfinder. The lens focuses the light on the film and can magnify the size of the object. The shutter allows light to enter the camera.

Section 12.2 Photographic film is coated with chemicals that are very sensitive to light. In a digital camera, tiny devices called "photosites" convert the light into an electrical charge. A movie camera uses the same components as a still camera. It photographs 24 pictures per second and creates an illusion of motion.

Section 12.3 Chemical developer is used to reveal the latent image in exposed film, creating a negative—areas that appeared light in the original scene appear dark, and dark areas appear clear. Information from a digital camera can be fed directly into a computer. The information does not have to be changed because it is encoded as electrical charges.

Review Content Vocabulary and Academic Vocabulary

1. On a sheet of paper, use each of these terms and words in a written sentence.

Content Vocabulary
- lens
- shutter
- focus
- aperture
- CCD
- CMOS
- latent image
- photosite
- developer
- negative

Academic Vocabulary
- compensate
- demonstrate
- undergo
- equip
- determine
- feature

Review Key Concepts

2. **List** a camera's main parts.
3. **Discuss** the similarities between film and digital cameras.
4. **Explain** how images are captured on film.
5. **Define** the term *latent image*.
6. **Describe** how digital cameras capture images.
7. **Summarize** how moving pictures are recorded.
8. **Describe** how photographic film is used to record a color image.
9. **Recall** the process of developing film negatives.
10. **Explain** how digital images are processed.

Real-World Skills

11. Evaluate Digital Cameras You want to buy a digital camera. Make a list of ways in which you intend to use it. Research the various cameras available. List their costs and features. Write a paragraph about the camera you would choose to purchase and why.

STEM Technology Skill

12. Make a Pinhole Camera You can make a simple camera using a discarded frozen orange juice can or an oatmeal box and some tape. Use the Internet to research how to build a pinhole camera.

a. Use your pinhole camera to take some pictures in bright daylight.

b. Write a set of instructions someone could use to build a pinhole camera. Include photographs or illustrations of the one you built.

Academic Skills

Social Studies

13. Write a two-page report about the earliest forms of image capturing and projection. When were the first images displayed on a screen? How were moving pictures first shown? Who are key inventors in photography?

STEM Mathematics

14. Lonnie's family stores their photos on a computer. They have 1,560 pictures stored. Lonnie's mother deletes ⅓ of the pictures. Then his father deletes 45. Write and solve an equation to find how many pictures are left.

Math Concept **Solving Equations** Do the math within the brackets first. Start with multiplication or division, and then perform addition or subtraction, left to right.

WINNING EVENTS

Photographer

Situation Develop a portfolio of digital photographs depicting the importance of family and friends.

Activity Use digital photography to capture moments that show the connections you have with your family and friends. Make a series of photographs. Select your best shots and edit them. Finally, arrange them in a portfolio.

Evaluation The portfolio and photographs will be evaluated by the following criteria:

- Originality
- Art—communicates the importance of family and friends
- Technical—effective use of the principles of photography

Go to **glencoe.com** to this book's OLC for information about TSA events.

Standardized Test Practice

Directions Choose the letter of the best answer. Write the letter on a separate piece of paper.

1. Where are images recorded in a digital camera?

A on a negative

B on an aperture

C on a video screen

D on a microchip

2. To record a moving image, video cameras capture 30 still frames per second.

T

F

Test-Taking Tip If each item on a test is worth the same number of points, do not spend too much time on questions that are confusing.

chapter 12

TECHNOLOGY LAB

Process Photographs

Film is developed in a darkroom using chemicals. Enlargers help you make big prints out of little negatives. Digital photos are processed by using the software that comes with your camera or by using graphics software such as *Photoshop*®, Google's *Picasa*® or *HyperStudio*®.

Tools and Materials

Film Processing

- ✓ A completed roll of black-and-white film
- ✓ Developing tank
- ✓ Photo enlarger
- ✓ Developing trays
- ✓ Photo paper
- ✓ Print dryer or hanging clips and wire
- ✓ Darkroom
- ✓ Developer
- ✓ Stop bath
- ✓ Fixer

Digital Processing

- ✓ Digital photos
- ✓ Computer and photo software
- ✓ Old damaged photo
- ✓ Scanner
- ✓ Printer

SAFETY Reminder

Pass a safety test before you work on the film processing part of this activity. Follow appropriate safety procedures and rules so you do not get hurt.

Set Your Goal

For this activity, you will develop and print a roll of black-and-white film. You will also use a computer either to process digital photographs or to repair a scanned image of an old, damaged photograph.

Know the Criteria and Constraints

In this lab, you will:

1. Select three of the best film photos to submit to the teacher. Indicate any changes you would make if you could alter the negative.
2. Select three of your best digital photos to submit and describe how you changed the picture on the computer to improve its appearance.
3. Turn in "before" and "after" prints that show how you removed scratches, cracks, and other defects from the damaged photograph.
4. Dispose of all chemicals properly when you have finished the lab.

Design Your Project

Follow these steps to complete this lab.

Film Processing

1. Set up the trays and other darkroom equipment following your teacher's instructions.
2. Load the film into a developing tank in darkness using only safelights.
3. Develop your negatives according to the manufacturer's instructions.
4. Place the film in the stop bath for the proper time.
5. Place the film in the fixer for the proper time.
6. Wash the film and dry it.
7. Follow your teacher's instructions for making prints.

Digital Processing

Request teacher approval before starting this activity. Your teacher will determine if you know proper procedures for the computer, scanner, and software that you will be using.

1. Scan a photo needing repair and save it as two computer files, one as it is originally ("before") and one with the changes you will make to it ("after").
   - Make repairs on the "after" version of the photo.
   - Print copies of the original photo and the retouched photo.
2. Download new photos into the computer.
   - Select three for processing.
   - Use the software tools to improve the photos by removing red eye, cropping unwanted areas, and fixing backgrounds.
   - Print copies of before and after versions.
3. Select a series of photos that you can "morph" together. Place the photos in the desired order and use the morphing tools to make your changes.
4. Show the results to other students.

SAFETY

Picturing Safety

Wear rubber gloves and OSHA-approved safety glasses, and use tongs when you work with photographic chemicals. Wash your hands with soap and water as soon as you leave the darkroom. Read labels before using chemicals. Follow your teacher's directions and shop safety rules.

Evaluate Your Results

After you complete the lab, answer these questions on a separate piece of paper.

1. What part of this lab did you find most interesting?
2. Describe any hardware or software problems that you experienced and the way you solved them.
3. If you were to repeat this activity, what would you do differently?

Academic Skills Required to Complete Lab

Tasks	English Language Arts	Math	Science	Social Studies
Learn to process photo film in darkroom using chemicals.	✓		✓	
Learn to use digital photography software.	✓	✓	✓	
Make a print in the darkroom.			✓	
Scan and retouch a photo on the computer.		✓	✓	
Show the results to others.	✓		✓	✓

chapter 13

Multimedia Technologies

Sections

What You'll Learn

- **Explain** how radio and TV signals are transmitted.
- **Describe** how discs are recorded.
- **Discuss** the use of computers for animation.
- **Explain** how film, video, and video games are produced.
- **Discuss** features of the World Wide Web.

Explore the Photo 

Movie Magic The crew for a film includes camera operators as well as makeup artists, directors, and many others. Once film or video has been shot, a post-production team will edit it and add music. *What else might be added during post-production?*

Create a Multimedia Web Page

At the end of this chapter, you will be asked to gather a variety of media and use software to develop a multimedia Web page. Get a head start by using this checklist to prepare for the Technology Lab.

PROJECT CHECKLIST

- ✓ Visit technology Web sites and study their designs.
- ✓ Think of some ideas for your own Web page.
- ✓ Make a list of the multimedia tools and applications you will use.

section

13.1 Audio and Video Technologies

Reading Guide

Before You Read **Preview** What is the difference between audio and video technologies?

Content Vocabulary

- multimedia
- video
- audio
- pixel
- animation
- digital compression

Academic Vocabulary

You will see these words in your reading and on your tests. Find their meanings at the back of this book.

- formula
- enable

Graphic Organizer

Draw the section diagram. Use it to organize and write down information as you read.

Requirements for Audio and Video Recording Systems

1. ____________________
2. ____________________
3. ____________________

Go to **glencoe.com** to this book's OLC for a downloadable graphic organizer and more.

TECHNOLOGY STANDARDS

STL 3 Relationships & Connections

STL 4 Cultural, Social, Economic & Political Effects

STL 17 Information & Communication Technologies

ACADEMIC STANDARDS

Science

NSES Content Standard E Understandings about science and technology

English Language Arts

NCTE 1 Read texts to acquire new information.

STL *National Standards for Technological Literacy*

NCTM *National Council of Teachers of Mathematics*

NCTE *National Council of Teachers of English*

NSES *National Science Education Standards*

NCSS *National Council for the Social Studies*

Multimedia Technologies

What kinds of multimedia do you use in your daily life?

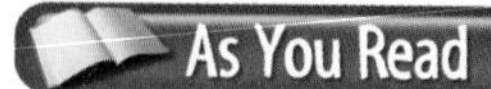

As You Read

Predict How are sounds and images recorded to DVDs?

Multimedia combines different media for text, sound, and images into one presentation to create a more enriched and entertaining message. Multimedia technologies include the machines and systems used to create and transmit these messages. You experience multimedia when you go to the movies, watch television, play video games, or explore Web sites on the Internet.

Radio

How does a remote control transmit a signal to your TV?

All wireless over-the-airwaves communication is based on radio transmission technology. Cell phones and TV remote controls are examples of devices that communicate by wireless radio.

The radio transmitter at a radio station or the one built into a wireless telephone converts sound into electrical signals. These signals are then carried by a radio carrier wave through the atmosphere to a receiving device. The radio on the receiving end converts the transmission back into its original form.

You can compare the joining of these two signals to a trip on a bus. The sound signals are like the passengers, and the radio carrier wave is like the bus. The bus carries passengers to their destination; the carrier wave carries the radio signal to your cell phone or wireless gaming device.

Television

How does a television work?

In your television, the **video** transmission and **audio** transmission are separate. The video signal contains the moving electronic images, and the audio signal contains the synchronized sound.

If your TV uses an antenna, audio and video signals reach it via radio carrier waves. If you have cable, your cable provider sends signals to your home through wire or fiber optic cables.

TV images consist of tiny dots of light called **pixels**. The full motion images you see are a series of quick-changing still images. These images are replaced 30 times each second, which is so fast your eyes are fooled into seeing a moving image.

High-Definition Television

In digital HDTV (high-definition television), the studio camera's image is converted into a compressed digital signal. This signal is then transmitted to your TV receiver. The computer in your HDTV converts this signal into moving images. The United States switched entirely to digital broadcasting in 2009. All new TVs manufactured after March 1, 2007 have digital tuners.

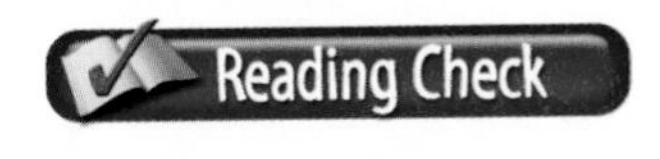

Summarize How are radio signals transmitted?

No Strings Attached Wireless and motion-sensitive game controllers have changed the way players experience video games. The wireless components in these game controllers are based on radio technology. *What are some other wireless devices?*

Audio and Video Recording

How are audio and video recorded?

All audio and video recording systems need to convert sound or images into a recordable signal. These signals must be stored on some type of material. The recording system must also have a way of converting the stored information back into sound and/or images in a playback unit.

Digital Video Recorders (DVRs) first convert sound and images into a digital electrical signal. This signal is then compressed (squeezed) so it will take up less space. The "write" head on the DVR's hard drive then converts this compressed signal into a magnetic pattern that it can record. Before digital technology, engineers recorded sounds and images onto magnetized tapes.

CD and DVD recorders use lasers instead of magnetism to record the same compressed digital file. During playback, the hard drive's magnetic head, or the CD or DVD's laser, "reads" the signal and sends it to the speakers and video screen.

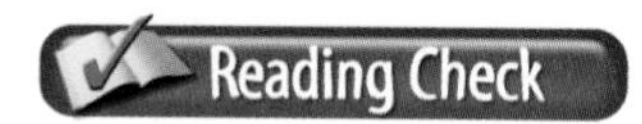

Explain How is audio recorded to CD and DVD?

Animation

How are individual drawings used to create animation?

Animation creates the illusion of movement of images. Each separate image or drawing is slightly changed to show movement. See **Figure 13.1**. The images move at 24 frames per second when played. This is too fast for your eyes to see one image at a time.

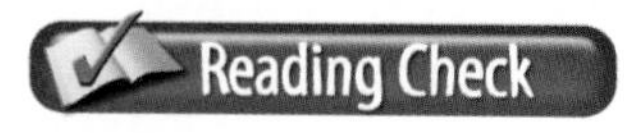

Define What is animation?

Ethics in Action

Facing the Music

When retailers first started selling music online, they protected it with DRM software. DRMs are like locks—they prevent people from illegally copying music files.

Changing Their Tune Now, retailers like iTunes are selling music without DRMs. Instead, when you purchase music, they record some of your personal data in the file. If you distribute one of these files illegally, the file can be traced back to you.

English Language Arts/Writing

Fighting Fire with Fire? Piracy has seriously hurt the music industry. However, many people do not like the idea of using personal data to combat piracy. They are concerned that hackers and thieves could access and abuse this information.

1. Do some research on anti-piracy in the music industry.
2. Hold a panel discussion to debate the pros and cons of this issue.

Figure 13.1 Animation

Illusion of Movement For animation, still drawings pass rapidly in front of your eyes and give the sensation of motion. *What happens if the drawings are shown at less than 24 frames per second?*

Computer-Drawn Animation

For computer-drawn animation, the beginning and ending images in a sequence are drawn first. The artist then saves the images to the animation program's memory. The computer fills in the images in between the beginning and the ending images. Feature-length animated films contain hundreds of thousands of separate digital images. Without the aid of computers, these films would take years to draw.

3D Animation

Animation that appears three-dimensional (3D) is created by using complex mathematical formulas. The process starts with a simple drawing that combines various shapes, lines, and arcs to create solid surfaces. This drawing is then *rendered*, which gives it depth and produces reflections. 3D animation makes video and computer games more realistic.

Video and Computer Games

In video and computer games, the graphics on the screen are always changing. Game software has many animation sequences and sounds stored in its memory. Each move you make with the controller tells the computer which sequence to run. As you play, you create a story.

The first video and computer games had simple controls and 2D graphics. Today's games have 3D graphics, movie-like music and animation, and wireless, motion-sensitive controllers. The computers and consoles that games are designed for are always being made faster and more powerful.

Academic Connections
Language Arts

Movie Talk Movies and TV shows seem more believable when the actors sound like real people. Their speech must be natural, not artificial.

Apply Write a conversation between two people who are trying to decide which of two TV shows is more interesting. Make their conversation realistic.

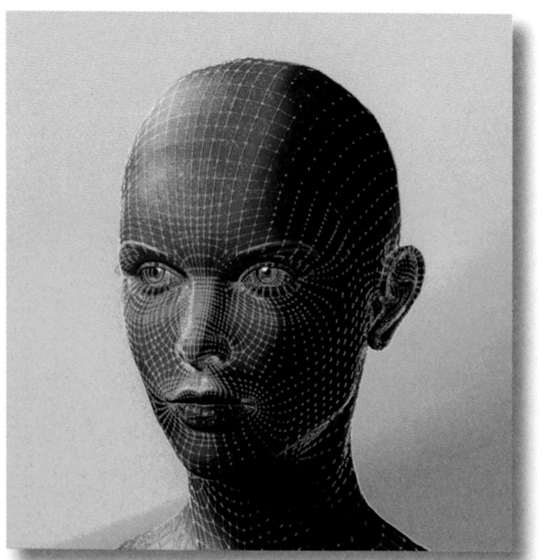

Keeping It Real Rendering and animation are used to make people and objects appear more realistic. *Why do designers render and animate new designs early in the design process?*

Presentations

Presentation software, such as PowerPoint® and HyperStudio®, helps create better presentations. Facts presented with graphics and sound are easier to remember. The programs **enable** you to add and organize text, graphics, sound, and other media.

Digital Compression

What is digital compression?

The images in an animated film or 3D video game contain too much data to store conveniently. **Digital compression** reduces the size of a digital file by removing bits of information that a computer can recreate when processed.

Digital compression is also used for ordinary audio and video files. The most common digital compression for video is called MPEG. The computer keeps the part of the video picture that remains the same and compresses only new data.

Audio compression drops out sound values that are beyond human hearing or that are drowned out by other sounds. The audio compression for music is commonly known as MP3. File compression reduces a digital file to 1/200 the original size.

section 13.1 assessment

After You Read Self-Check

1. Identify the form of transmission used for radio signals.
2. Explain the principle of animation.
3. Compare HDTV to ordinary TV.

Think

4. Some people think that computers have reduced the need for skilled artists in making animated films. Explain why you think this is or is not true.

Practice Academic Skills

Social Studies

5. Many viewers say that HDTV is so clear that they can see the individual blades of grass on a field. Research and write a one-paragraph report on the effect you think the switch from TV to HDTV has on people, the environment, the economy, and your favorite shows. Examine both positive and negative impacts.

STEM Mathematics

6. Marcus has $200 and wants to use it to subscribe to satellite radio. The radio costs $69.95, and the monthly subscription costs $12.95. Use estimation and mental math to figure out how many months he can subscribe before his money runs out.

Math Concept **Mental Math** Rounding numbers makes them easier to work with mentally.

1. Round the two types of costs to the nearest dollar.
2. Subtract the cost of the radio from Marcus's total, and divide the result by the monthly cost.

For help, go to **glencoe.com** to this book's OLC and find the Math Handbook.

section 13.2

Multimedia Applications

Reading Guide

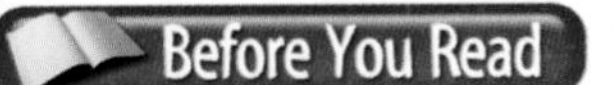

Before You Read **Connect** In what ways can technology be used to make videos and Web pages?

Content Vocabulary

- producer
- script
- director
- editing
- motion capture
- HTML
- URL
- browser
- search engine

Academic Vocabulary

- indicate
- link

Graphic Organizer

Draw the section diagram. Use it to organize and write down information as you read.

Producing a Video

Pre-Production	Production	Post-Production

Go to **glencoe.com** to this book's OLC for a downloadable graphic organizer and more.

TECHNOLOGY STANDARDS

STL 8 Attributes of Design

STL 9 Engineering Design

STL 11 Design Process

ACADEMIC STANDARDS

Science

NSES Content Standard F Science and technology in society

English Language Arts

NCTE 5 Use different writing process elements to communicate effectively.

STL *National Standards for Technological Literacy*

NCTM *National Council of Teachers of Mathematics*

NCTE *National Council of Teachers of English*

NSES *National Science Education Standards*

NCSS *National Council for the Social Studies*

Film, TV, and Video Production

What career skills must people have to be hired by a producer?

Feature films and TV programs cost a lot of money to create. These productions are run by a **producer**, who has financial and supervisory responsibility over every aspect of the project. This includes hiring the accountants, writers, the director, camera operators, performers, and the editor. In a small production, one person may do several jobs.

Producers and accountants use spreadsheets and other databases to track shooting schedules, props, costumes, and all financial aspects of a production.

As You Read

Predict Why are scripts important to film and television production?

Pre-Production

Every production, from a major motion picture to your classroom video assignment, can be divided into three phases: pre-production, production, and post-production. Pre-production includes all the jobs done before the cameras start rolling.

Scripts

If you think about the many movies you have seen, you will probably recognize that big-name actors and special effects cannot save a boring story. The **script** is the written version of the production. It lists the different characters and their dialogue. It also describes each set and indicates the action that takes place during each scene.

Some scripts are done in storyboard form. A storyboard contains sketches of what the scenes will look like from the camera's point of view. The sketches show the camera operators how to shoot each scene.

Other Pre-Production Tasks

The **director** controls and guides the camera work and helps the actors achieve their best performances. During pre-production, the director is in charge of rehearsals, which give the performers a chance to practice and learn their lines.

Many other tasks are also accomplished during pre-production. Set designers may create stage sets and backgrounds. Costume designers may make, rent, or purchase costumes and props. Cameras, lights, and microphones have to be set up. When everything is ready, production can begin.

Green Screen These actors are performing in front of a green screen. The audience will only see the final composite that will show them performing in a different location. *What type of shooting situations require a green screen?*

Production

Not all the scenes in a video or movie are shot in the same order in which you see them. Some scenes are shot in a studio. Other scenes are shot on location, off the studio lot. Scenes that require the same setting are all done together.

Green Screens

Directors can also shoot actors in front of a green screen. Later the editor can *composit* this footage, which means combining it with backgrounds or special effects to look real.

The Control Room

During live TV broadcasts, the director sits in a control room along with audio and video engineers. The director decides which camera feeds to send out to the public. The director can see what each camera sees by watching the different camera monitors.

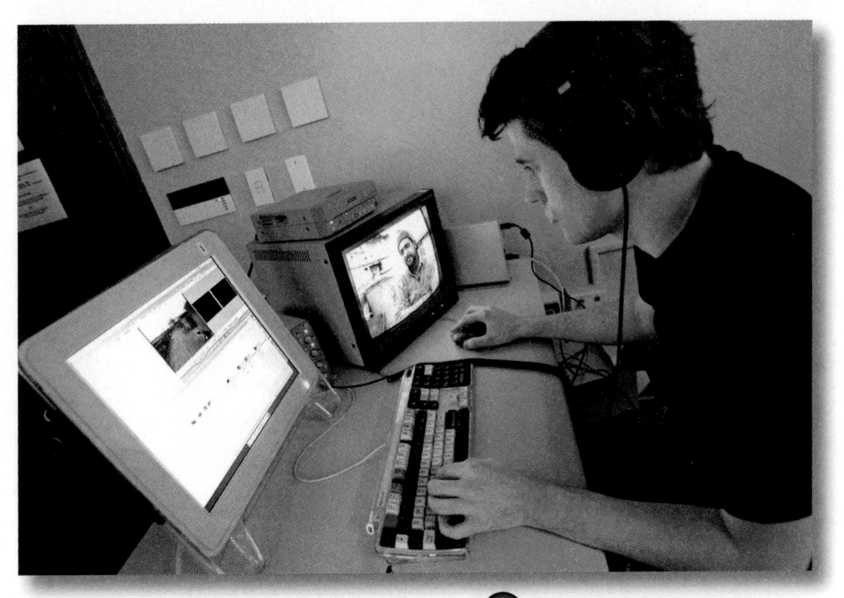

Compositing Post-production editors composit footage shot in front of a green screen with computer-generated landscapes and other special effects. By doing this, filmmakers can make scenes that are literally out of this world. *What other elements are added during post-production?*

Post-Production

When shooting is finished, post-production begins. This is the final phase. The most important post-production task is editing. **Editing** is the cutting and arranging of the taped material to decide its final order and content. Careful editing can make the difference between a successful production and a confusing one.

The editor copies scenes in the proper order onto a master file. He or she might add special graphics, music, and other elements at the same time, including the title and credits. When the editor is finished and the director is satisfied, the project is complete.

Producing a Video

Have you ever created your own video? All you need is a video camera and a computer. The right hardware and software can turn a computer into a desktop video studio, allowing you to do all of your editing directly on it. When the project is complete, you can burn it to DVD or upload it to your favorite Web site.

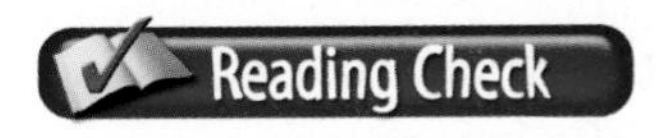

Explain How are TV programs, films, videos, and commercials produced?

Game Design

How do live actors help to produce video games?

Like TV or film production, computer and video game design usually requires many people. Game developers work with programmers, character and background artists, musicians, sound designers, and other technical experts.

A producer manages all of these people. If the game has a complicated storyline or scripted scenes, the producer might need to hire a writer. If the game will be sold in another country, translators will be essential.

Motion Capture The sensors attached to this actor's body record his (or her) movements and send the information to a computer. Designers or special effects artists use this information to create computer-generated characters. *What else can actors contribute to video games?*

Motion Capture

Game developers also use actors to provide characters with voices and for motion capture. **Motion capture** is the process whereby an actor performs physical actions that are recorded by sensors attached to his or her body. These sensors are connected to a computer, which uses the information to make animated characters move more realistically. This technology is also used to create computer-generated characters for feature films.

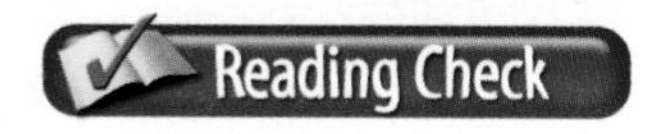

Compare What do game design and film and TV production have in common?

Web Design

What is the Internet?

The Internet is a research tool, meeting hall, shopping mall, and video game arcade all rolled into one. When you explore the Internet by clicking on a link or typing in an address, you move from computer to computer until you reach the site you want to find.

You will find many examples of multimedia technology on the Internet. Audio, video, text, graphics, photography, and animation all come together in an interactive format. (*Interactive* means that you can participate in the activity on a site.) Some people are concerned that this format can be addictive.

The World Wide Web

The World Wide Web is only one part of the Internet. It is a huge set of files linked in such a way as to make information easy to locate. That is why so many people and organizations use it.

Coding

The code language used for the Web is called **HTML**, which stands for HyperText Markup Language. HTML uses markers called "tags" to turn text, images, and video into a Web page.

Each site on the Web may have many "pages" and links to take you to other locations on the Web. Each Web page has its own address, called a **URL** (Uniform Resource Locator).

Browsers and Search Engines

Internet service providers (ISP) give you access to the Internet. Web **browsers**, such as Mozilla® Firefox® or Internet Explorer®, are software that allow you to view HTML documents.

Search engines such as Google™ help you search the Web. Web sites such as YouTube let you share videos. Social networking sites like Facebook provide a place for friends to communicate.

Web Designers

The Web sites of large companies or organizations are usually created by professional Web-page designers. These designers are skilled in combining different media to produce an exciting result. They may even write special software for pages with complex features. However, you do not have to be a professional designer to create your own Web page.

Building Your Own Web Page

You can design a page to share information with relatives, friends, or even strangers who share your interests. You can use text, audio, animation, links to other sites, e-mail links, and even video clips to create a page other people will enjoy.

Web-Site Building Programs

Software programs allow you to automatically convert your work into HTML. Most programs have preview buttons so you can see what your page will look like once it is converted.

Adobe® Dreamweaver®, Microsoft's Expression® Web, and Netscape Composer® are just a few of the many Web-site building programs that you might use.

Web for the People

The Green Wi-Fi Project is installing solar-powered networks around the world in an effort to give children in developing countries access to the Internet. The first prototype was made in 2006 in San Francisco.

Try This Make a list of items in your household or school that could be solar powered. Make a public service advertisement poster with illustrations of these items. Post it in your classroom.

Chad Hurley, Steve Chen, Jawed Karim

Co-Founders of YouTube

In 2005, Chad Hurley, Steve Chen, and Jawed Karim introduced a video sharing Web site called YouTube. Using Adobe Flash technology, they created an Internet phenomenon. According to one survey, over 100 million video clips are viewed every day on YouTube.

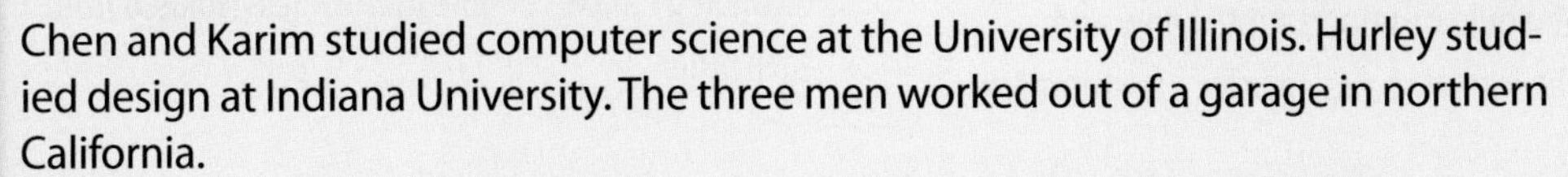

Chen and Karim studied computer science at the University of Illinois. Hurley studied design at Indiana University. The three men worked out of a garage in northern California.

Easy Money In November of 2006, Hurley, Chen, and Karim closed a deal with Google to sell YouTube for $1.65 billion, making them each multimillionaires. The sale came barely a year after the site was launched.

English Language Arts/Writing Have you ever posted or watched a video on YouTube? Write a few paragraphs explaining your experience with the Web site.

Go to **glencoe.com** to this book's OLC to learn about young innovators in technology.

Imagine This...

Non-Stop Internet

Researchers predict that by 2020, the Internet will be so common, we will not notice it anymore. You will even be able to turn on your dishwasher and other appliances over the Internet. Buildings and vehicles will be online. Enormous amounts of information will be available. *Will there be a downside? Can you think of some impacts?*

Go to **Glencoe.com** to this book's OLC for answers and to learn more about the future of the Internet.

Publishing Your Web Page

You can practice creating a Web site on a computer that is not connected to the Internet. However, to actually publish your page on the World Wide Web, you need an Internet connection.

Your Web-site building program can determine how you upload your page. After you upload your page, it will receive a URL. Enter this URL into a browser on any computer that has Internet access, and your page will appear. Many schools now provide space for students to upload their own Web pages. Most Internet providers have Web-page space set aside for their customers.

Describe What are some features of the World Wide Web?

Copyright Law

The elements that you use for your Web site can be of your own design or copied from sites that make them freely available. However, be cautious before you copy material that was designed by someone else. Copyright laws protect the material on the Web, and you may need permission to use it. You must assume that anything you find there is copyrighted. Many sites do make their graphics, templates, and animations available for anyone's use. But it is important that you respect copyright laws.

section 13.2 assessment

Self-Check

1. Name the source of dialogue in written form for a video production.
2. Identify what HTML stands for and explain its use.
3. Explain why a Web site needs a URL.

Think

4. Discuss why it would be important to know and understand the intended audience for a multimedia presentation.

Practice Academic Skills

English Language Arts/Writing

5. Develop a script for a one-minute radio commercial that uses two performers, music, and at least two sound effects. Indicate on the script where the music and sound effects will be inserted.

Social Studies

6. Thanks to the World Wide Web, you can shop, bank, and work without ever leaving your home. Some people feel this shift in communication has made the world increasingly impersonal. Write a few paragraphs telling whether you think that communication technology makes people more or less isolated from each other.

Exploring Careers in Technology

John Vechey

VIDEO-GAME DESIGNER

Q: *What is a typical day like for you on this job?*
A: Checking e-mails, prepping for meetings, playing our games to give feedback, and contemplating how to integrate our game with our Web site!

Q: *How did you train for your job?*
A: In 1997, Brian Fiete (one of the other co-founders of PopCap) and I created a game while we were in college. We sold that game to what is now Pogo.com, and then we went to work for Sierra Online. We worked there in various non-game development capacities before we quit to start PopCap Games.

Q: *What do you like most about your job?*
A: The fact that I get to work on great games like Bejeweled and Book Worm, and that the goal of the company is to only make the best! Furthermore, the casual-games side of the video-game industry is still emerging, so that most of the firms in this field are small and relatively informal. It's still a "garage," or cottage, industry in many ways.

Q: *How did you get interested in producing video games?*
A: Playing lots of games! That's the only way to get interested in game development. If you're in it for money, you're in the wrong industry.

English Language Arts/Writing

Making Predictions Today you can play video games on a cell phone or handheld gaming device. You can even play games online with players from all over the world.

1. Use the library or Internet to do research on early video game systems. How were they different from today's video games?
2. Think of ways that video games have changed in your own lifetime. Have video game controllers changed?
3. Write a few paragraphs explaining how video games may change in 20 years. Use the correct tense when talking about the future.

 Go to **glencoe.com** to this book's OLC to learn more about this career.

Real-World Skills

Creativity, problem-solving, organization

Academics and Education

English language arts, computer science, mathematics

Career Outlook

Growth faster than average for the next ten years

Source: *Occupational Outlook Handbook*

chapter 13 Review and Assessment

Chapter Summary

Section 13.1 Multimedia technologies use more than one medium in order to communicate. Audio devices include radios, CDs, and MP3 players. Video devices include TVs, VCRs, and DVD players. TV images consist of tiny dots of light called pixels. Animation creates the illusion of movement with a series of slightly different drawings or models.

Section 13.2 Pre-production, production, and post-production are all part of making feature films, TV programs, videos, and video games. The most important post-production task is editing, the cutting and arranging of material. The World Wide Web is one part of the Internet. The code language used for the Web is called HTML. Each page has its own URL, or Web address.

Review Content Vocabulary and Academic Vocabulary

1. On a sheet of paper, use each of these terms and words in a written sentence.

Content Vocabulary

- multimedia
- video
- audio
- pixel
- animation
- digital compression
- producer
- script
- director
- editing
- motion capture
- HTML
- URL
- browser
- search engine

Academic Vocabulary

- formula
- enable
- indicate
- link

Review Key Concepts

2. **Explain** how radio and TV signals are transmitted.
3. **Describe** how discs are recorded.
4. **Discuss** the use of computers for animation.
5. **Describe** how sound and images are reproduced on CDs or DVDs.
6. **Explain** how animation is used in video games.
7. **Describe** digital compression.
8. **Explain** how videos are produced.
9. **Discuss** features of the World Wide Web.
10. **Explain** the difference between pre-production, production, and post-production.

Real-World Skills

11. **Asking Permission** Before a company can use a photograph of someone, it must have the person's permission. With technology that allows people to take pictures or videos in public, receiving permission for those images is important. Create a document to be signed by someone who will give permission.

Technology Skill

12. **Film Production** Producing a film is a complex process with many steps.
    a. Write a script involving a conversation between two people. Include such information as where people should stand and how they should move.
    b. Using a video camera, make a short film based on your script. Write a paragraph describing how well the film followed your script.

Academic Skills

Social Studies

13. Research ways that multimedia technologies have changed. Pick one technology and write a few paragraphs about how the technology was introduced, how it changed, and its future.

Mathematics

14. Susan is producing an animated film for art school. She is drawing the individual frames by hand. She set the rate of the frames moving past your eyes at 25 frames per second. How many frames will she need to draw if her film is $5\frac{1}{2}$ minutes long?

Math Concept **Time** Convert units of time by using multiplication or division. If you need to convert from a large unit to a smaller unit, such as from hours to minutes, you multiply.

WINNING EVENTS

Electronic Game Designer

Situation Your design team must develop an E-rated electronic game. Your game must be exciting, intellectually challenging, and have educational and social value.

Activity Working as a team, brainstorm ideas for your game. Then determine the responsibilities of each team member. You are encouraged to use one of the free gaming engines available on the Internet.

Evaluation The game will be evaluated by the following criteria:

- Creativity and artisanship
- Technical skill
- Social value
- Flow, story, and overall appearance

Go to **glencoe.com** to this book's OLC for information about TSA events.

Standardized Test Practice

Directions Choose the letter of the best answer. Write the letter on a separate piece of paper.

1. What is the volume of a TV set that is 36 inches wide, 18 inches high, and 13 inches deep?
   A 8,420 cubic inches
   B 6,272 cubic inches
   C 8,424 cubic inches
   D 648 cubic inches
2. Digital compression removes bits of data that can be recreated later.
   T **F**

Test-Taking Tip Before you turn in your test, review your answers. Make changes you feel are important, but do not change an answer unless you have a good reason.

chapter 13 TECHNOLOGY LAB

Create a Multimedia Web Page

Web sites have many purposes. Commercial sites want to sell you their products. Government sites provide information about government policies and services. Consider who you want to view your Web site when you design a Web page.

Tools and Materials

- ✓ Computer
- ✓ Web-page design software

Optional

- ✓ Drawing materials or software
- ✓ Camcorder and editing equipment
- ✓ Photography equipment
- ✓ Music and sound recording equipment or digital audio files
- ✓ Props and/or costumes

SAFETY

Reminder

Be sure to follow your teacher's directions and general shop safety rules during this and all other lab activities. Follow your teacher's instructions on proper computer startup, software use, and computer shut down. If computer problems occur, ask your teacher for help before attempting to solve the problem on your own.

Set Your Goal

For this activity, you and the members of your team will create a Web page using at least three different media. Your page will inform viewers about some aspect of technology.

Know the Criteria and Constraints

In this lab, you will:

1. Use at least three of the following media to create your Web page:
   - Printed text
   - Photographs, drawings, or graphics
   - Animation
   - Speech, music, or sound effects
   - Video
2. Be aware that the purpose of your page will be to inform.
3. Choose your own topics and approach, but remember that the theme of your page will be "Exploring Technology".
4. Include links to other sites on your page, including sites developed by other teams in your class.

Design Your Project

Follow these steps to complete this lab.

1. Brainstorm with your team to select topics for your Web page and to decide on the approach you will take. A page entitled "Exploring Technology: Cars for the 21st Century" could include photos of new car designs with links to manufacturers' sites. You could inform readers what makes the cars special.
2. Research several topics before selecting those you will use. Consider what resources will be available to you.
3. Decide who your audience will be and which media will present your topics best.
4. Divide the work so that everyone on the team is responsible for some part of the project. Tasks will include designing the appearance of the page, finding links to use, creating audio or video sequences, taking photographs, or preparing artwork.
5. Follow the instructions from your Web-page design software to create your page.
6. Before you go live and connect your page to the Internet, obtain your teacher's approval.
7. Then show your page to your classmates.

Evaluate Your Results

After you complete the lab, answer these questions on a separate piece of paper.

1. What was the easiest part of creating the Web page? What was the most difficult? Why?
2. If you could add something more to your page, what would it be and why?
3. What other technology subjects would make interesting Web pages?

Academic Skills Required to Complete Lab

Tasks	English Language Arts	Math	Science	Social Studies
Brainstorm approach with team and research different topics.	✓		✓	✓
Complete assigned tasks.	✓		✓	
Use software to design Web page.	✓	✓	✓	
Put your Web page on the Internet.	✓		✓	
Show Web page to class.	✓		✓	

Technology Time Machine

Communication through Computers

Play the Game This time machine will travel to the past to show you events and inventions that made computer technology possible today. To operate the time machine, you must know the secret code word. To discover the code, read the clues, and then answer the questions.

Clue 1

3000 B.C.E. One of the first mechanical computers was the abacus. It was created in Babylonia, a country that existed in the Middle East about 5,000 years ago. People used it to calculate numbers by moving beads from side to side.

Clue 2

1642 The French mathematician Blaise Pascal built a wooden adding machine for his father, a tax collector. Pascal's machine added and subtracted numbers through the movement of wheels.

Clue 3

1830s The English mathematician Charles Babbage designed a machine that could perform complicated calculations. The machine operated by following instructions on punched cards. Babbage's "analytical engine" was never built, but computer designers of the 20th century adapted some of his ideas.

Clue 4

1944 The Mark I was the first computer powered by electricity. Professor Howard Aiken designed this large machine with many switches that opened and closed. This action formed a code that directed how the machine operated.

Clue 5

1946 The ENIAC was a computer that used vacuum tubes instead of mechanical switches. The ENIAC could perform 5,000 mathematical operations per second. Your school computer might perform more than 100 million operations per second.

Clue 6

1947 The transistor was invented at the AT&T Bell Labs. It was smaller, used less power, worked faster, and lasted longer than vacuum tubes. The first computer with transistors was built in 1956. By 1960, all computers were using transistors.

Clue 7

1960s Engineers put dozens of transistors onto a single chip called an "integrated circuit," which was used in calculators. Future integrated circuits may contain 10 billion transistors. The Intel Development Corporation produced the first programmable computer chip, which made the personal computer (PC) possible.

Clue 8

1979 A spreadsheet program called "VisiCalc" and a word processing program called "WordStar" went on sale to the public. These programs proved personal computers could do more than play games. Software guaranteed the PC revolution.

Crack the Code

On a piece of paper, write the answers to these questions:

1. What product guaranteed the PC revolution?
2. Who built a wooden adding machine in 1642?
3. What was one of the first mechanical computers?
4. In what products were integrated circuits with transistors first used?
5. What computer used vacuum tubes instead of mechanical switches?

Now write down the first letter of each answer. Put them together to discover the secret code word!

Hint Computers operate communication satellites located here.

Unit 3 Thematic Project

Shopping Online or In-Store

Past In Unit 3, you learned how communication technologies have changed our lives in many ways. One way is how we shop. In the early 20th century, small businesses in neighborhoods provided goods and services to the people living in those neighborhoods. You might walk to a nearby shop to buy groceries or dry goods, or to have your shoes repaired. After automobiles were invented, people began to drive to stores that were further away from home—and the shopping center developed. As people's incomes grew, shopping became recreational, and the shopping mall became a gathering place.

Present Today, with the Internet, people can stay home and use computers to purchase just about anything online, including groceries, clothing, books, music, and gifts. Because of communication technology, businesses have changed the ways in which they promote their goods and services.

This Project In this project, you will plan a retail Web site for a store.

Your Project

- Choose the store you represent.
- Choose and complete one task:
  1. Design an e-commerce Web site that allows consumers to order and pay for goods on the site.
  2. Design an informational Web site without e-commerce features.
- Write a report.
- Create a presentation with posters or presentation software.
- Present your findings to the class.

Tools and Materials

- ✓ Computer
- ✓ Internet access
- ✓ Newspapers
- ✓ Consumer magazines
- ✓ Word processing software
- ✓ Presentation software
- ✓ Posterboard
- ✓ Colored markers

The Academic Skills You'll Use

- Communicate effectively.
- Speak clearly and concisely.
- Use correct spelling and grammar when taking notes or writing presentations.
- Think about how businesses get their products to their customers.

English Language Arts

NCTE 4 Use written language to communicate effectively.

NCTE 12 Use language to accomplish individual purposes.

Social Studies

NCSS 7 Production, distribution, and consumption

Step 1 Choose Your Topic

You can choose any store as the topic for your Web-site project. Examples might include:

- Department stores
- Bicycle shops
- Gift shops
- Athletic shoe stores

Choose a store you would visit!

Step 2 Do Your Research

Research how to design the Web site and how your customers will use it. Think about the type of products the store offers and your customers' needs. Write answers to these questions:

- What features are on most retail Web sites?
- What products are likely purchased online?
- What are the advantages and disadvantages of purchasing online?
- What would your customers buy online?
- What would you like the most about online shopping? The least?

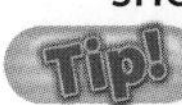

Find a Web site for a store like yours!

Step 3 Explore Your Community

Choose a local store that has a Web site. Ask the store's manager to share reasons for creating the Web site. Ask: What makes a good Web site? What are some mistakes made when designing an e-commerce Web site? Has it helped the business?

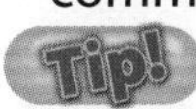

Remember to listen attentively!

Step 4 Create Your Project

Your project should include:

- 1 Web-site design plan
- 1 Web-site site map
- 1 report
- 1 presentation

Project Checklist

	Objectives for Your Project
Visual	✓ Make a poster or slide presentation to illustrate your Web site. ✓ Show how customers will use the Web site.
Presentation	✓ Make a presentation to your class and discuss what you have learned. ✓ Turn in notes from your interview and research to your teacher.

Step 5 Evaluate Your Presentation

In your report and presentation, did you remember to:

- Demonstrate your research and preparation?
- Be realistic?
- Show thorough evidence?
- Create quality content?
- Speak clearly?

Rubrics Go to **glencoe.com** to the book's OLC for a printable evaluation form and your academic assessment form.

GLOBAL TECHNOLOGY

Computer Circuit Boards

Through online stores, you can buy just about anything from anywhere in the world. E-commerce depends on computers, which are made with circuit boards. The most sophisticated equipment is used to build electronic circuit boards. Fiberglass, copper, and other components make up a circuit board. Many circuit boards for computers are assembled in countries such as China.

Critical Thinking *What other devices besides computers might use circuit boards?*

Mandarin Chinese

hello	ni hao
goodbye	zai jian
How are you?	Ni hao ma?
thank you	xie xie
You're welcome	Bu ke qi

Go to **glencoe.com** to the book's OLC to learn more and to find resources from **The Discovery Channel**.

unit 4

Biotechnologies

Chapters In This Unit:

Unit Thematic Project Preview

Discovering Careers in Biotechnology

As part of this unit, you will learn about how biotechnology has improved our way of life. You will also read about the different kinds of biotechnologies, including medical, industrial, agricultural, and marine, along with the interesting jobs in these fields.

As you read this unit, use this checklist to prepare for the project at the end of this unit:

PROJECT CHECKLIST

✓ Think of at least two interesting jobs each in medical and agricultural biotechnology. Ask your teacher if these jobs are appropriate.

✓ Explore at least two job-search Web sites.

✓ Make a list of people you know who might work in biotechnology.

WebQuest Internet Project

Go to **glencoe.com** to this book's Online Learning Center (OLC) to find the WebQuest activity for Unit 4. Begin by reading the Task. This WebQuest activity will help you find out about our genetic codes and the Human Genome Project.

Explore the Photo

Technology Meets Science Technologists are closely related to scientists, and sometimes they are the same. Innovations in medical biotechnology are improving the lives of many people. *What innovations in medical biotechnology are in the news?*

chapter 14

Medical Biotechnologies

Sections

What You'll Learn

- **Identify** disease prevention technologies.
- **Explain** the difference between pasteurization and irradiation.
- **Explain** how immunization works.
- **Define** genetic testing.
- **Describe** imaging technologies.
- **Identify** tests used to read electrical impulses of the human body.
- **Explain** how antibiotics work.
- **Describe** three types of surgery.
- **Define** genetic engineering.
- **Explain** the purpose of bionics.

Explore the Photo 

Exchanging Organs The human body is made up of many different organs. Doctors and medical technologists can transplant some of these organs, including kidneys. They are also learning how to replace hearts and other body parts with artificial ones. *Would this type of organ replacement have a major effect on medicine? Why?*

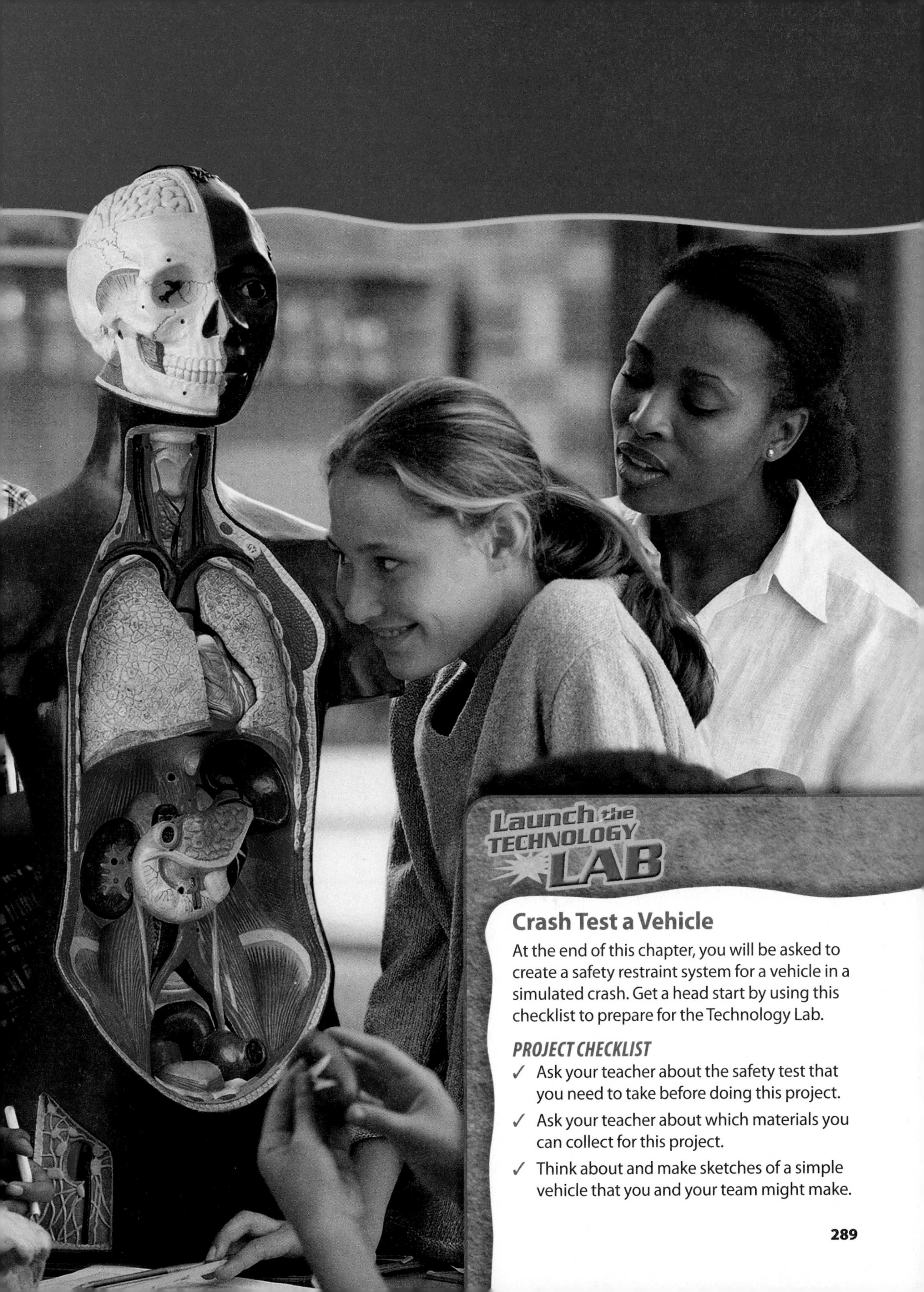

Crash Test a Vehicle

At the end of this chapter, you will be asked to create a safety restraint system for a vehicle in a simulated crash. Get a head start by using this checklist to prepare for the Technology Lab.

PROJECT CHECKLIST

- ✓ Ask your teacher about the safety test that you need to take before doing this project.
- ✓ Ask your teacher about which materials you can collect for this project.
- ✓ Think about and make sketches of a simple vehicle that you and your team might make.

section

14.1

Disease Prevention

Reading Guide

Before You Read **Preview** What do you already know about pasteurization?

Content Vocabulary

- pathogen
- pasteurization
- irradiation
- sanitation
- immunization
- vaccine
- ergonomics

Academic Vocabulary

You will see these words in your reading and on your tests. Find their meanings at the back of this book.

- seek
- ensure

Graphic Organizer

Draw the section diagram. Use it to organize and write down information as you read.

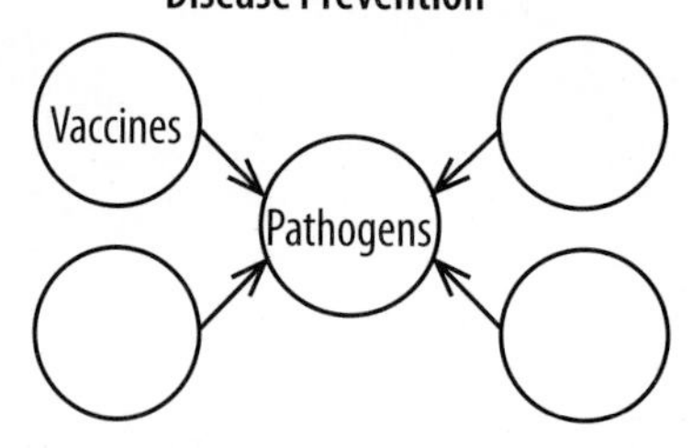

Go to **glencoe.com** to this book's OLC for a downloadable graphic organizer and more.

TECHNOLOGY STANDARDS

STL 5 Environmental Effects

STL 14 Medical Technologies

STL 15 Agricultural & Related Biotechnologies

ACADEMIC STANDARDS

Science

NSES Content Standard C Structure and function in living systems

English Language Arts

NCTE 1 Read texts to acquire new information.

STL	*National Standards for Technological Literacy*
NCTM	*National Council of Teachers of Mathematics*
NCTE	*National Council of Teachers of English*
NSES	*National Science Education Standards*
NCSS	*National Council for the Social Studies*

Medical Biotechnologies

What is medical biotechnology?

When you combine knowledge of technology with knowledge of biology, you get "biotechnology." Older biotechnologies use microorganisms and other biological substances to produce products. The newest biotechnologies use genetic engineering to create new biological agents not found in nature. Medical biotechnology uses biology and technology for disease prevention, diagnosis, and treatment. The first medical technologists used plant roots, herbs, and superstition to cure disease centuries ago.

As You Read

Identify What are the different ways to prevent disease?

Disease Prevention

How can disease prevention improve health?

If people never got sick, they would not need medicine and other treatments. That is the goal of disease prevention. Preventing disease involves many technologies. Some technologies, such as pasteurization and irradiation, sterilization, and water treatment, seek to remove or kill **pathogens**, which are organisms that can cause disease. Pathogens include bacteria, viruses, parasites, and fungi. Other technologies, such as immunization, can help the body stay strong so it can stay healthy.

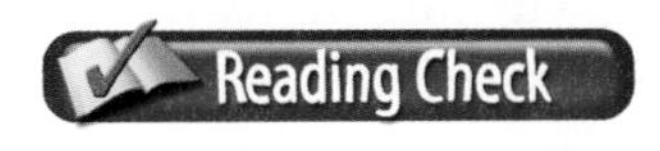

Identify What is the goal of disease prevention?

Pasteurization and Irradiation

How are pathogens destroyed in foods?

Pasteurization and irradiation kill pathogens in foods. **Pasteurization** is a heating process used to kill bacteria that turns milk sour and makes you sick. In 1864, a French chemist named Louis Pasteur developed this process.

Irradiation is also called "cold pasteurization," because it uses radiation to kill pathogens without heat. Exposing food to X-rays, ultraviolet rays, or gamma rays increases the shelf life of food by killing parasites, insects, mold, and bacteria. Some people fear irradiation might make the food radioactive or change its nutritional value. The FDA and the World Health Organization have 40 years of research that proves irradiation is a safe process.

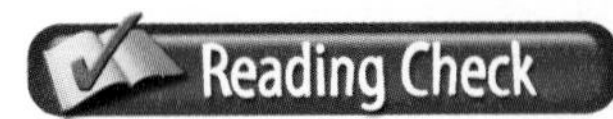

Connect What process sterilizes milk?

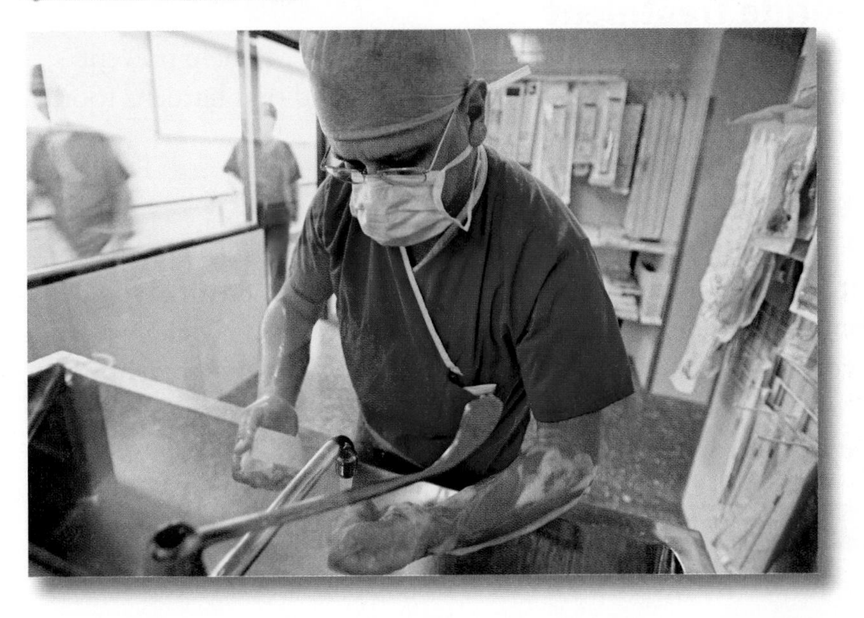

Keeping Clean Doctors and nurses must be as germ-free as possible before performing any medical procedure. *Why might the medical profession warn the public against overusing hand disinfectants?*

EcoTech

Aim Low

Growing livestock, such as cows and chickens, requires a tremendous amount of grain, water, and resources. Eating less meat products and more fresh unprocessed foods, such as fruits and vegetables, which are low on the food chain, can reduce your impact on the environment.

Try This Become aware of what you are eating. Is your meal made of fresh or processed foods? Some ready-made foods are more processed than others. Look at packages and read the ingredients to find out what you are eating.

Figure 14.1 Nano Filtration

Water for Life Nanotubes are tiny carbon atom tubes. Contaminated water or ocean water can flow out the tube as clean drinking water. *Why could this technology be a water purification breakthrough?*

Sterilization

When did cleanliness become important to health care?

Until the late 1800s, doctors performed surgery in their street clothes and did not always wash their instruments or their hands. Hospitals were seldom clean. Pathogens thrived, and many patients died from infections. Joseph Lister, an English surgeon, discovered that infections could be prevented by washing items with disinfectants—chemicals that kill pathogens. The disinfectants he developed saved many lives.

Today medical procedures require sterile (very clean) methods. Surgical instruments are put in an autoclave, a device that uses steam and pressure to kill pathogens. Doctors scrub their hands and wear latex gloves and special garments.

Water Treatment

How is water purified?

Clean drinking water is important to health. Therefore, most local community water supplies are purified before water can pass through the distribution system.

The water may be given several treatments, depending on the quality of the untreated water and the local regulations. Water is pumped through a filtering system that uses sand, gravel, and minerals to filter out any remaining impurities.

Fluoride Treatment

Some communities add fluoride to the water to help prevent tooth decay. Fluoride is a chemical compound that hardens tooth enamel and improves dental health.

Nanotube Technology

The newest water purification technology uses a silicon chip that contains billions of nanotube membranes made of carbon atoms. Each nanotube is 50,000 times thinner than a single strand of your hair. Each nanotube strand purifies water by preventing anything wider than six water molecules from passing through it. (See **Figure 14.1**.)

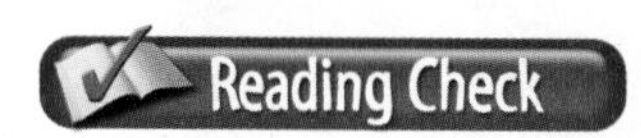

List What are some different water treatment methods?

Sanitation

What is sanitation?

Sanitation involves the removal of waste products that could cause disease or contaminate the environment. Public sewage systems dispose of waste from homes, factories, businesses, and public streets.

Consumed water that is used to wash your hands or take a shower is now labeled *grey water*. Some new buildings channel grey water to be reused to water lawns and gardens. Water from toilets is labeled as *black water* and must go through several treatment stages to make it safe for the environment. When waste water from sinks and toilets is not separated, it must all be treated as black water.

Sewage Treatment

At a sewage treatment plant (see the photo below), water is removed from the waste, purified, and released back into the environment. Solid waste is treated to make it safe to transport to landfills or to be burned.

Hazardous Waste

Hazardous (dangerous) wastes from medical facilities require special separate treatment to protect people from harmful organisms and disease. If possible, the waste is treated to make it harmless. Otherwise, it is burned or buried. The proper disposal of medical products contributes to medical safety.

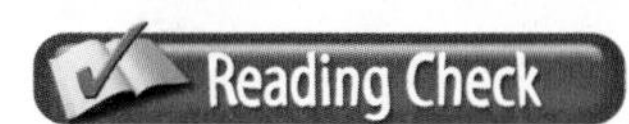

Explain Why is sanitation so important?

Waste Management This aerial view shows a sewage treatment plant. *Is grey water treated differently than black water when it enters this plant?*

BusinessWeek Tech NEWS

Scorpion Venom Fights Cancer

Researchers from Seattle Children's Hospital and Fred Hutchinson Cancer Research Center found a way to eliminate cancer cells using protein from the deadly scorpion. Chlorotoxin, found in a scorpion's venom, combined with fluorescent molecules light up cancer cells. This allows surgeons to clearly see and remove all cancerous material and tumors.

Critical Thinking *Why is it important to remove all cancerous material during surgery?*

Go to **glencoe.com** to this book's OLC read more about this news.

Immunization

How do vaccines prevent disease?

Immunization makes the body resistant to a disease by causing the immune system to attack the disease's pathogen. If you are immune to a disease, you cannot catch it or spread it.

History of Immunization

Before modern medicine, people recognized that if you survived certain illnesses, you would not catch the same disease again. Edward Jenner, an English doctor, noticed that people who caught cowpox, a mild disease, seemed to be safe from smallpox, which was often a fatal illness. In 1796, by using killed or weakened cowpox organisms, he created a vaccine for smallpox.

Vaccines do not cause disease, they just stimulate the body's immune system to recognize and attack the pathogen to prevent the disease. A vaccine cannot cure a disease you already have, so vaccines are usually given to you when you are young. Medical biotechnologists have recently developed a vaccine that can prevent cervical cancer.

Tech Stars

Dr. Jonas Salk

Developer of Polio Vaccine

The paralyzing viral disease known as *polio* was first recognized in 1840. By the 1940s and 1950s, the illness caused panic and dread as epidemics struck in the United States. Over 21,000 cases were reported in 1952. Its effects were crippling. In addition to paralysis, many people had difficulty breathing. Years of research produced no cure or vaccine. But in 1955, a physician and researcher named Dr. Jonas Salk developed a vaccine against the disease.

Dr. Salk became famous overnight for his discovery. He founded the Jonas Salk Institute for Biological Studies in La Jolla, California. There he continued his research on the causes, cures, and preventions of various diseases, including AIDS and cancer.

Not for Profit Dr. Salk did not seek to patent or profit from his discovery. His own words were: "Who owns my polio vaccine? The people! Could you patent the sun?"

English Language Arts/Writing Describe in a few paragraphs how important vaccines are in battling highly contagious diseases such as polio.

Go to **glencoe.com** to this book's OLC to learn about young innovators in technology.

Present and Future Immunization

Pharmaceutical (far-mah-SOO-tik-uhl) companies use special technologies to make current vaccines and develop new ones. Their goal is to find a weakened or dead strain, or even a small piece of a pathogen's structure that will cause an immune response.

The First Shot Children might not be anxious to go to a doctor to get an immunization. *What is a vaccine?*

Healthier Living

How does good nutrition help prevent disease?

Some diseases are not caused by pathogens. For example, heart disease may be related to unhealthy habits, such as smoking, eating diets high in certain types of fat, or lack of exercise. Proper nutrition helps keep the body strong and able to defeat or survive pathogens and other causes of disease. To keep medical costs down, laws have been passed to prohibit smoking in public places, eliminate trans fats in foods, and curb antibiotic and hormone use in our food.

Many food manufacturers add vitamins and other nutrients to their products to ensure that people receive adequate amounts. Major nutrients contained in processed foods are listed on food labels. Manufacturers also add preservatives to foods. Preservatives are chemicals that prevent spoilage.

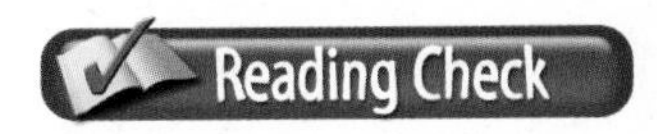

Identify What kinds of laws have been passed to promote healthier living?

Ergonomics

Why is comfort important to health?

Ergonomics, or human factors engineering, is the design of equipment and environments to promote human safety, health, and well-being. The people who work in this field design products considering people's limitations and comfort needs. See **Figure 14.2** on page 296. Have you noticed that certain tools, computer-game controllers, or kitchen utensils are comfortable to hold?

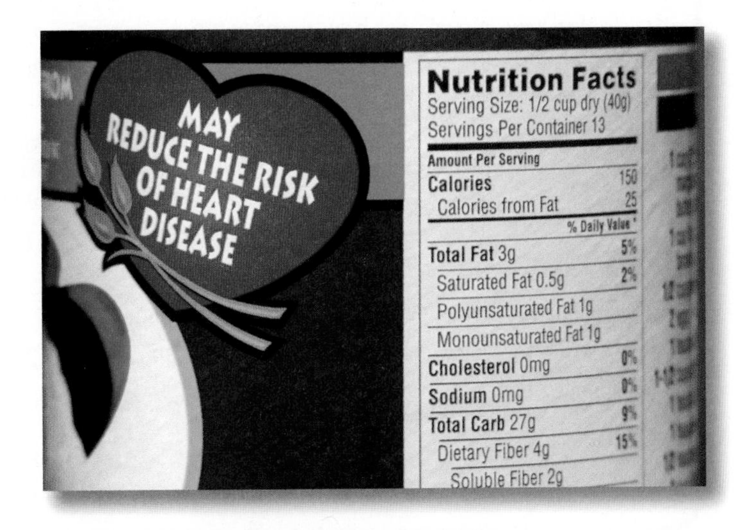

Knowing Your Nutrition Many food products are labeled with nutritional information. *Why do you think this information is important?*

Figure 14.2 Ergonomic Dimensions

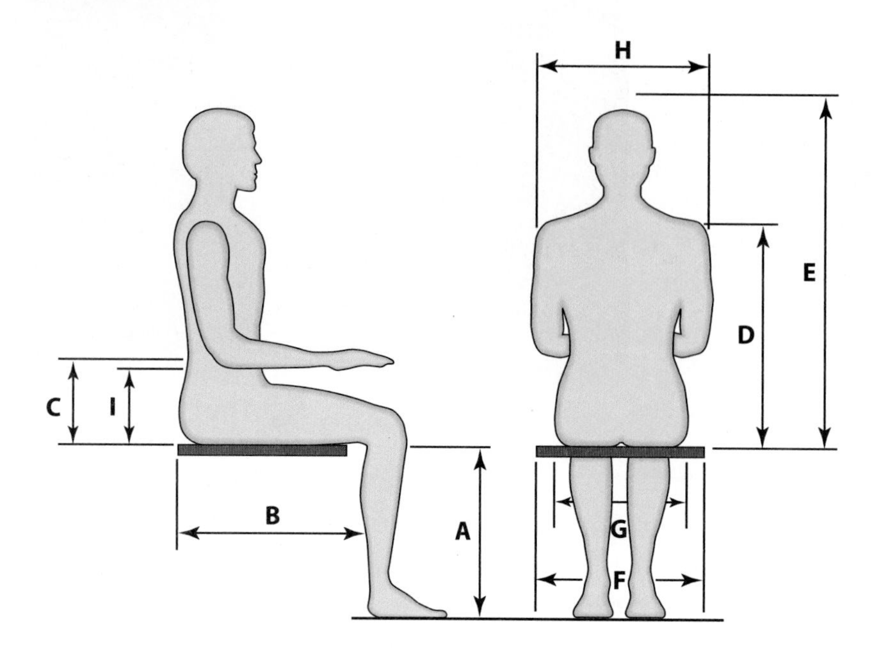

The Easy Chair Specific dimensions are used to ensure the comfort and usefulness of a chair. *What is the relationship between the price of a chair and its level of comfort?*

Ergonomic designers try to produce products that are comfortable to use and easy to control. This is especially important in the workplace where repetitive actions often lead to stress and injury.

Ergonomic engineers also help create life-sustaining environments that are needed by astronauts, deep-sea divers, soldiers, and extreme sports enthusiasts.

section 14.1 assessment

After You Read Self-Check

1. Describe a pathogen.
2. Compare pasteurization and irradiation.
3. Explain how immunization helps the body fight disease.

Think

4. Explain how ergonomics affect human health.

Practice Academic Skills

English Language Arts/Writing

5. Imagine you are camping in the woods near a stream. Describe how you would make sure you had a clean supply of water.

STEM Mathematics

6. A pharmaceutical firm produces a vaccine for animals. 250 vials are produced every day. If about 1 percent of the vials are damaged before they can be used, how many animals can be vaccinated using a 20-day supply of the vaccine?

Math Concept **Multi-Step Problems** Multi-step problems require extra attention to solve.

1. Make notes to help you organize the steps that need to be taken. Start by multiplying to find how many vials are made in 20 days.
2. Reduce that amount by 1 percent to find the number of animals.

For help, go to **glencoe.com** to this book's OLC and find the Math Handbook.

section 14.2

Diagnosis of Disease

Reading Guide

Before You Read **Preview** What is the meaning of the word *diagnosis*?

Content Vocabulary

- genetic testing
- CT scan
- ultrasound
- MRI
- endoscope

Academic Vocabulary

- conduct
- internal

Graphic Organizer

Draw the section diagram. Use it to organize and write down information as you read.

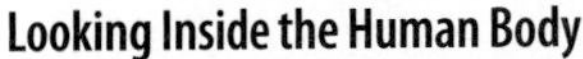

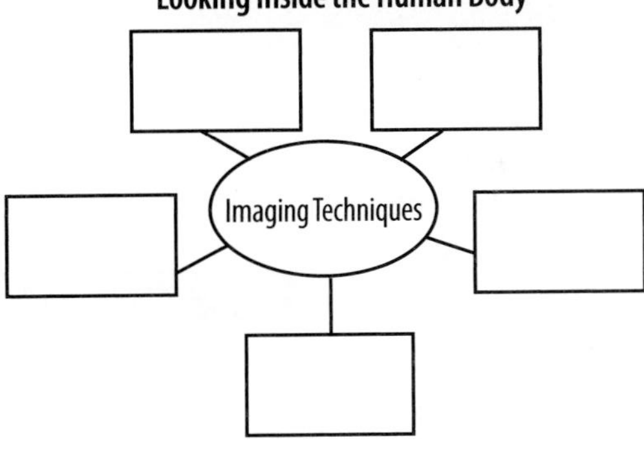

Go to **glencoe.com** to this book's OLC for a downloadable graphic organizer and more.

TECHNOLOGY STANDARDS

STL 2 Core Concepts of Technology
STL 3 Relationships & Connections
STL 4 Cultural, Social, Economic & Political Effects
STL 14 Medical Technologies

ACADEMIC STANDARDS

Science

NSES Content Standard E Understandings about science and technology

NSES Content Standard C Reproduction and heredity

STL *National Standards for Technological Literacy*
NCTM *National Council of Teachers of Mathematics*
NCTE *National Council of Teachers of English*
NSES *National Science Education Standards*
NCSS *National Council for the Social Studies*

Primary Physician

How do doctors diagnose an illness?

Disease prevention often starts with a general check-up with your primary physician. Your doctor will conduct some blood and other laboratory tests and use an 1816 medical invention called a "stethoscope" to listen to your heart, stomach, and lungs for abnormal sounds. A diaphragm in the stethoscope picks up the sound vibrations from your body and transmits them to the earpieces.

List What are the tools used by doctors to diagnose illnesses?

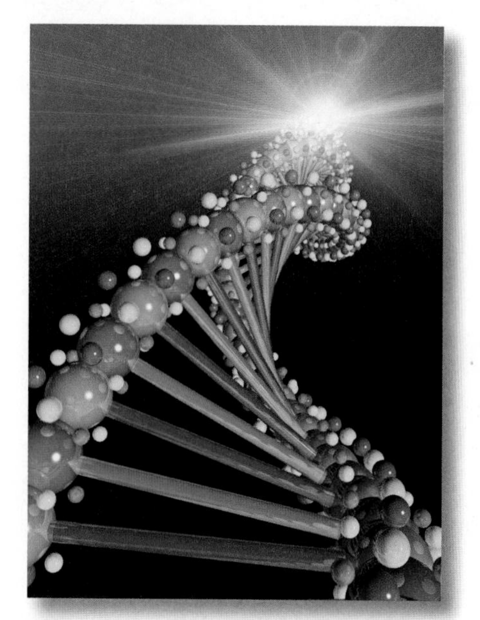

Life's Code A strand of your DNA carries a lot of information unique to only you. *When would a doctor recommend genetic testing?*

Laboratory Tests

How does the microscope help diagnose illnesses?

Microscopes make it possible to see pathogens and the cells that make up our bodies. Optical microscopes use lenses to magnify objects up to 2,000 times. Electron microscopes use a beam of electrons to magnify objects.

Using microscopes, doctors and technicians can examine body fluids for pathogens and other signs of disease. Along with other technical equipment, the microscope is an important laboratory testing tool.

Genetic testing is an evaluation to determine if your family has given you a gene that makes you more likely to get certain diseases. In genetic testing, DNA, the carrier of heredity, is examined. If test results say that you are at risk for a particular disease, you might be able to make certain lifestyle changes and reduce your risk of developing the disease.

Name What tool is used for lab tests?

Imaging Body Structures

How does an X-ray machine create a picture of broken bones in a person's arm?

Imaging machines can show doctors the inside of your body. The first imaging device was the X-ray machine. Fast-moving electrons give off X-ray radiation. This radiation passes through some parts of the body but not other parts.

Ethics in Action

Genetic Testing

Using just a few drops of blood, health care professionals can determine a person's risk for certain diseases. This is called "genetic testing," and it holds great promise for the future.

Disease Discrimination However, genetic testing has created new ethical questions. What if an employer does not hire a person because he or she is likely to develop a disease? What if an insurance company refuses to insure the person because treatment is too expensive?

English Language Arts/Writing

Facing the Future Each year, 4 million newborn babies in America undergo testing for genetic diseases. Some of the diseases do not have treatments. Should children be tested for diseases that have no cure? Parents and physicians have differing opinions.

1. Do some Internet research on current genetic testing.
2. Write your opinon in a paragraph and present it to your class.

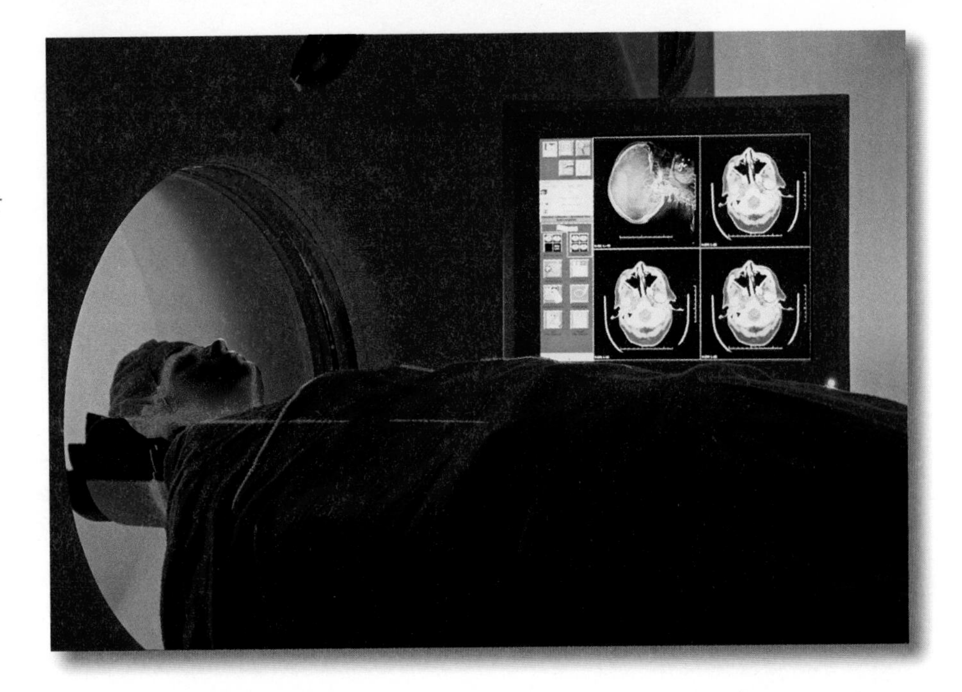

CAT Images This patient is undergoing a CT scan. *During a CT scan, what source of energy penetrates the body? What device converts the scan into images?*

The radiation not blocked by the bones of your body will expose the X-ray film. When the film is developed, your bones appear clear on the picture, and the areas where the film was exposed appear black.

To view places deep inside the body, a CT scan (computerized axial tomography) is used. In a **CT scan**, X-rays gather thousands of measurements of **internal** body structures. The measurements are then processed by a computer and are transformed into detailed images.

Ultrasound imaging bounces sound waves that you cannot hear off the structures inside your body. The reflected echo is interpreted by a computer and transformed into images. Ultrasound is popular for determining whether a fetus (an unborn baby) is developing properly. See the photo below. The newest form of ultrasound produces three-dimensional images.

For an **MRI** (magnetic resonance imaging), the patient is placed inside a magnetic field. As the magnetic field encounters internal body structures, it sends precise measurements to a computer. The measurements are interpreted, and then transformed into images.

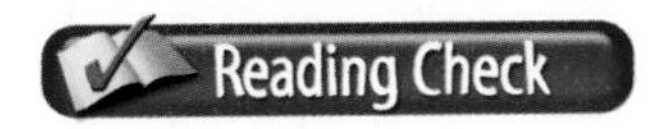

List What are some imaging methods?

Inside View An ultrasound image can see an unborn baby so doctors can check the baby's health. *What else can an ultrasound tell parents?*

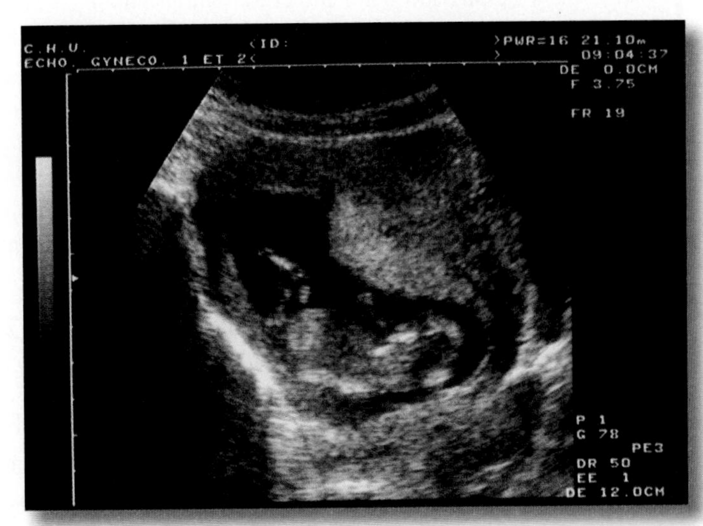

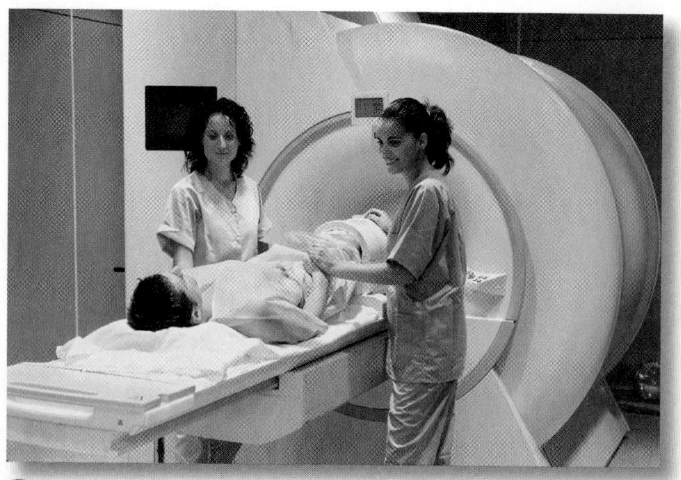

Magnetic Vision An MRI is another way for doctors to see what is going on inside the body. *When looking for problems in soft tissue, doctors might use an MRI instead of an X-ray. Why?*

To be sure of the diagnosis, doctors sometimes must examine a problem more directly. In these cases, they may use an **endoscope**. An endoscope is a flexible cable with a tiny camera and light. It is threaded into the body through a small incision or through the mouth. The doctor can view the image from the camera on a special screen. Endoscopes are commonly used to look inside a patient's stomach for ulcers and other problems.

Other Screening Methods

What are some other screening methods?

Some machines can read electrical impulses created by the body. One machine can produce an electrocardiogram this way. For an electrocardiogram (EKG), wires are attached to the patient's body, and a machine records the heart's electrical impulses. This information is printed out on a paper chart for doctors to read.

Another similar machine can create an electroencephalograph (EEG), a record of the activity of the brain. By reading the machine's EEG printout, doctors can tell if the brain is functioning properly.

section 14.2 assessment

After You Read Self-Check

1. Describe genetic testing.
2. Explain the difference between a CT scan and an MRI.
3. Describe an electrocardiogram.

Think

4. Discuss the pros and cons of being genetically tested for a disease.

Practice Academic Skills

English Language Arts/Writing

5. Research and write a short report about the first stethoscope. Then design, make, and test your own stethoscope. Present your report and demonstrate your stethoscope to the class.

STEM Mathematics

6. Josie went to a heart specialist for an evaluation. She walked on a treadmill while her heart was monitored. Her heart was beating 88 times a minute as she walked. If she walked on the treadmill for 35 minutes, how many times did her heart beat?

Math Concept **Using Equations** An equation can help you organize information to solve a problem.

1. Think of an equation as a math sentence that explains how one thing is equal to another.
2. Think through the steps. Then write an equation. Use a letter such as x for the unknown quantity.

For help, go to **glencoe.com** to this book's OLC and find the Math Handbook.

section

14.3 Treatment of Disease

Reading Guide

Connect What are some new medical treatments?

Content Vocabulary

- antibiotic
- laser surgery
- implant
- genetic engineering
- telemedicine
- bionics

Academic Vocabulary

- visual
- expert

Graphic Organizer

Draw the section diagram. Use it to organize and write down information as you read.

New Surgical Procedures

1. Laser surgery
2. ______
3. ______

Go to **glencoe.com** to this book's OLC for a downloadable graphic organizer and more.

TECHNOLOGY STANDARDS

STL 3 Relationships & Connections

STL 11 Design Process

STL 12 Use & Maintenance

STL 14 Medical Technologies

ACADEMIC STANDARDS

Social Studies

NCSS Content Standard 8 Science and technology in society

Science

NSES Content Standard F Science and technology in society

STL	*National Standards for Technological Literacy*
NCTM	*National Council of Teachers of Mathematics*
NCTE	*National Council of Teachers of English*
NSES	*National Science Education Standards*
NCSS	*National Council for the Social Studies*

Medicines

How were effective medicines first developed?

Some diseases can be cured with the help of medicines and other treatments. For non-curable diseases, doctors try to relieve symptoms and make patients more comfortable.

Bacteria, viruses, and fungi can cause many diseases. In 1928, Alexander Fleming, an English physician, discovered penicillin. Penicillin was the first antibiotic. An **antibiotic** is a medication that can kill bacteria and other germs. Penicillin is still effective against many life-threatening bacterial infections.

Define What is laser surgery?

Since 1928, many antibiotics have been created. However, because antibiotics have been overused, some strains of bacteria have grown resistant. This means the bacteria becomes too strong to be killed when exposed to the same antibiotics many times.

Some antibiotics can kill fungi, but no antibiotic works against viruses. Some common viral infections include colds, flu, and AIDS. Researchers have developed some antiviral drugs that suppress the infection. Some medicines, such as aspirin, relieve pain. Other medicines promote health, such as those that regulate blood pressure, reduce the spread of cancer cells, or control allergies.

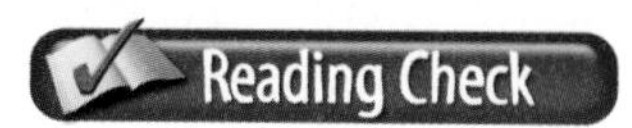

Connect Have you taken antibiotics? Why?

Surgical Procedures

What new surgical procedures are being used?

More than 40 million surgeries are performed in the United States each year. Some doctors use computer simulations to plan difficult surgeries. Using a computer and an MRI scan, the doctor enters surgical possibilities, and the computer analyzes the effects.

Some surgeries can be done by using very small incisions (cuts). A doctor can insert an endoscope with a cutting tool into the incision and guide it to the right location. Scars are smaller with this procedure, and patients require less healing time. Endoscopes are also being used to operate on babies in their mother's womb.

In recent years, lasers have been used to stop ulcers from bleeding and to correct vision. **Laser surgery** is surgery done with a laser beam instead of a scalpel (knife).Unlike scalpels, lasers do not cause bleeding. The surrounding tissue is vaporized. Because the laser is computer-controlled, cuts can be very precise.

Sound waves are also being used for surgery. Doctors use high-frequency sound waves to break up kidney stones in patients.

With robotic surgery, the surgeon uses controllers to tell the robot, located in the next room, where and how deep to cut, what to dissect, and where to suture (sew up the incision). The incision is small, and a patient's recovery is faster than with ordinary surgery. A robotic surgical arm can match the movement of a beating heart during surgery.

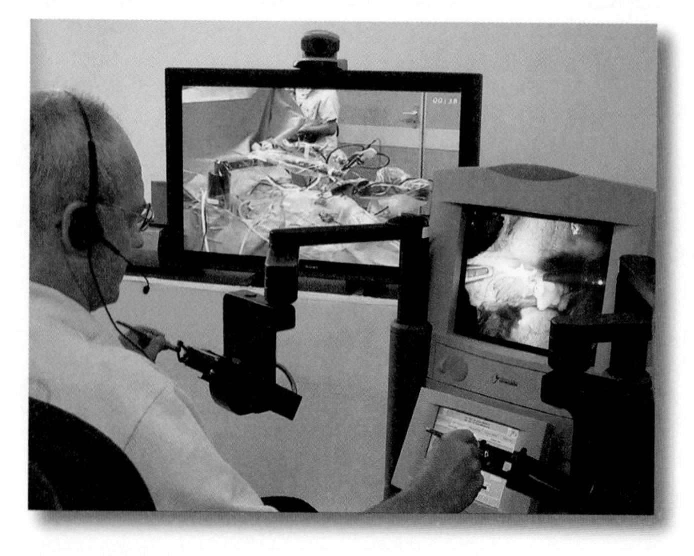

Precision Robotic Surgery Because a surgeon controls the robot via a computer, a robot surgeon can be very precise. *What kind of incision can a robot make?*

Reading Check **Identify** What are some surgical methods?

Electronic Implants

What are implants?

Implants are small devices inserted into the body to treat or solve medical problems. People may recover partial or complete hearing when an electronic implant is placed in the head. (See **Figure 14.3**.) Vision implants may be inserted into the eye or directly into the brain. The implant sends electronic signals to the visual cortex of the brain, which creates the image.

Pacemakers are implants used to stimulate a heart with electrical impulses. Researchers are working to develop pacemakers that also deliver medicine to the heart, radio for help in an emergency, and provide the doctor with a computerized report.

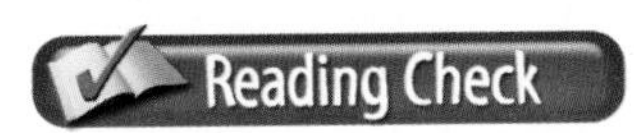

Identify What are some uses of implants?

Academic Connections
Science

Robotic Moves A robot's flexibility is measured in "degrees of freedom." Flexibility relates to how the robot's hand or arm can move—up or down and in or out. If there are more degrees, the robot is more flexible. Most robots have no more than six degrees.

Apply Research the joints in a robot's hand and wrist. Then compare them to the joints in a human's hand and wrist.

Genetic Engineering

What is genetic engineering?

Your genes can influence your risk for a particular illness. **Genetic engineering** is the process of altering or combining the genetic material in DNA to treat a disease or modify body characteristics. For example, if a person has a defective or missing gene, a normal gene could be inserted into a harmless virus that can take the new gene to the affected area. This is called "gene therapy." Gene therapy has been experimental. However, researchers hope that it will prove useful in treating illnesses such as cystic fibrosis and cancer.

Other genetic engineering research focuses on developing drugs for treating disease. Genes that can affect a particular disease are placed in other organisms. The organism produces substances that can treat the disease. The substances, or medicines, include vaccines and treatments for burns and some cancers.

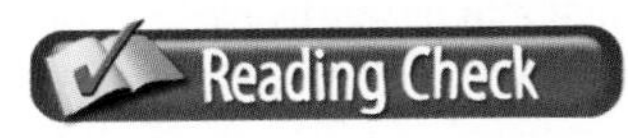

Explain What is gene therapy?

Figure 14.3 Hearing Implants

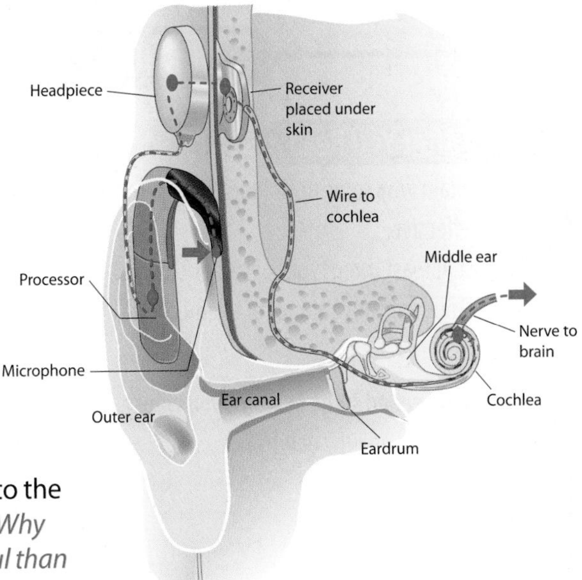

Hearing Signals Hearing implants send a signal to the middle ear. Nerves carry the information to the brain. *Why do you think hearing implants have been more successful than vision implants?*

Tissue Engineering

In an organ transplant, a kidney or a heart is removed from one person and placed in another person. But a patient's body may reject the organ. Tissue engineering may allow a new organ to grow in the lab, using the sick person's own cells. This would prevent rejection. Burn victims are treated with skin transplants. *Do you think bionic organs would be rejected? Explain.*

Go to **glencoe.com** to this book's OLC for answers and to learn more about tissue engineering.

Telemedicine

How can doctors treat patients who are far away?

Telemedicine is medicine practiced at a distance. Expert medical advice or treatment is given to patients through the use of computers, telephones, and/or Internet connections. A medical practitioner attaches electronic diagnostic equipment to the patient. The readings are then sent to the doctor who is usually miles away—sometimes in a distant country.

Bionics

What is bionics?

Medical engineers are developing systems that can replace many human body parts if they are missing, injured, or diseased. This medical area is called **bionics**. Regaining mobility or movement, due to loss of a limb, can help people have equality of opportunities.

Artificial organs are important because too few donor organs are available. However, all the materials to make the organs must be biologically neutral, so that the person's immune system will not attack what it considers to be a foreign body.

The new part must work in the environment of the human body for many years. It must duplicate the function of the part it replaces and not cause a breakdown in a body system. If a replacement heart valve causes blood clots, it cannot be used. If a replacement joint causes bones to break, it cannot be used. The AbioCor II artificial heart is a completely artificial heart that can be surgically implanted in place of a person's dying heart.

section 14.3 assessment

Self-Check

1. Explain why some antibiotics have become ineffective.
2. List at least three types of surgery.
3. Name some advantages of telemedicine.

Think

4. If you needed surgery, would you want a robot to do it? Explain why or why not.

Practice Academic Skills

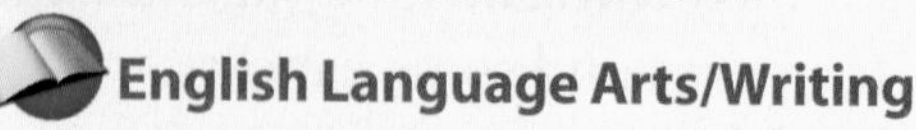

5. Research and write two paragraphs on any bionic device, such as an artificial heart. Make a poster or a model of the device and label its parts and functions.

STEM Science

6. Research how genetic engineering is helping to fight certain diseases. Write a few paragraphs describing the pros and cons of genetic engineering and the names of diseases that might be cured using it.

Exploring Careers in Technology

Kristen Dezzani

PHARMACEUTICAL ENGINEER

Q: *How did you get interested in your field?*

A: At 18, when I decided to major in chemical engineering, I wasn't sure what I wanted to do when I got out of college. My decision to become a chemical engineer was influenced by the fact that my father is a chemistry teacher and my brother is an engineer.

Q: *What do you do on a typical day?*

A: I work closely with my team to obtain status updates on our projects, as well as to make decisions on the path going forward. Most of my job has to do with interfacing with people, following up on projects with my team members, or reporting to upper management. The projects that we do are either facility- or equipment-related. We might bring in new storage tanks for solvents used in the manufacture of drugs or install a new process hot-water system to aid in the cleaning of manufacturing equipment.

Q: *What do you like most about your job?*

A: The challenges. Being a project manager is not as easy as it might seem. I have to meet deadlines and budgets. It's a very fast-paced environment that keeps me motivated and brings out the best in me. No two projects are alike, so I'm never bored. I also like interacting with people, which allows me to develop close personal and professional relationships. That's what keeps me coming back every day.

English Language Arts/Writing

Career Levels There are a variety of jobs that relate to most types of technology. Research different jobs related to chemical engineering or any other medical biotechnology:

1. Find a job for each of these levels of education: high school diploma, one-year certificate program, two-year degree, bachelor's degree, master's degree, and Ph.D.
2. Choose two of the jobs and find out the courses and training needed for these jobs. Then check job-search Web sites to find possible job openings.
3. Write a short report on your findings.

Go to **glencoe.com** to this book's OLC to learn more about this career.

Real-World Skills

Problem solving, adaptability, research, communication

Academics and Education

Chemistry, engineering, mathematics, English language arts

Career Outlook

Growth faster than average for the next ten years

Source: *Occupational Outlook Handbook*

chapter 14 Review and Assessment

Chapter Summary

Section 14.1 Medical technology is a biotechnology. Biotechnologies make use of information about living things. Prevention involves technologies that remove or kill pathogens. Governments remove pathogens from water supplies. Sanitation removes waste that causes disease or contaminates. Immunization helps the body fight disease. Ergonomics is the design of equipment and environments to promote safety, health, and well-being.

Section 14.2 Using microscopes, doctors and technicians can examine body fluids for pathogens and other signs of disease. In genetic testing, a person's genes are examined. Genes are the carriers of heredity. Imaging machines show doctors the inside of the body. The first imaging machine was the X-ray machine. Imaging technologies include CT scans, MRIs, ultrasound, and endoscopes.

Section 14.3 Penicillin was the first antibiotic. Many important medicines have been developed. In recent years, laser surgery has become important. Lasers do not cause bleeding. Implants are small electronic devices inserted in the body. People who are blind or deaf may recover partially or completely. Genetic engineering alters or combines genetic material to treat a disease or modify the body. Telemedicine is medicine at a distance.

Review Content Vocabulary and Academic Vocabulary

1. On a sheet of paper, use each of these terms and words in a written sentence.

Content Vocabulary

- pathogen
- pasteurization
- irradiation
- sanitation
- immunization
- vaccine
- ergonomics
- genetic testing
- CT scan
- ultrasound
- MRI
- endoscope
- antibiotic
- laser surgery
- implant
- genetic engineering
- telemedicine
- bionics

Academic Vocabulary

- seek
- ensure
- conduct
- internal
- visual
- expert

Review Key Concepts

2. **Discuss** disease prevention technologies.
3. **Explain** pasteurization and irradiation.
4. **Explain** how immunization works.
5. **Discuss** genetic testing.
6. **Describe** several imaging technologies.
7. **Identify** tests that read electrical impulses in the body.
8. **Explain** how antibiotics work.
9. **Explain** genetic engineering.
10. **Tell** the purpose of bionics.

Real-World Skills

11. Understanding Nutrition From package labels, collect the nutritional information from food you eat in a day. Make a chart listing the food and percent of the recommended daily amounts. Write a paragraph describing your diet. Discuss possible diet changes.

STEM **Technology Skill**

12. Imaging Systems There are many different types of imaging systems that doctors use to diagnose diseases.

a. Use the Internet to research different types of imaging systems. Research the diseases each system diagnoses.

b. Create a poster with an illustration of one of the systems. Include an image from the machine and text that explains the diseases.

Academic Skills

STEM **Science**

13. The design of products must consider a person's limitations and comfort level. Think of something you use often. Write a few paragraphs about its design.

STEM **Mathematics**

14. Courtney wants to know the age of her brother's wife. Her brother told her that his wife is one-eighth of their mother's age younger than him. If Courtney's mother is 64 and her brother is 35, how old is her brother's wife?

Math Concept **Algebra** Some problems can be easier to solve if you write an equation. An equation is like a sentence that explains how to solve a problem. The unknowns can be represented as letters such as x or y.

WINNING EVENTS

Scientist, Medical Technologist, and Ethicist

Situation You are part of a research team that will investigate a medical technology innovation.

Activity Brainstorm medical technology innovations. Choose one topic to study. Conduct research, prepare a report, develop sketches and drawings, and build a model and display of the technology. Give a formal presentation in class.

Evaluation The team's work will be evaluated by the following criteria:

- Report—well researched and written
- Sketches and drawings—accurate
- Display and model—attractive, accurate
- Presentation—informative, interesting

Go to **glencoe.com** to this book's OLC for information about TSA events.

Standardized Test Practice

Directions Choose the letter of the best answer. Write the letter on a separate piece of paper.

1. How much is $\frac{1}{16}$ of 438?

A 28.785 **C** 21.5

B 27.375 **D** 26.375

2. CT scans do not use a magnetic field to create images of body structures.

T

F

Test-Taking Tip Re-read all questions that include negative wording, such as *not* or *least*. Look out for double negatives used in a question.

chapter 14

TECHNOLOGY LAB

Crash Test a Vehicle

Your family car is designed to protect you in case of a crash. Its safety system includes: the front and rear bumpers, steel-reinforced door panels, lap and shoulder seatbelts, airbags, padded instrument panels, headrests, and a collapsible steering column.

The engineers who originally developed these systems considered the physical features of the human body. They tested their systems by using dummies in actual car crashes before a safety agency would approve the systems.

Tools and Materials

- ✓ Lumber for ramp and vehicle
- ✓ Technical drawing equipment or computer and CAD software
- ✓ 2 axles for each vehicle
- ✓ 4 wheels for each vehicle
- ✓ 2 eye hooks for each vehicle
- ✓ Uncooked eggs
- ✓ Ziploc®-style plastic sandwich bags
- ✓ String
- ✓ Protractor
- ✓ Glue
- ✓ Foam rubber
- ✓ Paper
- ✓ Rubber bands
- ✓ Balloons
- ✓ Camcorder

! SAFETY

Reminder

In this lab, you will be using tools and machines. You must pass a safety test. Follow appropriate safety procedures and rules so you and your classmates do not get hurt.

Set Your Goal

Your goal for this lab is to design and create a safety restraint system that can protect a raw egg from breaking during a vehicle collision. You and your team of three to five students will work together on this activity.

Know the Criteria and Constraints

In this lab, you will:

1. Be sure vehicles are the same so only the safety system is tested.
2. Include anything in your design that will directly or indirectly protect the egg. But you may not interfere with the vehicle's speed.
3. Be sure the egg or its shell is *not* strengthened in any way, including by hard-boiling it or coating it with nail polish.
4. Place all eggs in plastic bags before testing.
5. Send the vehicle containing the egg down a ramp at a 75 degree angle. Then it will crash into a wall. If the egg "survives" the crash, the design is successful!

Design Your Project

Follow these steps to design your project and complete this lab.

1. In teams, design a vehicle.
2. Build the vehicle from the design.
3. Design a ramp.
    - Build the ramp long enough so you can increase vehicle speed by starting the cars higher up on the ramp.
    - Build a protractor into the design so you can increase the ramp angle by specific degrees. A higher angle increases the vehicle's speed at impact.

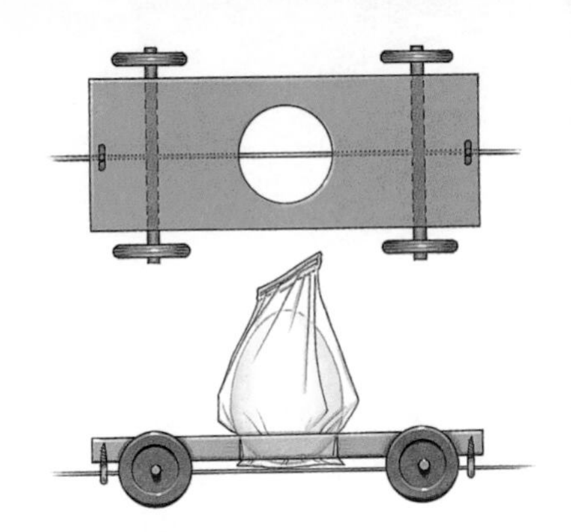

4. Place a string guide through eye hooks on the vehicle to guarantee it runs the full track.
5. As a class, select the vehicle that works the best.
6. Mass-produce this vehicle so that every team has a vehicle of the same design for testing their safety system.
7. Along with your teammates, design a safety restraint system.
8. Place your egg in a Ziploc®-style bag for easy cleanup.
9. Run a test of your restraint system and improve its design if necessary.
10. Hold a class-wide competition to see which team's restraint system allows the egg to survive the most violent crash.
11. If possible, record the competition on film.

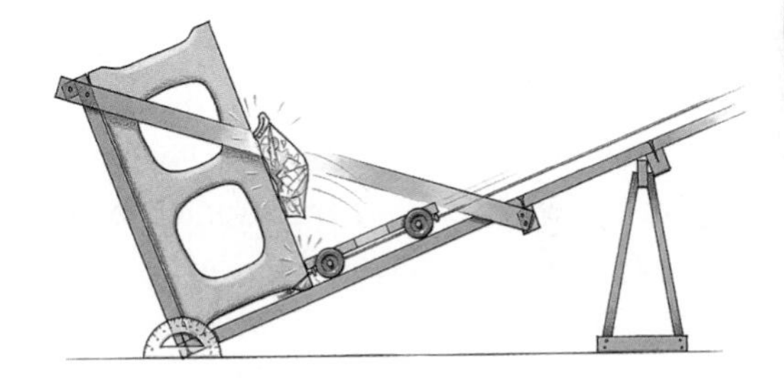

Evaluate Your Results

After you complete the lab, answer these questions on a separate piece of paper.

1. Did your egg survive the crash? What part of your design seemed to be most effective?
2. Did any team's egg survive more than one crash? What method did that team use to protect it?
3. Did the protections correspond to those used in an automobile? Explain.

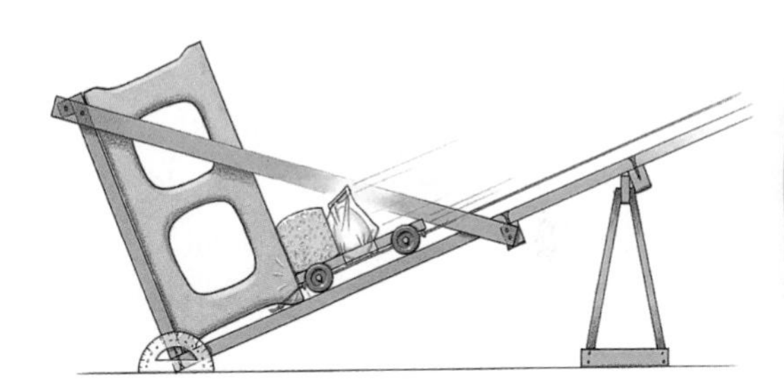

Academic Skills Required to Complete Lab

Tasks	English Language Arts	Math	Science	Social Studies
Design a vehicle as a team.	✓	✓	✓	
Build vehicle and test it on ramp.	✓	✓	✓	
Choose the best vehicle design as a class.	✓		✓	
As a team, design a restraint system and test it to see if your egg "survives."	✓	✓	✓	
Record the competition with camcorder.	✓		✓	

chapter 15

Agricultural Biotechnologies

Sections

What You'll Learn

- **List** highlights in the evolution of farming.
- **Explain** how animals are traditionally bred.
- **Discuss** methods used in plant and animal maintenance.
- **Discuss** genetic engineering in agriculture.
- **Describe** the cloning process.
- **Identify** the roles of the USDA, FDA, and EPA in approving new agricultural technologies.
- **Discuss** biosynthesis and pharming.
- **Explain** the purpose of bioremediation.
- **Describe** different kinds of artificial ecosystems.

Explore the Photo 

Mechanical Farm Hand One driver runs this harvester that picks and separates cotton for cotton bales. *What other things might advanced machines do to be more efficient?*

Construct a Hydroponic System

At the end of this chapter, you will be asked to build and operate a hydroponic system used to grow plants. Get a head start by using this checklist to prepare for the Technology Lab.

PROJECT CHECKLIST

✓ Go to the Internet and do a search using the keywords *hydroponic systems*. Find an example of a simple system.

✓ Your teacher will provide many of the materials needed for the lab, but you can begin to collect markers, water, and a pH testing kit.

✓ Ask your teacher to review safety procedures to follow before doing this lab.

section

15.1 Farming Technologies

Reading Guide

 **Before You Read** **Connect** Have you heard the word *hybrid*? What might it mean?

Content Vocabulary

- hybrid
- irrigation
- fertilizer
- monoculture farming
- dehydrate

Academic Vocabulary

You will see these words in your reading and on your tests. Find their meanings at the back of this book

- monitor
- income

Graphic Organizer

Draw the section diagram. Use it to organize and write down information as you read.

Outcomes of Farming

	Fertilizers	Antibiotics
Positive		Promotes Growth
Negative		

 Go to **glencoe.com** to this book's OLC for a downloadable graphic organizer and more.

TECHNOLOGY STANDARDS

STL 7 Influence on History

STL 13 Impact of Products & Systems

STL 15 Agricultural & Related Biotechnologies

ACADEMIC STANDARDS

Science

NSES Content Standard F Science and technology in society

Mathematics

NCTM Problem Solving Solve problems that arise in mathematics and in other contexts.

STL *National Standards for Technological Literacy*

NCTM *National Council of Teachers of Mathematics*

NCTE *National Council of Teachers of English*

NSES *National Science Education Standards*

NCSS *National Council for the Social Studies*

Agriculture's Biotechnology Roots

Why is agriculture classified as a biotechnology?

Agriculture is the science of cultivating the land to produce crops and raise livestock. As a biotechnology, it uses knowledge of biology to increase crop yields and improve livestock health. It also uses industrial processing to convert living organisms into processed foods, textile fibers, fuels, and medicine.

New agricultural biotechnologies use genetic engineering to modify species. These new animals and plants become healthier foods. They also become living chemical factories that produce medicines, plastics, and fuels, and can digest oil spills.

 **As You Read**

Predict How has farming changed since ancient times?

The Evolution of Farming

What tools and equipment did the first farmers use?

Low-Tech Plows Early plows were usually guided by a farmer who walked behind the animals that pulled the plow through the soil. *Why is the plow considered to be such a critical farming invention?*

Cave dwellers obtained food by hunting and gathering. They traveled with the seasons and followed animal migrations. Eventually they began to domesticate animals and plant crops. They worked the soil with sticks and primitive tools. Then, around 3000 B.C.E., people developed the plow, which enabled them to work larger areas of land more efficiently.

The Industrial Revolution produced a real revolution in agriculture. After 1750, people began to leave the countryside to work in factories in the cities. With fewer workers available, farmers began to rely on machinery to help them prepare the soil, plant, and harvest crops.

The tractor was introduced to farming at the beginning of the 20th century. Now, at the start of the 21st century, it is undergoing important changes. It is becoming part of the Information Age. Some tractor models are equipped with robotic control systems. Guidance information comes from the Global Positioning System (GPS). GPS is a series of satellites orbiting the earth that tell the farmer where the tractor is located. Computers also help monitor crop quality and make planting the next crop more efficient.

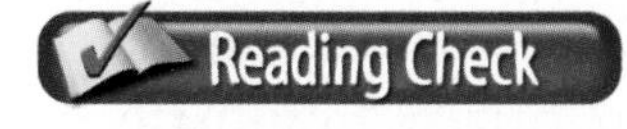

Name What was the name of the era when people began to work in factories?

Soil Mapping The GPS system, combined with a computer on board the tractor, allows the farmer to map each field to determine how much fertilizer and pesticide each area will need before planting. *How will this knowledge save the farmer time and money?*

Tech Stars

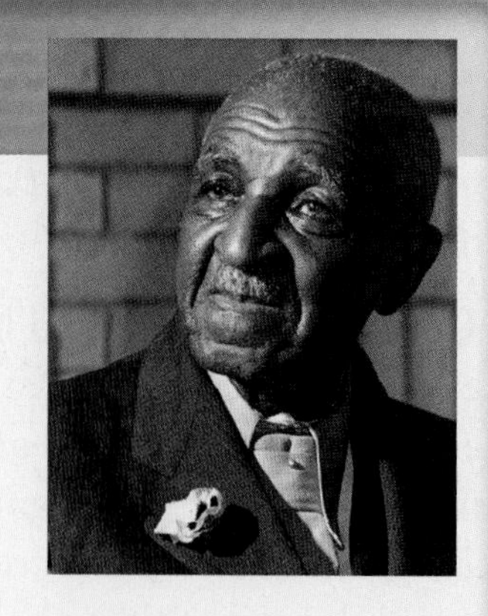

George Washington Carver

Agricultural Chemist

George Washington Carver was born into slavery on a Missouri farm in 1864. After the Civil War, when slavery was abolished, Carver pursued an education in horticulture. He eventually earned his master's degree. His research focused on developing byproducts from plants such as peanuts, soybeans, pecans, and sweet potatoes. He also researched improved soil conditions for agricultural purposes.

Among Carver's many contributions were recipes and improvements to hundreds of byproducts. These included cereals, oils, dyes, ink, instant coffee, shaving cream, carpeting, synthetic marble, and food substitutes. He owned three U.S. patents.

His Epitaph The words that are inscribed on his memorial say, "He could have added fortune to fame, but caring for neither, he found happiness and honor in being helpful to the world."

English Language Arts/Writing Write an article for your school newspaper discussing at least five items that you use on a daily basis that are byproducts from plants.

Go to **glencoe.com** to this book's OLC to learn about young innovators in technology.

Breeding Plants and Animals

How are plants and animals traditionally bred?

The first farmers probably saved seeds from wild plants and planted them. Over time, seeds from crops with desirable characteristics, such as hardiness or tastiness, were prized. Eventually certain plants were crossed with other plants to improve crop quality. Pollen from one plant was transferred to the flowers of another plant. When it worked, the seeds produced plants with the desired characteristics.

The first livestock were found in the wild and raised for human needs. Herders learned to select the best mates for breeding bigger, healthier, and easier to control animals. When animals of two different species or varieties are bred, the resulting animal is called a **hybrid**.

These breeding processes, though controlled by humans, are actually natural processes. The animals themselves have not been altered. More advanced breeding technologies are discussed in Section 15.2.

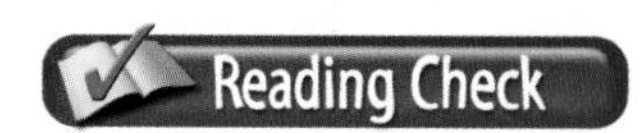

Identify How did some plants crossbreed with other plants?

Plant and Animal Maintenance

What is necessary to maintain plants and animals?

After seeds have been planted and animals have been purchased or bred, they must be cared for, or maintained. They need water and food, and conditions have to be right for their growth. They must also be kept free of disease.

Irrigation

Without water, rich farm soil quickly dries out and destroys the crops. Famines have been caused by too little rainfall. Too much water causes flooding, which can also destroy crops.

To offset dry spells, modern farms use **irrigation** systems to pump and sprinkle water where it is needed. To offset wet spells, the land on these farms is graded so that excess rainwater will flow away from the crops into canals and ponds.

Fertilizing

When a plant grows, it takes nutrients out of the soil. If you plant the same crop on the same plot of land repeatedly, crop yields will decrease. So, before the development of modern agriculture, farmers learned to rotate their crops. They might plant corn one year in a field and soybeans in that same field the next year. Some fields were left unplanted.

Then, in 1912, Fritz Haber, a German chemist, developed a process for making nitrogen fertilizer. **Fertilizer** is a chemical compound that restores nutrients to the soil. Mixing fertilizer that contains nitrogen, phosphorus, and potassium into the soil makes it possible to grow plentiful crops year after year without rotating them in the same fields.

A Mule is Not a Donkey A mule is a cross between a horse and a donkey. It is a hybrid animal. *Before off-road vehicles were invented, why did people prefer to ride mules instead of horses as pack animals in hot, mountainous terrain?*

Water When Needed This tractor irrigation system is mobile. A farmer can quickly set it up to provide watering where needed. *When would an underground sprinkler system that pipes water to certain locations be more practical?*

Ethics in Action

Organic Foods

More and more people are concerned about the food they eat. They want groceries that do not contain antibiotics, pesticides, or GMOs, which are genetically modified organisms. So, many consumers are turning to fresh food grown by organic farmers.

Going Organic Organic farmers let animals roam freely. They use natural feed and no antibiotics. Instead of using synthetic fertilizers, they use manure. By bringing back traditional techniques like crop rotation, farmers do not have to use pesticides.

English Language Arts/Writing

Good Business Pretend you and a partner are selling apples. Your partner wants to say they are "organic," even though they are not completely organic. By doing this, you can sell them at a higher price.

1. Think about what you would say to your partner and why.
2. Write a short role play for you and your partner.
3. Present your role play to the class.

Fertilizers have made **monoculture farming** cost effective. Monoculture is a farming practice that involves growing only one crop or one plant species on many acres of farmland.

Pros and Cons of Fertilizing

Fertilizer has caused some unexpected negative results. It has produced big crops of weeds. Runoff of fertilizer into lakes and oceans has affected wildlife and increased the growth of algae. On a monoculture farm, pests or disease can quickly spread through the single crop and completely destroy it, which negatively affects the farmer's income.

Use of Antibiotics

To keep costs low, many farm animals are raised in crowded conditions where diseases can spread very quickly. To prevent disease and promote animal growth, farmers started to give healthy animals antibiotics.

This technology, too, has negative outcomes. When antibiotics were given to healthy animals, small numbers of mutated pathogens that were antibiotic-resistant survived and multiplied. Now antibiotic-resistant pathogens cause diseases that can no longer be cured by using older, less expensive antibiotics.

The overuse of antibiotics by the general population, plus ingesting antibiotics when eating meat and poultry, has helped create diseases that are resistant to antibiotics. Also, antibiotics in animal waste have entered the soil and waterways.

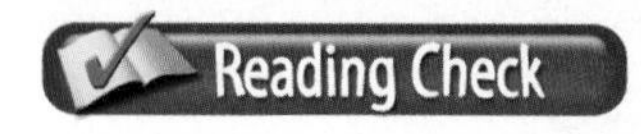

List What are three important processes for growing plants?

From Farm to Consumer

What processes are used during harvesting?

At harvest time, large machines go into the fields and pick cotton, corn, beans, and other crops. Waste, such as stalks, is separated out, and the rest is sent to a market or a factory for processing. Livestock, such as hogs and cattle, are shipped to processing plants where they are butchered for meat.

Cold Processing

Today most food is processed in very cold warehouses or factories to keep it fresh. Many food products are shipped in refrigerated trucks and stored in refrigerated cases at supermarkets. Freezing is also used for some food products.

Dehydration

Some foods, such as rice and other grains, must be kept very dry to maintain freshness. Other foods are **dehydrated**, which means the water they normally contain is removed. For example, raisins are dried grapes.

section 15.1 assessment

After You Read Self-Check

1. Define a hybrid.
2. Explain irrigation.
3. Name at least three methods for preserving foods.

Think

4. Discuss how tractors have become part of the Information Age.

Practice Academic Skills

English Language Arts/Writing

5. Write a short report about the possible solution to the overuse of antibiotics in meat-producing animals. Discuss how farmers might continue to keep their animals disease-free and also promote growth.

STEM Mathematics

6. Jennifer wanted to plant a vegetable garden. She made 4 rows of plants. In the first row, she planted 8 cucumbers. In the second row, she planted 12 carrot plants. She planted 5 eggplants in the third row and 3 tomato plants in the last row. If the seeds for each plant were sold in packs of 20, how many extra seeds would she have?

Math Concept **Choosing Operations** Some problems can be solved in more than one way. Think about the operations you will use to solve the problem and the order in which you will use them.

1. First, use multiplication to find out how many seeds Jennifer had in all. Then add up the number of seeds she used and subtract the sum from the total.
2. Another way to solve this problem is to subtract the number of each type of seed planted from 20, and then add up the extras.

For help, go to **glencoe.com** to this book's OLC and find the Math Handbook.

section

15.2

New Breeding Technologies

Reading Guide

Before You Read **Preview** What do you know about cloning?

Content Vocabulary

- gene
- DNA
- transgenic organism
- cloning

Academic Vocabulary

- principle
- convince

Graphic Organizer

Draw the section diagram. Use it to organize and write down information as you read.

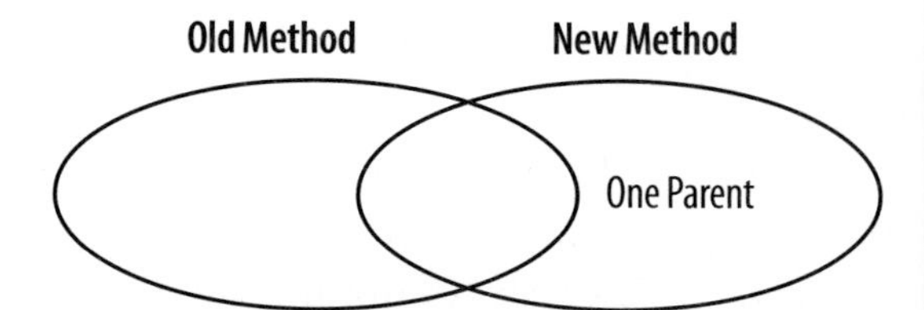

Go to **glencoe.com** to this book's OLC for a downloadable graphic organizer and more.

TECHNOLOGY STANDARDS

STL 6 Role of Society

STL 14 Medical Technologies

STL 15 Agricultural & Related Biotechnologies

ACADEMIC STANDARDS

Science

NSES Content Standard F Science and technology in society

NSES Content Standard C Reproduction and heredity

STL *National Standards for Technological Literacy*

NCTM *National Council of Teachers of Mathematics*

NCTE *National Council of Teachers of English*

NSES *National Science Education Standards*

NCSS *National Council for the Social Studies*

Genetics

What happens when one plant is crossbred with a different plant?

In 1865, the Austrian monk Gregor Mendel discovered the basic principles of genetics by crossbreeding pea plants. He discovered that each plant had hereditary elements that are now called **genes**. Genes contain all the information needed to reproduce an organism. Genes are small sections of **DNA** (deoxyribonucleic acid), which are the molecules that contain the genetic information that determines inherited characteristics.

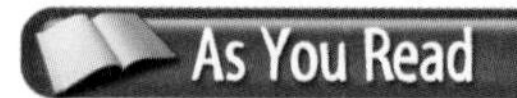

As You Read

Identify How is genetic engineering used today?

In the past, farmers crossbred the best of breed to produce the most favorable characteristics. Today genetic engineering and cloning are used to create new organisms. Many people question these processes and need to be convinced of their safety.

Genetic Engineering

How does genetic engineering relate to agriculture?

In the 1990s, researchers learned how to directly alter and recombine the genetic material in DNA to produce desired characteristics or remove undesirable ones. For example, a **transgenic organism** occurs when genes from one organism of a different species are transplanted into another.

GM Food Approval

It was very difficult to get the first genetically modified (GM) food, a tomato, approved. The Federal Drug Administration (FDA) regulates food additives. The U.S. Department of Agriculture (USDA) oversees food safety and agricultural research. The Environmental Protection Agency (EPA) is charged with protecting the environment. Before the "Flavr Savr" tomato could go on sale, these organizations had to sign off on its safety. Today approval of genetically modified species is mostly in the hands of the USDA.

To be approved, the offspring of the altered plants or animals must first be raised in a laboratory to see if the new characteristics show up and can be passed on. Tests are done to learn if the altered plant or animal could be dangerous to the environment. Other tests determine if food produced from these plants or animals is safe for human consumption.

Advantages of GM Crops

Improving plants genetically has produced crops that need less water, can grow in salty soil, are immune to certain diseases, can kill pests, can grow and ripen faster, and can stay fresh longer after harvesting. Also, improving livestock genetically has produced animals that can grow faster, are immune to certain diseases, and are larger and leaner than other livestock. Their meat tastes better and stays fresh longer.

Growth of GM Crops

A USDA biotech report indicated that, in 2007, a majority of different crops grown in the United States germinated from genetically engineered crops:

- 82 percent of corn crops
- 89 percent of soybean crops
- 86 percent of cotton crops

The current estimate is that 75 percent of all processed foods eaten in the United States contains GM ingredients. Although researchers claim these foods are safe for humans to eat, many people have doubts.

Academic Connections
Social Studies

What Came First? Throughout history, the "how" of technology often came before the "why" of science. For thousands of years, people used special "seeds" to make bread rise. Scientists later discovered that the seeds were actually yeast, a type of fungus.

Apply Obtain a packet of yeast used to make bread. Sprinkle it into a bowl of lukewarm water and add a pinch of sugar. Observe what happens over the next 20 minutes. Explain why the result you observe might occur.

Genetic Control Since 1997, many crops have been genetically modified. *In what way is each one considered superior to its non-modified species?*

Disadvantages of GM Crops

Many people fear that the pollen or seeds from altered plants could escape into the wild and affect other plants. For example, could the genes of a plant that is resistant to weed killer find their way into the weeds themselves? If so, the weeds might then be unstoppable. To prevent this, some seed companies have developed a "terminator" gene. This type of gene prevents a plant from reproducing. Its seeds will not grow. Any genetic changes would die with it. However, this situation might be even more dangerous. If the terminator gene escapes into the wild, all affected plants would die out in one generation.

Other Applications

You should not assume that genetic modification only modifies plants and animals for human consumption. In July of 2007, a research project, sponsored by the U.S. Defense Advanced Research Project Agency (DARPA), developed a biosensor that contains a genetically engineered yeast that has the ability to sniff out explosives. The researchers infused the specific rat genes for identifying odors into the yeast's DNA. They combined the special-function genes from the yeast and the genes from the rat to create a new organism biosensor that could smell and identify dangerous chemical odors.

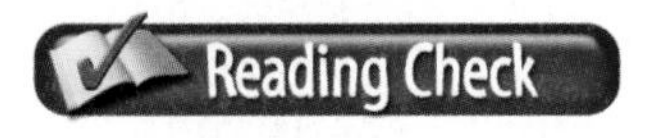

Recall What are some possible advantages and disadvantages of genetic engineering?

Cloning

What is cloning?

Cloning is a process that produces an identical copy of a plant or animal. Some clones, such as identical twins, are natural. A single cell divides in the mother's womb and produces two identical people. The older method of animal cloning uses a medical procedure to cause the egg in a pregnant animal to split. This produces identical twins by mimicking the natural accident that causes twins.

Cloning of some species of plants has been common for a very long time. For example, a plant cutting is placed in soil; it takes root, and a new plant grows. See **Figure 15.1**.

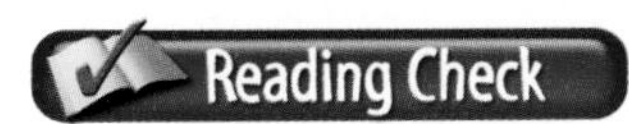

Explain How does cloning take place?

In 1996, a sheep named "Dolly" was cloned in a new way. This method creates a clone of an adult animal. The single parent and baby have identical DNA, even though their birthdays are years apart. A cell is taken from the animal to be cloned and placed in a weak nutrient culture where it will grow. The lack of all the needed nutrients causes the cell to stop dividing and switch off its active genes. Then an unfertilized egg cell is taken from a female animal, and its nucleus is removed. Although its genes are gone, the egg cell still has all the other things needed to produce a baby. The nucleus from the first cell is removed and fused into the egg cell with a spark of electricity. Another spark "wakes up" the sleeping genes in the nucleus, and the cell begins to divide. After a few days, the embryo is placed in the womb of another female animal and allowed to develop.

Figure 15.1 **Cloning Plants**

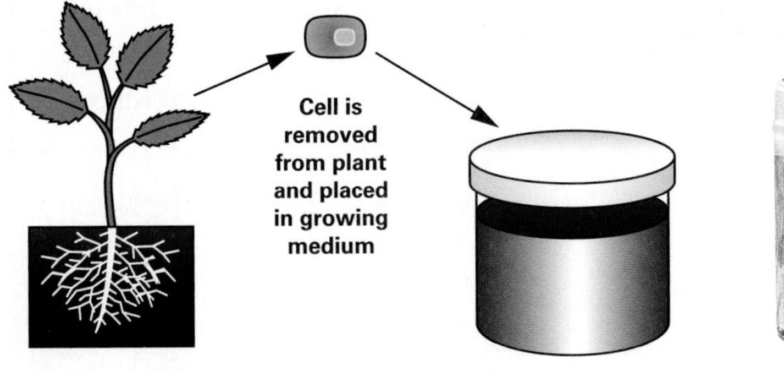

Two of a Kind In plant cloning, one piece of a plant or even one cell from the plant is used to grow a new plant. *What makes these two separate plants clones?*

The First Cloned Sheep Many animals have been cloned since Dolly was cloned in 1997. *Do you think it is all right to clone any living creature? Why or why not?*

Genetic engineering and cloning can be combined. A new genetically engineered cow cannot catch mad cow disease because the natural cow protein that gets infected with this disease has been eliminated. Altered cows and bulls with the genetic change were cloned into a herd of safe cows. Cows bred from this herd would remain immune to the disease because the protein needed for the disease no longer exists. Could this procedure be used to clone animals back into existence after they have become extinct? A Texas A&M University project, Noah's Ark, is freezing tissue of endangered animals in the hope that some day clones of them can be created.

section 15.2 assessment

After You Read Self-Check

1. Define a transgenic animal or plant.
2. Explain a clone.
3. Describe a terminator gene.

Think

4. The U.S. Department of Agriculture must approve new technologies that affect the foods we eat. Write a paragraph discussing the pros and cons of this mandate.

Practice Academic Skills

English Language Arts/Writing

5. The McIntosh apple is a cloned species. Do research on the Internet and/or at the library to find out the location of the parent tree and when the apple was first cloned. How many copies have been made, and where is it grown? Also, find out what other fruits have been cloned. Write a report on your findings.

STEM Mathematics

6. A field of soybeans is long and narrow. It measures 145 yards by 76 feet. What is the area of the field in square feet?

Math Concept **Measurement Conversion** Converting measurements involves multiplication or division.

1. Determine which measurement needs to be converted.
2. To solve this problem, remember that there are three feet in a yard.

For help, go to **glencoe.com** to this book's OLC and find the Math Handbook.

section 15.3

Other Agricultural Technologies

Reading Guide

Before You Read **Preview** What is aquaculture?

Content Vocabulary

- biosynthesis
- pharming
- bioremediation
- artificial ecosystem
- hydroponics
- aquaculture
- agroforestry

Academic Vocabulary

- source
- adequate

Graphic Organizer

Draw the section diagram. Use it to organize and write down information as you read.

Artificial Ecosystems

1. Hydroponics	2. Aquaculture	3. Agroforestry
		Fast-Growing

Go to **glencoe.com** to this book's OLC for a downloadable graphic organizer and more.

TECHNOLOGY STANDARDS

STL 5 Environmental Effects

STL 7 Influence on History

STL 15 Agricultural & Related Biotechnologies

ACADEMIC STANDARDS

Social Studies

NCSS Content Standard 8 Science, technology, and society

Science

NSES Content Standard C Population and ecosystems

STL *National Standards for Technological Literacy*

NCTM *National Council of Teachers of Mathematics*

NCTE *National Council of Teachers of English*

NSES *National Science Education Standards*

NCSS *National Council for the Social Studies*

Biosynthesis

Could a living organism produce a piece of fabric?

Biosynthesis is the making of chemicals by using biological processes. Technologists are working to develop genetically altered living organisms that will produce chemicals that were once produced in factories. For example, one microbe can manufacture polyester used to make clothing. Another can produce silk. A British company has developed biopolymers, which are plastics produced from living organisms.

As You Read

Compare What is the difference between *farming* and *pharming*?

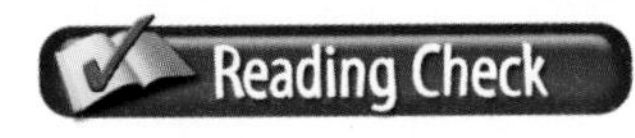

Reading Check **Recall** What is biosynthesis?

EcoTech

Pest-Free Naturally

Traditional pesticides can be as harmful to humans or pets as they are to the pests that they eliminate. Nontoxic alternatives to pesticides include natural repellants, such as herbs or soap, and predators like ladybugs.

Try This To get rid of pests, try using liquid dish soap, garlic, or household products such as borax. They can often eliminate or repel insects without harming people, pets, and your environment.

Pharming

How can biosynthesis create medicines?

Some crops, animals, and microbes have been genetically modified to produce medicines. This process is called **pharming** (*pharmaceutical* and *farming*). Plants, animals, and microbes become bio-factories that convert food and water into medicine.

Certain transgenic animals, such as cows, are able to produce compounds in their milk that are active against diabetes, arthritis, hemophilia, emphysema, and gastrointestinal infections. A tobacco plant can produce vaccines and human growth hormones. Other transgenic organisms produce human blood components and anti-tooth decay compounds.

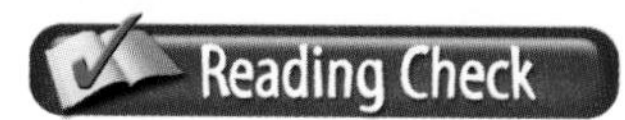

Recall What are bio-factories?

Bioremediation

What is bioremediation?

Bioremediation is the use of bacteria and other organisms to clean up contaminated land and water. Ordinary microbes work in landfills (dump sites) to break down garbage. Petroleum waste can be biodegradable by mixing it with soil, nutrients, and microbes. Engineered microbes clean up oil spills by eating the oil. Today other microbes remove dangerous materials from toxic (poisonous) wastes.

Biofuels

What kinds of fuels can agriculture produce?

Biofuels are fuels made from agricultural products. Biofuels are a source of renewable energy. Materials such as corn, crop wastes, and lumber wastes can all be used to manufacture fuel alcohol. When this alcohol is mixed with gasoline, it is sold as "gasohol."

Plant and animal wastes can be mixed with bacteria to produce methane gas, which is similar to natural gas and propane fuel. Methane gas can also power vehicles. Some farms sell their waste products to power plants that can recycle the wastes to produce electricity.

A Way to Go! Gasohol is a combination of gasoline and alcohol made from corn and crop waste that can substitute for plain gasoline. *Why has gasohol affected the price of food?*

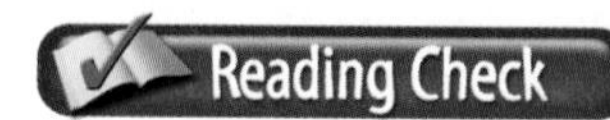

Connect What biofuels could run a car?

Water Culture Sprouts grow as the conveyor moves in this hydroponic hot house. *How are these plants watered?*

Artificial Ecosystems

Why are artificial ecosystems developed?

An **artificial ecosystem** is a human-made, controlled environment that is built to support humans, plants, or animals. Some artificial ecosystems include hydroponic farms, aquacultures, and agroforests. They copy some aspects of the natural environment. Astronauts in space live in an artificial ecosystem that supports their need for air, food, and water.

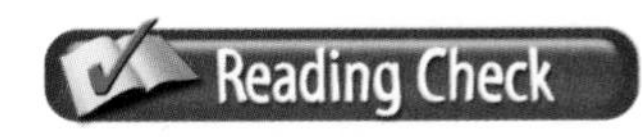

Name What are examples of some artificial ecosystems?

Hydroponics

Only about 6 percent of the earth's surface is adequate for traditional farming in soil. With **hydroponics**, plants are grown in nutrient solutions without soil.

Some hydroponic farms plant their seedlings on a long, wide, slow-moving conveyor system. The system moves at the same speed as the plants grow. When the plants reach the end of the conveyor system, they are ready for harvest.

Hydroponics re-circulates water and fertilizer. This means the fertilizer that is used is not released into the environment to pollute lakes and streams. Insects cannot reach the plants easily. Weather conditions are not important, so crops can grow in any climate. All these factors make some people think hydroponics could be used to help relieve world hunger.

Imagine This...

Breathing Easy

NASA has created an artificial ecosystem on the International Space Station. It is also working on systems that could sustain life during space journeys to distant planets. One engineer spent 15 days sealed in a chamber, breathing oxygen produced by wheat plants. All air, food, water, and waste would be recycled for journeys that take many years. *Write a paragraph on how air, food, water, and waste could be recycled on board a spacecraft.*

Go to **glencoe.com** to this book's OLC for answers and to learn more about artificial ecosystems.

Aquaculture

When fish, shellfish, or plants that naturally grow in water are grown in artificial water ecosystems, the process is called **aquaculture**. Aquaculture farms can be a totally enclosed indoor environment or a pen in a lake, pond, river, or ocean. Predators that would destroy or consume the plants or fish cannot enter these enclosures protected by nets.

From Seed to Paper Paper-mill reforestation programs use very fast-growing tree saplings to grow enough new trees to replace harvested trees. *If using computers results in a paperless society, do you think reforestation will still be needed?*

Agroforestry

Agroforestry is turning forests into controlled environments dedicated to the replacement of trees. Under natural conditions, trees grow very slowly. When clear cutting is done, even if the entire forest is replanted with seedlings, it will be many years before the forest will return to its former state.

Hybridization and genetic engineering help to create fast-growing softwood and hardwood trees. New trees are quickly planted to rebuild the stock for future cutting seasons. Biotechnologists have developed a fast-growing tree species for papermaking.

section 15.3 assessment

After You Read Self-Check

1. Define bioremediation.
2. Identify products produced from pharming.
3. Name three kinds of artificial ecosystems.

Think

4. Discuss the advantages of biofuels versus fossil fuels.

Practice Academic Skills

English Language Arts/Writing

5. Write a radio commercial for a new biofuel. Choose a current biofuel being used today. Describe the advantages of using it. Also think of a catchy new name for the fuel.

STEM Science

6. Research cloning technology. Write a few paragraphs explaining the process of cloning and what achievements have been made in cloning. Discuss the ethical issues related to types of cloning. Include your opinion of cloning and what types of cloning, if any, should be allowed.

Exploring Careers in Technology

Sonia Barker

NATURALIST

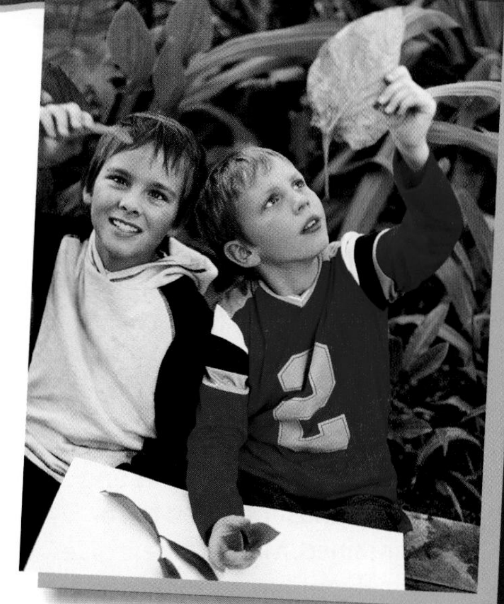

Q: *What got you interested in being a naturalist?*
A: I didn't even know what a naturalist was until I was a junior in college. The director of the arboretum at my college gave a special course on being a naturalist. I was studying research biology at the time. Being a naturalist appealed to me because it didn't deal with just one tiny part of science—you get to learn about all of it.

Q: *What do you do?*
A: I work as a resident naturalist at an elementary school. My first responsibility is to take care of the animals. I feed them and clean their cages. Then the students come down to my area, which is called "The Inquiry Zone." We start each class with a focus point. That could be an animal or an artifact that I bring from the Bell Museum of Natural History, my employer, who partnered with the school to make this job possible.

Q: *What do you like most about your job?*
A: I feel that even if I teach the same class 100 times, it's new every time, especially with younger kids. They have a fascination that catches me every time. Now that the kids know me, they bring things to my office. It could be something they found in their yard or maybe a bug that bit them. We get out books and discover what it is. They come and find out more, and it's great one-on-one time.

English Language Arts/Writing

Your Environment Find a familiar plant in your yard at home or in a nearby park. Find out the name of it. Do research on the Internet or at the library and write a report. Find out this information:

1. Does this plant grow naturally in your part of the country?
2. If not, where did it come from? If so, where else does it grow?
3. Can this plant be used for anything besides decoration?

Go to **glencoe.com** to this book's OLC to learn more about this career.

Real-World Skills

Communication, research, organization

Academics and Education

Biology, chemistry, natural history

Career Outlook

Growth faster than average for the next ten years

Source: *Occupational Outlook Handbook*

chapter 15 Review and Assessment

Chapter Summary

Section 15.1 Agriculture is a biotechnology because it is related to living things. It applies the principles of biology to create commercial products and processes. During the Industrial Revolution, people left the countryside to work in factories in the cities. Farmers began to rely on machinery. Breeding animals is part of agriculture. Plants and animals must be cared for, or maintained. Many farmers give animals antibiotics to protect them from disease.

Section 15.2 Gregor Mendel discovered the basic principles of genetics by careful breeding of pea plants. In the past, improving a species of plant or animal was a slow process. Today improvement is more rapid with genetic engineering and cloning. In some cases, genes from one organism are transplanted into another. The result is a transgenic plant or animal. Cloning is a process that produces an identical copy of a plant or animal.

Section 15.3 Some crops, animals, and microbes have been genetically modified to produce drugs. This process is called "pharming." Bioremediation is the use of bacteria and other organisms to clean up contaminated land and water. Biofuels are made from agricultural products, which can be a source of renewable energy. Artificial ecosystems include hydroponic, aquaculture, and agroforestry farms.

Review Content Vocabulary and Academic Vocabulary

1. On a sheet of paper, use each of these terms and words in a written sentence.

Content Vocabulary

- hybrid
- irrigation
- fertilizer
- monoculture farming
- dehydrate
- gene
- DNA
- transgenic organism
- cloning
- biosynthesis
- pharming
- bioremediation
- artificial ecosystem
- hydroponics
- aquaculture
- agroforestry

Academic Vocabulary

- monitor
- income
- principle
- convince
- source
- adequate

Review Key Concepts

2. **List** three highlights in the history of farming.
3. **Explain** how animals are traditionally bred.
4. **Discuss** methods of plant and animal maintenance.
5. **Give an example** of genetic engineering in agriculture.
6. **Define** cloning.
7. **Identify** the role of the USDA, FDA, and EPA with agricultural technologies.
8. **Describe** biosynthesis.
9. **Explain** the process of bioremediation.
10. **Identify** different types of artificial ecosystems.

Real-World Skills

11. **Organic Foods** Go to your grocery store and look at organic foods. Compare the foods labeled "organic" with similar foods not labeled "organic." Write a few paragraphs describing what you find, including the price and appearance.

STEM Technology Skill

12. **Farming Technology** The tools used in farming and gardening are constantly changing. Some are simple, and others are very complex.
    a. Research some of the different tools used in the agriculture industry.
    b. Make a model of one of the tools. Use available materials or sketch the tool. Change the design if you think it will improve the tool.

Academic Skills

Social Studies

13. Many people in the world suffer from hunger and even famine. Research and find areas of the world that suffer from famine. Write a few paragraphs describing the reasons for famine and ways to eliminate it.

STEM Mathematics

14. A farmer is changing fertilizer. The old fertilizer cost $140.80 for a 55-gallon drum. The new fertilizer costs $86.70 for a 30-gallon drum. Which fertilizer costs less per gallon and by how much?

Math Concept **Multi-Step Problems** Some problems require two steps. Take notes to organize information given. Write out what you are solving for to determine which step to take first.

WINNING EVENTS

Agricultural and Biological Scientists

Situation You are part of a research team which includes scientists, technologists, and ethicists who will investigate an agricultural or biotechnology discovery or innovation.

Activity Working in a team, brainstorm contemporary agricultural or biotechnology discoveries and innovations. Choose one topic. Conduct research, prepare a report, develop drawings, and build a model of your topic. Then give a formal presentation.

Evaluation The team's work will be evaluated by the following criteria:

- Report—well researched, well written
- Drawings—complete, accurate
- Display and model—complete, attractive
- Presentation—well organized, informative

Go to **glencoe.com** to this book's OLC for information about TSA events.

Standardized Test Practice

Directions Choose the letter of the best answer. Write the letter on a separate piece of paper.

1. What is the volume of a box that is 13.2 inches wide, 10.5 inches tall, and 24.4 inches deep?
   A 3,351.84 cubic inches
   B 4,800.1 cubic inches
   C 48.1 cubic inches
   D 3,381.84 cubic inches
2. Bioremediation involves genetically modifying plants and animals to produce medicine.
   T
   F

Test-Taking Tip Answer the simple questions first to help build up your confidence for answering the harder questions.

chapter 15

TECHNOLOGY LAB

Construct a Hydroponic System

Hydroponics is a good method of farming for regions where rainfall is scarce, soil is poor, insects are difficult to control, or farmland is limited. To grow plants, you need seeds or young plants, water, sunlight, air, nutrients, and a way to hold the plants in place. In hydroponic greenhouses, plants grow faster, take up less space, and produce larger crops. Water and nutrients are recycled.

Tools and Materials

- ✓ One 6-inch diameter plastic pipe, 2 feet long
- ✓ Two ¼-inch-thick plastic pieces, 6½-inches square
- ✓ One 4-foot length of ⅜-inch flexible plastic tubing
- ✓ Two adaptors with rubber washers and nuts to fit tubing
- ✓ Two ⅝-inch clamps
- ✓ 32-ounce plastic pails
- ✓ Drill set
- ✓ Electric drill
- ✓ 2¼-inch hole saw
- ✓ Marking pen
- ✓ Drill press
- ✓ All-purpose plastic cement
- ✓ Hot glue gun
- ✓ Hot glue
- ✓ Chlorine-free water
- ✓ pH testing kit
- ✓ Baking soda
- ✓ Vinegar
- ✓ Sterilized sand
- ✓ Fine plastic screening
- ✓ Plant food
- ✓ Four small potted plants
- ✓ Wood blocks
- ✓ 18″ × 10″ × ¾″ plywood

Set Your Goal

You will build and operate a hydroponic system used to grow plants.

Know the Criteria and Constraints

In this lab, you will:

1. Keep a daily record of solution mixtures and any experiments.
2. Monitor your system daily for water and to check for leaks.

Design Your Project

Follow these steps to design your project and complete this lab.

1. **Build Your Hydroponic System**
   - Stand the pipe (the planter) on end in the center of the plastic square. Draw a circle using the pipe as a template.
   - Mark the locations for the tubing adaptor on the plastic square and on the pail. Drill holes in both places.
   - Insert one tubing adaptor in the holes in the plastic square and the pail.
   - Place rubber washers on both sides of the adaptor to prevent water leaks. Tighten the nuts to hold in place.
   - On the inside of this plastic square, place a circle of hot glue around the tubing adaptor. Place a piece of screen on it.

- Place the pipe on plywood to keep it raised. Use stop blocks to keep it from rolling.
- Coat both pipe ends with plastic glue. Attach one plastic square to each end, making sure that the bottom edges rest flat on the table.
- Hold or clamp the parts together as the glue sets.
- Carefully set the project aside to dry.
- After glue is dry, mark spots for plant holes on pipe.
- Drill holes in the pipe using the hole saw in the drill press.
- Clamp tubing to the planter and pail.
- Fill the planter with water to check for leaks.
- Partially fill the planter with sand.

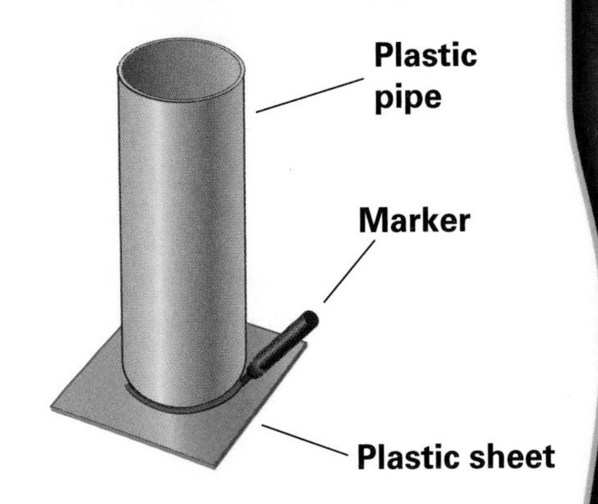

2. **Add plants to planter**
   - Remove plants from containers and wash away soil.
   - Carefully place your plants into the planter, filling in sand.
   - Test the pH of the water you set aside. It should read 6 or 7 for best growth. Add a few drops of baking soda if pH is low or a few drops of vinegar if pH is high. Retest.

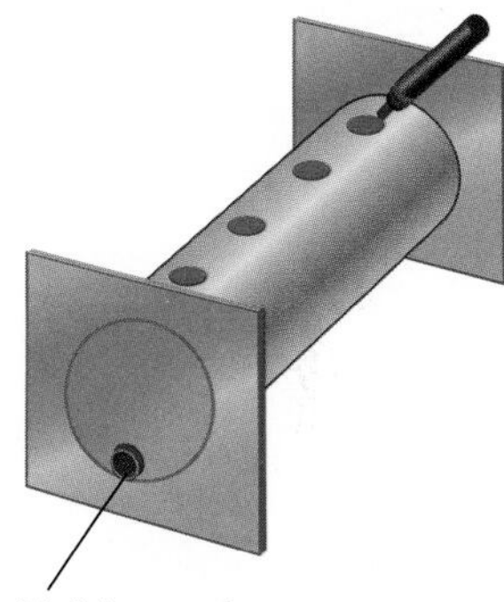

3. **Care for the Plants**
   - Mix plant food with water according to instructions. Place the pail higher than the planter so the mixture will fill the planter.
   - Lower the pail so the extra solution drains.
4. **Follow-Up**
   - Check plants daily and keep the sand moist. Add water to the pail to replace evaporated water. Keep a record.

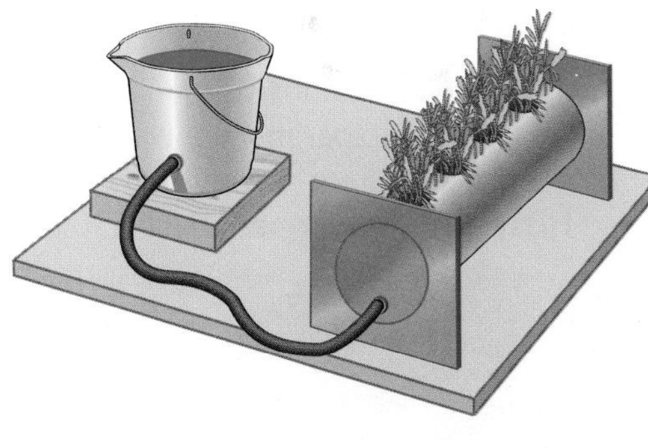

Evaluate Your Results

After you complete the lab, answer these questions on a separate piece of paper.

1. Which nutrient mixtures were most effective for plant growth?
2. If you repeated this activity, what would you do differently and why?

Academic Skills Required to Complete Lab

Tasks	English Language Arts	Math	Science	Social Studies
Build hydroponic system.	✓	✓	✓	
Test pH of water and adjust.		✓	✓	
Learn to care for plants.	✓		✓	
Keep track of water needed daily.		✓	✓	
Perform experiments on plants.	✓	✓	✓	

Technology Time Machine

Understanding Genetics

Play the Game This time machine will travel to the past to show you that genetics is the science of heredity. It can explain how specific traits are passed on from generation to generation. To operate the time machine, you must know the secret code word. To discover the code, read the clues, and then answer the questions.

Clue 1

1857 The field of genetics began with an Austrian monk named Gregor Mendel. He studied what happened when plants with one set of characteristics were crossed with plants with another set of characteristics. He theorized that cells from the parent plants contained certain elements that influenced heredity. Today these elements are called "genes."

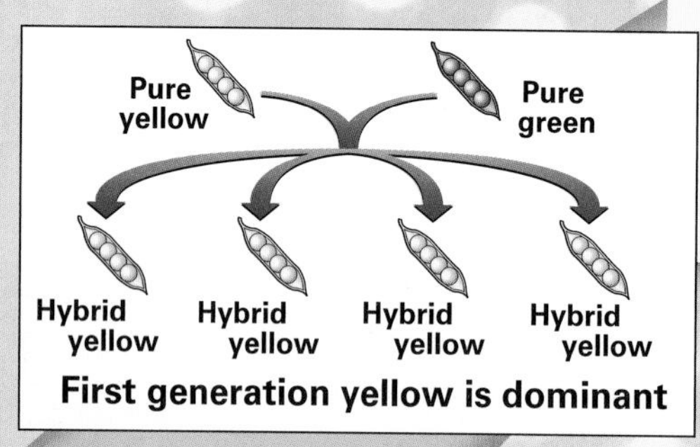

First generation yellow is dominant

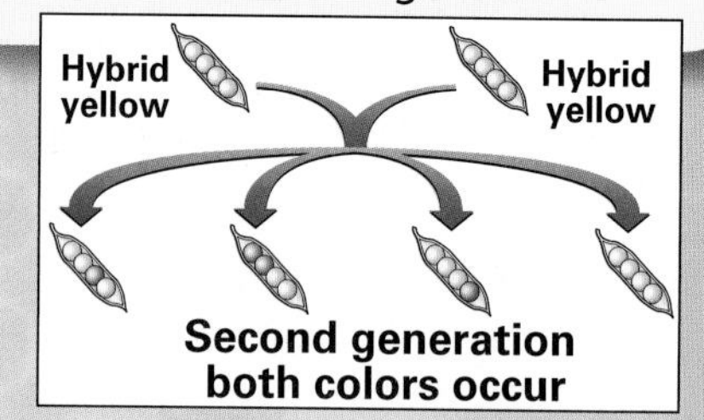

Second generation both colors occur

Clue 2

1907 Thomas Hunt Morgan used fruit flies for his gene experiments because they are simple creatures that reproduce quickly. His research confirmed that each gene is responsible for a particular characteristic.

Clue 3

1920s Researchers developed corn plants that increased crop yields by pollinating them for desired characteristics.

Clue 4

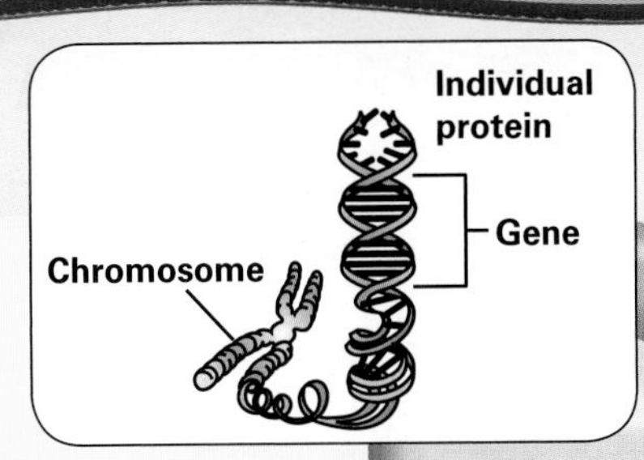

1953 Genetic researchers worked out the structure of DNA (deoxy-ribonucleic acid). Within each gene, DNA is a long chain of four molecules twisted into a coil that looks like a double helix. DNA molecules carry the genetic information.

Clue 5

1995 The bacterium *Haemophilus influenzae* was the first free-living organism whose DNA was successfully analyzed. In the United States, the Human Genome Project is an effort to analyze human DNA.

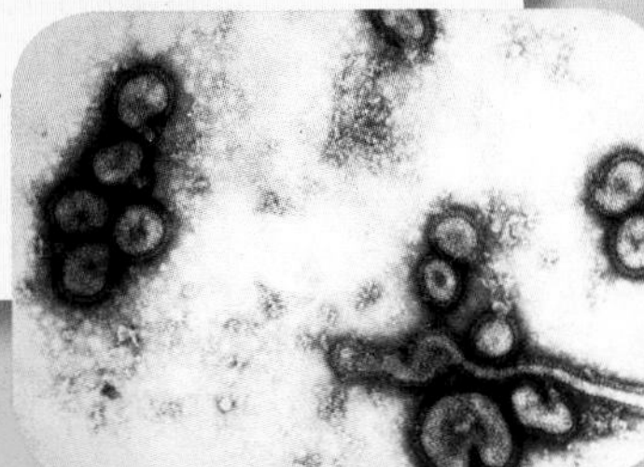

Clue 6

2000s Genetic codes may have been formed more than 3 billion years ago. Some researchers believe humans can be traced back to an original parent they have nicknamed "Eve."

Crack the Code

On a piece of paper, write the answers to these questions:

1. What type of molecule carries genetic information?
2. What did Gregor Mendel theorize cells from parent plants could do?
3. Who hopes to discover secrets in the genetic code?
4. What is the nickname of a possible original parent?
5. Where was Gregor Mendel from?
6. Morgan used fruit files because they were ________.
7. What did Morgan and Mendel do to make their discoveries?

Now write down the first letter of each answer. Put them together to discover the secret code word!

Hint Pathogens are organisms that cause this in living beings.

unit 4 Thematic Project

Discovering Careers in Biotechnology

Meeting Human Needs In Unit 4, you learned how biotechnology has improved our way of life. You also read about careers in biotechnologies, including medical, industrial, agricultural, and marine.

The Colors of Biotechnology

Red—Medical biotechnology offers careers in fields as varied as pharmaceuticals, genetic testing, gene therapy, and X-ray technology.

White—In industrial biotechnology, chemists design organisms to produce useful chemicals. Lab technicians use enzymes to destroy hazardous chemicals.

Green—Agricultural biotechnology touches all aspects of food production. Farmers grow crops, and chemists genetically engineer plants.

Blue—Marine and aquatic applications of biotechnology include restoring and preserving endangered marine species and aquaculture, also known as fish farming.

This Project In this project, you will research a career in biotechnology that you think would be interesting.

Your Project

- Choose a "color of biotechnology".
- Research career possibilities in that field.
- Choose and complete one task:
  1. Create a checklist of educational requirements for this career.
  2. Design a fictional day-in-the-life blog entry.
  3. Develop a fictional résumé.
- Write a report about what you learned.
- Create a presentation.
- Present your findings to the class.

Tools and Materials

- ✓ Computer
- ✓ Internet access
- ✓ Trade magazines
- ✓ Word-processing software
- ✓ Presentation software
- ✓ Posterboard
- ✓ Colored markers

The Academic Skills You'll Use

- Communicate effectively.
- Speak clearly and concisely.
- Use correct spelling and grammar when taking notes or writing presentations.
- Think about what it takes to be a biotechnologist.

English Language Arts

NCTE 4 Use written language to communicate effectively.

NCTE 12 Use language to accomplish individual purposes.

Science

NSES Content Standard E Science and Technology: Understandings about Science and Technology

Step 1 Choose Your Topic

You can choose any career in the biotechnology field. Examples might include:

- Biopharmaceutical engineer
- Bioinformatics scientist
- Farmer
- Marine biotechnician

Choose a job that interests you!

Step 2 Do Your Research

Research your project. Your fact-finding may include several activities. Answer these questions:

- What does someone with that job do during a day?
- What college courses are required?
- What do companies' online job postings say they require in a biotechnologist?
- What do old and new articles in libraries and online sites say about your topic?

Remember to list all your resources!

Step 3 Explore Your Community

Find someone in your community who knows something about your topic. Ask him or her in person or on the phone how his or her job connects to your career. Does biotechnology make the job easier? Is biotechnology controversial? Why or why not?

Thank the person afterward!

Step 4 Create Your Project

Your project should include:

- 1 research project (checklist, blog, or résumé)
- 1 report
- 1 presentation

Project Checklist	
	Objectives for Your Project
Visual	✓ Make a poster or slide presentation to illustrate your project.
Presentation	✓ Make a presentation to your class and discuss what you have learned.
	✓ Turn in research and names from your interview to your teacher.

Step 5 Evaluate Your Presentation

In your report and/or presentation, did you remember to:

- List your sources?
- Explain ideas using clear examples?
- Show thorough evidence?
- Practice before presenting?
- Write clearly?

GLOBAL TECHNOLOGY

Health Tourism

There are jobs for people who work in medical biotechnology all over the world. Tourists visit Thailand to see ancient temples and pristine beaches. But its state-of-the-art medical science and reputation as "The Land of Smiles" make it popular for health tourism. Patients from Asia, Europe, and North America travel there for vacations—and less expensive treatment.

Critical Thinking *What patients should probably not take part in health tourism?*

Thai	
hello	sa-wat dee
goodbye	sa-wat dee!
How are you?	Sabai dee rue?
thank you	khorb koon
You're welcome	Mai ben rai

Go to **glencoe.com** to the book's OLC to learn more and to find resources from **The Discovery Channel.**

unit 5

Manufacturing Technologies

Chapters In This Unit:

Unit Thematic Project Preview

Making Things Real

As part of this unit, you will learn about how manufacturers use technology to create products in mass quantities. These manufacturing technologies help produce products in the more efficient way possible.

As you read this unit, use this checklist to prepare for the project at the end of this unit:

PROJECT CHECKLIST

✓ Think of some products you use that are mass-produced.

✓ Look up factories that make the products in your area.

✓ Find out if the factories use computers and/or robots.

WebQuest Internet Project

Go to **glencoe.com** to this book's Online Learning Center (OLC) to find the WebQuest activity for Unit 5. Begin by reading the Task. This WebQuest activity will help you understand the history of manufacturing and its effect on our lives since the Industrial Revolution began in the 18th century.

Explore the Photo

Mass Production Making enough products to sell to people all over the country or world requires manufacturing technology. Imagine you got an order to make 10,000 globes in one month. How would you do it? Old and new manufacturing methods can help. *What types of machines do you think help make many items quickly?*

chapter 16

Manufacturing Systems

Sections

What You'll Learn

- **Discuss** how manufacturing evolved.
- **Explain** the difference between durable and non-durable goods.
- **Discuss** the importance of assembly lines and division of labor.
- **Describe** manufacturing systems using the universal systems model.
- **Explain** the concept of added value.
- **Describe** several manufacturing tools and processes.
- **Describe** general steps in setting up and running a small factory.
- **Explain** the function of market research, quality assurance, and just-in-time delivery.

Explore the Photo 

Waterjets Waterjet cutting uses water forced through a tiny hole under very high pressure. Openings on the dashboards on many automobiles are cut with waterjets. *What other kinds of technology can be used for cutting?*

Launch the TECHNOLOGY LAB

Manufacture a Bookstand

At the end of this chapter, you and two teammates will set up a production system to manufacture adjustable bookstands. Get a head start by using this checklist to prepare for the Technology Lab.

PROJECT CHECKLIST

✓ Take a look at some bookstands at a furniture store.

✓ Bring a pencil and paper with you and make sketches of bookstands.

✓ Gather the materials you will need for the project.

section

16.1 The Evolution of Manufacturing

Reading Guide

Before You Read **Connect** How have manufacturing systems evolved?

Content Vocabulary

- durable good
- non-durable good
- craft
- Industrial Revolution
- scientific management
- time and motion study
- division of labor
- assembly line

Academic Vocabulary

You will see these words in your reading and on your tests. Find their meanings at the back of this book.

- period
- establish

Graphic Organizer

Draw the section diagram. Use it to organize and write down information as you read.

Manufactured Products

Durable Goods	Non-Durable Goods

Go to **glencoe.com** to this book's OLC for a downloadable graphic organizer and more.

TECHNOLOGY STANDARDS

STL 2 Core Concepts of Technology

STL 3 Relationships & Connections

STL 4 Cultural, Social, Economic & Political Effects

STL 19 Manufacturing Technologies

ACADEMIC STANDARDS

Social Studies

NCSS Content Standard 2 Time, continuity, and change

English Language Arts

NCTE 2 Read literature to build an understanding of the human experience.

STL *National Standards for Technological Literacy*

NCTM *National Council of Teachers of Mathematics*

NCTE *National Council of Teachers of English*

NSES *National Science Education Standards*

NCSS *National Council for the Social Studies*

Manufacturing and the Modern World

Why is manufacturing essential to our way of life?

As You Read

Predict Think of some products that should not be produced in an assembly operation.

Most of the things we use every day were manufactured in factories. Manufacturing makes it possible to have a modern way of life. Without factories, we would not have cars, clothing, televisions, breakfast cereals, books, microwave ovens, or cell phones.

Products and Consumers

Manufactured products are sent to stores. Consumers are people who buy and use the products. We consume products by using soap, wearing clothes, and reading books. Without consumers, there would be no manufacturers.

Durable and Non-Durable

All products are either durable or non-durable. The federal government defines **durable goods** as those expected to last three years or longer. Television sets, hammers, and bicycles are all examples of durable goods. **Non-durable goods** are expected to last less than three years. Calendars and packaged foods are examples.

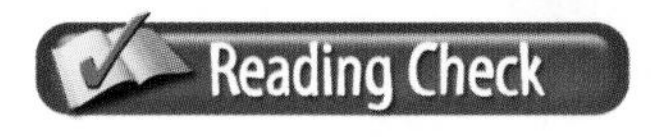

Contrast What is the difference between durable and non-durable goods?

Early Manufacturing

How has the evolution of manufacturing changed society?

The products that people used in the past were not always made in factories. People had to make their own fabric for clothing, farming tools, candles, wagons, furniture, and even their children's toys. It took a long time to make each item.

Not everyone could do a good job at making so many different things. Some people could make good wagons but not good tools. Others were better at making cloth. People soon began to specialize in just one type of **craft**, such as wagon making or shoemaking.

The Industrial Revolution

Around 1750, great changes started to take place around the world because of new methods of manufacturing. Because new methods were so important, that period of time is called the **Industrial Revolution**. A revolution brings about great changes. Goods once produced by hand were then produced by machines in factories.

Cities grew in size when factories drew people from farms to cities. The Industrial Revolution was a great turning point in the history of the world, and it caused many social changes. It changed the U.S. economy from an economy based on farming to one based on industry. However, factories were wasteful of workers' efforts.

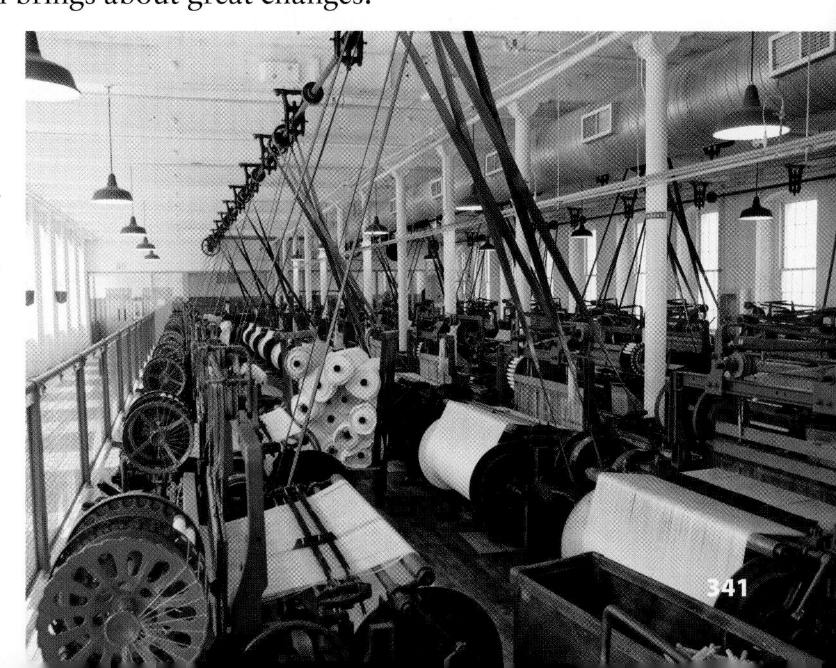

Producing Change Factories like this one were built during the Industrial Revolution, which was a great turning point in world history. *What kind of economy did the U.S. have before the Industrial Revolution?*

Figure 16.1 Division of Labor

Efficiency With division of labor, fewer workers are needed to manufacture larger quantities of high-quality products. *How does a flow chart help in establishing a division of labor?*

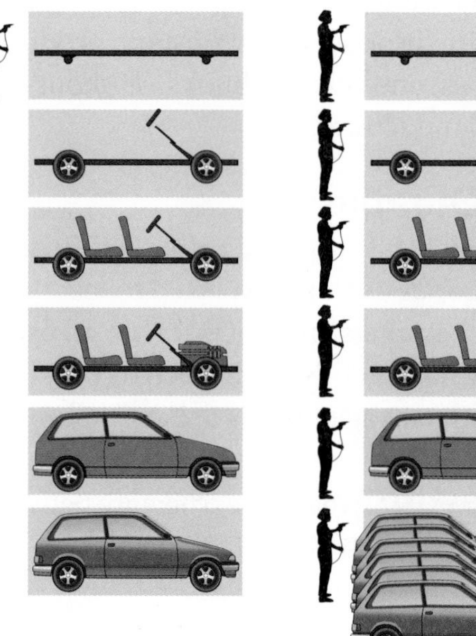

One worker doing all six required steps in manufacturing a product can make one unit.

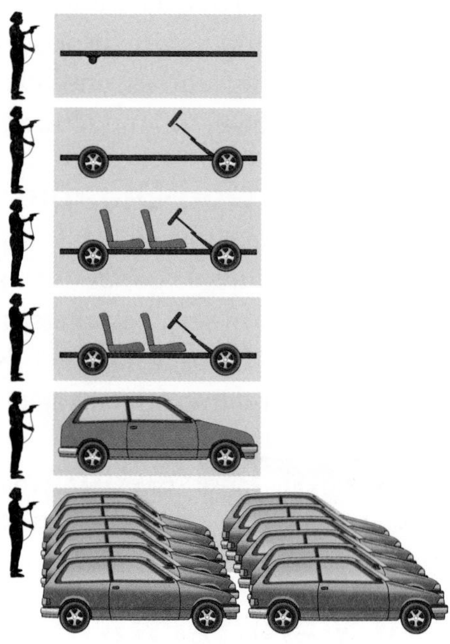

Six workers, each specializing in one of six steps, can make twelve units in the same amount of time.

Improvements in Manufacturing

As factory production increased, people began to look for better ways to do things. Frederick Taylor was the first person to develop standard ways of doing things. He made it easier for workers to manufacture products. Taylor's method is called **scientific management** and is an important part of modern factories.

The husband-and-wife team of Frank and Lillian Gilbreth also worked on better ways of doing things. Like Taylor, they worked in the new field of **time and motion study**. Time and motion study is a method of finding the way to do a job in the shortest time with the fewest movements. The Gilbreths improved factory working conditions and developed many techniques to measure factory efficiency.

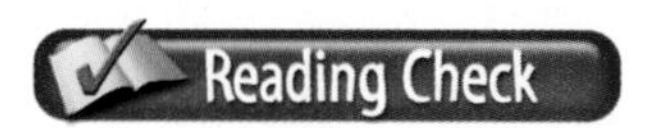

Summarize How has manufacturing evolved?

Modern Manufacturing

How can all factories be different and similar at the same time?

Early experts provided the foundation for establishing factories to manufacture products. However, each product is different and requires different materials and methods.

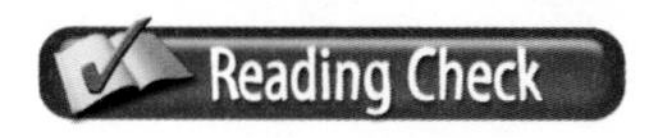

Predict What is division of labor?

Division of Labor

Many companies use a **division of labor**. This means dividing the work into smaller steps done by certain groups of people. Workers develop specific skills. See **Figure 16.1**. A person who is good at electrical wiring will wire while someone good at painting will paint. Sometimes a flow chart is drawn showing the steps for producing the product and how they relate.

Assembly Lines

Many companies use **assembly lines** or some type of assembly operation. In an assembly line, the workers often stay in one place and work on the product as it passes by. Automobiles and vacuum cleaners are put together this way.

Some products are assembled automatically by machines. For example, most medicines and foods are automatically processed and packaged. Other products do not lend themselves to an assembly operation. Examples include gas, paint, and tissue paper.

Quality Control

During the manufacturing operation, each company makes many checks on the product's quality. A company making canned soup has to keep the machinery and the food very clean. The product has to taste good and be nutritious. Such companies also have strict governmental regulations they must follow regarding content, safety, and portion size.

Academic Connections
Social Studies

Sales of Durable Goods To determine the state of the economy, government officials often use the most recent data for the sales of durable goods. Durable goods are usually more expensive than non-durable goods. If people are willing to invest in cars, televisions, and washing machines, they may be confident about the economy.

Apply Find out what kinds of durable goods are being manufactured. Make a list of the top five. Compare the list to the top five from 10 or 20 years ago.

section 16.1 assessment

After You Read **Self-Check**

1. Identify examples of durable and non-durable goods.
2. Explain effects of the Industrial Revolution.
3. Define a division of labor.

Think

4. Do you think all goods should be durable? Explain.

Practice Academic Skills

English Language Arts/Writing

5. Read a novel that takes place in England or the United States during or after the Industrial Revolution. Write a short report describing everyday life during the time in which the novel is set. Some examples of novels to choose are *David Copperfield* and *Oliver Twist* by Charles Dickens and *The Jungle* by Upton Sinclair.

STEM **Mathematics**

6. Renée works at a tire shop. Her main responsibility is repairing flat tires that people bring to the shop. One day she repaired 12 tires during a 7-hour period. If she is paid $6.75 an hour, how much was she paid for each tire she repaired?

Math Concept **Multi-Step Problems** Multi-step problems require extra attention to solve.

1. Figure out how much Renée made during her work day.
2. Next divide her wages for the day by the number of tires she repaired.

For help, go to **glencoe.com** to this book's OLC and find the Math Handbook.

section

16.2 Organizing a Manufacturing System

Reading Guide

Before You Read **Preview** How is a manufacturing system organized?

Content Vocabulary

- added value
- abrading
- resin

Academic Vocabulary

- distribute
- purchase

Graphic Organizer

Draw the section diagram. Use it to organize and write down information as you read.

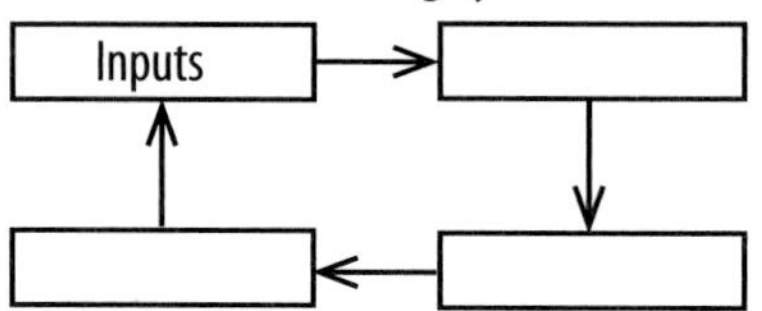

Go to **glencoe.com** to this book's OLC for a downloadable graphic organizer and more.

TECHNOLOGY STANDARDS

STL 2 Core Concepts of Technology

STL 17 Information & Communication Technologies

STL 19 Manufacturing Technologies

ACADEMIC STANDARDS

Science

NSES Content Standard E Understandings about science and technology

English Language Arts

NCTE 3 Apply strategies to interpret texts.

STL *National Standards for Technological Literacy*

NCTM *National Council of Teachers of Mathematics*

NCTE *National Council of Teachers of English*

NSES *National Science Education Standards*

NCSS *National Council for the Social Studies*

Inputs

What inputs are needed by manufacturing systems?

All manufacturing companies use the same system inputs to produce their products. They are people, materials, tools and machines, energy, information, capital, and time.

People

Larger companies have many engineers and technologists. Some people design the products the companies make. They are called "design engineers" or "designers/drafters." Production engineers decide the best way to manufacture the products. Quality control engineers and technicians inspect the product. Many people operate manufacturing machinery, while others set up or adjust the machines. Then people distribute and sell it.

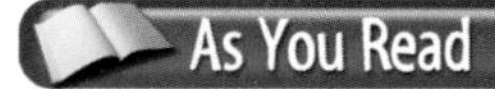
As You Read

Connect Think of some examples of manufacturing tools.

Production Materials

The materials companies use to make their products are called production materials, or engineering materials. They are different from raw materials. Raw materials are materials in their natural state, such as iron ore, trees, and raw cotton. Most production materials have already been processed to some degree. For example, trees have already been cut into lumber and dried. Materials are chosen for their different properties. Plastic used to make balloons, for example, has to be soft and stretchy.

You've Got Nails The steel wire used to make these nails is more valuable now. *What has been added to the wire?*

Added Value

Companies stay in business by adding value to production materials. **Added value** is the increase in a material's worth after it has been processed into a finished product. For example, a nail is a one-piece product made from a roll of strong steel wire. A nail-making machine puts a point at one end and flattens the other end. Companies add value to steel wire by changing it into nails.

Combining Materials

Other companies combine several different materials or make a product with many different parts. An ordinary flashlight has ten parts: plastic base, top, lens, switch, lamp (bulb) holder, spring, two flat pieces of metal, lamp reflector, and metal lamp conductor. The two batteries and lamp are purchased from other companies. See **Figure 16.2.**

Some companies make products that have hundreds, or even thousands of parts. You probably have some complicated products right in your own home.

Figure 16.2 **The Parts of a Flashlight**

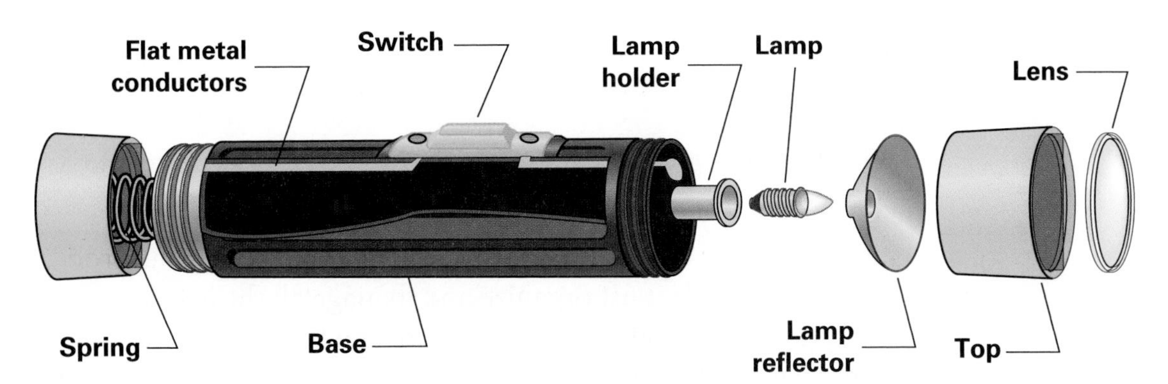

Fitting Together A flashlight has parts made from different materials. All the parts have to fit well and work together. *What parts of a flashlight are purchased from other companies?*

Power Tools Portable power tools are useful for many kinds of manufacturing processes. *What are some other examples of portable power tools?*

Tools and Machines

During manufacturing, tools and machines change the shape of materials and fasten them to other materials. Hand tools, portable power tools, and machine tools and their many purposes are discussed in Chapter 3.

Machine Tools

Machine tools are bolted to the floor and operated by electric motors. They are some of the most important tools used in a factory. Some are as large as a room. Others are smaller, such as a drill press. Machine tools are used for such jobs as drilling holes in jet engines and bending steel for car doors.

Portable Power Tools

A portable power tool uses a small motor for power and is usually handheld. Some power tools, such as a table saw, cannot be held in your hands or easily moved. People use power tools for such jobs as sanding flat surfaces on wooden furniture and assembling electronic parts.

Other Inputs

Energy is needed to run machines, heat buildings, and provide light. Capital (money) is needed to build factories, buy production materials, and pay workers. Of course, time is needed to do all the different jobs. Another important input is information. Workers need information in the form of drawings and instructions to make products.

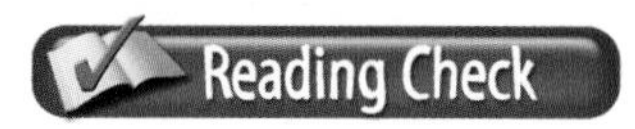

Explain How do companies add value to production materials?

Processes

What are some manufacturing processes?

Almost everything that takes place during manufacturing involves processes. Someone has to design and develop the product. Other people must organize and manage all the jobs that are done, make the product, and distribute and sell it. After the consumer has bought the product, another process involves servicing the product or making any repairs to a product that is returned.

Changing Size and Shape

Processing often involves changing the shape or size of a material. The shape or size of materials is often changed by cutting, bending, casting, forging, or extruding, which are processes discussed in Chapter 3. **Abrading** means to scrape or rub off small pieces. Filing, sanding, and grinding remove material by abrading it.

Chemical Processing

Conditioning is another process discussed in Chapter 3. Conditioning changes materials by using chemicals. Chemicals are used for etching the circuit boards in your computer.

Plastics are often processed from chemicals. The basic material is called a **resin**. Other ingredients are added to produce special properties, such as strength.

Fastening

Fastening is combining two or more parts together. Furniture, for example, is fastened together with nails, screws, and glue. Nails and screws are mechanical fasteners. Many companies use screws because they have more holding strength than nails. Glue is an adhesive, which is a chemical fastener.

George C. Devol, Jr. and Joseph Engelberger

Developers and Promoters of the First Robot

In 1954, George C. Devol, Jr. developed and patented Unimate I,the first industrial robot. He then joined forces with Joseph Engelberger in 1956 to form the company Unimation, Inc., which built and promoted the robots.

Devol's invention combined computer processing, electricity, and hydraulics. The Unimate I was able to respond to 200 commands. A self-made engineer, Devol also made contributions to the fields of industrial automation in machine vision and bar coding. He has over 40 U.S. patents.

Spreading the Word Engelberger was also an engineer, but it was his skills as an entrepreneur that helped launch Unimation. He was a vocal advocate of robotic technology in fields such as the service industry, manufacturing, health care, and space exploration.

English Language Arts/Writing Write a short essay explaining what types of commands you would want a robot to be able to perform.

Go to **glencoe.com** to this book's OLC to learn about young innovators in technology.

> **EcoTech**
>
> **Packaged with Care**
>
> Wasteful packaging from the past includes long cardboard boxes for CDs and Styrofoam for take-out food. Today you can promote sustainable living by buying products that are packaged with reusable, recyclable, or biodegradable materials.
>
> **Try This** When wrapping gifts, reuse materials such as newspaper, shopping bags, or used gift-wrap.

Plastic parts can be fastened with snaps. A small extension on one part snaps into a hole in another and holds everything together. This method allows the parts to be easily taken apart and recycled.

Plastic and metal parts can also be fastened by melting. Some plastic CD or DVD cases are heated and melted together. No mechanical fasteners or adhesives are necessary. Many metal parts of a car body are welded (melted) together with high heat.

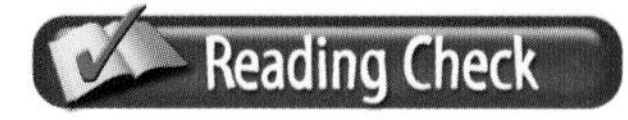

List Give some examples of manufacturing processes.

Outputs and Feedback

What outputs and feedback are part of a manufacturing system?

The products themselves are the main outputs of a manufacturing system. Another is providing jobs for a community. Waste or pollution might be another.

Feedback can also come in many forms. High demand or low demand for the product is part of feedback. The manufacturer must review what went right or wrong to determine if changes need to be made to the product and system. What changes might be needed if a product is *too* successful?

section 16.2 assessment

After You Read Self-Check

1. Compare a design engineer to a production engineer.
2. Identify the difference between production materials and raw materials.
3. Define the term *added value*.

Think

4. Think of a common manufactured item, such as a bicycle. Use the systems model to describe the system used to make the item.

Practice Academic Skills

Social Studies

5. If you had been a small child 150 years ago, what kinds of toys would you have played with? Think of one example. How do you think that toy would have been manufactured without the use of electricity and power tools? How would you make the same toy today? Write a one paragraph response.

STEM Mathematics

6. A flashlight consists of many different parts. A complete flashlight weighs 1.23 pounds. The lamp in the flashlight weighs 0.058 pounds. What percent of the total weight of the flashlight does the lamp represent?

Math Concept **Percents** Think of percents as parts of the whole.

1. Divide the part by the whole to get a decimal equivalent of the percent.
2. Multiply the decimal by 100 to get the percent.

For help, go to **glencoe.com** to this book's OLC and find the Math Handbook.

section

16.3 Running Your Own Factory

Reading Guide

Before You Read **Connect** How could a teenager start a factory?

Content Vocabulary

- market research
- supplier
- just-in-time delivery
- quality assurance
- marketing
- commission
- profit

Academic Vocabulary

- item
- crucial

Graphic Organizer

Draw the section diagram. Use it to organize and write down information as you read.

Market Research

Step 1	Step 2	Step 3

Go to **glencoe.com** to this book's OLC for a downloadable graphic organizer and more.

TECHNOLOGY STANDARDS

STL 2 Core Concepts of Technology
STL 8 Attributes of Design
STL 9 Engineering Design
STL 10 Troubleshooting & Problem Solving
STL 19 Manufacturing Technologies

ACADEMIC STANDARDS

Science
NSES Content Standard F Science and technology in society

Social Studies
NCSS 7 Production, Distribution, and Consumption

STL *National Standards for Technological Literacy*
NCTM *National Council of Teachers of Mathematics*
NCTE *National Council of Teachers of English*
NSES *National Science Education Standards*
NCSS *National Council for the Social Studies*

Market Research

How can you find out what products people will purchase?

Suppose you have an idea for a new board game for two to four players. In addition to the board, the game will include a colored button for each player and cards with questions written on them. You are on a summer break, so you have time to manufacture and sell the game yourself.

1. You plan the game and make sketches. By putting your idea on paper, you can show others what you are thinking. The drawing is part of your design.

As You Read

Predict What do you think is necessary to set up your own factory?

BusinessWeek Tech NEWS

The Chips Have It

In 2007, four of the top 10 fastest-growing tech companies manufactured semiconductors and materials used to make them. Finishing fourth was Cypress Semiconductor, which makes the chips [for] Apple's iPod click wheel. The chip notes how the finger is moving and translates the information into a computer command, such as "more volume" or "next song."

Critical Thinking *What makes a manufacturer successful?*

Go to **glencoe.com** to this book's OLC to read more about this news.

2. You want opinions from other people, so you talk to your friends and teachers. Each person gives you suggestions for how to improve your game. This is market research. **Market research** is a way to find out what people will purchase.
3. You make one complete game to test with friends. After you make some changes, you are satisfied.

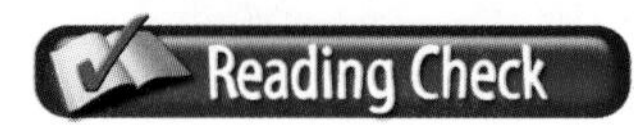

Describe What is the purpose of market research?

Manufacturing the Product

Who are the suppliers in a manufacturing process?

4. You want your playing board to fold up like other board games. You can draw the playing surface on paper, but you cannot make the heavy cardboard back. You talk with the owner of a frame shop. He says that he can make 100 cardboard backs for $1.20 each.
5. A friend of yours has a printer that can print on lightweight cardboard. She agrees to print the game's question cards for $1.50 per set. She also agrees to print the instructions. Both the picture framing shop and your friend are your suppliers. A **supplier** is a person or company that provides something for making your manufactured product.

Ethics in Action

Outsourcing

U.S. companies are employing more and more people out of the country. This is called "outsourcing." Wages are lower in some countries, so the companies are able to save money. In turn, they can sell products, such as cameras, cell phones, and shoes, at lower prices.

The Downside Some people say that outsourcing takes jobs away from Americans. They also point out that working conditions in some developing countries are poor and unsafe.

English Language Arts/Writing

Child Labor Laws Federal law prohibits you from working at most jobs until you are 14. This is for your protection. In some countries, children work long hours in dangerous conditions and cannot go to school.

1. Do research and find out what the law says. What jobs can you do now? How does the law protect you?
2. Write a brief report.

6. Of course you also want to sell the game. You check with four stores in your town. Three owners agree to offer your game for sale.
7. The framing shop requires an order of at least 100 cardboard game backs. So you make out a budget based on manufacturing 100 games:

100 cardboard game backs, $1.20 each	$120
100 packs of question cards, $1.50 per pack	150
400 colored buttons, $0.10 each	40
100 plastic bags to package games, $0.25 each	25
Colored pens, paper on which to draw the playing surface, rubber cement, and other items	65
TOTAL	**$400**

Your parents help you borrow $400 for the project. That money becomes your capital.

8. You ask your suppliers to send you ten game backs and ten sets of question cards each week for ten weeks. You do not have room to store much at home, so you want to use **just-in-time (JIT) delivery** and receive those items just in time to do the work. JIT is used by many manufacturers. With JIT, they do not need extra warehouse space to manage items.
9. Before you draw the playing surface, you get your production materials together. This includes the paper, pens, ruler, and other items necessary to draw the playing surface. Now you go into production and draw the playing surface for ten boards. You have added value to the materials.
10. You carefully inspect each playing surface you finish. This inspection is called **quality assurance** or quality control. It means you will meet the requirements, or standards, set for the games. You glue the ten playing surfaces to the cardboard backs. You check the paper to be sure that you did it well. This is also quality assurance. Quality assurance is crucial to manufacturing. It enables you to use information about your system to change it as needed.

Making Sketches You might sketch out your game on a big piece of paper and then get input from your teachers and friends. *Why would you ask your teachers and friends to give you input on your game?*

Self-Shifting Bicycle

Imagine a bicycle that shifts gears for you. The bicycle company Trek is introducing a three-speed bicycle with an automatic transmission. A computer shifts the gears as wheel speed changes. The rider can focus on traffic conditions and not on operating the bicycle. Market research suggested that city dwellers who do not usually ride bicycles might use this bike.

What are some ways to perform quality assurance on a self-shifting bicycle?

Go to **glencoe.com** to this book's OLC for answers and to learn more about self-shifting bicycles.

11. Now it is time to assemble the game. You ask a friend to help. You place four different-colored buttons in a small plastic bag. You add a pack of question cards. Then you pass the bag to your friend. Your friend places the game board, instructions, and smaller plastic bag into a larger plastic bag. This is like working on an assembly line in a factory.

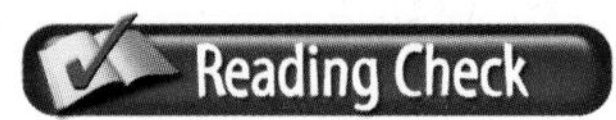

Explain Why is quality assurance important?

Marketing and Sales

What is involved in marketing your product?

12. Next, you have to plan the **marketing** for your game. In other words, you have to tell people about it. One way to do this would be to create flyers using your computer. You can then print them on a color printer and post them where they will be seen.
13. You drop off two games at each of the stores that agreed to sell them. The store owners will charge you one dollar for each game they sell. This is their commission. A **commission** is payment based on sales.
14. You price the games at ten dollars each. How much profit will you make? **Profit** is the money left over when all your bills are paid. Remember that the loan, too, must be repaid. If you sell all of the games, how much money will you make?

section 16.3 assessment

After You Read Self-Check

1. Define the term *market research.*
2. Identify the meaning of JIT.
3. Describe the purpose of quality assurance.

Think

4. Name some ways that a saw mill adds value to trees and a furniture company adds value to lumber.

Practice Academic Skills

English Language Arts/Writing

5. Suppose several customers returned the board games discussed in this section. They report such problems as missing cards or buttons, a loose playing surface, and instructions that are hard to follow. Outline a maintenance system for inspecting and servicing these products so they will function properly.

STEM Science

6. The Industrial Revolution brought manufacturing into factories, especially after Henry Ford developed a moving assembly line. Research some of the other changes the manufacturing process has seen over the ages. Write a few paragraphs describing what you find. Include a discussion of where you think the manufacturing process is headed.

Exploring Careers in Technology

Catherine Andrae

PROCESS ENGINEER

Q: *What do you do on a typical day?*

A: My company manufactures custom color and additive masterbatches used in the production of plastics. Each day, I review incomplete tasks from the previous day. I talk to the supervisors, operators, and mixers to evaluate the different products processed during the evening shift and weigh concerns, comments, or suggestions. I participate in the facility-wide, walk-around meeting to keep up with the larger-scale concerns and projects. Lately I also deal with reformulating several products to improve processing and reduce variation. Each day I review products and adjust the process settings to reflect optimal conditions.

Q: *What kind of training and education did you need to get this job?*

A: Polymer and fiber engineering was my major at Georgia Tech. I learned about polymer processing, completed hands-on lab experiments for colorants and additives, and mastered technical data about antioxidants, waxes, and processing aids. I also completed three internships, which really drove home several key areas of knowledge I need to do my job.

Q: *What do you like most about your job?*

A: I really like the freedom I have to tackle problems, as well as the positive feedback I receive from the manufacturing floor when I improve a process or product.

English Language Arts/Writing

Evaluating Choices What kind of career is best for you?

1. Think of at least two different careers that you find interesting. Research the two careers on the Internet, at a library, or in your school's career center.
2. Make a list of your personal characteristics, including your talents, personality, and subjects and activities you enjoy.
3. Compare the information about the careers with the data you gathered about yourself. Write a 50-word paragraph about the best match.

Go to **glencoe.com** to this book's OLC to learn more about this career.

Real-World Skills

Observation, speaking and listening, organization

Academics and Education

Chemistry, physical science, mathematics

Career Outlook

Growth as fast as average for the next ten years

Source: *Occupational Outlook Handbook*

chapter 16 Review and Assessment

Chapter Summary

Section 16.1 Factories make two kinds of goods: durable and non-durable goods. In factories, production materials are converted into finished products. Workers develop special skills partly because the companies use a division of labor. Workers put everything together using an assembly line.

Section 16.2 To start a company or make a new product, a person first needs an idea and creates a design. The final product will be made from production materials and parts provided by suppliers. The manufacturing process adds value to raw materials.

Section 16.3 Market research is a way to find out what people will purchase. Suppliers provide materials or products to help you manufacture your own products. Quality assurance is a method for being sure products meet quality standards.

Review Content Vocabulary and Academic Vocabulary

1. On a sheet of paper, use each of these terms and words in a written sentence.

Content Vocabulary

- durable good
- non-durable good
- craft
- Industrial Revolution
- scientific management
- time and motion study
- division of labor
- assembly line
- added value
- abrading
- resin
- market research
- supplier
- just-in-time delivery
- quality assurance
- marketing
- commission
- profit

Academic Vocabulary

- period
- establish
- distribute
- purchase
- item
- crucial

Review Key Concepts

2. **Summarize** the history of manufacturing.
3. **Compare** durable and non-durable goods.
4. **Explain** why having assembly lines and a division of labor is important.
5. **Use** the universal systems model to describe manufacturing systems.
6. **Define** added value.
7. **Name** several tools and processes used in manufacturing.
8. **List** several steps in setting up and running a small factory.
9. **Discuss** the purposes of market research, quality assurance, and just-in-time delivery.
10. **Explain** what is involved in marketing a product.

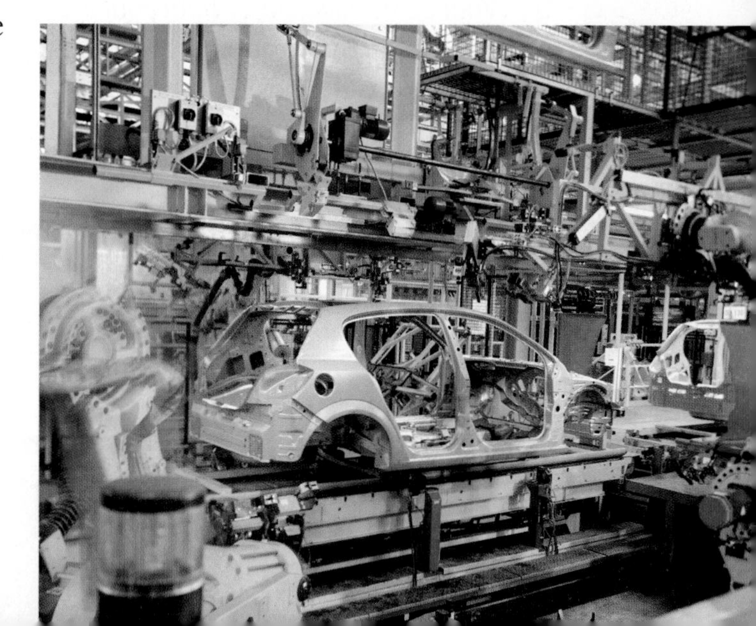

Real-World Skills

11. Time and Motion Choose a task you and a classmate do on a regular basis, such as getting ready for school. Keep a record of the steps you take to accomplish the task and the time it takes you to do each step. Compare your record with your classmates to see who accomplishes the task more efficiently.

STEM Technology Skill

12. Using Presentation Software Use presentation software to create a slide show about how something is manufactured. You might choose a product such as steel, ceramic tile, or lumber. Research the manufacturing process.

a. Create a step-by-step slide show explaining how the material is manufactured.

b. Include photographs, charts, and diagrams in your slide show.

WINNING EVENTS

Industrial Engineer

Situation Your team will design and mass produce a product made from discarded material solicited from a business.

Activity Develop assembly drawings for your product. List the parts, tools, and machines you will need. Develop a flowchart of the production process. Set up your assembly line. Make a series of trial runs, and make changes.

Evaluation The production process will be evaluated by these criteria:

- Design of process—thoughtful, complete
- Assembly drawings—accurate, complete
- Materials list—complete, appropriate
- Flowchart—appropriate, clear
- Production run—safe, efficient

Go to **glencoe.com** to this book's OLC for information about TSA events.

Academic Skills

Social Studies

13. Many companies have been accused of using foreign sweatshops. A sweatshop is a factory that has poor working conditions and pays low wages. Research the issue of sweatshops and write a few paragraphs about what you find.

STEM Mathematics

14. Jason needs to repair the clutch on his car. A mechanic says that the fan belt and flywheel also need work. The total cost will be $409.20. Replacing the fan belt will cost $32.50. The flywheel repair will cost ¼ of the total cost. How much will the clutch cost?

Math Concept **Algebra** You can solve some problems with an equation. An equation is like a sentence that explains what you are trying to solve. The unknowns can be represented as letters x or y.

Standardized Test Practice

Directions Choose the letter of the best answer. Write the letter on a separate piece of paper.

1. How much is $\frac{1}{16}$ of 368?

A 16 **C** 32

B 22 **D** 23

2. Abrading is the process of changing the shape of a material by rubbing off small pieces.

T **F**

Test-Taking Tip Be alert for multiple ideas or concepts within a true/false statement. All parts must be true or the statement is false.

chapter 16 TECHNOLOGY LAB

Manufacture a Bookstand

The manufacturing process includes designing, developing, making, and servicing the product. Being organized is important. It may make all the difference in your ability to have a successful manufacturing business.

Tools and Materials

- ✓ Three pieces of wood, each about 1 inch thick, 6 inches wide, and 18 inches long
- ✓ Six pieces of wood, each about 1 inch thick, 6 inches wide, and 8 inches long
- ✓ Twelve ⅜-inch-diameter dowels, 1½ inches long
- ✓ Ruler
- ✓ Pencil
- ✓ Glue
- ✓ Hammer
- ✓ Hand drill with ⅜-inch drill bit
- ✓ Fine sandpaper
- ✓ Can of water-based clear finish
- ✓ Paint brush

! SAFETY

In this lab, you will be using tools and machines. Be sure to always follow appropriate safety procedures and rules so you and your classmates do not get hurt.

Set Your Goal

For this activity, you and two teammates will set up a production system to manufacture adjustable bookstands. When you are finished, each of you will have a bookstand to use.

Know the Criteria and Constraints

In this lab, you will:

1. Use an assembly line and division of labor for part of your operation. After the bookstand is complete, you will turn in a flow chart showing how the materials moved along your assembly line.
2. Establish a method of controlling quality and of servicing products that do not meet quality standards. Turn in a paragraph describing the process you used.
3. The first bookstand you produce will be a prototype. Study it for ways to improve your methods. Keep a record of changes you make.
4. Recycle or discard any waste safely.

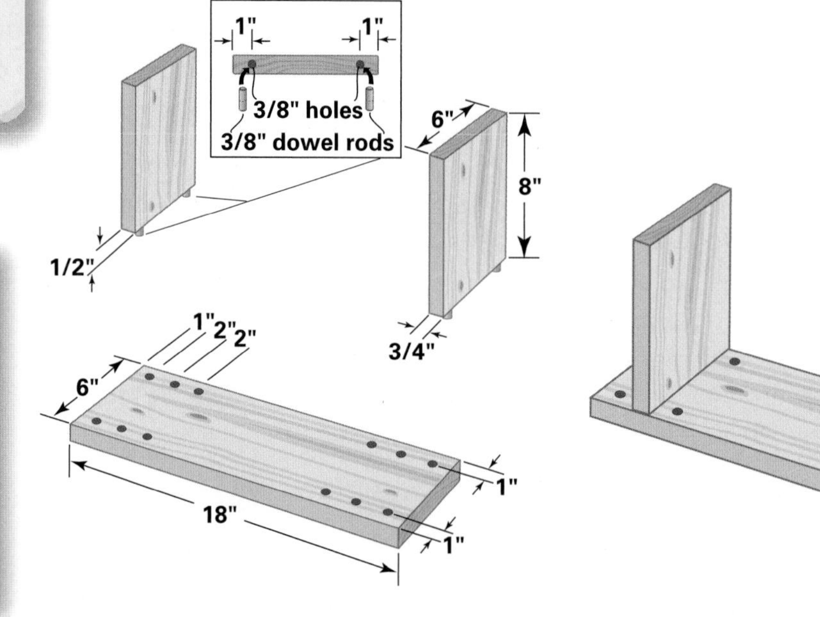

Design Your Project

The following steps are only for making the bookstands. You will determine your own steps for setting up the system.

1. Look at the assembly drawings.
2. Establish methods of controlling quality and of servicing products.
3. Take one of the 6-inch x 8-inch pieces of wood.
   - Drill two 3/8-inch-diameter holes, 1 inch deep, into one end. The center of each hole should be 1 inch from the edge.
   - Repeat for the other five bookstand ends.
4. To attach dowel to bookstand ends:
   - Place a small amount of glue on the end of a 1½-inch-long dowel and insert the dowel in a hole, glue-side down.
   - Tap the dowel with the hammer until it goes into the hole.
   - Repeat for the other 11 dowels and holes.
5. Drill twelve 3/8-inch diameter holes in the 18-inch-long pieces of wood. Drill each hole a little deeper than ½ inch—say, 5/8 inch.
6. After glue is dry, insert the dowel in the holes in the bookstand base.
7. Repeat for making the other two bookstand bases.
8. When your bookstands are complete, sand the surfaces and edges.
9. With proper ventilation, paint the bookstand with clear finish.
10. Write a brief paragraph describing your quality-control process.
11. Create an assembly-line flow chart and write your servicing plan.

Evaluate Your Results

After you complete the lab, answer these questions on a separate piece of paper.

1. What could you have done to improve the quality of your bookstand?
2. What would you do to make and sell 30 adjustable bookstands?

Academic Skills Required to Complete Lab

Tasks	English Language Arts	Math	Science	Social Studies
Set up assembly line for bookstands with teammates.	✓	✓	✓	✓
Build prototype of bookstand.	✓	✓	✓	
Improve method for producing other bookstands.	✓		✓	
Create a flow chart that describes your assembly line.	✓		✓	
Write about the plan for quality control and servicing products.	✓	✓		✓

chapter 17

Manufacturing in the 21st Century

Sections

What You'll Learn

- **Explain** the purpose of research and development.
- **Discuss** how products are designed.
- **Describe** a virtual factory.
- **Compare** CNC, CAM, and CIM.
- **Examine** the use of industrial robots and e-manufacturing.
- **Summarize** the roles of quality assurance and safety in modern manufacturing.
- **Define** the purpose of the marketing department.
- **List** several forms of advertising.
- **Identify** the difference between a wholesaler and a retailer.

Explore the Photo 

Virtual Factory A computer-generated virtual factory allows the manufacturing engineers to try various layouts. *What is the advantage of setting up a virtual factory before setting up a real one?*

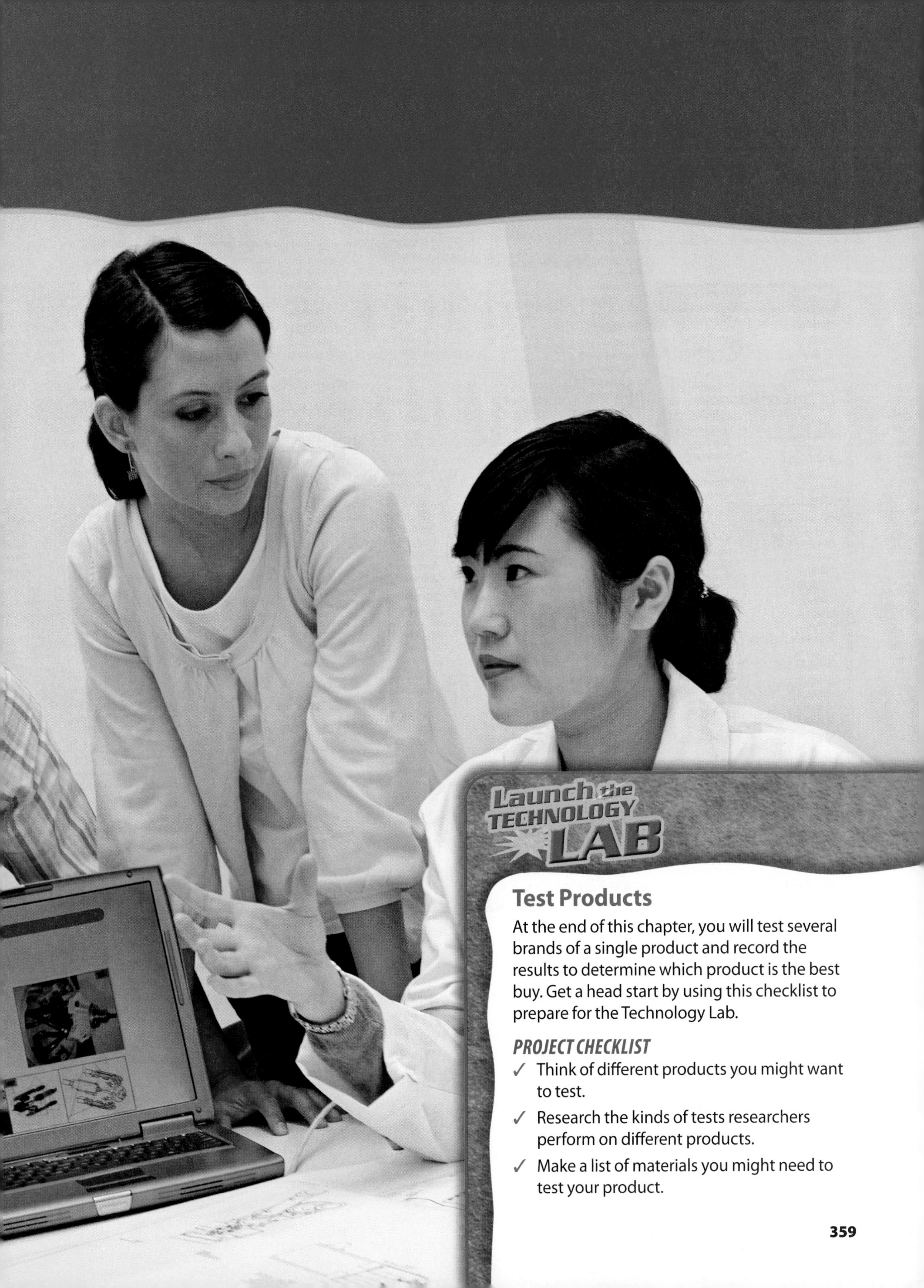

Launch the TECHNOLOGY LAB

Test Products

At the end of this chapter, you will test several brands of a single product and record the results to determine which product is the best buy. Get a head start by using this checklist to prepare for the Technology Lab.

PROJECT CHECKLIST

- ✓ Think of different products you might want to test.
- ✓ Research the kinds of tests researchers perform on different products.
- ✓ Make a list of materials you might need to test your product.

section 17.1

Preparing for Manufacturing

Reading Guide

Connect How do companies prepare for manufacturing?

Content Vocabulary

- R&D
- virtual factory

Academic Vocabulary

You will see these words in your reading and on your tests. Find their meanings at the back of this book.

- concept
- acquire

Graphic Organizer

Draw the section diagram. Use it to organize and write down information as you read.

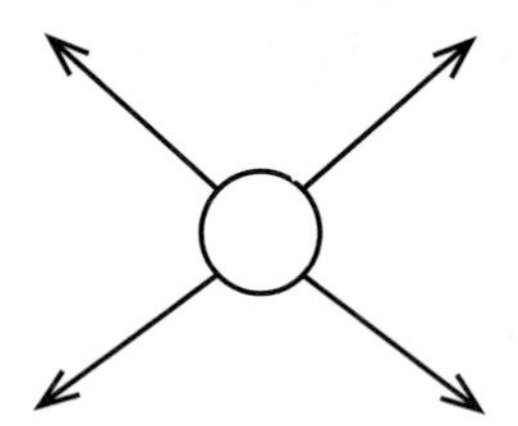

Go to **glencoe.com** to this book's OLC for a downloadable graphic organizer and more.

TECHNOLOGY STANDARDS

STL 8 Attributes of Design

STL 10 Troubleshooting & Problem Solving

STL 11 Design Process

STL 19 Manufacturing Technologies

ACADEMIC STANDARDS

Social Studies

NSES Content Standard F Science and technology in society

English Language Arts

NCTE 1 Read texts to acquire new information.

STL *National Standards for Technological Literacy*

NCTM *National Council of Teachers of Mathematics*

NCTE *National Council of Teachers of English*

NSES *National Science Education Standards*

NCSS *National Council for the Social Studies*

Manufacturing in the United States

How has the United States become a leader in manufacturing?

The United States has more than 400,000 factories. About 15 million people work in them and make everything from toys to locomotives. The United States produces more manufactured products than any other country in the world. Advances in technology have enabled American factories to manufacture large amounts of high-quality products. New technologies and products result from the demands, values, and interests of businesses and consumers.

Predict What is the purpose of a virtual factory?

Something Old, Something New Research engineers look for new products or new ideas for old products. *Can you think of an example of how an older product evolved into a newer product?*

Research and Development

Where do new products and advancements in manufacturing come from?

Before manufacturing can begin, decisions must be made. What product will be manufactured? How will it be designed? How will the factory be set up to make it?

A company's **R&D** (research and development) department is where invention and innovation take place. That is where new products and methods are usually born. Research looks for new concepts. Development uses the research to create the new products and processes. The people who work in the research and development department enjoy trying new things and considering all ideas. They also consider that products have a life cycle of introduction, growth, maturity, and decline. They might often solve technological problems through experimenting.

R&D Case Study

Suppose, for example, that the research group at Z. Z. Zipper Company is told that people might be interested in buying a zippered folding boat. They look for a new soft plastic that could be made into a leak-proof zipper. The development group makes sure that the new plastic is strong enough. They choose a specific nylon fabric for the boat. The research group and the development group work together closely.

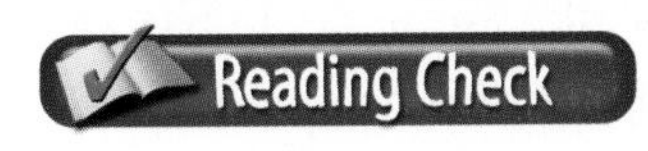

Describe What is the purpose of research and development?

Henry Ford

Pioneer Automobile Manufacturer

Henry Ford was born in Dearborn, Michigan, in 1863. As a young boy, he enjoyed tinkering with various machines found on his family's farm. Part-time work in a Detroit machine shop and later at Westinghouse Engine Company, allowed him to further experiment with machinery. In 1896, Ford built his first horseless carriage, and by 1903, he founded the Ford Motor Company.

Ford's Model T debuted in 1908. It sold for $950. During its 19 years in production, more than 15 million were sold. In order to keep costs down and cut manufacturing time, Ford invented the modern assembly line. This marked the beginning of the Motor Age. Cars evolved from being luxury items for the wealthy to an essential form of transportation for everybody.

Cranking Out Cars Ford's mass-production techniques allowed his company to manufacture one complete Model T every 24 seconds. Ford died in 1947, but his company remains one of the largest manufacturers of automobiles today.

English Language Arts/Writing Write an article for your school newspaper explaining how assembly lines work.

Go to **glencoe.com** to this book's OLC to learn about young innovators in technology.

Product Design

What are some of the jobs performed by the engineering design department?

The design department works closely with the R&D department. Design engineers help decide what requirements a product should meet, such as how big it should be. They identify the criteria and constraints involving the color, the material, the shape, and everything else that is part of a new product's design. They also determine what trade-offs may be necessary among competing values. Values include such things as cost, availability of materials, desirability of product features, and waste.

The engineers at Z. Z. Zipper acquire the R&D department's information. They have to answer questions like:

- What should be the length of the boat?
- The nylon fabric will wrap around a frame, much like an umbrella. Should the frame be made of composite plastic or aluminum?
- Should the boat frame fold up or come completely apart for storage?

In today's factories, the engineers draw their plans with computer-aided design (CAD) and drafting equipment. The plans are stored in the computer's memory and are later used by the production department.

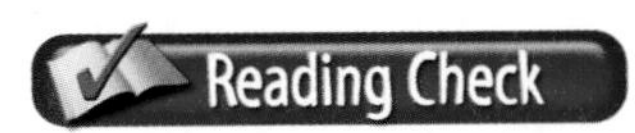

Explain How are products designed?

EcoTech

Appliance Upgrades

You can replace an old vehicle with a newer model that gets better gas mileage. In the same way, you can also upgrade old appliances. Newer televisions, refrigerators, and ovens are more efficient than old ones, and save energy and money.

Try This Identify appliances at home and school to see where energy can be saved by replacing them with new models.

Planning the Factory

In what ways does product design affect factory design?

After a new product has been developed and designed, the factory itself must be planned. Some products are very complicated to assemble. A new automobile model, for example, might have 15,000 to 20,000 parts. The parts include everything from tiny screws to large door panels. It is sometimes difficult to figure out the order in which everything should be put together.

That is where the virtual factory comes in. A **virtual factory** is a three-dimensional image that appears only on a computer screen. It is not real. Virtual factory software allows manufacturers to "try out" different layouts. Animation lets them observe the work flow before they actually set up the factory or change the process. It is just one of the ways computers have changed manufacturing.

section 17.1 assessment

After You Read Self-Check

1. Explain the difference between research and development.
2. Name four factors involving a product's design that are the responsibility of the design department.
3. Identify the kind of factory that appears only on a computer screen.

Think

4. Suppose you are a design engineer working with outdoor furniture. Think of at least four questions you might have to answer about your product.

Practice Academic Skills

STEM Science

5. How innovative can you be? Recycle a plastic milk container into a useful item. Perhaps you can convert it into a planter for new seeds. Another idea might be to make a tiny hole to drip out water and use a float to make water clock.

STEM Mathematics

6. The design of a wheelchair requires the front casters to have a radius of 2.5 inches. What is their circumference?

Math Concept **Geometry** Geometry problems can often be solved using formulas that show how the different features of a geometric share are related.

1. The formula for the circumference of a circle is its diameter multiplied by pi: $C = \pi D$
2. Use 3.14 to represent *pi*.

For help, go to **glencoe.com** to this book's OLC and find the Math Handbook.

section 17.2

Producing the Product

Reading Guide

Before You Read **Preview** What goes on during production in a modern factory?

Content Vocabulary

- schedule
- CNC
- CAM
- CIM
- troubleshooting
- standard
- robotics
- e-manufacturing
- OSHA
- NIOSH

Academic Vocabulary

- target
- contact

Graphic Organizer

Draw the section diagram. Use it to organize and write down information as you read.

Producing the Product

Acronym	Complete Phrase
CNC	
CAM	
CIM	

Go to **glencoe.com** to this book's OLC for a downloadable graphic organizer and more.

TECHNOLOGY STANDARDS

STL 10 Troubleshooting & Problem Solving
STL 11 Design Process
STL 17 Information & Communication Technologies
STL 19 Manufacturing Technologies

ACADEMIC STANDARDS

Science

NSES Content Standard E Understandings about science and technology

Social Studies

NCSS 8 Science, technology, and society

STL *National Standards for Technological Literacy*
NCTM *National Council of Teachers of Mathematics*
NCTE *National Council of Teachers of English*
NSES *National Science Education Standards*
NCSS *National Council for the Social Studies*

Computers and Manufacturing

How are computers used in modern factories?

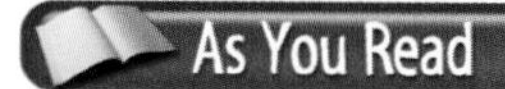

As You Read

Connect What do you think are the most common uses for robots?

The production department is responsible for actually making the company's product. It must be a quality product and manufactured to a specific schedule. A **schedule** is a plan for what must be done by a certain time. Some production activities use computers to keep on schedule.

Figure 17.1 Just-in-Time Delivery System

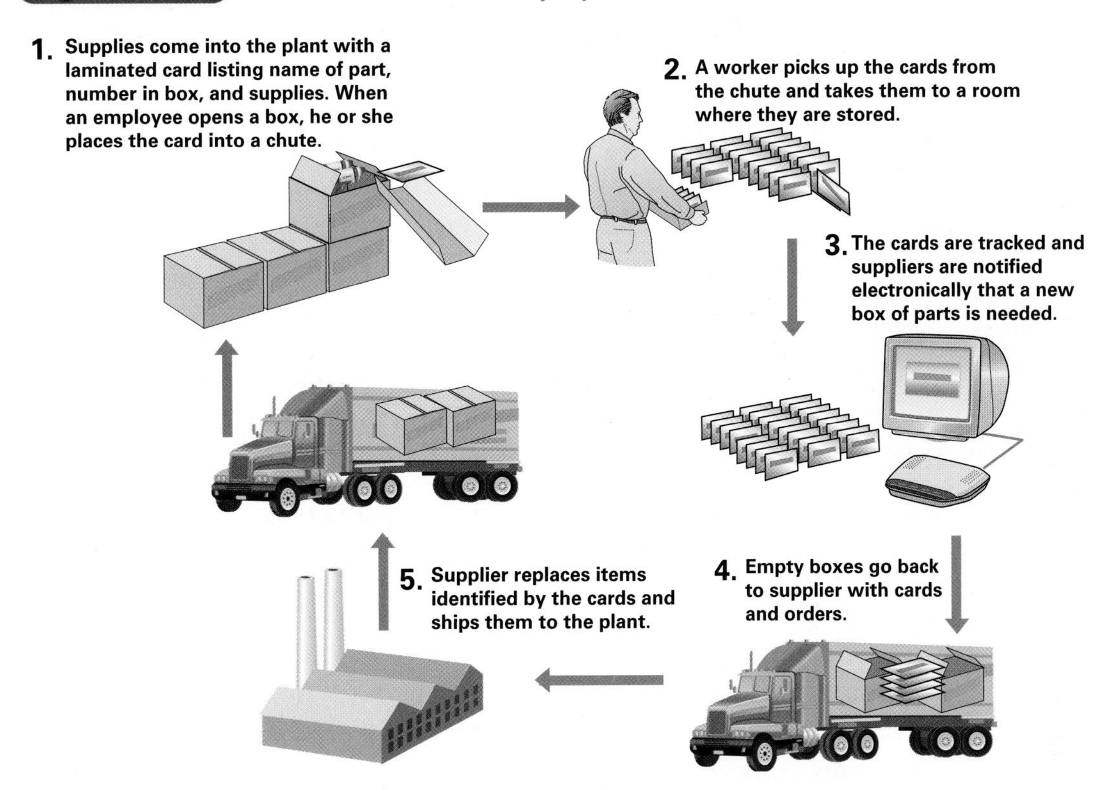

Delivered as Needed Toyota's JIT delivery concept means its products are made and delivered only when they are needed. *Why do you think this process is efficient?*

CNC and CAM

CNC stands for computerized numerical control. With CNC, machine tools operate by commands from a computer. The operator types in the instructions, and the machine does the work. With **CAM**, or computer-aided manufacturing, machine-tool operators program computers to operate all the machinery.

CIM

CIM is computer-integrated manufacturing. All the computers in a company are linked together, or integrated. Design and production departments can communicate instantly. The purchasing department can tell just-in-time (JIT) suppliers when to deliver materials. See **Figure 17.1**. The marketing department can plan when to start advertising. Management can direct the entire company from one location. CIM is a company-wide process.

Lean Manufacturing

Lean manufacturing, or smart manufacturing, uses JIT delivery from suppliers. The company makes only what it can sell quickly and keeps only a few products in the warehouse.

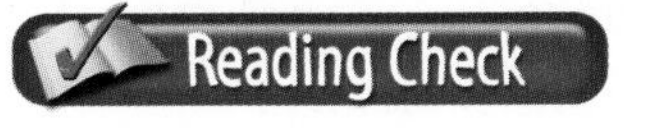

Compare How do CNC, CAM, and CIM differ?

I, Robot Engineers can program robotic arms to perform many tasks in a factory. *What are some tasks a robot could do better than a human?*

Quality Assurance

Can computers help guarantee quality?

Many companies use computers and special instruments to build quality products. One computer-controlled machine measures parts and compares the sizes with dimensions given on product plans. Microscopes and X-ray machines may be used as well.

Troubleshooting

If too many products are defective, a problem may exist in the system. **Troubleshooting** is the method used to identify the malfunction. Different processes are tested until the cause is found.

Standards

A **standard** is a rule or guideline. Standards are used so certain products conform to a particular size, shape, or level of quality. The American National Standards Institute (ANSI) publishes standards for screws and other materials.

Important international standards are published by the International Standards Organization (ISO) in Geneva, Switzerland. ISO 9000 standard targets quality assurance.

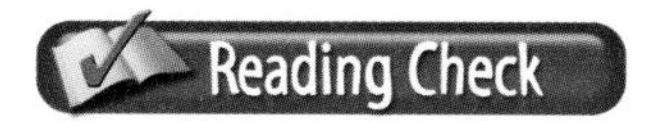

Summarize What is the role of quality assurance?

Robots

What does an industrial robot look like?

Robotics is the technology of industrial robots. Industrial robots are being used more and more in manufacturing. An industrial robot usually has one mechanical arm and is classified as a machine tool. The end of the robot's arm might have a gripper to hold and move items, or it might have a tool or welding tip. Welding and painting are the most common functions for robots.

E-Manufacturing

What is e-manufacturing?

The term **e-manufacturing** means using electronic information in the manufacturing of a product. In one type of e-manufacturing, all machine tools are linked to the Internet. A person can connect to the machine with a laptop computer, a personal digital assistant (PDA), a cell phone, or other device from anywhere.

Almost all manufacturers have suppliers. Some have hundreds. With e-manufacturing, a company can stay in immediate contact with them. The company can manage different suppliers of the same product and keep inventories up to date using the Internet.

Safety

How are factories made safer?

Nothing is more important in a factory than the safety of the people who work there. Safety means freedom from injury or any danger of injury.

Protective Equipment

When you use a hammer, metal pieces might chip off. A face shield or safety glasses will protect your eyes. Other types of protective equipment include hard hats, earplugs, gloves, and safety shoes with steel toes. When the air contains dust or vapors, you should wear special filters over your nose and mouth.

OSHA and NIOSH

Many factory safety rules are required by federal or state laws. OSHA, the Occupational Safety and Health Administration, establishes safety rules and checks up on companies. NIOSH, the National Institute of Occupational Safety and Health, approves protection equipment such as hard hats and safety glasses.

Academic Connections
Language Arts

Robots in the Slow Lane? Many people think robots are faster than humans, but this is not necessarily true. Most robots work at about the same speed as a human works. However, robots do not become bored and distracted, and their work is more accurate for repetitive tasks.

Apply Write a short story about a robot in a factory. Use at least ten of the terms you have learned in this chapter.

section 17.2 assessment

After You Read Self-Check

1. Describe e-manufacturing.
2. Tell what the acronyms OSHA and NIOSH mean.
3. Discuss industrial robots and why they are classified as machine tools.

Think

4. Companies sometimes claim, "Quality is built into our products." Explain what you think that means.

Practice Academic Skills

Social Studies

5. Count the number of minutes devoted to advertising during an hour of television. Make a pie graph showing how much time is devoted to advertising and to actual programs. Is this an appropriate distribution of time? Write a paragraph on your opinion.

STEM Mathematics

6. A company is trying to control its expenses for rents and leases. Company policy is that the monthly cost to lease one of their plants should be no more than 8 percent of the revenues generated by the plant. If the average monthly revenues of the plant are $106,250, what should the rent be?

Math Concept **Percents** Percents can be thought of as parts of the whole.

1. Divide the part by the whole to get a decimal equivalent of the percent.
2. Multiply the decimal by 100 to get the percent.

For help, go to **glencoe.com** to this book's OLC and find the Math Handbook.

section

17.3 Selling the Product

Reading Guide

Before You Read **Connect** Have there been advances in the way products are marketed?

Content Vocabulary

- advertising
- wholesaler
- retailer

Academic Vocabulary

- potential
- automate

Graphic Organizer

Draw the section diagram. Use it to organize and write down information as you read.

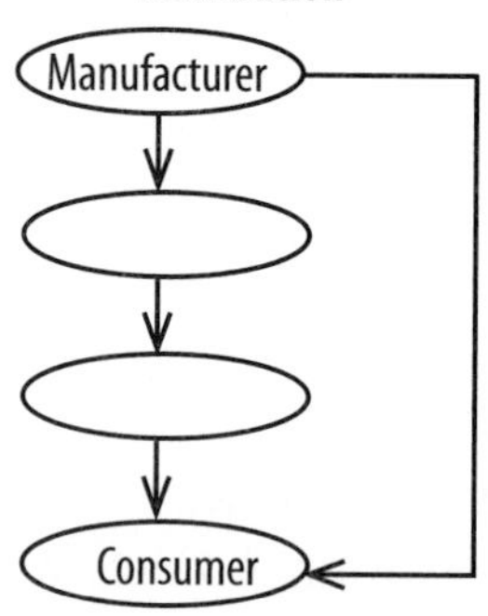

Go to **glencoe.com** to this book's OLC for a downloadable graphic organizer and more.

TECHNOLOGY STANDARDS

STL 2 Core Concepts of Technology

STL 11 Design Process

STL 17 Information & Communication Technologies

STL 19 Manufacturing Technologies

ACADEMIC STANDARDS

Science

NSES Content Standard F Science and technology in society

English Language Arts

NCTE 6 Apply knowledge of language structure and conventions to discuss texts

STL *National Standards for Technological Literacy*

NCTM *National Council of Teachers of Mathematics*

NCTE *National Council of Teachers of English*

NSES *National Science Education Standards*

NCSS *National Council for the Social Studies*

Marketing

How is advertising done?

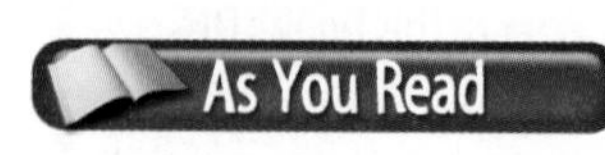

As You Read

Predict How are sales and marketing related?

The final steps in manufacturing are handled by the marketing department. Marketing is telling potential customers about the company's products and services in such a way as to make them eager to buy.

Advertising

Advertising is making a public announcement that your product is available. You have probably seen thousands of advertisements during your lifetime. You see ads on television, hear them on the radio, and read them in newspapers and magazines. Advertising can help create demand for a product.

Outdoor Advertising You have probably seen advertisements like this one on billboards and at bus stops. *Why do you think ads are placed at these locations?*

Outdoor and Internet Ads

Companies also advertise outdoors. You can see it on the sides of trucks and other vehicles, and on clothing. Do you have a pair of designer jeans with the designer's name on them? If so, you are a walking advertisement for that designer's products.

The Internet is another way that companies can advertise their products. Some companies have their own Web sites. Others insert ads on related Web sites. Still others send e-mails to potential customers.

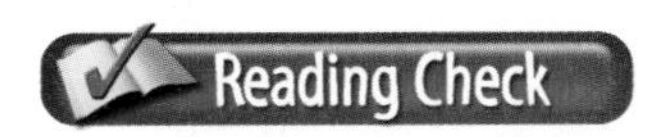

Name What are ways to advertise products?

Sales

What is the difference between direct and indirect sales?

Salespeople are responsible for closing the deal with a customer. Some sales are made directly to the person or people who will use the products. Other sales are made to a store of some kind.

Ethics in Action

Viral Marketing

Teens are some of the people advertisers want to reach. But teens have grown wise to their methods. They skip commercials with TiVo. They watch video clips on the Internet. But marketers are finding new ways to reach them.

Marketing in Cyberspace One method is called "viral marketing." Marketers use social networking sites like MySpace and Facebook to promote new movies, music groups, and products. They also use Internet chat rooms and forums to get teens interested.

English Language Arts/Writing

Word of Mouth When you see a movie because your friend recommends it, you are listening to "word of mouth." With viral marketing, marketers try to create word of mouth artificially.

1. Pretend you are an Internet marketer promoting a new movie.
2. Think of a movie to advertise.
3. Write a two-paragraph blog entry about the movie aimed at teenagers.

A Historic Yet New Ship

Imagine building a new ship out of an old building. A U. S. Navy ship being built in New Orleans will use 24 tons of steel reclaimed from the World Trade Center Towers. Named the *USS New York*, it may have a crew of 360 sailors. The steel was placed in the bow (front) so it will lead the way as the ship carries out missions. *Do you think this is a good use for reclaimed steel? Explain.*

 Go to **glencoe.com** to this book's OLC for answers and to learn more about the *USS New York*.

For example, remember the Z. Z. Zipper Company? The production department assembled all the zippered boat parts and put them into a strong cardboard box. Twenty packed boxes of boats went to the company's warehouse every day. To sell the boats, the marketing department of the Z. Z. Zipper Company contacted the B. B. Boat Sales Company. This kind of selling is an example of indirect sales because it involves selling to a store.

Wholesale and Retail

B. B. Boat Sales Company is a wholesaler. A **wholesaler** is a company that purchases large numbers of products from a manufacturer and then sells smaller numbers those of products to retailers. A **retailer** is a company or store that sells products to consumers. Target and Wal-Mart are two examples of large retailers. Distributors move the products from the manufacturer to the retailer who sells it to the consumer.

Online Sales

The Internet has changed the way selling is done. Many steps have become **automated**. Other steps have been eliminated altogether. Stores and manufacturers have Web sites where you can order and buy products. Other sites, like Amazon.com, do all their business on the Internet. They do not have a store that you can visit. Another common sales technique uses the U.S. Postal Service mail to send information to possible customers.

section 17.3 assessment

After You Read Self-Check

1. Discuss the role of the marketing department.
2. Identify the difference between a wholesaler and a retailer.
3. Describe how the Internet has changed the way products are sold.

Think

4. Infer why manufacturers might want to sell directly to customers.

Practice Academic Skills

English Language Arts/Writing

5. Advertisements appear on TV, on the radio, in newspapers and magazines, and over the Internet. The nature of advertising is continually changing. As media technology changes, so does the way ads appear on it. Choose a product that you use regularly and create an ad for it. Write a few sentences describing the audience you want to reach.

STEM Science

6. Design and build your own protective container for shipping an egg. To test it, drop the container holding a fresh egg from a height of 6 to 10 feet onto the floor. Some recommended rules: (a) The container must be no larger than $4 \times 4 \times 4$ inches. (b) Everyone cleans up his or her own mess. (c) The egg will be put into the container before the test, so the container cannot be built around the egg. (d) Every container must have solid sides; wrapping it with foam rubber is not allowed. (e) To promote creative packaging techniques, no padding should be used.

Exploring Careers in Technology

Una Kim

SHOE COMPANY OWNER

Q: *What do you do?*
A: I am the CEO of a women's footwear and apparel company called "Keep." I oversee the operations of the business and creatively direct our products and our marketing.

Q: *What kind of training and education did you need to get this job?*
A: I have a bachelor's degree in economics and a master's degree in business administration. School provides an important backbone, but getting out there and working on projects is the best preparation. This can mean working at other jobs that will give you relevant skills, but it can also mean your own projects. I worked in marketing, but I also organized music festivals, played in bands, and contributed to zines! Keeping up with the latest technology is important.

Q: *What do you enjoy most about your job?*
A: I love working with people I respect and who inspire me every day. I feel proud when I see my shoes and clothing come into being.

Q: *How did you become interested in your field?*
A: I've always loved bridging creativity with business. The process of taking an idea and bringing it to life is a fulfilling and challenging journey. Business is an avenue to bring cool ideas into the world to share with other people.

English Language Arts/Writing

Present a Product Write a description of new footwear using selling points such as technology and style.

1. Using a word-processor, write a product description of footwear of your choice, listing materials and other selling points.
2. On paper, draw the shoes, depicting their points of interest.
3. Using presentation software, show the footwear model to the class, presenting it as you would to a potential buyer.

Go to **glencoe.com** to this book's OLC to learn more about this career.

Real-World Skills

Speaking, listening, problem-solving

Academics and Education

Mathematics, English language arts, industrial design, marketing

Career Outlook

Growth slower than average for the next ten years.

Source: *Occupational Outlook Handbook*

chapter 17 Review and Assessment

Chapter Summary

Section 17.1 In a company organization, the research and development department looks at new ideas and creates new products. After the company decides to make a new product, design engineers decide how it will look. After all the testing is finished, the production department decides how to manufacture the product.

Section 17.2 With CNC, machine tools operate by commands from a computer. When computers are programmed to operate all the machinery, CAM is being used. CIM stands for computer-integrated manufacturing. The completed product is inspected by the quality assurance department.

Section 17.3 The marketing department determines how to advertise the product. A wholesaler sells to retailers. Some manufacturers sell directly to consumers on the phone, through the mail, and over the Internet.

Review Content Vocabulary and Academic Vocabulary

1. On a sheet of paper, use each of these terms and words in a written sentence.

Content Vocabulary

- R&D
- virtual factory
- schedule
- CNC
- CAM
- CIM
- troubleshooting
- standard
- robotics
- e-manufacturing
- OSHA
- NIOSH
- advertising
- wholesaler
- retailer

Academic Vocabulary

- concept
- acquire
- target
- contact
- potential
- automate

Review Key Concepts

2. **Summarize** the importance of research and development.
3. **Explain** how products are designed.
4. **Identify** some of the advantages of using a virtual factory.
5. **Define** CNC, CAM, and CIM.
6. **Discuss** the use of industrial robots and e-manufacturing.
7. **Describe** the role quality assurance plays in modern manufacturing.
8. **Identify** the purpose of the marketing department.
9. **Name** several forms of advertising.
10. **Compare** wholesalers and retailers.

Real-World Skills

11. Persuasion Advertisements try to persuade you to buy a product. Pick an issue that you feel strongly about. Write a short speech about your topic in which you try to persuade your classmates to agree with your point of view. After you present your speech, ask if anyone changed their opinion.

Technology Skill

12. Just-In-Time Delivery Just-in-time delivery keeps only the supplies that are needed immediately in inventory.

a. Search the Internet for companies that use just-in-time techniques. Use the search words "just in time" + (your state)."

b. Determine which companies in your state use these techniques. Write a paragraph explaining the advantages and disadvantages of using JIT.

Academic Skills

Social Studies

13. Research artificial intelligence versus human decision-making, and impacts of AI on society. Write a paragraph.

STEM **Mathematics**

14. Terry has a job hanging billboard advertisements. The paste he uses comes in gallon cans. He is working on a billboard 16 yards long and 13 feet high. A can of the paste covers 100 square feet. How many cans will he need?

Math Concept **Area Measurement** To determine the area of a rectangle, multiply its length by its width.

1. Express measurements using the same unit of measure.
2. The result of multiplying two numbers is the product. Express the product of length and width in square units.

WINNING EVENTS

Manufacturing Entrepreneur

Situation You are part of a group of entrepreneurs who work together to design, manufacture, package, and sell a product.

Activity Working with your team, make contact with two other teams outside your community who wish to participate. Determine what you want to produce and sell. Develop a plan and a timeline for each team.

Evaluation Your activities and product will be evaluated by these criteria:

- Communication skills
- Team skills
- Organizational skills—effective
- Product—well designed, packaged

Go to **glencoe.com** to this book's OLC for information about TSA events.

Standardized Test Practice

Directions Choose the letter of the best answer. Write the letter on a separate piece of paper.

1. What is the surface area of a cube with each edge measuring 4 feet?
   A 16 square feet
   B 96 square feet
   C 64 square feet
   D 48 square feet

2. OSHA is the agency that approves the use of protective equipment, such as safety glasses.
   T **F**

Test-Taking Tip Skim through the test before you begin, so you know how to pace yourself.

chapter 17

TECHNOLOGY LAB

Test Products

A product manufactured by one company is usually a little different from the same product made by another company. Even though they all may meet the same standards, one company's product might be better than others. *Consumer Reports* tests and evaluates all kinds of products, including cars, door locks, and toothpaste, and publishes its recommendations in a magazine and on its Web site, consumerreports.org.

Tools and Materials

- ✓ Flashlight
- ✓ Three sets of the number of batteries that fit your flashlight: one set each of brand A, brand B, and brand C. Make sure they all have the same ratings printed on the outside (for example, "Size D Alkaline" or "Size AA Heavy Duty").
- ✓ One or two plastic grocery bags per experiment
- ✓ Clock

Optional

- ✓ Calculator
- ✓ Computer with spreadsheet software

SAFETY

Reminder

Be sure to always follow appropriate safety procedures and rules so you and your classmates do not get hurt.

Set Your Goal

For this activity, you will choose different brands of a single product and test them. You will record the test results and determine which product is the best buy.

Know the Criteria and Constraints

In this lab, you will:

1. Test flashlight batteries. Or, with your teacher's permission, choose a different product and test it.
2. Determine in advance what qualities of the product you will test. If the product comes with an instruction manual, read it to learn the correct way to use it.
3. Keep a record of the test results.

Design Your Project

Follow these steps to design your project and complete this lab.

1. Record the cost of each brand. Note the battery expiration dates. Write down all your information or use a spreadsheet.

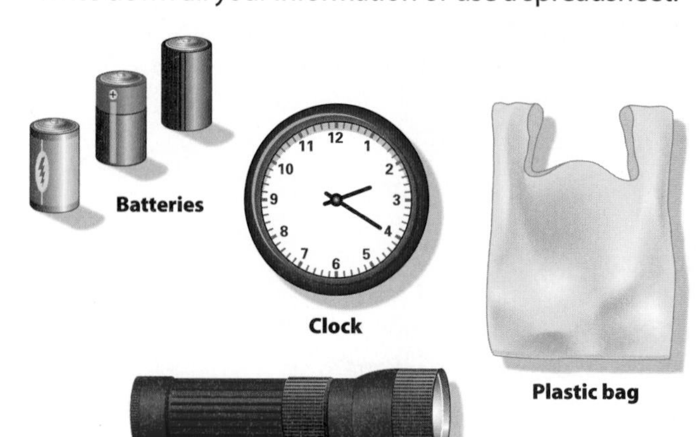

2. Determine your test procedure.
   - Your first test can be to find out which set of batteries produces the strongest light. One way to do this is to use white or light brown plastic bags from a grocery store.
   - See how many layers of plastic it takes to completely block the light from a flashlight that has brand A batteries inside.
   - Keep folding the plastic. You may need 20 or more layers.
   - Repeat the test using brands B and C.
   - Write down all your information or use a spreadsheet.

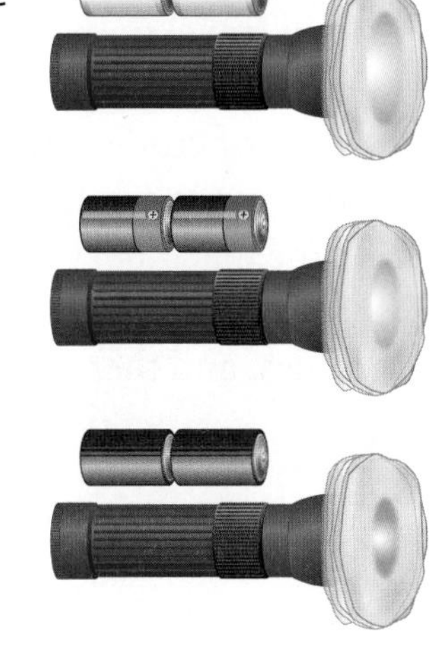

3. Another test is to find out how long it takes for the batteries to wear out.
   - Put new brand A batteries in the flashlight and turn it on. Write down the time you start. Keep the flashlight on until it no longer puts out much useful light.
   - Repeat for brands B and C. Write down all your information or use a spreadsheet.
   - When you are done, set the batteries aside for later disposal at a household hazardous waste facility. Do not throw them in the trash.

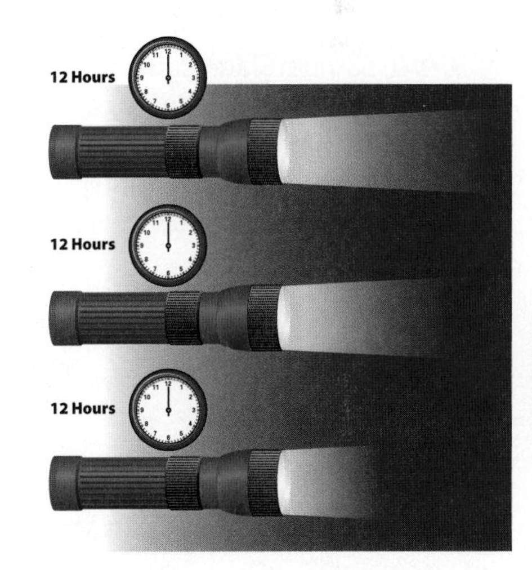

Evaluate Your Results

After you complete the lab, answer these questions on a separate piece of paper.

1. Based on your testing, which brand provided the most value? Why?
2. What other tests could you have conducted on the batteries?
3. Did you have any trouble reading the information you wrote down? How important was it to clearly write down all the information?

Academic Skills Required to Complete Lab

Tasks	English Language Arts	Math	Science	Social Studies
Record the cost of items to be tested.		✓		✓
Determine testing procedure.	✓		✓	
Perform a test on three different brands of a product.	✓	✓	✓	✓
Analyze results.	✓	✓	✓	✓
Find out how to dispose of batteries.	✓		✓	✓

Technology Time Machine

Evolution of the Factory

Play the Game This time machine will travel to the past to show you how manufacturing has changed from the days of crude cotton spinners to today's computer-run factories. To operate the time machine, you must know the secret code word. To discover the code, read the clues, and then answer the questions.

Clue 1

1700s The factory system was founded in Great Britain with engine-driven machinery producing various goods. The development of factory organization led to a huge increase in the production of goods—and the Industrial Revolution.

Clue 2

1790 Samuel Slater came to the United States with plans for a machine that would spin cotton. However, labor in this country was scarce, making the growth of factories slow at first.

Clue 3

1798 Another important innovation was the use of interchangeable parts. Prior to this time, parts were made individually. Eli Whitney and Simeon North, makers of small weapons, designed parts that could be used on any weapon they produced. Also, their workers were trained to specialize in one job. Both methods speeded up production.

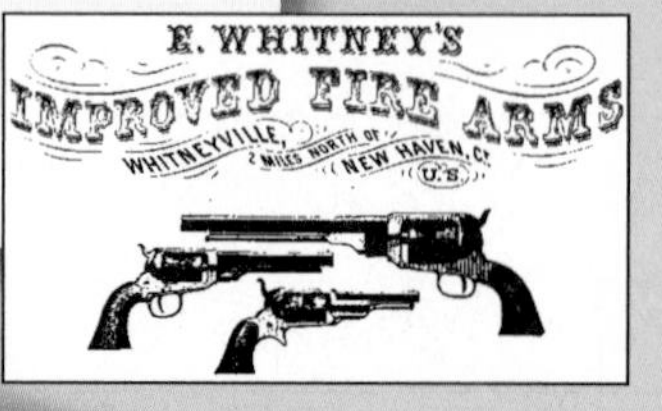

Clue 4

1914 With interchangeable parts and special training came the moving assembly line. It was introduced by automobile maker Henry Ford. Workers in the assembly line added parts as the product moved past them on a conveyor belt. This was a major improvement.

Clue 5

1950s With automation, manufacturing methods changed again. The automated restaurant was an early form of the vending machine. Automatic controls over machines were also used to make products. Automation relied on early computers. One result was industrial robots replaced human workers for certain tasks.

Clue 6

1970s In the United States, machinery and methods used in factories became outdated. Meanwhile, factories in Germany and Japan were using technology along with modern labor methods. They began to out-produce the United States, particularly in the automobile industry.

Clue 7

1980s Manufacturers in the United States introduced new factory systems using computers in the production process. Factories became more flexible and required fewer workers. Many products improved. These changes may lead to the factory of the future.

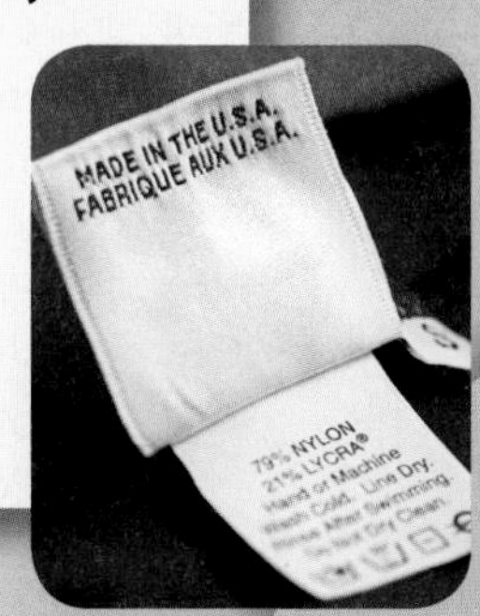

Crack the Code

On a piece of paper, write the answers to these questions:

1. What system began in Great Britain in the 1700s?
2. What relied on early computers in the 1950s?
3. Name the products that Eli Whitney and Simeon North produced.
4. What helped Germany and Japan out-produce the United States?
5. What was weapon-maker Whitney's first name?
6. What replaced humans for certain industrial tasks?

Now write down the first letter of each answer. Put them together to discover the secret code word!

Hint The assembly line meant manufacturers could make products like this.

unit 5 Thematic Project

Making Things Real

Manufacturing Products In Unit 5, you learned about how manufacturers use technology to create products. Taking classes in physics, machine shop, and CAD (Computer-Aided Design) in high school can lead you to a career in manufacturing engineering technology.

Things to Know To run a successful manufacturing company, you need knowledge in many fields. You also must answer questions about "logistics": Where do you store parts? How do you get them to the production line? Where do you store the finished product? What do you do with waste?

The Faster Way When you mass-produce anything, it is faster and easier to use an assembly line. In today's world robots are doing some of the more dangerous and/or repetitive jobs.

This Project In this project, you will research and design a manufacturing plant to make a product, such as a toy, T-shirt, or something of your choice.

Your Project

- Choose a product.
- Research how the product is manufactured.
- Design a manufacturing plant to make the product. Include storage, production lines, and traffic areas.
- Complete this bonus task for extra credit: Produce a toy your class or group could donate to a family shelter.
- Write a report on your product and plant.
- Create a presentation with posters, video, or presentation software.
- Present your ideas to the class.

Tools and Materials

- ✓ Computer
- ✓ Internet access
- ✓ Trade magazines
- ✓ Word-processing software
- ✓ Presentation software
- ✓ Video camera, digital camera, or cell phone with video
- ✓ Posterboard
- ✓ Colored markers

The Academic Skills You'll Use

- Communicate effectively.
- Speak clearly and concisely.
- Use correct spelling and grammar.
- Conduct research using a variety of resources.
- Incorporate reading, writing, and speaking with viewing, representing, and listening.

English Language Arts

NCTE 8 Use information resources to gather information and create and communicate knowledge.

Social Studies

NCSS 8 Science, Technology, and Society

Step 1 Choose Your Topic

You can choose any manufactured product to research or a simple toy to produce. Examples include:

- Canned beans
- Printed greeting cards
- Crayons
- Wooden tops
- Bean-bag games
- Pom-pom dolls
- Simple puzzles

If you are having trouble finding information, choose something else.

Step 2 Do Your Research

Research your project. Your fact finding may include finding answers to the questions below:

- What types of machines are used to mass-produce this product?
- Could computers help the creation of this item? If so, how?
- How much will wages be for your workers?
- How can you keep the process eco-friendly?

Ask adult relatives and friends who know about your topic.

Step 3 Explore Your Community

Find someone in your community who knows something about your topic. Ask how his or her job connects to your product or manufacturing process. Ask for feedback on your process.

Use any suggestions that will help you!

Step 4 Create Your Project

Your project should include:

- 1 research project (design, blueprint, or model)
- 1 report
- 1 presentation

Project Checklist

	Objectives for Your Project
Visual	✓ Make a poster, blueprint, model, or slide presentation to illustrate your project.
Presentation	✓ Make a presentation to your class and discuss what you have learned.
	✓ Turn in research and names from your interview to your teacher.

Step 5 Evaluate Your Presentation

In your report and/or presentation, did you remember to:

- Demonstrate your research and preparation?
- Present plans for a manufacturing plant?
- Label your designs or test your process?
- Use facts and evidence to back up your ideas?
- Speak clearly and engage your audience?

Rubrics Go to **glencoe.com** to the book's OLC for a printable evaluation form and your academic assessment form.

GLOBAL TECHNOLOGY

Globe Making

At a globe factory, production revolves around world events. These spherical maps have to be constantly revised because war and politics can change borders. Nova Rico is a small Italian map-making company that has been making globes for more than 50 years. It is one of the biggest globe makers in the world, but the only one in Italy. It once received an order from a leader with instructions on where to place his country's borders!

Critical Thinking *How might a customer's special instructions affect the manufacturing of a certain product?*

Italian

hello	ciao
goodbye	arrivederci
How are you?	Come stai?
thank you	grazie
You're welcome	Prego

Go to **glencoe.com** to the book's OLC to learn more and to find resources from **The Discovery Channel.**

unit 6 Construction Technologies

Chapters In This Unit:

Unit Thematic Project Preview

Planning a Green Shelter

As part of this unit, you will learn about construction systems for homes and commercial buildings. You will also find out how construction systems evolved from log homes to the skyscrapers of today.

As you read this unit, use this checklist to prepare for the project at the end of this unit:

PROJECT CHECKLIST

✓ Look at magazines and Web sites for examples of green buildings.

✓ Find out what features are important for green buildings.

✓ Check the Yellow Pages to find builders who specialize in green construction.

WebQuest Internet Project

Go to **glencoe.com** to this book's Online Learning Center (OLC) to find the WebQuest activity for Unit 6. Begin by reading the Task. This WebQuest activity will help you learn about bridge designs and how they work. You will also learn about the history of bridge building and some famous bridges.

Explore the Photo

Constructing Our World Buildings come in all shapes and sizes. From the smallest house to the most magnificent bridge and skyscraper, technology plays a part in design and planning for usefulness, beauty, and safety. *What materials do you think are used the most in construction?*

chapter 18

The World of Construction

Sections

What You'll Learn

- **Discuss** how construction systems evolved.
- **Name** some important structures.
- **Explain** the purpose of different construction materials.
- **Discuss** building codes and safety.

Explore the Photo

Super Tall The *Burj Al Arab* at Jumeirah Beach in Dubai, United Arab Emirates, is a unique sail-shaped building. It is one of the world's tallest hotels. *What is the disc-shaped "balcony" near the top floor?*

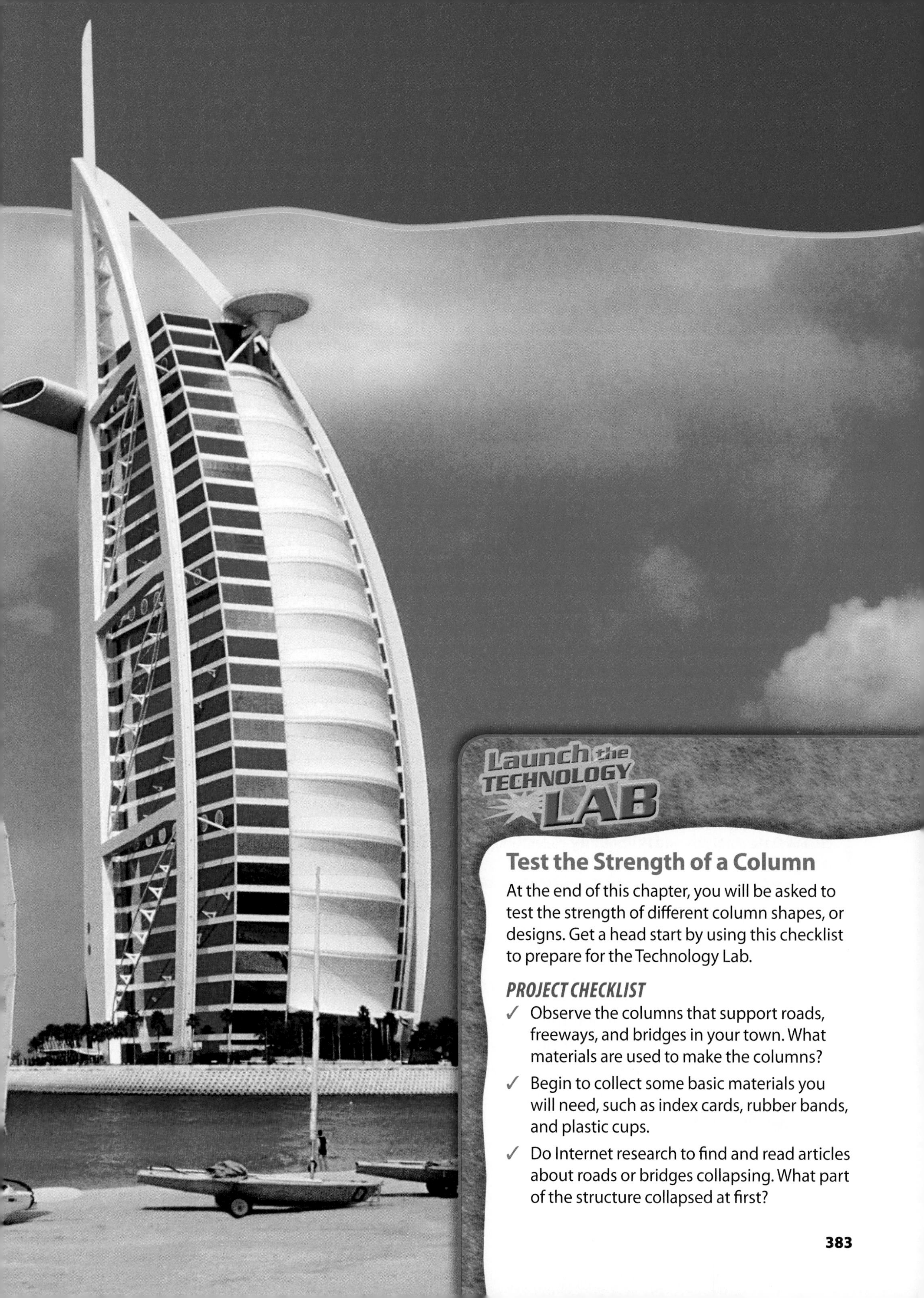

Test the Strength of a Column

At the end of this chapter, you will be asked to test the strength of different column shapes, or designs. Get a head start by using this checklist to prepare for the Technology Lab.

PROJECT CHECKLIST

- ✓ Observe the columns that support roads, freeways, and bridges in your town. What materials are used to make the columns?
- ✓ Begin to collect some basic materials you will need, such as index cards, rubber bands, and plastic cups.
- ✓ Do Internet research to find and read articles about roads or bridges collapsing. What part of the structure collapsed at first?

section 18.1

The Evolution of Construction

Reading Guide

Before You Read **Predict** What kinds of construction have all civilizations needed?

Content Vocabulary

- stick construction
- skyscraper
- surfacing

Academic Vocabulary

- maintain
- require

Graphic Organizer

Draw the section diagram. Use it to organize and write down information as you read.

Types of Construction

1. Residential	2. Commercial	3. Civil
Apartments		
Houses		

Go to **glencoe.com** to this book's OLC for a downloadable graphic organizer and more.

TECHNOLOGY STANDARDS

STL 6 Role of Society

STL 7 Influence on History

STL 11 Design Process

STL 20 Construction Technologies

ACADEMIC STANDARDS

Social Studies

NCSS 2 Time, continuity, and change

Mathematics

NCTM Data Analysis and Probability Select and use appropriate statistical methods to analyze data.

STL *National Standards for Technological Literacy*

NCTM *National Council of Teachers of Mathematics*

NCTE *National Council of Teachers of English*

NSES *National Science Education Standards*

NCSS *National Council for the Social Studies*

Types of Construction

Besides homes and office buildings, what are some other structures?

When the weather is severe, you can stay safe and dry inside a well-built building. Perhaps you are inside a house or apartment building. You might be inside a school, store, restaurant, bus terminal, or hospital. The buildings we use every day for shelter were assembled by people who work in the construction industry.

Other important construction projects include building roads, bridges, and tunnels. Other large construction projects include dams, canals, and even space stations.

As You Read

Connect What materials would you use to build a house?

Construction systems have existed for as long as people have built structures. Technology enables construction. There are three basic types of construction: residential, commercial/industrial, and civil. (See **Figure 18.1**.)

1. **Residential construction** This type provides places where people live, including apartment buildings and single-family homes.
2. **Commercial/industrial construction** This type includes office structures, shopping malls, and factories, as well as churches and other houses of worship.
3. **Civil construction** This type creates large structures for public use. Hospitals and schools are examples, as are roads, bridges, tunnels, and dams. Most civil construction is paid for with tax money collected by the government.

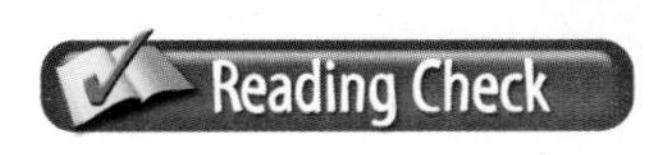

Connect What type of construction is used for your home?

EcoTech

Constructive Construction

A "sustainable design" considers the natural environment and ecosystem around a building to affect them as little as possible. It includes healthier and more efficient models of construction, renovation, operation, maintenance, and demolition.

Try This Look at buildings in your neighborhood and identify examples of sustainable design.

Buildings

What kinds of shelters did early Americans construct?

During the American colonial period, many trees grew in the eastern United States. Log houses were popular because they could be quickly put together. Log houses, however, were wasteful because so much wood was used. Their construction also required a great deal of strength to position the heavy logs. A new type of house construction started to appear in the 1840s. Instead of using logs or large wooden beams, these new houses were made of lightweight pieces of wood. This method was called **stick construction**.

Figure 18.1 Types of Construction

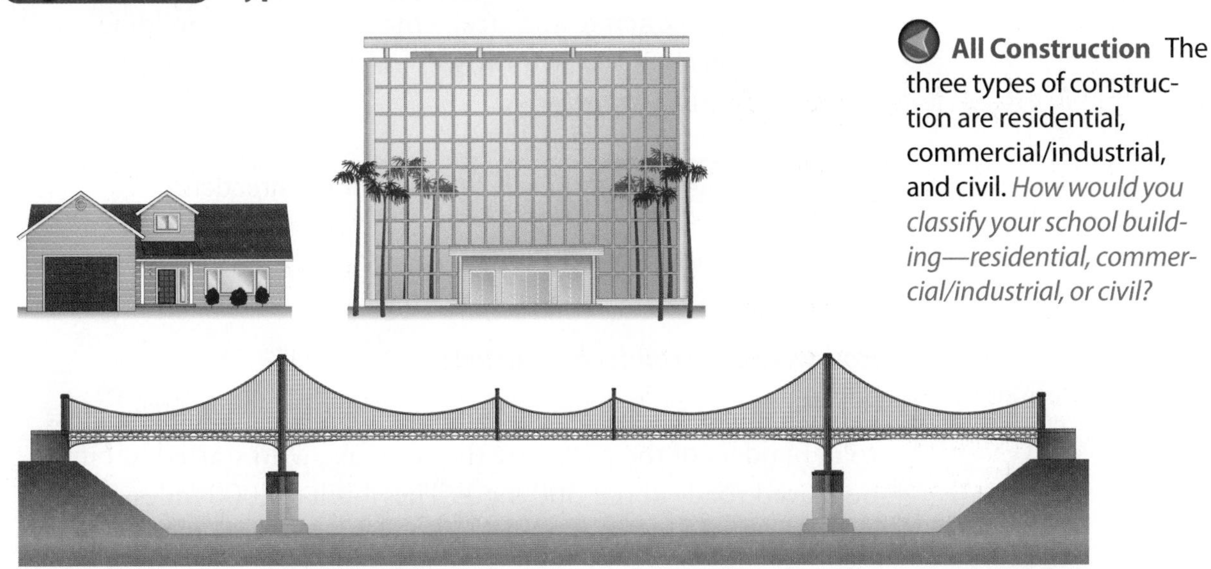

All Construction The three types of construction are residential, commercial/industrial, and civil. *How would you classify your school building—residential, commercial/industrial, or civil?*

The First High-Rise The 1885 Home Insurance Building in Chicago was the world's first skyscraper. It utilized stick construction with steel instead of wood. *Why was Chicago the place where skyscrapers first appeared?*

The frames went up quickly and provided both safe and strong dwellings. Almost all modern houses are still built this way. Even **skyscrapers** can be built with a type of stick construction that uses steel.

The first skyscraper was the 1885 Home Insurance Building in Chicago. At ten stories high, it was not as tall as some other buildings. However, it was the first to use a metal frame as a basic part of its design. The outside walls were connected to the metal frame. The walls did not support the building as in log houses.

One of the world's tallest skyscrapers is Taipei 101 in Taipei, Taiwan. It is 1,676 feet tall. Tall and strong modern skyscrapers are made with concrete and steel. One of the tallest buildings in the United States is the Sears Tower in Chicago, standing at 1,450 feet and 110 stories.

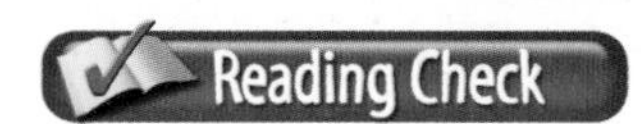

Recall What is stick construction?

Roads

How were early roads constructed?

Most ancient roads were little more than dirt paths. The great road builders of the past were the Romans, who started to build improved roads about 300 B.C.E. They built 50,000 miles of roads, more than America's interstate highway system.

Types of Road Surfaces

Roads can more easily support heavy loads if the roads are covered with a strong and durable material. This covering process is known as **surfacing** a road.

Early Roads

The ancient Romans used flat stones for road surfaces; some early American roads were surfaced with logs or planks placed crosswise. Both types of roads were difficult to **maintain**. They were also so bumpy that people had to travel on them very slowly. The first section of the National Road in the eastern United States, which is now U.S. Highway 40, or Interstate 70, was made from logs!

McAdam Roads

Around 1800, George McAdam from Scotland developed a method for making a road surface smooth. He used tar, which comes from crude oil. The tar is heated and spread over a thick layer of crushed rocks on the ground. It is still a common way of surfacing roads, driveways, and parking lots. Americans call this surfacing material "asphalt" or "blacktop." The British call it "macadam," or they use the brand name "Tarmac."

Today's Roads

Today high-speed highways are usually surfaced with concrete because concrete lasts longer than asphalt. Most of America's approximately 47,000 miles of interstate highways have concrete surfacing.

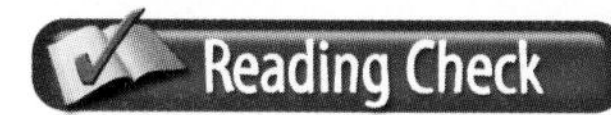

Compare What materials were used to make the first roads and the most recent roads?

McAdam's Method Road workers lay down asphalt for a highway. *Why do you think asphalt is heated before it is used for road surfacing?*

Academic Connections
Social Studies

Reach for the Sky Many countries seem to compete to see who can construct the tallest skyscraper.

Apply Research the world's tallest buildings and make a list of the top five, arranged by height. Write down the year each one was completed. Copy a world map and label where each is located with its year.

Figure 18.2 The Arch

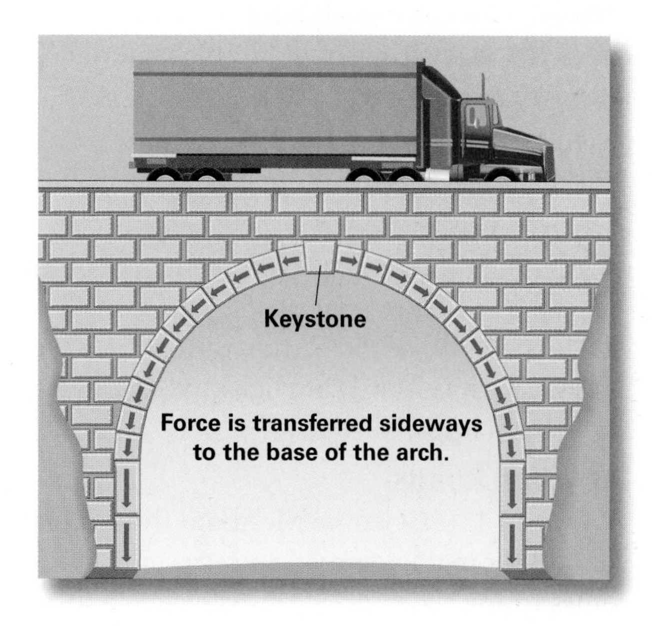

Simply Strong An arch transfers the truck's weight from the center to the two sides. *Could a bridge have two or more arches?*

Bridges

When were the first metal bridges made?

The Romans of 300 B.C. gave us long-lasting roads, but they also devised strong, well-designed bridges. They developed the arch—using wedge-shaped stones arranged and locked in a curve. The arch shape distributed weight sideways as well as downward. See **Figure 18.2**. Like their roads, some Roman bridges have lasted for many centuries.

During the late 1700s, improvements in metal manufacturing greatly reduced the price of iron. That was when people started to make bridges out of iron. One hundred years later, bridge builders began to use steel, which is a much stronger material. In 1874, the world's first major all-steel bridge, the Eads Bridge, was built across the Mississippi River at St. Louis, Missouri. It was named for its designer and chief engineer, James Buchanan Eads.

Modern bridges are still made of steel and supported by concrete. The modern bridge with the longest distance between supports is the *Akashi-Kaikyo* Bridge, which is near Kobe, Japan.

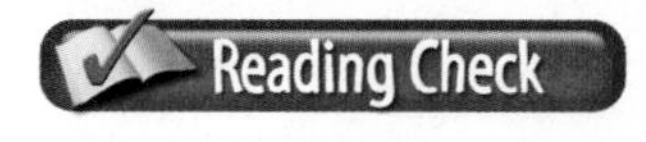

Identify What material(s) is used to build the strongest bridges?

Tunnels

What is the purpose of a tunnel?

Bridges take people over obstacles, while tunnels take them through or under them. Tunnels are less noticeable than bridges and can be less inviting. They often make us think of mysterious caves.

Rebuilding New Orleans

In August 2005, Hurricane Katrina devastated the city of New Orleans. Eighty percent of the city was flooded. Now, the people of New Orleans are trying to rebuild their homes. They are getting help from governments, agencies, businesses, and individuals.

Planning to Preserve But it is not easy. Most people want to protect the city's landmarks. They also want to improve conditions for everyone. This requires good urban planning, which is the *design* of cities and communities.

English Language Arts/Writing

Maintaining Technology Why did the flood protection system in New Orleans fail? One reason is that the design was flawed. Another reason is that the system was not properly maintained.

1. Make a checklist of things you can do to properly maintain your bike, computer, or other device.
2. Write reminders in your calendar to do regular maintenance.

The first major tunnel in the United States was constructed through the Hoosac Mountains in western Massachusetts. The railway tunnel is almost five miles long, took over 20 years to complete, and opened in 1875. It is still used today.

The world's longest tunnel is the Seikan Railway Tunnel in Japan. It is more than 33 miles long, but more than 14 miles are under water.

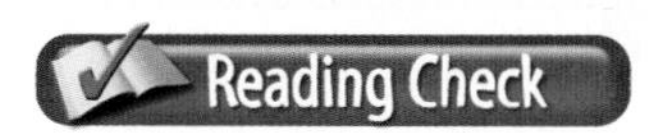

Explain What is the purpose of a tunnel?

Other Construction

What are some other large types of construction projects?

There are several other types of large construction that are not buildings or houses. These include canals, dams, and construction of structures used in space.

Canal Construction

Our ancestors used small canals to bring water to their crops. Canals are human-made waterways. Ships carrying passengers and products once had to travel a long, dangerous route around the southern tip of South America to get from the Atlantic Ocean to the Pacific Ocean. One of the largest waterways was opened in 1914, the Panama Canal in Central America. It was built by American engineers to connect the Atlantic and Pacific Oceans. Since then trade and travel has increased.

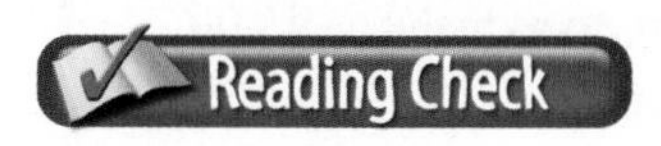

Recall What is a canal and what is the purpose of a canal?

 **Super Canal** The United States completed the Panama Canal, but ownership of it was later transferred to the country of Panama. *What was the purpose of the canal?*

Dam Construction

Dams divert the flow of water or cause it to form a pool. Sometimes this can prevent floods. Dams often improve water supplies and the economy of an area, but that can compete with other needs. Animal habitats can be changed or destroyed by human-made dams; people can be displaced from their homes; and areas downstream can become too dry. Preventing these problems requires a proper balance. The Chinese government is dealing with these problems as they build the Three Gorges Dam on the Yangtze River in China.

Construction for Space

However, the world's most unusual and advanced construction project is taking place above the earth—in space. This project is the International Space Station, being built by the United States and other nations. It circles, or orbits, around our planet every 90 minutes. Astronauts take turns living on the space station.

section 18.1 assessment

After You Read Self-Check

1. List the kinds of structures built by the Romans.
2. Identify the new type of house construction that began in the 1840s.
3. Name the type of construction that can make a bridge strong.

Think

4. You have been asked to build a dam. Name at least three problems you would have to overcome.

Practice Academic Skills

English Language Arts/Writing

5. Suppose you have been asked to build a tunnel. Write a paragraph describing three problems you must overcome.

STEM Mathematics

6. The five longest spans on suspension bridges include the *Akashi-Kaikyo* in Japan (6,529 feet), the *Great Belt Bridge* in Denmark (5,328 feet), the *Humber* in the United Kingdom (4,626 feet). Two bridges in China complete the list: the *Runyang Bridge* (4,888 feet) and the *Jiangyin Suspension Bridge* (4,543 feet). What are the mean and median of the spans of the bridges?

Math Concept **Measure of Central Tendency** Two ways to describe a collection of data are to compute the mean (average) of the elements and the median. The median is the middle number of the data listed from least to greatest.

1. Determine the mean. Add the numbers and then divide by the number of elements.
2. To determine the median, list the numbers in order from least to greatest.

 For help, go to **glencoe.com** to this book's OLC and find the Math Handbook.

section 18.2

Design Requirements

Reading Guide

Predict How are construction systems like other systems?

Content Vocabulary

- lumber
- curtain wall
- concrete
- building code
- conservative design

Academic Vocabulary

- objective
- route

Graphic Organizer

Draw the section diagram. Use it to organize and write down information as you read.

Ingredients of Concrete

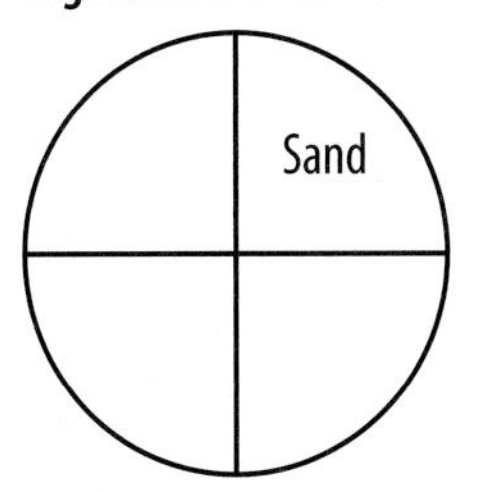

Go to **glencoe.com** to this book's OLC for a downloadable graphic organizer and more.

TECHNOLOGY STANDARDS

STL 4 Cultural, Social, Economic & Political Effects

STL 7 Influence on History

STL 10 Troubleshooting & Problem Solving

STL 20 Construction Technologies

ACADEMIC STANDARDS

Science

NSES CONTENT STANDARD F Science and technology in society

Social Studies

NCSS 8 Science, technology, and society

STL *National Standards for Technological Literacy*

NCTM *National Council of Teachers of Mathematics*

NCTE *National Council of Teachers of English*

NSES *National Science Education Standards*

NCSS *National Council for the Social Studies*

Understanding Requirements

Why must construction projects meet certain requirements?

Like other technology systems, construction systems include inputs, processes, and outputs. *Inputs* include the seven resources—people, materials, tools and machines, information, energy, capital, and time. *Processes* include designing the structure and putting it together. *Outputs* include the structure itself. Malfunctions of any part of this construction system may affect the way the system works and the quality of the outputs. For example, if low-quality steel is used, the bridge may collapse.

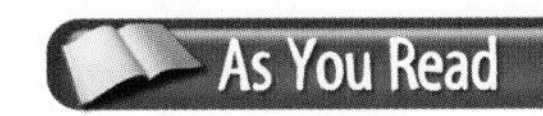

Identify What materials are used in construction?

Engineers and governments place many requirements on construction projects to ensure safety and long life. For example, the floor in your school classroom must be strong enough to support you and your classmates. In comparison, the floor in a single-family home must support a smaller load or weight. So, a building code may require your classroom floor to be made of concrete, but the floor in your house could be made of wood.

Although there are no perfect designs, engineers and designers can meet many requirements during the creative design process before building. Some requirements involve materials, building codes, and safety considerations.

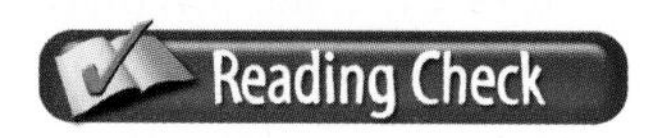

Identify What are the elements in a construction system?

Materials

What are the best materials to use in construction?

Centuries ago, before people acquired scientific knowledge, they built their homes and other structures from whatever materials were available.

Tech Stars

I.M. Pei

Master Architect

Award-winning architect I.M. Pei was born in China in 1917. He was educated in Hong Kong and Shanghai before moving to the United States at age 18. He studied architecture at the University of Pennsylvania, MIT, and later Harvard University. Pei was from a prominent family whose residence was surrounded by many rock sculptures, carved naturally by water. This led to a keen interest in how nature combined with buildings and how light and shadow mixed.

Career Building Pei is a master of high-modernist architecture. He works in abstract forms, using concrete, glass, stone, and steel. His many designs include the Mile High Center in Denver, the Four Seasons Hotel in New York City, the Pyramids of the Louvre in Paris, the Rock and Roll Hall of Fame in Ohio, the John F. Kennedy Library in Boston, and the Chinese Embassy in Washington, D.C. Pei is one of the most successful architects of his time. Yet despite his many awards, immense wealth, and high acclaim, he remains a humble man.

English Language Arts/Writing Describe in a short essay what types of architecture you find interesting.

Go to **glencoe.com** to this book's OLC to learn about young innovators in technology.

Today scientists and engineers work to develop new and better materials, such as:

- Wood
- Steel
- Concrete
- Other materials

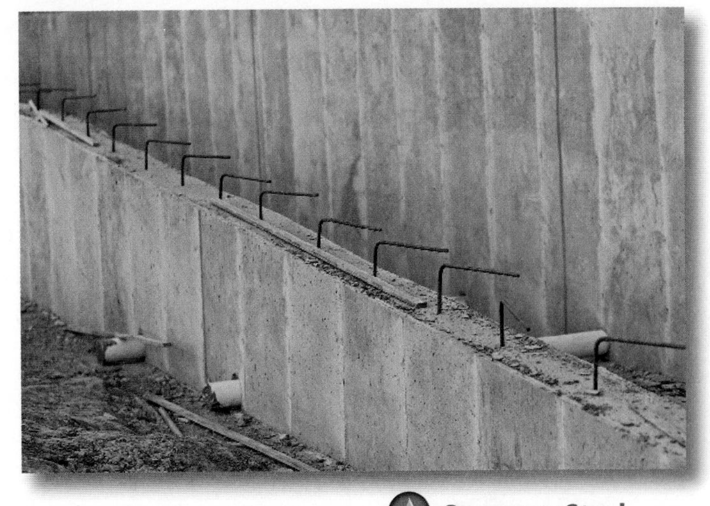

Strong as Steel Steel reinforcing bars make concrete much stronger. *Why?*

Wood

Trees growing on commercial tree farms produce wood in its natural state. People cut this wood to make **lumber**. Most single-family homes are made from lumber because it is readily available, easy to work with, and economical.

Wood is also a renewable resource. After trees are cut down, new ones can be planted in their place. However, trees take many years to grow, so it is important to "stretch" our current supply of wood. To do this, crooked trees, sawdust, and other wood wastes are used to create engineered wood materials, such as beams, plywood, and hardboard. Some engineered wood beams are even stronger than beams made from natural wood.

Steel

Steel is a mixture of iron and small amounts of carbon. It is used primarily to support large structures. Skyscrapers, for example, are built with a steel framework. Then floors and interior walls are added. The outside walls are hung on the metal frame. The walls do not help support the building. These exterior walls are called **curtain walls**. They can be made of many different materials, including glass.

Commercial buildings are sometimes made with steel supports and painted steel walls. These buildings might be airport terminals, grocery stores, and factories. Many suspension bridges are made of steel.

A Glass Tower? This office building has glass curtain walls. *With all that glass, what holds up the building?*

Paving the Way These workers are screeding the concrete for a driveway. *What do you think* screeding *means?*

Concrete

Made of a mixture of dry cement, sand, stones, and water, **concrete** is the most adaptable construction material. Builders use it for large buildings, highways, dams, bridges, and foundations of houses. Nearly every structure contains some concrete. Some people call concrete "cement," but cement is just one ingredient in a concrete mixture. It is the binder that holds everything together.

Concrete begins as a mixture that looks like ordinary mud. It is poured into forms or molds that hold it in place while it hardens. These forms give the concrete its final shape. Depending on the shape, it can be used for many purposes. Steel reinforcing rods or other steel shapes can be placed in the wet concrete to help strengthen it.

Other Materials

Many construction methods use materials besides wood, steel, or concrete for specific applications. For example, asphalt is an important surfacing material, made of crude oil and other substances, used for road construction and repair, driveways, and walkways.

Composite Materials

Composite materials are used for some bridges. A composite is a combination of two or more materials, such as fiberglass and carbon. Composite materials are much lighter than concrete and are weather-resistant, unlike steel. Composites are used for products such as bathtubs and roofing materials.

Bridge Materials This bridge is made of composite materials that were created in a factory and shipped to the site to be used. *Is it better to build a concrete bridge onsite or to have it shipped from the factory?*

Materials for Subsystems

Builders also use other materials in construction, such as masonry (bricks and stone), window glass, vinyl exterior wall coverings, fiberglass insulation, copper wire, and plastic water pipes. These materials are usually part of a structure's subsystems.

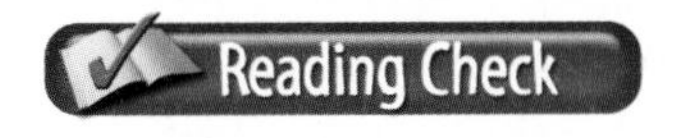

List What are the materials used in construction?

Regional Requirements

Are all buildings built the same way?

Many regions in the United States have different types of terrain and weather. Structures designed for one region might not be suitable for another region. For example, flatter roofs commonly appear on houses in the south, where there is little snow. However, people in the north must be concerned about the weight of snow on their roofs in the winter. If flat roofs were used, the snow would not slide off easily, and the roofs might collapse. The people in California and some other states must build structures to withstand earthquakes. People in some central states experience tornadoes. Other regions must be ready for floods and rising water, and/or landslides.

Building Codes

Local governments establish rules regulating the types of structures that can be constructed in their areas. Rules are part of the information inputs used by construction systems. They are called **building codes**.

Construction specialists agree that there must be certain standards for each structure. Commercial buildings must have enough exits in case of an emergency. Houses must have safe water supplies. Bridges and tunnels must support heavy loads. These are just a few of the many objectives of building codes.

Building codes can vary by state, county, or city. They are modified as new materials and methods of construction are developed. Before construction begins, builders contact a local inspection office at city hall or the county courthouse to apply for a permit. The builders must follow the regulations in the building code books.

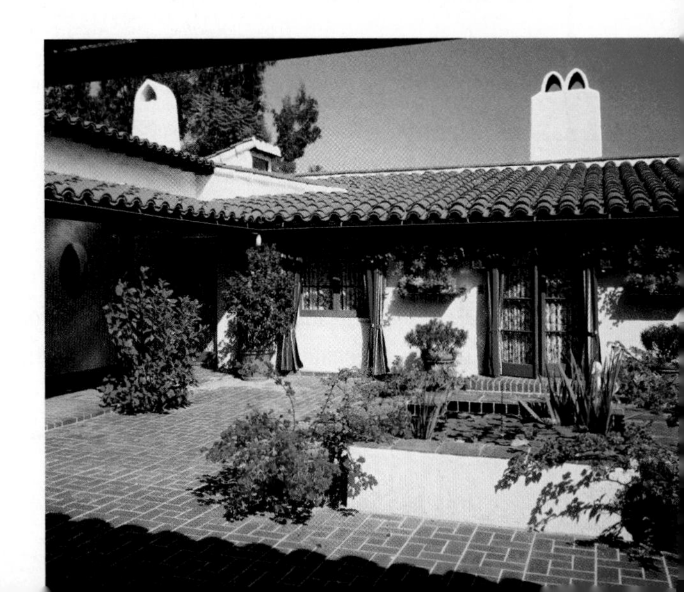

Rooftop Construction Tile roofs with little slope are common in the southern states. *Why are steeper roofs common in the northern states?*

House Roofs of Metal

Many residential houses have roofs made of asphalt shingles. But metal roofs are increasing in popularity. They keep houses cooler in the summer because they reflect more of the sun's energy. They are also safer because they do not burn. Some aluminum roofs are made from recycled beverage cans and resemble cedar shingles. The more popular metals include painted aluminum, steel, and copper. *Do you think a metal roof is a good choice for the house in which you live?*

Go to **glencoe.com** to this book's OLC for answers and to learn more about new roofing materials for homes.

For example, a builder might not be allowed to construct a tall building in a town's historic district because people want to preserve the historic appearance of the area. In another case, the building codes might permit building a house near a lake, river, or seashore if the house is on strong poles to keep it above water.

Building codes are designed to protect people's safety and the environment. They cover electricity, plumbing, energy use, and most other aspects of construction. That is why there is a lot of paperwork associated with building a new structure.

Safety

Buildings, bridges, and other structures are designed to support much more weight than would be placed on them. Engineers call this **conservative design**. Any major structural failure usually appears on the national news because it is a rare event. The failure of an interstate highway bridge in Minneapolis, Minnesota, in 2007, caused loss of life and disrupted traffic, but it was unusual. Nature can produce disastrous volcanic eruptions, tornadoes, tsunamis, and hurricanes, such as Hurricane Katrina in 2005, that result in severe damage.

Things happen that nobody can predict. So, engineers design escape **routes** and safety devices. All tall buildings have elevators, but they also have stairways in case electricity is cut off. Sprinklers put out small fires and keep them from spreading. Entrance doors and emergency exits in public buildings open outward so people can leave quickly. When accidents or other problems occur, engineers troubleshoot to prevent problems from happening again. New innovative technologies have sometimes resulted.

Section 18.2 Assessment

After You Read Self-Check

1. Name the four ingredients of concrete.
2. Define composite material. Give an example.
3. Identify the standards for building.

Think

4. You have been asked to design a house built on the side of a hill and a house built in an area that has tornadoes. Explain the differences in your designs.

Practice Academic Skills

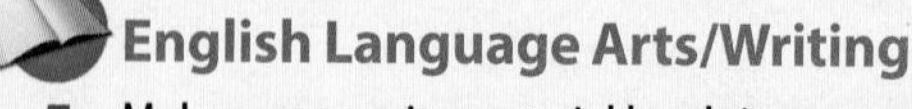

English Language Arts/Writing

5. Make a composite material by gluing a small section of window screening between two index cards. Write a paragraph evaluating the product for strength.

STEM Science

6. Using products that do not harm the environment or waste natural resources is known as "going green." Construction companies have been using materials that are friendly to the environment. Research some materials that construction companies use when they build a structure that is friendly to the environment. Write a few paragraphs about these materials and the pros and cons of using them.

Exploring Careers in Technology

Sue Tsoi

LAND SURVEYOR

Q: *What is a typical day like in your job?*
A: Each day varies, but it mostly involves talking with clients, researching and analyzing data, generating maps, writing legal descriptions, working on boundary resolution, and managing projects.

Q: *How did you get interested in your job?*
A: In my junior year, our high school offered its first environmental science class, and that got me hooked on science. I started going to nature camps and decided to major in forestry. In college, I was required to take some land-surveying classes and decided to major in the field.

Q: *What kind of training and education did you need to get this job?*
A: I took many science classes in college, such as botany, geology, and soil science, and received a degree in land surveying. After college, I worked for different surveying firms and earned my Land Surveying intern certificate and my Professional Land Surveyor license.

Q: *What do you like most about your job?*
A: I enjoy working in a profession that is considered behind the scenes. Most land surveyors are the first and last people who work on a construction project.

English Language Arts/Writing

Connect Construction Write a one-page report on how land surveyors are integral to the field of construction, and include the following information:

1. Do research online and at the library to find out what is required to receive a Land Surveying Intern certificate.
2. Write a step-by-step plan for a high school student who wants to become a land surveyor.
3. Go to a job-search Web site to find jobs advertised by construction companies seeking land surveyors.

Go to **glencoe.com** to this book's OLC to learn more about this career.

Real-World Skills

Communication, problem solving, adaptability

Academics and Education

Biology, geology, statistics, mathematics, English language arts

Career Outlook

Growth faster than average for the next ten years

Source: *Occupational Outlook Handbook*

chapter 18 Review and Assessment

Chapter Summary

Section 18.1 Construction occurs at a building site. The first skyscraper was the Home Insurance Building in Chicago. It had a metal frame design. Roads are construction projects. The Romans were the first great road builders. The British developed a material called "macadam" for surfacing roads. The material is called "asphalt" in the United States. The first major all-steel bridge was built across the Mississippi River in St. Louis, Missouri. The first major American tunnel was the Hoosac Tunnel in Massachusetts.

Section 18.2 Many structures are built of lumber, which is wood cut into useful shapes. Tall buildings are built with a steel framework. The outside walls, or curtain walls, are hung on the framework. Concrete is a construction material made of cement, sand, stones, and water. Because it can take on any shape, concrete is the most adaptable material used in construction. Structures built in different regions have different requirements. Building codes protect the environment and people's safety. They vary by region and are regulated by local governments.

Review Content Vocabulary and Academic Vocabulary

1. On a sheet of paper, use each of these terms and words in a written sentence.

Content Vocabulary
- stick construction
- skyscraper
- surfacing
- lumber
- curtain wall
- concrete
- building code
- conservative design

Academic Vocabulary
- maintain
- require
- objective
- route

Review Key Concepts

2. **List** some important structures.
3. **Discuss** the purpose of a tunnel.
4. **Describe** the first types of shelter.
5. **Explain** how early roads were once constructed.
6. **Tell** when and why metal bridges were first used.
7. **Describe** how wood is often used in building.
8. **Discuss** why steel is popular.
9. **Explain** how concrete is made.
10. **Explain** the purpose of having building codes.

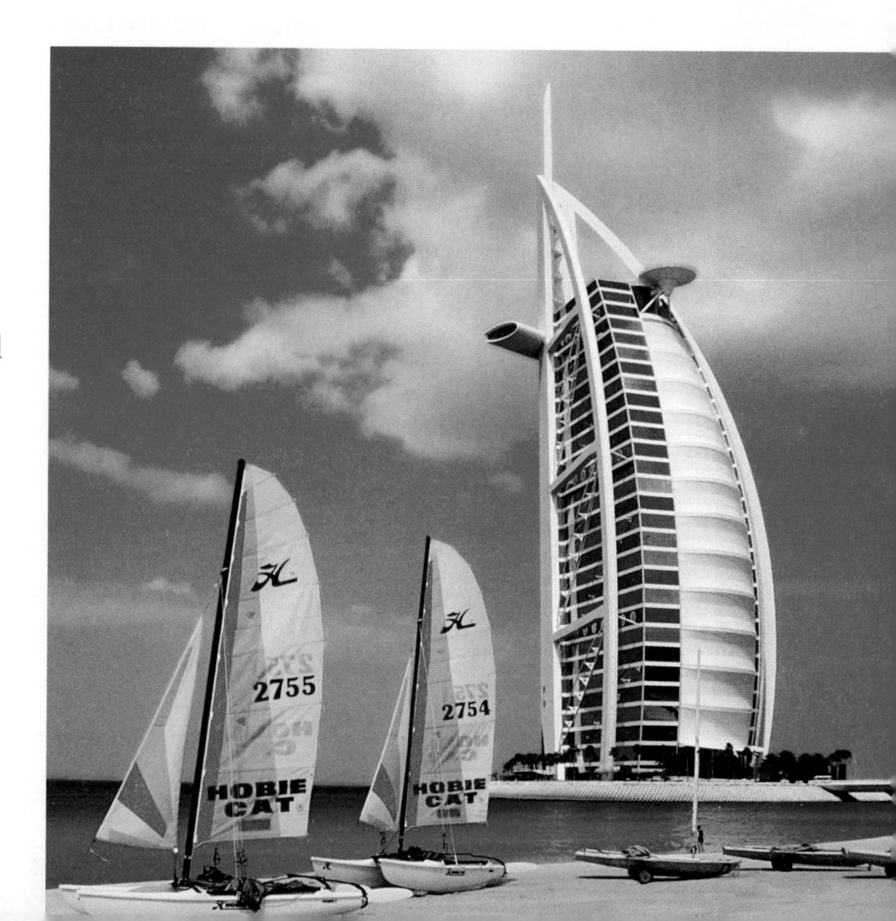

Real-World Skills

11. **Building Materials** Research the types of materials that are used for homes and businesses. Look at your own house and school and notice the types of materials used. Write a few paragraphs summarizing your findings.

Technology Skill

12. **Construction Safety** The construction industry is a dangerous field, despite numerous types of safety equipment to protect construction workers.
    a. Research the hazards at a construction site, and technologies for improving safety.
    b. Write a paper outlining some of the hazards and the safety equipment. Discuss gaps in safety equipment and how to solve the problem.

Academic Skills

Social Studies

13. Research construction practices of different countries. Choose a country from each of the continents. Write a few paragraphs explaining the construction practices, for both residential and commercial buildings, in each country.

Mathematics

14. Telephone poles are placed every 40 yards on a country road. How many poles would there be on a 2-mile stretch of this road?

Math Concept **Measurement** When solving word problems involving measurements it is necessary to convert the measurements to like units.

WINNING EVENTS

Urban Planner

Situation Your city council has proposed a four-lane highway through your neighborhood to reduce traffic congestion. Your team must identify the location and positive and negative impacts on your neighborhood.

Activity Work with your team and use a map of your neighborhood to determine where the highway should be located. Identify its impacts upon the environment, your neighbors, and the area's economy.

Evaluation Your assessment will be evaluated by these criteria:

- Report—well researched
- Impacts clearly identified for:
    - Neighborhood environment
    - Neighborhood citizens
    - Neighborhood economy

Go to **glencoe.com** to this book's OLC for information about TSA events.

Standardized Test Practice

Directions Choose the letter of the best answer. Write the letter on a separate piece of paper.

1. What is the area of a triangle with a base of 18 and a height of 12?
   A 108 **C** 54
   B 216 **D** 30

2. A curtain wall is an exterior wall that does not help support the building.
   T
   F

Test-Taking Tip Answer all questions on a multiple-choice test in order. Identify doubtful answers by marking them in the margin. Recheck these if you have time after you have answered all questions.

chapter 18

TECHNOLOGY LAB

Test the Strength of a Column

Look at the concrete columns supporting highway bridges and overpasses, particularly on an interstate highway. Some columns are cylinders, and others have square corners like boxes. They might be wide or they might be narrow. Their shape depends upon the load that the bridge or overpass has to carry. The shape of the column affects its strength.

Tools and Materials

- ✓ 3 × 5-inch index cards, four per experiment group
- ✓ Thin rubber bands
- ✓ Lightweight plastic cups
- ✓ Small weights, such as metal washers or bolts
- ✓ Scales for weighing the cup and the weights
- ✓ Plastic tubs
- ✓ Calculators

Set Your Goal

For this activity, you will fold index cards into different shapes to discover how much of a load each one can carry before it collapses.

Know the Criteria and Constraints

In this lab, you will:

1. Make paper folds that are sharp and uniform.
2. Keep a record of the maximum loads carried by each of the columns.
3. Observe the influence that each shape has on the column's load-carrying ability.

Design Your Project

Follow these steps to design your project and complete this lab.

1. Look at the drawing of column shapes. Call them V (one fold), N (two folds), M (three folds), and O (tube). Your folded cards will look like the letters when viewed from the top.

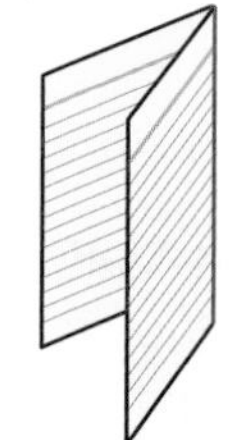

V Column

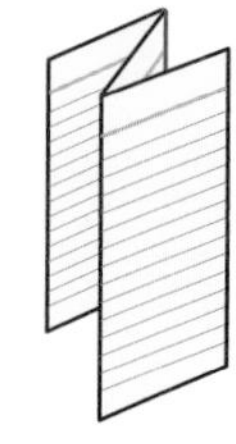

N Column

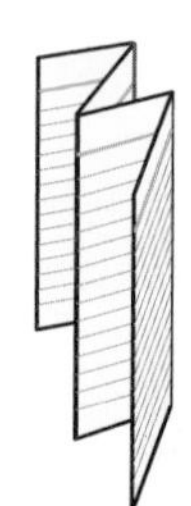

M Column

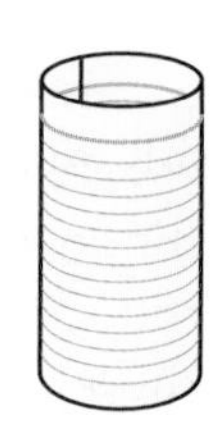

O Column

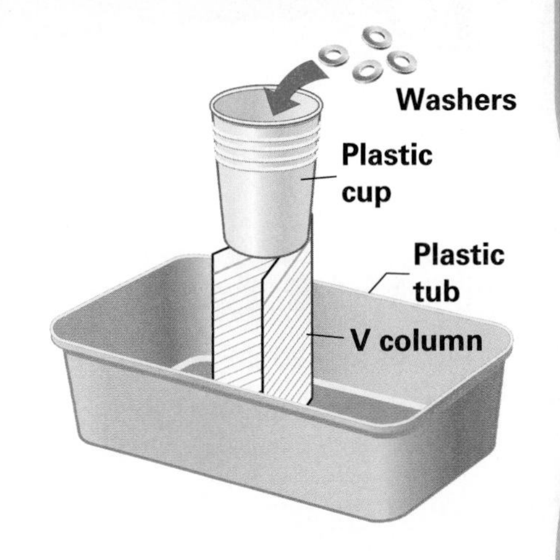

2. Weigh the plastic cup on the scale. Then weigh ten washers or bolts. After you know what ten of them weigh, you will be able to calculate how much 20, 35, or any other number will weigh. Do not mix types of weights. Use only washers or only bolts.
3. Now you are ready to start the experiment.
   - Fold one card into a V and place it in the plastic tub.
   - Carefully set the plastic cup on top of the folded card. If necessary, use a rubber band to help your V column keep its shape.
   - Slowly and carefully, add weights to the cup as shown in the illustration.
   - Eventually, the column will collapse and scatter the washers or bolts in the tub.
   - Collect and count them, and then calculate the maximum load supported by the V column. It will be a small amount, but add in the weight of the plastic cup.
4. Repeat the experiment for the N column and the M column.
5. Make the O column using a rubber band to hold its shape. Repeat the experiment.

Evaluate Your Results

After you complete the lab, answer these questions on a separate piece of paper.

1. Of the V, N, and M columns, which one carried the largest load? Why?
2. Did the O column carry a smaller or larger load than the folded columns carried?
3. What conclusions can you draw about column design from this lab experiment?

Academic Skills Required to Complete Lab

Tasks	English Language Arts	Math	Science	Social Studies
Weigh washers, bolts, and cup.		✓		
Do experiment, counting the number of weights each column could hold.		✓	✓	
Calculate the actual weight each paper column could hold before breaking.		✓		
Consider the relationship between column shape and strength.	✓		✓	

chapter 19

Building a House

Sections

19.1 Homes for People

19.2 House Construction

What You'll Learn

- **Identify** different residential dwellings.
- **Explain** the differences between manufactured houses and site-built houses.
- **Describe** how a building site is chosen.
- **Explain** how a house is assembled.

Explore the Photo 

Building Homes and Lives Habitat for Humanity enables people of all ages to help families build their own home. *Why is this man working safely?*

Build a Model House

At the end of the chapter, you will be asked to design and build a model of a house. Get a head start by using this checklist to prepare for the Technology Lab.

PROJECT CHECKLIST

- ✓ Do research on the Internet or go to a local hobby store to find examples of miniature model houses.
- ✓ Begin to collect materials you will need to do the project, such as marking pens, white glue, and wax paper.
- ✓ Ask your teacher to review the safety reminder for this lab.

section

19.1 Homes for People

Reading Guide

 **Before You Read** **Connect** What are some different types of homes?

Content Vocabulary

- residential building
- building site

Academic Vocabulary

You will see these words in your reading and on your tests. Find their meanings at the back of this book.

- community
- associate

Graphic Organizer

Draw the section diagram. Use it to organize and write down information as you read.

Advantages of Prefabriated Houses

1. Requires less labor
2. ____
3. ____
4. ____

 Go to **glencoe.com** to this book's OLC for a downloadable graphic organizer and more.

TECHNOLOGY STANDARDS

STL 12 Use & Maintenance

STL 20 Construction Technologies

ACADEMIC STANDARDS

Science

NSES Content Standard F Science and technology in society

Mathematics

NCTM Algebra Use mathematical models to represent and understand quantitative relationships.

STL *National Standards for Technological Literacy*

NCTM *National Council of Teachers of Mathematics*

NCTE *National Council of Teachers of English*

NSES *National Science Education Standards*

NCSS *National Council for the Social Studies*

The Structure of Homes

What covers the insides and outsides of homes?

Look at your arm. You can see your skin—the outer covering of your arm—but you do not see the bone that is part of its internal support. You see only what is on the outside, not what is inside. A house is similar.

The outside of a house may be covered with materials such as brick or wood. The walls inside may be covered with paneling, wallpaper, or paint. The floors may be covered with carpeting, wood or wood laminate, tile, or other materials. They make the rooms appear pleasant and comfortable. They also hide the inner structure of the house.

As You Read

Connect Have you ever seen the inside of an attic or a basement of a house or apartment? What did you notice there?

The inner structure holds up the walls, the ceilings, the roof, and all the other parts of a house. It has to be strong enough to support the loads the house must carry, but lightweight enough to be easy to construct and maintain.

Basic Types of Homes

What are the most lived-in types of homes in the United States?

There are three basic types of buildings: commercial, public, and residential. Commercial buildings are used by businesses, such as stores. Public buildings, such as a city hall, are for public use. **Residential buildings** are the buildings in which people live. This chapter is about the different kinds of housing and how a house is built. They include multiple-family housing and single-family housing. Single-family homes make up about two-thirds of the households in the United States.

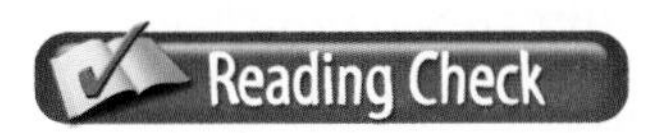

Name What are the basic types of buildings?

Multiple-Family Housing

What is multiple-family housing?

The cost of land in some communities is quite high. Large buildings that can house many families require a smaller area of land per person than single-family homes. They include apartments, townhouses, and condominiums.

Buildings with ten or more housing units in them provide about 13 million units in the United States. They are often built using techniques similar to those used for commercial buildings.

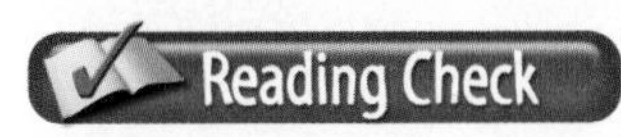

Describe What is multi-family housing?

Environmentally Friendly Bricks

Often red in color, clay bricks are used on the outside of brick houses and other buildings. They are baked in ovens that use a lot of fuel. But there is a new brick made from power-plant ashes and water that uses much less energy. The bricks are formed under high pressure in a low-temperature steam bath. They are smoother and more uniform, which makes them easier for workers to use at the construction site. Plus they cost less to make. *Explain why these new bricks are considered "green."*

Go to **glencoe.com** to this book's OLC for answers and to learn more about new materials for building.

Modern Living In large cities, many people live in high-rise apartment buildings. *Why are such tall buildings more popular in large cities than in smaller communities?*

Academic Connections
Social Studies

This Old House Many communities seek to preserve old buildings with historical value. Historical societies, museums, and community Web sites often feature them.

Apply Identify an old, historical house in your community. Find out when it was built and who the owners have been. Research interesting events or history regarding the house. Share your information with interested neighbors and classmates.

Single-Family Housing

What kind of construction is used for most single-family homes?

A single-family house is free standing and designed for the members of one family. The property on which it is built is called the **building site**, or construction site.

Stick-built construction is used for most single-family homes. Workers assemble pieces of lumber resembling sticks over a foundation (base) of some kind. However, other styles and methods are also used to build single-family houses.

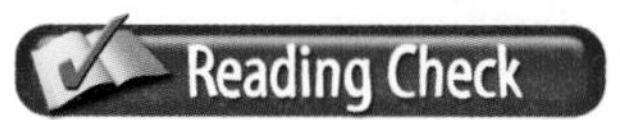

Define What is stick-built construction?

Log Houses

Many people associate log houses with the American pioneers. However, they were originally designed for the forests of Switzerland, Germany, and Scandinavia. Immigrants from those countries brought this construction knowledge with them to the United States.

Their links to the past and pleasing appearance have made log homes popular again. Today log houses are usually sold in kits, and the owner can hire a builder to assemble the house.

Rustic Trends Many people like the appearance of a log house. Designs for these homes are more elaborate than designs for the original log cabins. *Why do you think there is there no bark on the logs?*

Manufactured Houses

Most stick-built houses are built on the site where they will be located. Manufactured houses, or parts of these houses, are made in factories. They are also called "prefabricated houses." Factory production of houses uses less energy, less raw material, and less labor than building them on the site. Factories also can make more houses in less time.

The houses are made in sections and are delivered by trucks to the site where a concrete foundation has been installed. Large cranes position them on the foundation, and everything is bolted together.

Mobile homes and trailers are also manufactured houses. They have their own wheels and can be easily moved. Some New Orleans-area families who were displaced by the 2005 Katrina hurricane were housed in trailers. Including all types, there are about 9 million manufactured homes in the United States.

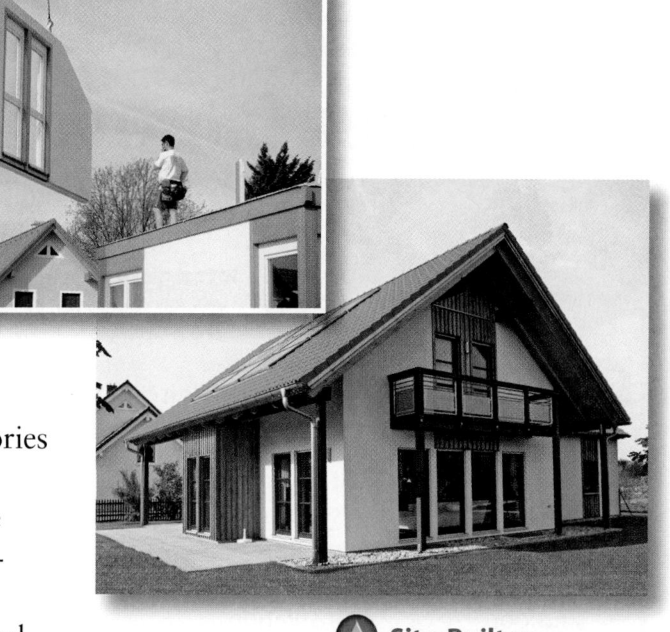

Site-Built or Manufactured? When a manufactured house is completed, it is almost impossible to tell it from a conventionally built house. *What are some advantages of a manufactured house?*

section 19.1 assessment

After You Read Self-Check

1. Name the three basic types of buildings.
2. List three examples of multiple-family housing.
3. Think of three possible advantages and three disadvantages of living in an apartment as compared to a single-family home.

Think

4. How could you tell the difference between a manufactured and a site-built house?

Practice Academic Skills

Social Studies/History

5. Research and write a one-page report on the history of the log cabin, tracing it from Europe to the United States.

STEM Mathematics

6. Madison lives in a high-rise apartment building with her father. Her father told her that he hid something for her on a certain floor. The number of the floor is three times her age, minus half of his age, plus the difference in their ages. Madison is nine, and her father is 38. On what floor will she find what he hid?

Math Concept **Writing Equations** Writing equations can help you solve problems with many different values.

1. Write out in word form the operations that need to be performed.
2. Replace the word form of the quantities with their numerical values and solve.

For help, go to **glencoe.com** to this book's OLC and find the Math Handbook.

section

19.2 House Construction

Reading Guide

Before You Read **Preview** Is it difficult to build a house? Why or why not?

Content Vocabulary

- foundation
- footing
- mortar
- subfloor
- stud
- sheathing
- gable roof
- truss
- insulation
- drywall

Academic Vocabulary

- restrict
- secure

Graphic Organizer

Draw the section diagram. Use it to organize and write down information as you read.

Questions When Choosing a Location

1. ______	2. How much is property tax?
3. ______	4. ______

Go to **glencoe.com** to this book's OLC for a downloadable graphic organizer and more.

TECHNOLOGY STANDARDS

STL 20 Construction Technologies

ACADEMIC STANDARDS

Science

NSES Content Standard E Understandings about science and technology

Social Studies

NCSS 8 Science, technology, and society

STL *National Standards for Technological Literacy*
NCTM *National Council of Teachers of Mathematics*
NCTE *National Council of Teachers of English*
NSES *National Science Education Standards*
NCSS *National Council for the Social Studies*

Early Housing

How did the pioneers build their homes?

During the 1800s, there were no construction companies in many regions. Pioneer farmers constructed shelters for themselves and their animals. Neighbors helped, and everyone worked together. They had no power tools, few metal nails, and no printed building plans. However, many of their buildings are still standing. Their basic construction techniques are still used today.

As You Read

Connect If you could build your own house anywhere, what location would you choose and why?

Choosing a Location

Can you build a house anywhere?

Some building sites are better than others. Before choosing one, there are some questions that need to be answered.

- Are electricity lines nearby? How about city water, natural gas for heating, telephone lines, and sewers? Without sewers, the house must have a septic tank.
- How much will property taxes cost?
- How much dirt will have to be moved? Houses on hillsides are more expensive to build than houses that are located on level land.

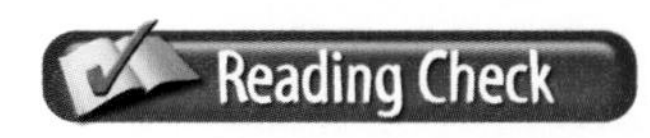

Identify What services should be available to build a house?

Selecting a House Design

Why do we have different house designs?

Selecting a house design depends on several factors besides personal preferences. These factors include building laws and codes, cost, and climate. Other factors may include the way in which the house will function. For example, will it be used for a growing family or a retired couple?

Some local laws restrict the type of houses that can be constructed in areas. These laws protect the homeowners already there. They can be sure the neighborhood will appear about the same.

After choosing the general design, the next step is to determine the size of the house. House size is calculated by its floor area. Small houses have about 1,000 square feet of living area. An average-size house might have 1,400 to 1,600 square feet.

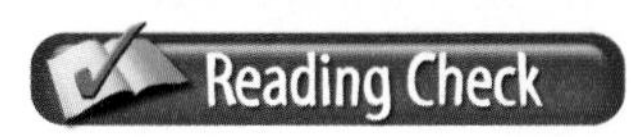

Recall How do you choose a house design?

EcoTech

Sustainable Flooring

The trees used to make traditional hardwood flooring can take 30 to 100 years to grow, and are not sustainable. In contrast, bamboo is a grass that can be used after five to seven years of growth. Also, it regrows after being cut.

Try This Use the Internet or your local library to find out about other sustainable flooring materials and how and where they are being used. Make a poster using this information, and then share it with your class.

Urban Sprawl

In many cities it is getting harder to buy a house. Buyers have to look further away from cities to find homes they can afford. New housing developments bring more freeways, shopping malls, and traffic. This is called "urban sprawl."

Urban sprawl can have a negative impact on the environment. It can also affect the quality of life. People might have to spend more time in traffic and less time at home.

English Language Arts/Writing

Smart Growth Drive Time One alternative to urban sprawl is "smart growth" communities that are designed to be compact, so it is easier to go places by walking, riding a bike, or taking public transportation.

1. You are a real estate agent. Do online research on smart growth.
2. Write an advertisement for a house in a smart growth community.

Building on a Strong Foundation Wet concrete is poured into forms. When it hardens, the concrete will make a strong foundation for the house that will be built on top of it. *What makes up a concrete mixture?*

The Foundation

How is a foundation constructed?

Buildings contain a variety of subsystems. The **foundation** is the part of the house that rests on the ground and supports the structure. The foundation starts with a trench dug where the outside walls will be located. Its depth depends on the local building code. Concrete is poured into the trench to create **footings**.

The foundation usually includes a low wall, sized for a full basement or a crawl space. The foundation wall is made from poured concrete or pre-cast concrete blocks with large holes to reduce weight. The blocks are positioned on the footing and fastened together with **mortar**, which is similar to concrete but lacking stones.

Describe What is the foundation?

Figure 19.1 Floor Joists

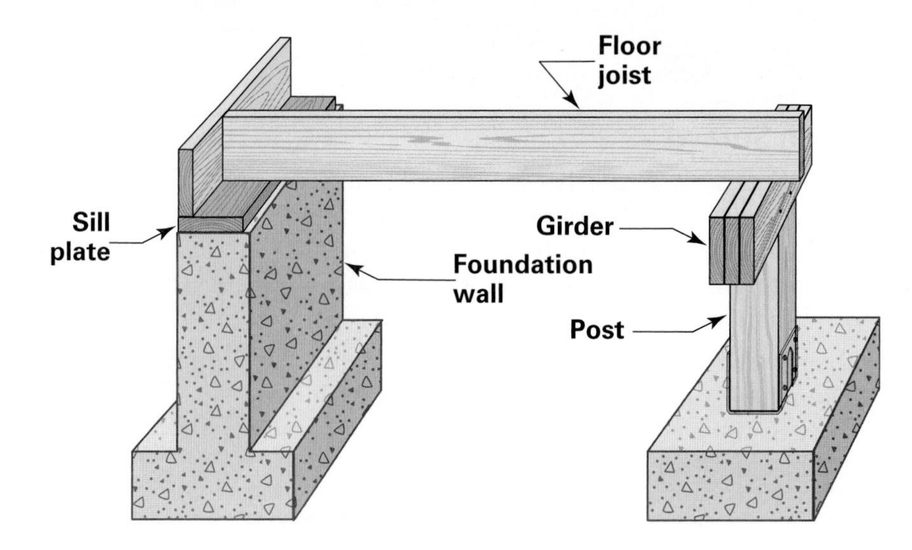

Floor Construction Floor joists usually rest on a sill plate and girder on the foundation of a house. *Where would you go to inspect the floor joists in a house?*

The Basic Structure

What holds up the house?

Wooden supports in the walls, floor, and roof make up the inner structure of a house. Large lumber is used because it must hold the structure's weight.

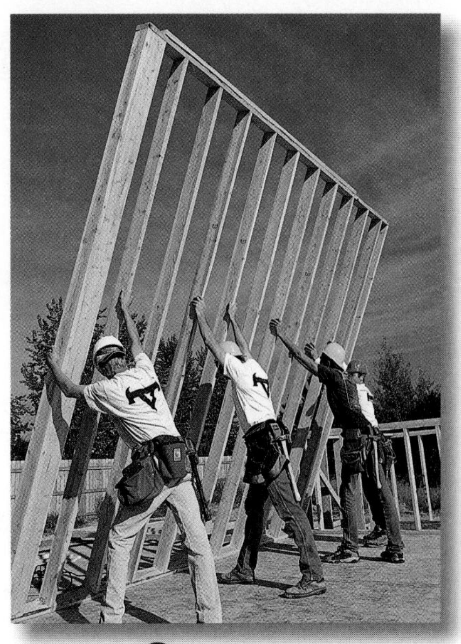

Raising the Wall After a wall frame has been nailed together on the ground, workers raise it into position and brace it to keep it straight. *What size lumber is normally used in a framed wall?*

The Floor

After the foundation is completed, the floor is built. (See **Figure 19.1.**) Bolts are placed in the top of a poured foundation wall. In a concrete block wall, holes in the blocks are filled with mortar, and the bolts are placed in the wet mortar. After the mortar hardens, a piece of lumber is bolted to the foundation. Floor supports are nailed to this lumber, which is the sill plate. The floor is supported underneath by floor joists. Joists are boards that extend from the front of the house to the rear. When the foundation area is large, a wooden or steel center beam, a girder, is installed in the middle.

A **subfloor** is nailed to the floor joists. The subfloor is usually made of plywood sheets. During construction, the subfloor takes a great deal of abuse. Tools are dropped on it, and nails are hammered into it. Later a finished floor will be installed over it.

Walls

Next the wall framework is put together on the subfloor, with openings for windows and doors. The vertical wall supports are called **studs**. When the framework is finished, it is raised into position by two or more people and nailed to the floor.

Each wall is nailed at the corners to walls that are already up. The walls are strengthened by ceiling joists connecting the top of the front wall to the top of the back wall. Then **sheathing**, made of engineered wood material, encloses the structure. Inside walls dividing the house into rooms are assembled and nailed in place.

The Roof

The roof must keep water out when it rains. It also must be strong enough to support heavy snow loads in cold climates. Most roofs have two sloping sides that meet at the center board, or ridge. This style is called a **gable roof**. The shape makes a strong and secure cover for the house.

Long pieces of lumber called "rafters" are nailed to the top of the walls and meet at a peak. The ridge board at the peak acts like a central support for the rafters. The rafters are covered with sheets called "roof decking or sheathing."

Many roofs are prefabricated triangular frameworks called **trusses**. Trusses include the rafters and ceiling joists.

Figure 19.2 The House Structure with Gable Roof

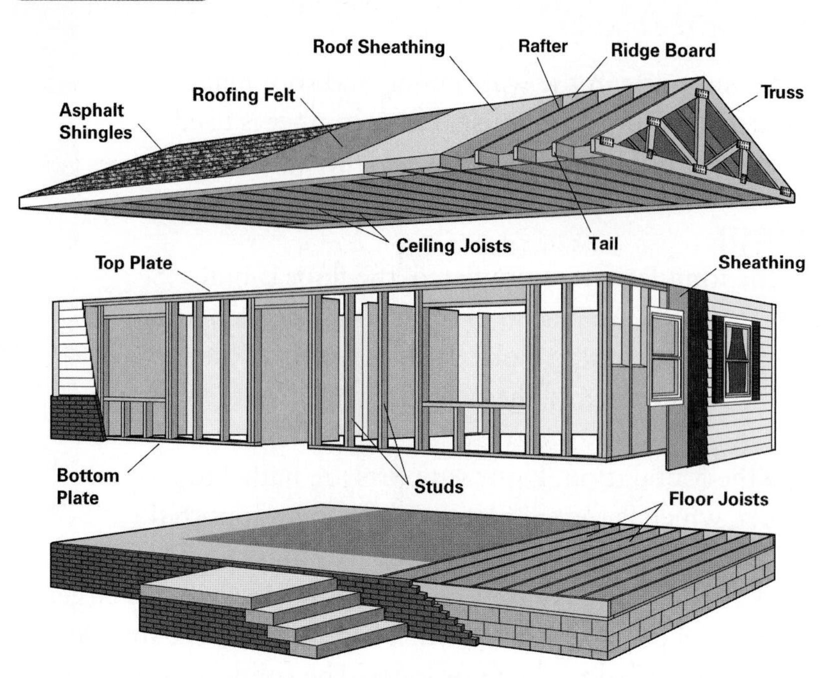

Whole House This drawing shows the major structural elements of a house. *Why might a house with framed rafters have more useful attic storage than a house with roof trusses?*

Wall and Roof Coverings

The external appearance of a house is what gives it character. Some houses just seem to look better than others. It might be because the builders or owners carefully picked siding and roofing material. Siding covers the walls. Roofing covers the roof.

Many different kinds of siding material can be fastened to the outside walls of a house. Wood, plastic, metal, and brick are only a few types. They are attached to the sheathing.

On many houses, asphalt shingles are nailed to the roof decking over a layer of heavy asphalt paper called "underlayment," or "roofing felt." The two layers provide a seal against rain and snow.

After the windows and doors are installed, the outside is almost completed. A person walking by might think the house could be occupied. (See **Figure 19.2**.)

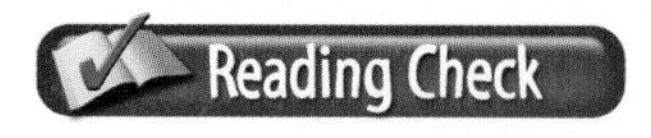

List What are the basic components of a roof?

The Interior

How is the interior of a house completed?

The interior of a house is completed after the exterior is finished. Workers have the advantage of being protected from bad weather. They install things such as insulation, utilities, and interior wall coverings.

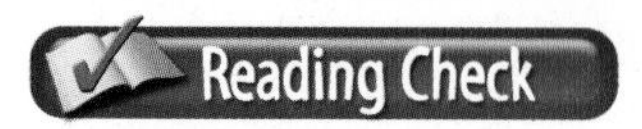

Recall When is a house interior completed?

Insulation

Insulation is like a blanket tucked inside the walls and ceilings. It helps keep a house warm in the winter. A popular insulation is a fluffy type of fiberglass made in long rolls that fit between floor joists, ceiling joists, and wall studs.

Cracks appear in a house during its construction. They can let in cold air. Caulking is used to seal them up. Caulking looks like toothpaste and is applied with a caulking gun.

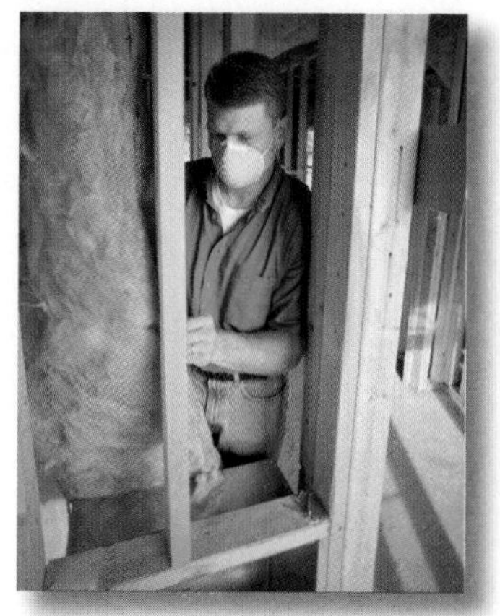

Safety First A worker must always use the proper safety equipment when installing insulation. *Why is this worker wearing a breathing mask?*

Utilities

Buildings contain a variety of subsystems referred to as utilities: electricity, natural gas, water, and sewage disposal. These services are provided by businesses called "public utilities." Electricity is used for heating, lighting, cooking, air conditioning, and other purposes. Natural gas is used for heating and cooking. Water is used for drinking and cleaning. Sewage disposal removes wastes from the house. Subcontractors install these subsystems.

Electric Systems

Electricians install wiring, outlets, and light fixtures in the house. Electricity enters the house through a heavy insulated wire. A meter located outside measures how much electrical power is used.

Heating and Cooling Systems

Heating and cooling systems require furnaces, air conditioners, pipes, and ducts. Heating systems usually use natural gas, electricity, or fuel oil. Air conditioning is usually powered by electricity.

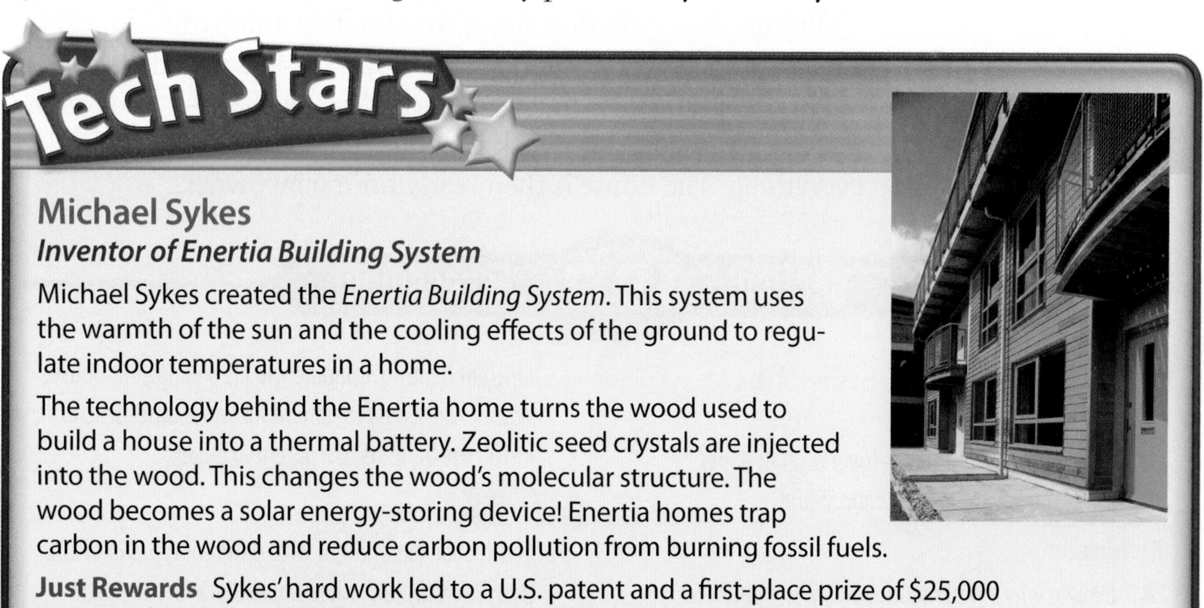

Tech Stars

Michael Sykes

Inventor of Enertia Building System

Michael Sykes created the *Enertia Building System*. This system uses the warmth of the sun and the cooling effects of the ground to regulate indoor temperatures in a home.

The technology behind the Enertia home turns the wood used to build a house into a thermal battery. Zeolitic seed crystals are injected into the wood. This changes the wood's molecular structure. The wood becomes a solar energy-storing device! Enertia homes trap carbon in the wood and reduce carbon pollution from burning fossil fuels.

Just Rewards Sykes' hard work led to a U.S. patent and a first-place prize of $25,000 in the "2007 History Channel Modern Marvels Invent Now" contest.

English Language Arts/Writing Write a few paragraphs describing ways you can protect the environment within your own home.

Go to **glencoe.com** in this book's OLC to learn about young innovators in technology.

Drywall Specialist Drywall can be screwed to the walls and ceilings with a screw gun. *Why is this worker wearing stilts?*

Water and Sewage Systems

These systems are installed by plumbing subcontractors. They are responsible for the network of pipes that carry water to the kitchen, laundry room, and bathrooms, and sewer pipes that carry wastes out of the house.

Water and sewage services are usually provided by the same public utility company. Pressurized water is delivered to the house through underground pipes. Inside the house, pipes branch out to where water is used. Some houses have a water well. Those houses might also have a septic system. The waste water is treated in a large underground tank and slowly returns to the ground.

The main water shutoff valve is on the water line where it enters the house. You should know where it is located.

Inside Wall Coverings

Drywall, or sheetrock, is made of plaster sandwiched between two layers of sturdy paper. The drywall sheets are nailed or screwed to the walls and ceiling as covering. A plaster-like material fills in the cracks between the sheets and covers the nail heads.

Finishing Touches

After the walls and ceilings are finished, the floor is completed. Floors can be finished with wood or wood laminate, stone, tile, or carpeting.

Trim is installed around the edges of floors, around doors and windows, and sometimes where ceilings and walls meet.

Interior doors are then hung, and kitchen and bathroom cabinets are added. Shelves and countertops are also installed.

Landscaping—planting trees, grass, and bushes—is done as the house is finished. After work is completed, the contractor checks everything. The house is then ready for a new owner.

section 19.2 assessment

After You Read Self-Check

1. Define the word *truss*.
2. Explain why roof shingles cover asphalt paper.
3. List some advantages of wood siding.

Think

4. Explain why a wall requires only 2 × 4 lumber.

Practice Academic Skills

English Language Arts

6. If a home buyer is concerned about heating costs, the seller might mention the home's insulation and efficient heating. Imagine you are selling your house. Write a mock advertisement for your home, describing all the benefits to a potential buyer.

STEM Science

5. Conduct an experiment with different types of insulation. Paint three or more boxes black on the outside. Leave one box empty; line the walls of each remaining box with a different type of insulation. Put thermometers inside each box and place the boxes in the sun. Record the results and analyze your findings.

Exploring Careers in Technology

Tiffanie Turner

ARCHITECT

Q: *What do you do?*

A: During the last 14 years, I've worked on large train stations, schools, and a high-tech "green" house. I am now working on renovations and additions to Victorian and mid-century homes in San Francisco and Marin County, California. I like the challenge of modernizing older buildings while respecting the Victorian style.

Q: *What kind of training and education did you need?*

A: I attended a 5-year bachelor of architecture program, followed by many years of interning at different architecture firms. I then took the licensing exam.

Q: *What do you enjoy most about your job?*

A: My favorite part of the job comes after I meet with a client, understand what he or she wants, and start to design. The process takes months and months. My second favorite part is construction administration, which happens when the construction really gets moving.

Q: *How did you become interested in your field?*

A: I took a high school design course. It was a great opportunity to use my love of art and drawing in a more technical, professional field. But I realized I had already been interested in architecture: I used to wish the world would stop so I could go exploring other people's houses!

English Language Arts/Writing

Write a Proposal Choose a building that you would like to renovate or add on to. Then write a proposal.

1. Using a word-processing document, describe how you would like to renovate or what you would like to add to an existing building.
2. Using a spreadsheet program, create a table and estimate the costs for the project. Include expenses such as materials, labor, and licenses.
3. Using a presentation program or visual aids that you create, present and explain your ideas to the class.

Go to **glencoe.com** to this book's OLC to learn more about this career.

Real-World Skills

Speaking, listening, interpersonal, problem-solving

Academics and Education

Mathematics, English language arts, marketing

Career Outlook

Growth as fast as average for the next ten years

Source: *Occupational Outlook Handbook*

chapter 19 Review and Assessment

Chapter Summary

Section 19.1 Apartments, townhouses, and condominiums house many families. Most houses are built at a specific location or site. Parts for manufactured houses are prefabricated in a factory. Log houses are usually sold in kits, and the owner hires a builder. Houses that face south are sometimes less expensive to heat than houses facing north.

Section 19.2 The foundation walls of a house are made from poured concrete or concrete blocks. The floor is supported by joists that rest on the foundation. The subfloor absorbs wear during construction and supports a finished floor. The walls are nailed together on the floor and raised into place. Ceiling joists help hold them in position. The roof structure can be made with rafters or trusses and is covered with shingles. Inside walls are covered with drywall. The utilities in a house can include electricity, natural gas, and water. Subcontractors install pipes, wires, and duct work. The finished floor, trim, cabinets, and landscaping are done last.

Review Content Vocabulary and Academic Vocabulary

1. On a sheet of paper, use each of these terms and words in a written sentence.

Content Vocabulary
- residential building
- building site
- foundation
- footing
- mortar
- subfloor
- stud
- sheathing
- gable roof
- truss
- insulation
- drywall

Academic Vocabulary
- community
- associate
- restrict
- secure

Review Key Concepts

2. **Identify** different residential dwellings.
3. **Explain** how manufactured houses and site-built houses differ.
4. **Explain** multiple-family housing.
5. **Describe** the type of construction used for most single-family homes.
6. **Discuss** how log houses are sold.
7. **Describe** how a building site is chosen.
8. **Explain** how a house is assembled.
9. **Discuss** the reasons for different house designs.
10. **Describe** how the interior of a house is completed.

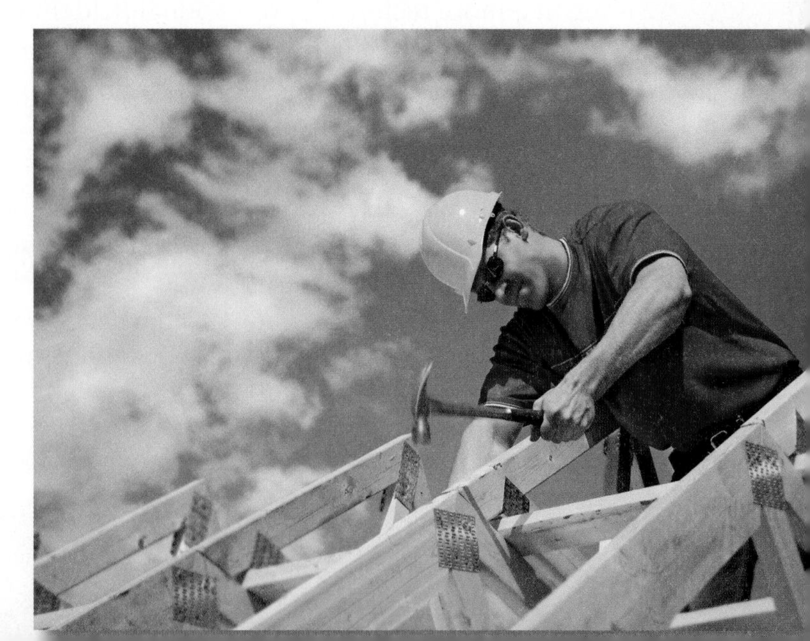

Real-World Skills

11. Disaster Relief When a disaster such as an earthquake or hurricane strikes, homes can be destroyed. Some companies make portable homes that can be used after a disaster. Research the different types of modular homes that could be used. Write a few paragraphs on what you find.

STEM Technology Skill

12. Floor Plan A floor plan is an illustration showing the layout of a structure. A floor plan should indicate where all the important places, such as electrical boxes and gas lines, are located.

a. Find the main water valve and electrical breaker switch in your home.

b. Use a computer illustration program, CAD software, or hand sketching to draw the floor plan of your home.

Academic Skills

STEM Science

13. Research different geodesic domes and their uses. Write a paragraph about the pros and cons of different domes.

STEM Mathematics

14. The average price of homes in a suburb of Chicago rose by 18.3 percent in a year. If the average price of homes in this area was $93,150 at the beginning of the year, what is the average price after the increase?

Math Concept **Percents** A percent represents a part of a whole. When determining the nominal value of a percent, convert the percent to a decimal by dividing by 100. If you are figuring a percentage increase of something, add 1 to the decimal equivalent and multiply.

WINNING EVENTS

Landscape Architect

Situation Your team will design and build an area at school where students can sit during breaks. The area must be functional, inviting, beautiful, and compatible with the environment and architecture.

Activity Develop a series of sketches. Make a three-dimensional model of your proposal. Include models of benches. Landscape the area and build and place the seating.

Evaluation Your project will be evaluated by these criteria:

Landscape	Seating
• Functional	• Functional
• Attractive	• Attractive
• Appropriate	• Well made

Go to **glencoe.com** to this book's OLC for information about TSA events.

Standardized Test Practice

Directions Choose the letter of the best answer. Write the letter on a separate piece of paper.

1. What percentage of 245 is 98?

A 52% **C** 0.40%
B 46% **D** 40%

2. The truss is a layer of material between house framing and the outer covering.

T **F**

Test-Taking Tip If you encounter a question that you think you cannot answer, leave it and answer it last—but do answer it.

chapter 19

TECHNOLOGY LAB

Build a Model House

Many different house designs exist because people have different needs and like different designs. When architects design a house, they often will build a model of it first. The model helps people visualize what the house will look like when it is built.

Tools and Materials

- ✓ Poster board
- ✓ White glue
- ✓ Marking pens
- ✓ Waxed paper
- ✓ Razor knife
- ✓ Metal ruler
- ✓ Wooden cutting surface
- ✓ House floor plan

Set Your Goal

Your goal for this lab is to make a model of a house that you think is interesting or attractive.

Know the Criteria and Constraints

In this lab, you will:

1. Make a model out of poster board.
2. Choose a fairly simple house design.
3. Create a gable roof for your house.
4. Use a scale of ½ inch equals 1 foot.
5. Label the rooms and any special features of the house.
6. Submit your sketches along with your model.

Design Your Project

Follow these steps to design your project and complete this lab.

1. Do some research on house plans. Sunday editions of newspapers may show plans for houses. Some magazines also include plans.
2. Select a house design that interests you. A single-story house will be easier to construct than one with two stories.
3. Make sketches of your model's floor plan and outer walls. Use a scale of ½ inch equals 1 foot. If a dimension on the original plan is 10 feet, it will be 5 inches on your model.

SAFETY Reminder

In this lab, you will be using cutting tools. Be sure to always follow appropriate safety procedures and rules so you and your classmates do not get hurt.

4. Place your drawings on the poster board.
   - Use a razor knife to cut out the front wall.
   - Draw the windows and doors with your marker. Also, draw any desired details.
5. Repeat for the other three walls. Make one wall at a time.
6. Place a small piece of waxed paper on your work surface to protect it. Hold the front wall against one of the side walls and glue the corners together along the inside of the walls. Repeat for all the other walls.
7. After the glue has dried, glue the four walls onto another piece of poster board.
8. Draw the entire roof on poster board.
   - Cut it out. Draw a line down the middle.
   - Use razor knife and metal ruler to score a line down the middle of the roof.
   - Then fold it in the middle.
   - Place the roof on the walls but do not glue it.
9. Remove the roof and add interior walls. Place labels in the rooms.
10. You can add a porch or a deck if you want.
11. Add landscaping like bushes and trees cut from poster board.
12. Label the features of your house that you especially like.

Evaluate Your Results

After you complete the lab, answer these questions on a separate piece of paper.

1. Write a paragraph explaining why you chose this house design.
2. Discuss whether the models helped you visualize the house.

Academic Skills Required to Complete Lab

Tasks	English Language Arts	Math	Science	Social Studies
Research house plans.	✓			✓
Draw model to scale.		✓		
Build outside of house model and attach to poster board floor.		✓		
Add interior walls.		✓		
Label house model.	✓			

chapter 20 Heavy Construction

Sections

What You'll Learn

- **Compare** the construction of large buildings to the construction of houses.
- **Describe** basic methods used in building skyscrapers.
- **Explain** why asphalt and concrete are preferred materials for roadways.
- **Identify** the ways in which bridges are supported.
- **Discuss** methods used to build tunnels.
- **List** the three main parts of a dam.
- **Tell** the purpose of a monument.
- **Explain** how construction in space is different from that on Earth.

Explore the Photo 

For Art's Sake The Guggenheim Museum in Bilbao, Spain, was designed by American architect Frank Gehry. It is a remarkable feat of engineering. Complex computer-simulation programs helped make construction possible. *What kind of construction project is the Guggenheim Museum: residential, commercial, or civil?*

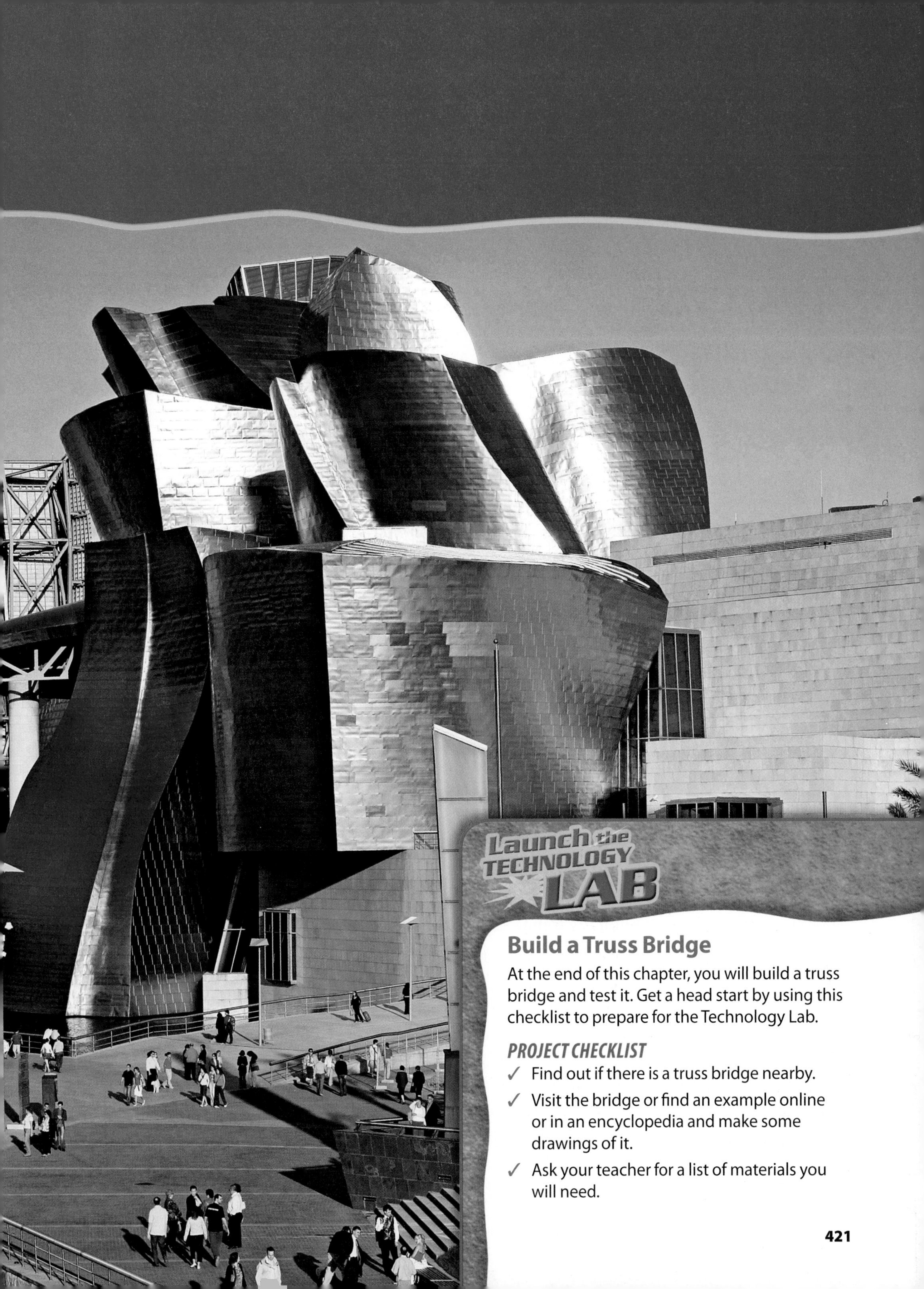

Build a Truss Bridge

At the end of this chapter, you will build a truss bridge and test it. Get a head start by using this checklist to prepare for the Technology Lab.

PROJECT CHECKLIST

✓ Find out if there is a truss bridge nearby.

✓ Visit the bridge or find an example online or in an encyclopedia and make some drawings of it.

✓ Ask your teacher for a list of materials you will need.

section

20.1 Large Buildings

Reading Guide

Before You Read **Preview** How does safety affect the construction of large buildings?

Content Vocabulary

- excavation
- pile
- pier
- crane

Academic Vocabulary

You will see these words in your reading and on your tests. Find their meanings at the back of this book.

- integral
- control

Graphic Organizer

Draw the section diagram. Use it to organize and write down information as you read.

Safety Issues for Tall Buildings

1. ______________________
2. ______________________
3. ______________________

Go to **glencoe.com** to this book's OLC for a downloadable graphic organizer and more.

TECHNOLOGY STANDARDS

STL 3 Relationships & Connections

STL 11 Design Process

STL 12 Use & Maintenance

STL 20 Construction Technologies

ACADEMIC STANDARDS

Science

NSES Content Standard F Science and technology in society

English Language Arts

NCTE 1 Read texts to acquire new information.

STL *National Standards for Technological Literacy*

NCTM *National Council of Teachers of Mathematics*

NCTE *National Council of Teachers of English*

NSES *National Science Education Standards*

NCSS *National Council for the Social Studies*

Working with Heavy Construction

What are some examples of heavy construction projects?

Structures built with heavy construction methods include skyscrapers, dams, tunnels, monuments, and bridges, which may be the most attractive structures built by modern people. Because most of these structures will be a part of the daily lives of numerous people, safety is integral to their design.

Heavy construction requires many workers with different skills. Sometimes thousands of people work on a single project. The designs for these massive projects come from many qualified engineers and architects who specialize in specific areas of technology. Specialization can drive technological improvements.

As You Read

Connect How is the construction of large buildings similar to the construction of houses?

The Burj of Greatness Located in the United Arab Emirates, the Burj Dubai in Dubai may be the tallest building in the world. *What do you think "Burj" means?*

Skyscrapers

What is the tallest skyscraper in the United States?

Some large buildings are spread out over a great deal of land but are not very tall. They may be built where land is inexpensive or set aside for public use. Other large buildings, called "skyscrapers," are very tall, but they do not cover a lot of land area. They are usually built in large cities where land is scarce and expensive.

Skyscrapers in the United States

Of the five tallest skyscrapers in the United States, Chicago has three. They include the Sears Tower at 1,454 feet, which is the tallest in America, the Amoco Building at 1,136 feet, and the John Hancock Center at 1,127 feet. New York City has the other two tallest skyscrapers. They are the Empire State Building at 1,250 feet and the Chrysler Building at 1,046 feet.

New York's Manhattan Island is mostly rock. That provides a strong foundation support. This is one reason why so many other skyscrapers are located there. Manhattan's 21-story Fuller Building, also known as the *Flatiron Building*, because its shape resembles a clothes iron, is the oldest skyscraper in the world that is still standing. It was built in 1902.

Malls vs. Stand-Alone Stores?

Imagine a world without shopping malls. Today shoppers and store designers seem to prefer smaller, more open designs to large enclosed malls. Some stores are avoiding malls completely, preferring their own space. Others do their business via the Internet. *Why might businesses build their own stores?*

Go to **glencoe.com** to this book's OLC for answers and to learn more about the future of shopping.

Ethics in Action

The World's Tallest Building

The world's tallest building is called the Burj Dubai in the United Arab Emirates. But taller projects are already underway. The first skyscrapers were built because there was not enough open land in large cities. Why are they being built today?

Top This Cities take pride in skyscrapers. They bring visitors from around the world; they inspire people to dream of even taller structures. But how tall is too tall? Some believe that the contest to build the world's tallest building is a waste of resources.

English Language Arts/Writing

No Contest Competition is essential in sports. Players practice hard because they want to win. But at other times, success is the result of cooperation.

1. Discuss competition in a small group. Is it always good? What is good about it? What is bad?
2. Choose one person to take notes. Choose someone else to share your thoughts with the rest of the class.

Foundations

The foundation of a skyscraper must support a huge load. A deep **excavation** is dug, and **piles** made of concrete or steel are driven deep into the soil until they hit solid rock. Rock gives the building the proper support. Foundation walls are built, and extra support is added in the form of strong **piers**, or columns, made of reinforced concrete.

Frameworks

The inner framework of a skyscraper is usually made of steel. Workers build it floor by floor. As each floor is completed, large **cranes** lift the steel parts to the next floor level. Workers weld, bolt, or rivet the parts in place. Then the curtain walls are lifted into position and attached. These curtain walls merely hang on the structure and provide little support.

Safety

It is not unusual for people to be concerned about their safety in tall buildings. Modern construction procedures, building codes, and strict rules cover fire protection and ways of escape. Tall buildings have restrictions and controls about the use of fire-resistant construction materials. Even lightning conductor systems are built into modern tall buildings. All skyscrapers must sustain strong winds without swaying. In areas around the world that have earthquakes or violent storms, builders use special supports and designs to withstand movement.

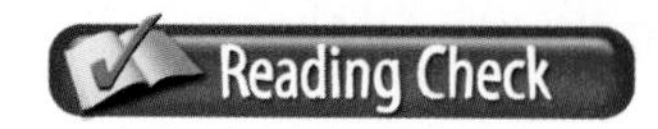

Summarize How are skyscrapers built?

Other Large Buildings

How is your school built differently than a skyscraper?

Some buildings spread out over large areas of ground. In the suburbs more land may be available and may be less expensive than in the city. Buildings constructed on these suburban sites might be supermarkets, shopping malls, restaurants, office buildings, and schools.

Like skyscrapers, these structures often have steel frames and many concrete parts, unlike residential buildings. Your school may be built in this way.

However, the shape and appearance of these large buildings are usually quite different. Your school, for example, would probably have different outer walls than walls for a grocery store or a place for religious worship.

Builders usually cannot use column-and-beam framing for very wide buildings, such as auditoriums. Instead, they may use built-up girders to support large areas. They might also use trusses and arches for support.

Green Paint

Many common house paints contain lead or other toxic additives. These chemicals are added to kill mildew. However, these chemicals are also harmful to you, your family, and your pets. If you pour them down the drain, they may contaminate the water supply.

Try This Find out the brand names of non-toxic paint and stain removers, as well as cleaning agents.

section 20.1 assessment

After You Read Self-Check

1. Name one of the tallest skyscrapers constructed in the United States, and also name the city in which it can be found.
2. Compare the foundation of a skyscraper to the foundation of a house.
3. Describe how it is possible for skyscrapers to have walls of glass.

Think

4. Which category or categories of buildings do skyscrapers belong to: residential, commercial/industrial, or public? Explain your answer.

Practice Academic Skills

5. Research the world's tallest skyscrapers at the library or on the Internet. Draw scaled representations of the top five skyscrapers. Use a scale of 1 inch = 100 feet. Label your drawings, including the name of the skyscraper, its height, the year it was built, and the city and country where it can be found.

STEM Mathematics

6. The Empire State Building is 1,250 feet tall. The Chrysler Building is 1,046 feet tall. By what percentage would the Chrysler Building need to be raised to equal the height of the Empire State Building?

Math Concept **Percents** To determine a percentage divide the part by the whole.

1. Find the number of feet difference between the two buildings by subtracting.
2. Divide the number of feet difference by the total height of the Chrysler Building.

For help, go to **glencoe.com** to this book's OLC and find the Math Handbook.

section

20.2 Roadway Projects

Reading Guide

Before You Read **Connect** What are some typical roadway projects?

Content Vocabulary

- subgrade
- abutment
- span
- suspension bridge
- cable-stayed bridge
- truss bridge
- cantilever bridge
- beam bridge
- shield

Academic Vocabulary

- allocate
- consist

Graphic Organizer

Draw the section diagram. Use it to organize and write down information as you read.

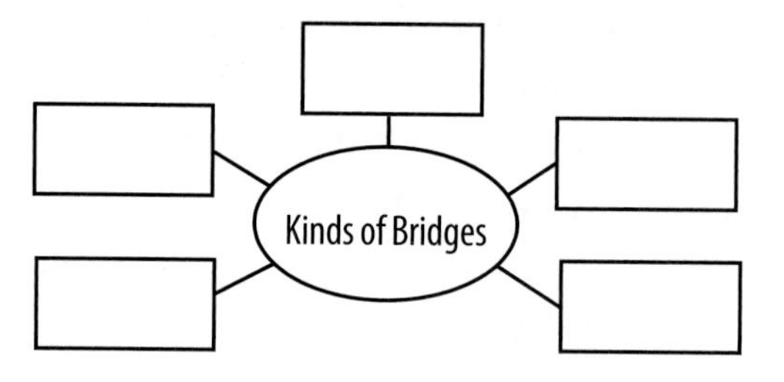

Go to **glencoe.com** to this book's OLC for a downloadable graphic organizer and more.

TECHNOLOGY STANDARDS

STL 11 Design Process

STL 12 Use & Maintenance

STL 17 Information & Communication Technologies

ACADEMIC STANDARDS

Science

NSES Content Standard E Understandings about science and technology

NSES Content Standard F Science and technology in society

STL *National Standards for Technological Literacy*

NCTM *National Council of Teachers of Mathematics*

NCTE *National Council of Teachers of English*

NSES *National Science Education Standards*

NCSS *National Council for the Social Studies*

Roads and Highways

What is below the surface of a highway?

Did you know that roadway projects include streets, highways, bridges, and tunnels? Bridges allow highways or railways to cross rivers and valleys. Tunnels allow roads or railways to run *through* obstacles, such as mountains, rather than go over or around them.

As You Read

Explain Why are bridges and tunnels also considered roadway projects?

Roadways in the United States

The United States has millions of miles of roadways. Our interstate highway system covers more than 47,000 miles in length. Federal, state, and local governments **allocate** billions of dollars every year to maintain and construct roadways.

Road Construction

All paved roads are made in three layers. See **Figure 20.1**. The **subgrade** is the natural soil along the roadway. If it is not level or firm enough, heavy machines scrape and pack the soil. Next comes the base. A common base is sand or gravel. It provides support and keeps water from collecting underneath, which could freeze and break the pavement. Finally, the surface material is added. The surface is smooth and higher at the middle to drain off water.

Figure 20.1 Paved Roads

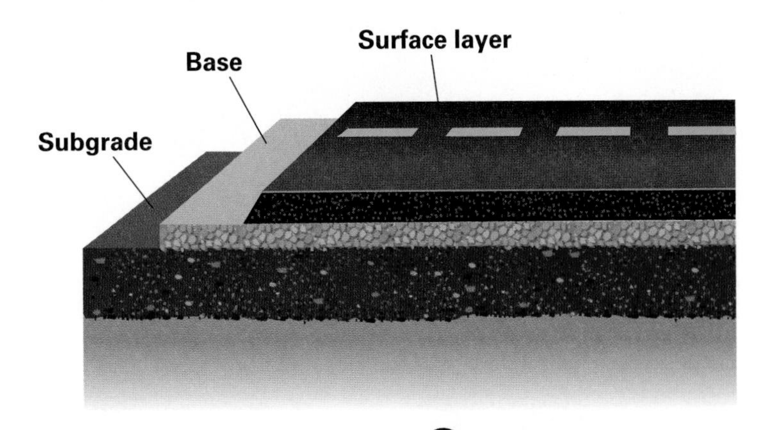

Hit the Road You have probably seen highways being constructed in your area. *Highway surface layers have to be thicker than on the roads in residential neighborhoods. Why?*

Concrete

The surface material chosen depends on the type of traffic that will use the road. Highways, which are main roads, are usually surfaced with asphalt or concrete. Asphalt is a brownish-black, flexible material made from crude oil and other substances. Concrete is often preferred because it is easy to mix, does not have to be heated like asphalt, and dries to a hard, durable surface. Concrete can be strengthened with steel bars or steel mesh placed into it when it is wet. That is why it is used where the heaviest traffic is expected. Airport runways and interstate highways are surfaced with strengthened concrete that is about ten inches thick.

Asphalt

Asphalt is used for less important roads. It is also used to repair worn or damaged concrete roads. Asphalt is flexible and sticks better to old concrete than fresh concrete sticks to it.

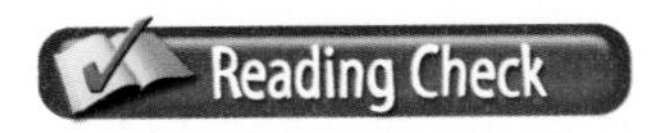

Identify Why are asphalt and concrete the preferred materials for making roadways?

The Golden Gate The Golden Gate Bridge in San Francisco, California, is one of the most famous bridges in the world. It was built in 1937. *What kinds of obstacles do bridges help people to cross?*

Figure 20.2 Bridges

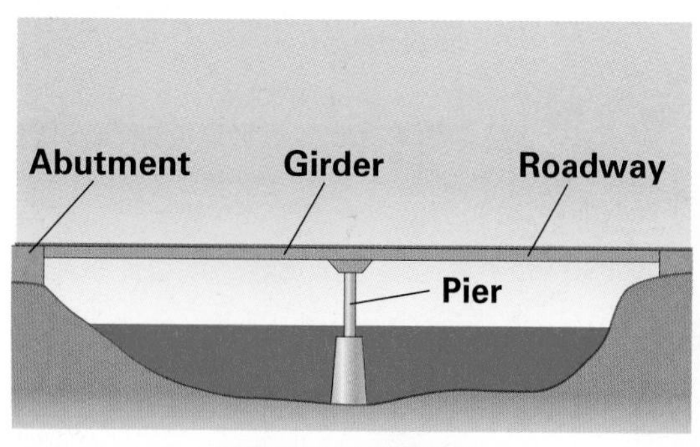

GIRDER BRIDGE

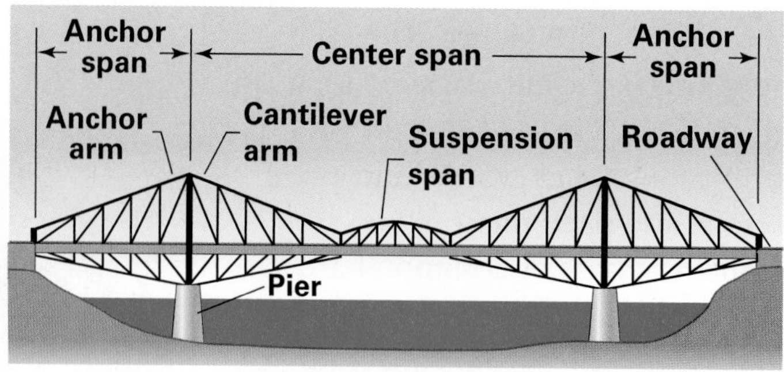

CANTILEVER BRIDGE

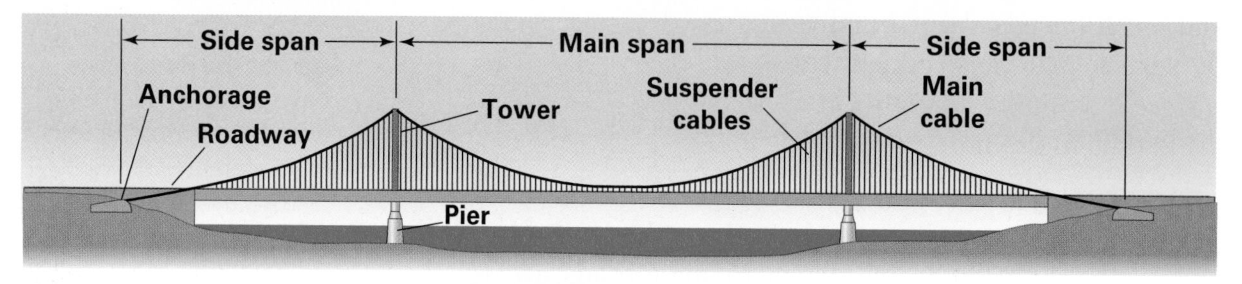

SUSPENSION BRIDGE

Types of Bridges Bridges can be constructed in several ways. *Which type of bridge might be the most expensive to construct?*

Bridging the Gap The Gateshead Millennium Pedestrian Bridge crosses the River Tyne in northeast England. It rotates to allow ships to pass. *Do you think its designers thought appearance was important?*

Bridges

What are the different kinds of bridges?

The roadway approach to a bridge is an important part of its design. At the ends of a bridge where it meets the land, **abutments** support both the bridge and the earth. If the bridge is long, piers may support the roadway between the abutments.

The distance between supports is called a **span**. (Sometimes the entire length of a bridge is also the span.) If the earth beneath the bridge is not stable, piles may also be used.

Suspension Bridges

The Golden Gate Bridge in San Francisco is stunning example of a suspension bridge. **Suspension bridge** roadways hang from large cables and cross wide spans. When it opened in 1937, the Golden Gate Bridge had the world's largest span at 4,200 feet, and the highest supporting towers at 746 feet.

Another type of suspension bridge is known as the **cable-stayed bridge**. Inclined cables, called "stays," connect the roadway to tall support towers. Cable-stayed bridges are cheaper and easier to construct.

Truss and Cantilever Bridges

A **truss bridge** is a bridge held together with steel beams. Beams are fastened together in the shape of triangles. The Eads Bridge in St. Louis, Missouri, is a **cantilever bridge** strengthened with trusses. A cantilever is a self-supporting beam fastened to the ground at one end. Two cantilevers meet in the middle to make the bridge. See **Figure 20.2**. This strong design resists high winds.

Beam Bridges

A **beam bridge** has a simple structure. The roadway rests on girders laid across the span. Beam bridges are frequently supported by piers partway along the span. Many interstate highway bridges use this design.

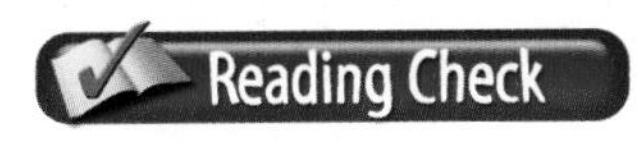

Summarize What are some of the ways bridges are supported?

Tunnels

How are tunnels made?

Bridges go over natural barriers, but tunnels go under or through them. Tunnels are constructed through mountains and under water. Modern tunneling methods consist of blasting with explosives and drilling with huge machines.

Patricia Billings
Inventor of Geobond

Patricia Billings was a trained artist. She used plaster of Paris for sculpture. But she also wanted a material that would be more resistant to breaking. Billings discovered that mixing a special cement additive with gypsum and concrete made an almost unbreakable plaster.

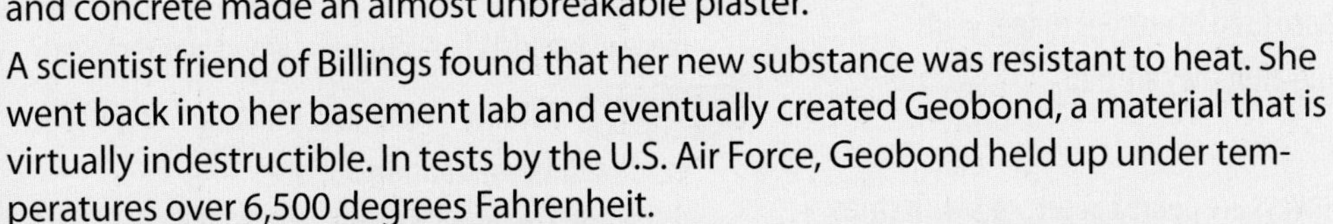

A scientist friend of Billings found that her new substance was resistant to heat. She went back into her basement lab and eventually created Geobond, a material that is virtually indestructible. In tests by the U.S. Air Force, Geobond held up under temperatures over 6,500 degrees Fahrenheit.

Safer Sculpting Billings created Geobond for her artistic works. However, the nontoxic material has now replaced cancer-causing asbestos, and is used to build aircraft, bridges, and other structures. Billings was awarded a U.S. patent for Geobond in 1997.

English Language Arts/Writing Research a few common items that are made with Geobond and write a short essay about them.

Go to **glencoe.com** to this book's OLC to learn about young innovators in technology.

A large metal tube called a **shield** fits inside the tunnel as it is drilled. For immersed tunnels, pre-built sections are sunk into an excavation. Then workers connect the sections together.

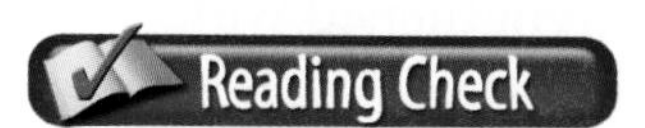

Identify What is a shield?

Tunnels in the United States

At 1.7 miles, the Eisenhower Memorial Tunnel in Colorado is the longest highway tunnel in the United States. It is at an altitude of 11,000 feet and is among the highest tunnels of its type.

The Fort McHenry Tunnel in Baltimore is an immersed tunnel and was not built like others. Workers first dug an undersea trench to accommodate sealed twin-tube steel and concrete sections. They floated sections over the trench and pumped in concrete to make each section settle to the bottom. The tunnel has 32 concrete tube sections and runs 1½ miles long.

The Chunnel

The most expensive private construction project in history is the 32-mile-long tunnel under the English Channel between England and France. It was a joint venture between British and French companies and cost more than $13 billion.

The project started in 1986, and service began in 1994. The English Channel Tunnel, or "Chunnel," has 23.6 miles of its length underwater and is the largest undersea tunnel ever built. Its depth varies from 90 to 480 feet below the bottom of the seabed.

Academic Connections — Social Studies

The Bridges of New York City New York City includes several islands. Eleven major bridges help people get from place to place. Six of the bridges were designed by Othmar Ammann (1879–1965).

Apply Using a large map of New York City, identify its bridges and the regions they connect. Use other references to identify the bridge designs used. Can you find out which ones Ammann designed?

section 20.2 assessment

After You Read — Self-Check

1. Explain why roadway surfaces are higher in the middle.
2. Compare a suspension bridge to a truss bridge.
3. Name the device used to prevent tunnel walls from collapsing after they are dug.

Think

4. Concrete highways are poured in sections with strips of tar between them. Explain the tar strips and why the concrete sections are not joined together.

Practice Academic Skills

English Language Arts/Writing

5. Do an Internet search for Web sites featuring famous roads, bridges, or tunnels. Some possibilities include the Silk Road or the Hoosac Tunnel. Make a poster-board display with drawings and descriptions.

STEM Mathematics

6. A ¾ inch drill was used to bore a hole through a 3-inch piece of wood. What volume of wood was removed?

Math Concept **Volume** Volume is a measure of the space inside a three-dimensional shape.

1. A drill cuts out a cylinder when it is bored through a board with parallel sides.
2. The volume of a cylinder is found by using the formula $V = \pi r^2 h$ where $\pi = 3.14$, r is the radius, and h is the height.

For help, go to **glencoe.com** to this book's OLC and find the Math Handbook.

section

20.3 Other Structures

Reading Guide

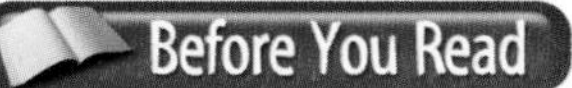

Preview What other structures are built using heavy construction?

Content Vocabulary

- embankment
- outlet works
- spillway

Academic Vocabulary

- benefit
- unique

Graphic Organizer

Draw the section diagram. Use it to organize and write down information as you read.

Main Parts of a Dam

1. Embankment	2. ________	3. ________

Go to **glencoe.com** to this book's OLC for a downloadable graphic organizer and more.

TECHNOLOGY STANDARDS

STL 3 Cultural, Social, Economic & Political Effects

STL 5 Environmental Effects

STL 6 Role of Society

ACADEMIC STANDARDS

Social Studies

NCSS Content Standard 8 Science, technology, and society

Science

NSES Content Standard F Science and technology in society

STL *National Standards for Technological Literacy*

NCTM *National Council of Teachers of Mathematics*

NCTE *National Council of Teachers of English*

NSES *National Science Education Standards*

NCSS *National Council for the Social Studies*

Dams

What is the purpose of a dam?

Heavy construction methods are used to build many other structures, such as dams. A dam is built across a river to block the flow of water. It is usually done for one of three reasons: 1) to provide a water supply for nearby communities and farms; 2) to provide flood control; or 3) to provide electrical power.

Most dams have three main parts: the embankment, the outlet works, and the spillway. The **embankment** is the large section that blocks the flow of water. The **outlet works** contains gates that allow a certain amount of water to flow through the dam.

Connect How is construction in space different from construction on Earth?

BusinessWeek Tech NEWS

Preventing Disaster

Following Hurricane Katrina, 300 top scientists and engineers developed a threat-assessment model that measures risks to infrastructure (public systems such as transportation, water, and power). Although the tool was developed for New Orleans, it may be used for other areas. This includes California, where an earthquake or flood could devastate the 2,600-mile long levee system that runs from San Francisco Bay to Sacramento.

Critical Thinking *What do you think is the meaning of "threat assessment" and why might it be necessary?*

Go to **glencoe.com** to this book's OLC to read more about this news.

When too much water builds up behind the dam, the **spillway** allows the excess to flow around or through it and prevents the dam from breaking.

Hydroelectric dams have a power station. As water flows through the outlet works, it turns the blades of a turbine, which turns a generator, and makes electricity.

Dams in the United States

The construction of a dam usually takes many years. The Hoover Dam was opened in 1936 after about eight years of work. The project was too large for any one company, so six companies banded together to build the dam. There are numerous smaller concrete and earth dams in the United States. An earth dam is made of carefully selected soil that is hauled to the site. Layer upon layer of the soil is compacted with heavy rollers to form a watertight mass.

The Three Gorges Dam

Dams have many benefits, but they also impact the environment. Wildlife habitats are often lost, and there are many other trade-offs. The gigantic Three Gorges Dam in China is an example. It is the largest dam in the world. It controls severe flooding and produces as much electricity as 15 large power plants. At the same time, the dam caused the loss of more than 1,500 cities, towns, and villages, as well as uncounted wildlife habitats. The river valley is now under almost 200 feet of water.

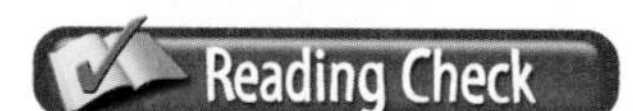

Identify What are three main parts of a dam?

Trade-Offs The Three Gorges Dam in Hubei province, China, is designed to provide electrical power for nine provinces and at least two cities. *What are some of the dam's negative impacts?*

Monuments

What is the purpose of a monument?

Monuments appeal to the human spirit. They are designed to honor the past and look to the future. The Statue of Liberty, the St. Louis Gateway Arch, and other grand monuments often inspire feelings of reverence and pride.

The Lincoln Memorial in Washington, D.C., the Eiffel Tower in Paris, France, and the Taj Mahal in Agra, India, are also examples of monuments. Some are so huge and unique that they required more careful design and construction than most traditional structures.

The Statue of Liberty

The Statue of Liberty that greets all ships entering New York City's harbor was a gift from France in 1886. It is meant to celebrate the personal freedoms that Americans enjoy. Its internal structure of iron beams resembles the metal cage of a skyscraper. See **Figure 20.3**. The framework is covered with molded copper sheets. As the copper oxidizes, the statue turns green in color.

The St. Louis Gateway Arch

The St. Louis Gateway Arch is in the middle of a national park on the banks of the Mississippi River in St. Louis, Missouri. It was built in memory of the Louisiana Purchase of 1803 and the westward expansion that followed. The arch is constructed of stainless steel, and its foundation extends 60 feet into the ground.

The arch was built starting from both ends at the same time. The two halves were connected at the top in 1965 after four years of construction. The Arch is 630 feet tall, which makes it the tallest monument in the United States.

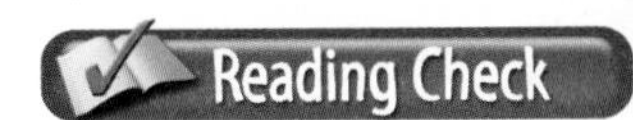

Explain Why do people build monuments?

The International Space Station

How does construction in space differ from construction on Earth?

The United States, Russia, and other nations are building the International Space Station (ISS) that orbits around the earth. The station is permanently occupied by rotating international crews.

Figure 20.3 **The Statue of Liberty**

Supporting Liberty The Statue of Liberty is supported by a strong interior metal frame. *Which nation gave the Statue of Liberty to the United States as a gift?*

The ISS The International Space Station orbits the earth every 90 minutes. *What does the International Space Station use as a source of power?*

The ISS serves primarily as a laboratory where researchers can do experiments. They learn about the effects of very low gravity on materials and processes. They are also learning about what people need to live and work in space.

Construction of the ISS

Since construction began in 1998, all assembly of the station has taken place in space. Modules (sections), materials, and equipment are brought up from Earth in American space shuttles or Russian spacecraft. All maintenance is also done in space.

The framework of the ISS is a series of trusses. Living quarters and other modules are attached at various points. As the astronauts and cosmonauts work, their tools and equipment must be tied down to prevent them from drifting away.

The ISS has huge solar collectors that draw energy from the sun and recharge its electrical batteries. The collectors extend about 350 feet, which is longer than a football field.

section 20.3 assessment

After You Read Self-Check

1. List three reasons for constructing a dam.
2. Name the material used to construct the St. Louis Gateway Arch.
3. Identify the type of energy used to power the International Space Station.

Think

4. Hydroelectric dams create less pollution than many other power sources, yet they are used to produce only about five percent of America's electrical power. Why you think this is so? Explain your answer.

Practice Academic Skills

English Language Arts/Writing

5. Some monuments inspire feelings of patriotism; others were built to commemorate great leaders, thinkers, or artists. Imagine that your community has constructed a new monument. You have been asked to write a dedication. Your dedication will be engraved in the monument in stone. Write a two-paragraph dedication for the monument that explains why it was built and what it stands for.

Social Studies

6. The English Channel tunnel, or "Chunnel," is 32 miles long and runs from England to France, underneath the English Channel. In some places, it is 480 feet under the water. The safety requirements, both in its construction and operation, are unique. Research some of the Chunnel's safety requirements. Summarize your findings in two or three paragraphs.

Exploring Careers in Technology

John Chan
ARCHITECT

Q: *What do you do?*
A: I design buildings for an architectural firm. The projects I work on range from the really small ones to big ones, like schools, subway stations, and university buildings. However, no matter how big or small, each structure is designed for the experience and pleasure of the people who use them.

Q: *What kind of training and education did you need?*
A: I have always liked to draw and make things. After high school, I applied to an architecture school at a university. I also visited a lot of buildings. I looked at them, studied them, and sketched them.

Q: *What do you like most about your job?*
A: I enjoy making something and putting it in this world for others to enjoy. It also gives me great pleasure to collaborate with other people. When I visit a construction site and see something that I drew take physical form, I feel a sense of happiness that is hard to compare!

Q: *How did you get interested in your job?*
A: When I graduated from architecture school, I wanted to work in an office that really explored creative solutions for buildings in this world. I set out looking for my dream job, and I found something very close to it!

English Language Arts/Writing

Propose a Building Describe a building complex you would like to construct. It can be anything from retail to entertainment to housing.

1. Using a word-processing program, write a description of the buildings, including their placement, materials, style, and other details.
2. On paper, draw the buildings from various angles.
3. Create a catalogue of the buildings and present it to your class.

Go to **glencoe.com** to this book's OLC to learn more about this career.

Real-World Skills

Speaking, listening, problem-solving

Academics and Education

Physics, mathematics, structural engineering, English language arts

Career Outlook

Growth as fast as average for the next ten years

Source: *Occupational Outlook Handbook*

chapter 20 Review and Assessment

Chapter Summary

Section 20.1 Major projects like skyscrapers are heavy construction projects. The word *skyscraper* has come to mean a tall building with an interior support frame. Skyscraper safety is an important concern. Building codes and other forms of legislation lay down strict rules for modern construction.

Section 20.2 All paved roads are made in three layers. The top layer is called the "surface." Many bridges are strengthened with trusses. A truss is made of steel beams fastened together in the shape of many triangles. A metal tube called a "shield" fits inside a tunnel as it is constructed.

Section 20.3 Dams have three main parts: the embankment, the outlet works, and the spillway. Monuments are large structures that appeal to the human spirit. The International Space Station orbits the earth. Its framework is a series of trusses.

Review Content Vocabulary and Academic Vocabulary

1. On a sheet of paper, use each of these terms and words in a written sentence.

Content Vocabulary

- excavation
- pile
- pier
- crane
- subgrade
- abutment
- span
- suspension bridge
- cable-stayed bridge
- truss bridge
- cantilever bridge
- beam bridge
- shield
- embankment
- outlet works
- spillway

Academic Vocabulary

- integral
- control
- allocate
- consist
- benefit
- unique

Review Key Concepts

2. **Compare** the construction of large buildings to building a house.
3. **Summarize** the process of building a skyscraper.
4. **Discuss** why construction differs in other large buildings.
5. **Describe** the use of asphalt and concrete in road construction.
6. **List** ways in which bridges are supported.
7. **Identify** methods used to build tunnels.
8. **Name** a dam's three main parts.
9. **Discuss** monuments and why people build them.
10. **Compare** construction in space to construction on Earth.

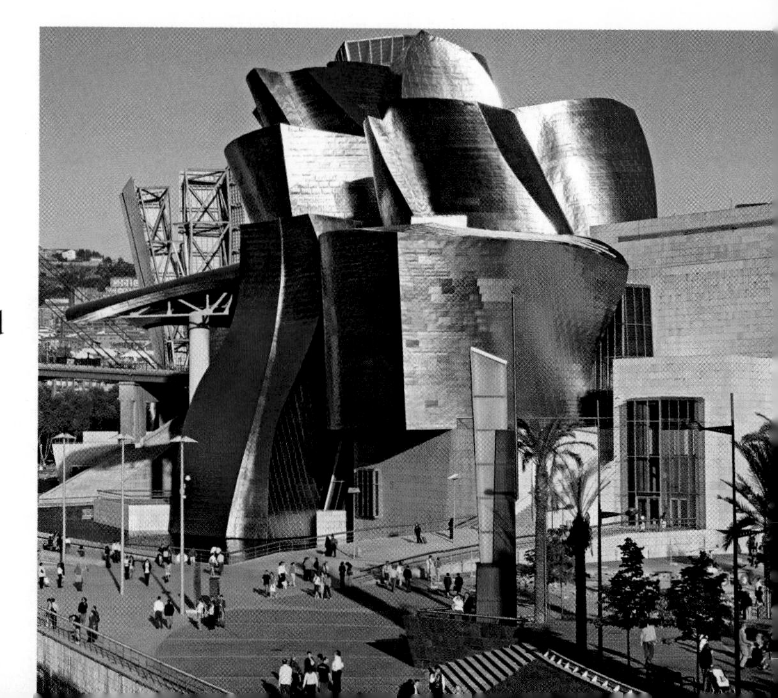

Real-World Skills

11. Construction Safety Research safety concerns at a high-rise construction site. Write a summary of what you find. Discuss the safety measures and the equipment used.

STEM Technology Skill

12. Monuments Many monuments in this country were built a long time ago. The technology used to build them differs greatly from technology today.

a. Use the Internet to research U.S. monuments.

b. Summarize what you find. Compare the technology used in the past to today's technology.

Academic Skills

Social Studies

13. Do some research at the library on different bridges around the world. Write a few paragraphs describing some of the bridges you research. Include a discussion of types and their history.

STEM Mathematics

14. Susan walks her two dogs every day. The path she takes is 2.3 miles long. If she completes her walk in 45 minutes, what is her average speed in miles per hour?

Math Concept **Average Speed** When calculating average speed, pay attention to the units. To convert minutes to the decimal equivalent of hours, divide the minutes by 60. There are 60 minutes in 1 hour.

WINNING EVENTS

Civil Engineer

Situation You have been asked to design a fire watch tower for a local forest. The tower will be on high ground and must comfortably accommodate two live-in rangers, permit 360° viewing, and be structurally sound. The observation deck must be between 36′ and 37′ above the ground.

Activity Working with your team, develop preliminary sketches. Agree on a design and construct a rough model. Test it for structural stability (wind tunnel) and strength (compression). Make any necessary changes. Finally, construct the final model to scale with 1/4 inch = 1 foot. Present your design to your class.

Evaluation Your project will be evaluated by these criteria:

- Functional—360° viewing
- Structurally sound—withstands tests
- Living environment—comfortable, attractive

Go to **glencoe.com** to this book's OLC for information about TSA events.

Standardized Test Practice

Directions Choose the letter of the best answer. Write the letter on a separate piece of paper.

1. How many minutes are in 4 days?

   A 2,870 **C** 5,760

   B 7,375 **D** 3,600

2. The section of a dam with gates that allow water to flow through is called the outlet works.

   T

   F

Test-Taking Tip Getting a good night's sleep before the test can reduce test anxiety.

chapter 20

TECHNOLOGY LAB

Build a Truss Bridge

Much of the strength of a bridge comes from its design. Truss bridges were developed in the 1500s. The parts of a truss are arranged in the form of many triangles.

Tools and Materials

- ✓ Six 36-inch-long pieces of 3/16-inch balsa strips
- ✓ Quick-drying adhesive
- ✓ Pencil
- ✓ Paper
- ✓ Ruler
- ✓ Cutting blade
- ✓ 2- to 3-gallon plastic bucket with handle
- ✓ Sand
- ✓ Nylon cord
- ✓ Small piece of hardwood
- ✓ Scale
- ✓ Safety goggles and long-sleeved shirt

Set Your Goal

For this activity, you will build a truss bridge and test it.

Know the Criteria and Constraints

In this lab, you will:

1. Construct a bridge that measures 18 inches long, 4 to 6 inches tall, and 4 to 6 inches wide. It must be able to accommodate the piece of hardwood used in the test.
2. Use balsa wood to construct your bridge. Do not use any metal.
3. Make all joints flush. No joints may overlap.

Design Your Project

Follow these steps to complete this lab.

1. Research truss bridge designs.
2. Make a full-scale drawing of a bridge that is exactly 18 inches long, between 4 and 6 inches tall, and about 4 to 6 inches wide.
3. Construct the bridge using 3/16-inch balsa wood strips. To improve stiffness of the main horizontal supports, glue two strips together. See illustration. No overlapping joints are allowed.

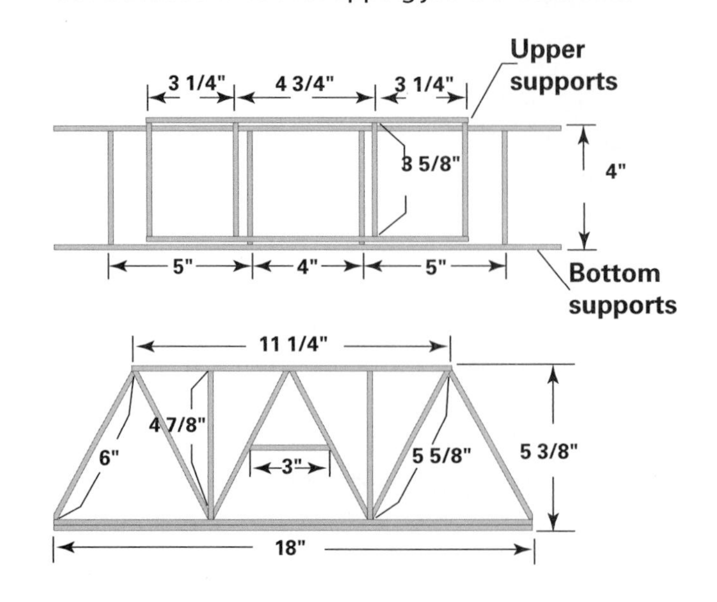

SAFETY

Reminder

In this lab, you will be using cutting tools and must wear safety glasses. Be sure to always follow appropriate safety procedures and rules so you and your classmates do not get hurt.

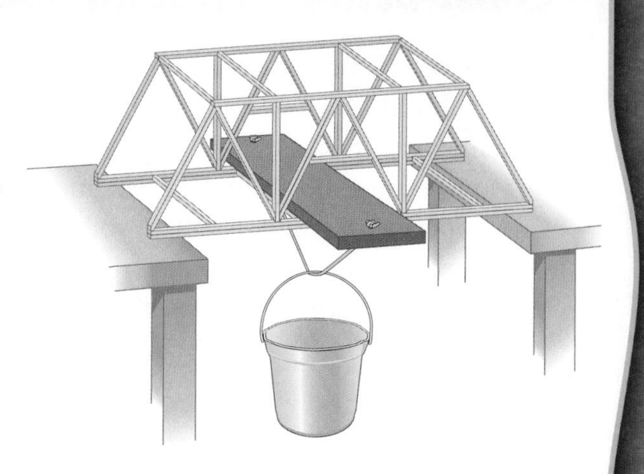

4. Your design must be able to fit the piece of hardwood used to test the bridge. Leave an opening that measures at least 1½ × ½ inches across the middle of your bridge. This is where the hardwood and bucket will fit.
5. After the adhesive has dried, record the weight of your bridge.
6. Arrange two tables so they are 16 inches apart. Place your bridge across the gap between the tables. Put the hardwood on the bridge and tie the bucket to it with nylon cord. The bucket's bottom should be about 4 to 6 inches above the floor. See illustration.
7. Perform the experiment:
   - Put on your safety glasses and a long-sleeved shirt.
   - If possible, make a video to show how long your bridge lasts.
   - Slowly pour sand into the bucket until the bridge breaks; otherwise, it may fail quickly and could scatter broken wood and sand.
   - Make sure the bucket of sand does not tip over when it drops.
   - Weigh the bucket to see how much load your bridge carried.
   - Record the results.

Evaluate Your Results

After you complete the lab, answer these questions on a separate piece of paper.

1. How much weight did your bridge carry in comparison to its own weight?
2. Where did your bridge begin to fail: at the center, the edges, or the top? How could you change it so that it would carry a heavier load?

Academic Skills Required to Complete Lab

Tasks	English Language Arts	Math	Science	Social Studies
Research truss bridge design.	✓		✓	✓
Make full-scale drawing of your bridge.		✓	✓	
Construct bridge with balsa wood.		✓	✓	
Conduct experiment to see how much weight the bridge will hold.		✓	✓	
Calculate ratio of weight of bridge to weight of load.		✓		

Technology Time Machine

Construction by Design

Play the Game This time machine will travel to the past to show you that all great achievements in design and construction require careful planning and engineering. To operate the time machine, you must know the secret code word. To discover the code, read the clues, and then answer the questions.

Clue 1

2500 B.C.E. Egypt's three great Pyramids of Giza were built about 4,500 years ago. The largest pyramid contains 2,300,000 stone blocks, each averaging 2½ tons. The mystery of how the heavy blocks were put in place has been studied for generations. The builders used surprisingly accurate mathematics.

Clue 2

221–206 B.C.E. Construction for the world's longest structure, the Great Wall of China, began during the Qin dynasty to keep out invaders. The wall was built by hand with earth and stones held in wood frames. During the Ming dynasty (1368–1644), the wall was given its present form. The main wall is about 1,500 miles long, and the total length is nearly 4,500 miles.

Clue 3

1895 The Biltmore Estate outside Asheville, North Carolina, was inspired by the great mansions of Europe. Created for George Washington Vanderbilt III, the house features 250 rooms with 34 bedrooms, 3 kitchens, 43 bathrooms, 65 fireplaces, a bowling alley, a gym, and an indoor swimming pool.

Clue 4

1931 New York City's Empire State Building was completed in 16 months. At 1,250 feet in height with 102 stories, it was the world's tallest building until the 1970s, when the now-destroyed World Trade Center was built. In 1950, it measured 1,472 feet with antennas.

Clue 5

1944 Construction of the U.S. Highway System was delayed until 1956, when President Dwight D. Eisenhower authorized moving the project forward for transport of the military. Interstate highways carry more than 20 percent of all traffic, but represent only about 1 percent of the nation's roadways.

Clue 6

1954 R. Buckminster Fuller received a patent on the geodesic dome, one of the lightest, strongest, and most cost-effective structures. Domes are assembled from interlocking polygons. They enclose more space without internal supports than any other structure. A well-known geodesic dome is Epcot Center at Walt Disney World in Florida.

Clue 7

2001 The leading "green" agency built the "greenest" high-rise in the country. The 25-story Joe Serna Jr. Building in Sacramento, California, is home to the state's Environmental Protection Agency. It includes recycled-content ceiling tiles and worm-composting bins.

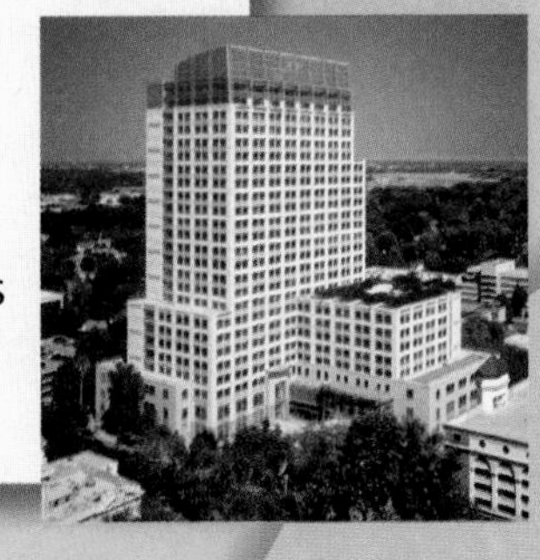

Crack the Code

On a piece of paper, write the answers to these questions:

1. Which president authorized the U.S. Interstate Highway System?
2. What is the tallest building in New York City?
3. What type of finished blocks were used to build the great pyramids?
4. Geodesic domes are assembled from this type of polygon.
5. Name one of the lightest and strongest structures.
6. Where is the Empire State Building located?

Now write down the first letter of each answer. Put them together to discover the secret code word!

Hint It is important to choose this prior to building a home or other structure.

unit 6 Thematic Project

Planning a Green Shelter

Construction Technology In Unit 6, you learned how construction systems for homes and shelters evolved from log houses to contemporary skyscrapers. Today the problem of global warming is affecting how we build shelters. We look for ways to reduce our use of fossil fuels. Sustainable development is influencing construction practices that reduce negative impacts on the environment.

Building Green A green building is a sustainable building. It is a structure designed, built, and operated in a way that uses and reuses energy, materials, and water efficiently.

This Project In this project, you will plan a green shelter.

Your Project

- Choose a type of shelter.
- Complete these tasks:
  1. Research one or more aspects of the design of your green shelter:
     - Materials Selection
     - Heating and Cooling
     - Appliances and Fixtures
     - Conservation and Pollution Control
     - Interior Design
  2. Draw a picture or create a 3D model.
- Write a report about your research and model.
- Create a presentation with posters or presentation software.
- Present your findings to the class.

Tools and Materials

- ✓ Computer
- ✓ Internet access
- ✓ Books and magazines about home and shelter building
- ✓ Word-processing software
- ✓ Presentation software
- ✓ Posterboard
- ✓ Colored markers

The Academic Skills You'll Use

- Communicate effectively.
- Speak clearly and concisely.
- Employ correct spelling, grammar, and usage in a written report.
- Conduct research using a variety of resources.
- Incorporate reading, writing, and speaking with viewing, representing, and listening.

English Language Arts

NCTE 4 Use written language to communicate effectively.

Science

NSES Content Standard F Science in Personal and Social Perspectives: Science and technology in society

Step 1 Choose Your Topic

You can choose any type of shelter for your project. Examples include:

- New or remodeled house
- Dog house for cold winters and hot summers
- Garden shed with an attached greenhouse
- Barn for horses with a wash rack, indoor arena, and viewing stands

 *Choose a shelter you would like to use!*

Step 2 Do Your Research

Research different types of shelters people or animals live in around the world. Look for ideas to use. Find out about techniques builders use in green construction. Answer these questions:

- How can you reuse and recycle construction and demolition debris?
- How can you limit damage to existing trees and other plants at the building site?
- What "green" materials will you use?
- How can you maximize energy efficiency?

 *Ask a contractor where to find information!*

Step 3 Explore Your Community

Find a local architect, builder, or general contractor who has worked on environmentally friendly buildings. Interview him or her about the positive and negative aspects of green construction.

 *Remember to listen attentively!*

Step 4 Create Your Project

Your project should include:

- 1 research report
- 1 model of your shelter
- 1 written report about your project
- 1 presentation

Project Checklist

	Objectives for Your Project
Visual	✓ Make a scale drawing or 3D model.
	✓ Show the aspects of your shelter that make it environmentally friendly.
Presentation	✓ Make a presentation to your class.
	✓ Include the notes from your research.

Step 5 Evaluate Your Presentation

In your report and/or presentation, did you remember to:

- Demonstrate your research and preparation?
- Engage your audience?
- Back statements with facts and evidence?
- Use visuals effectively?
- Speak slowly and enunciate clearly?

 **Rubrics** Go to **glencoe.com** to the book's OLC for a printable evaluation form and your academic assessment form.

GLOBAL TECHNOLOGY

Reaching for Green

Skyscrapers are getting taller every year, but taller is not always "greener." However, in Mumbai, India, architects designed The India Tower. It includes residences, a hotel, and offices. It might be the tallest building in the world with a certified green rating. Sustainable features include a solar chimney to generate electricity, waste-water reclamation, "daylighting" design for use of sunlight, natural ventilation, and rainwater harvesting.

Critical Thinking *Why are this building's features considered sustainable and green?*

Hindi

hello	**namaste**
goodbye	**achcha**
How are you?	**Ap kaise hain?**
thank you	**sukriya**

 Go to **glencoe.com** to the book's OLC to learn more and to find resources from **The Discovery Channel.**

unit 7

Transportation Technologies

Chapters In This Unit:

Unit Thematic Project Preview

Advancing Transportation

As part of this unit, you will learn about the different types of engines and motors that make vehicles go. You will also explore the many transportation systems that allow you to move around your neighborhood, your country, and the world!

As you read this unit, use this checklist to prepare for the project at the end of this unit:

PROJECT CHECKLIST

✓ Look for transportation magazines and Web sites.

✓ Find out what kinds of transportation problems are in your community.

✓ Consider whether you will use materials or a computer to make a model.

WebQuest Internet Project

Go to **glencoe.com** to this book's Online Learning Center (OLC) to find the WebQuest activity for Unit 7. Begin by reading the Task. This WebQuest activity will help you learn about space exploration, especially to the planet Mars. You will find out about past missions to Mars as well as current and future travel to that planet and beyond.

Explore the Photo

Moving into the Future As operating gasoline-fueled engines becomes more expensive and causes more pollution, technologists are inventing new ways to get around. Solar, electric, and alternative fuel powers are not just science-fiction—they are real today! *What kind of transportation do you imagine you will use when you grow up?*

chapter 21

Transportation Power

Sections

What You'll Learn

- **Explain** how an external combustion engine works.
- **Compare** a steam engine to a steam turbine.
- **Tell** how an internal combustion engine works.
- **Define** four-stroke and two-stroke engine cycles.
- **Describe** the purpose of a crankshaft.
- **Identify** differences among various engines.
- **Examine** how electric motors are used to power locomotives.
- **Discuss** how a hybrid automobile is powered.
- **Summarize** the advantages and disadvantages of hybrid and fuel cell cars.

Explore the Photo

Power Until the invention of the steam engine, transportation power came directly from natural sources, such as wind and flowing water. *What might be the source of power shown in this picture?*

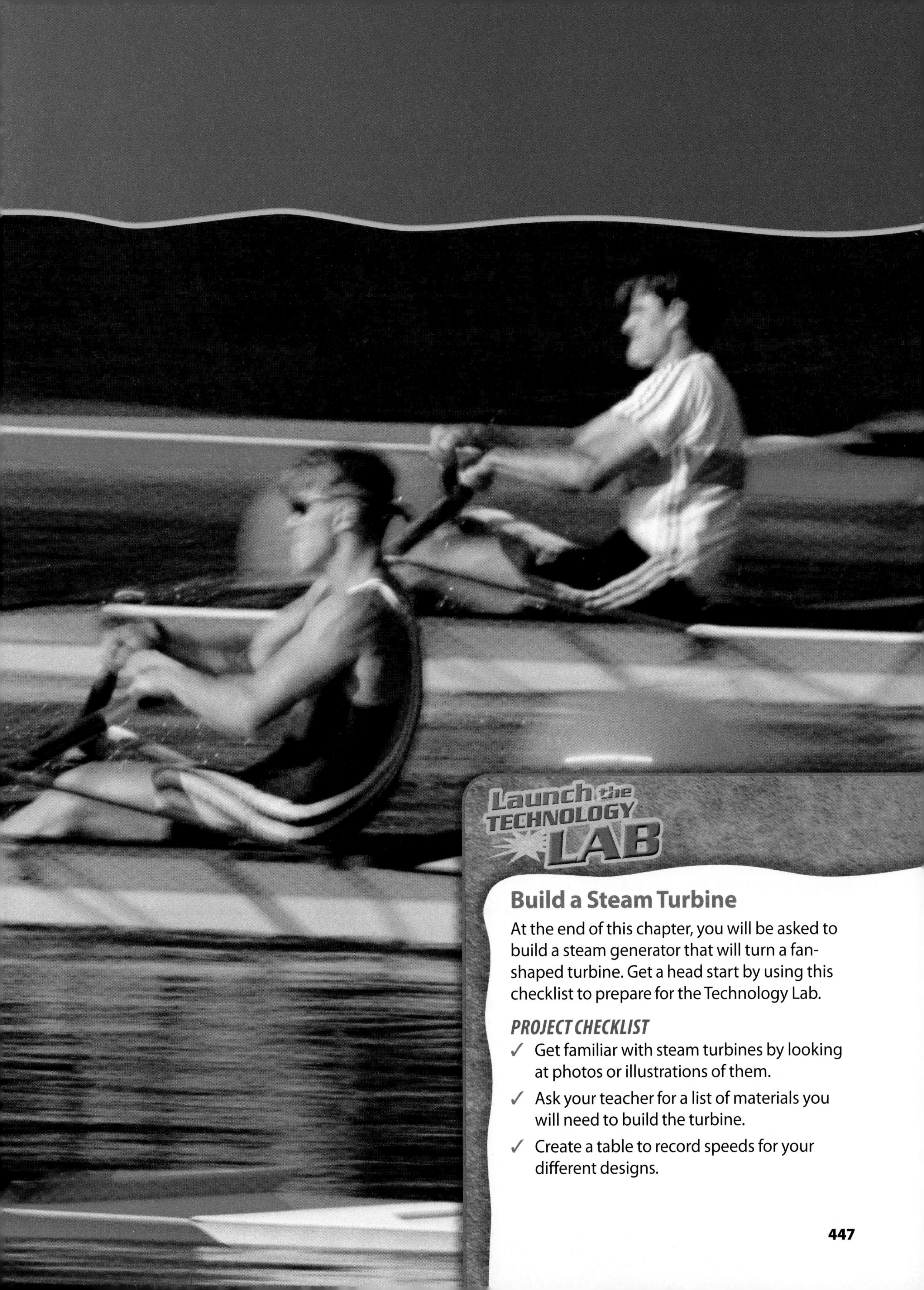

Build a Steam Turbine

At the end of this chapter, you will be asked to build a steam generator that will turn a fan-shaped turbine. Get a head start by using this checklist to prepare for the Technology Lab.

PROJECT CHECKLIST

- ✓ Get familiar with steam turbines by looking at photos or illustrations of them.
- ✓ Ask your teacher for a list of materials you will need to build the turbine.
- ✓ Create a table to record speeds for your different designs.

section

21.1 External Combustion Engines

Reading Guide

Before You Read **Preview** What are some types of transportation power?

Content Vocabulary

- external combustion engine
- piston
- turbine

Academic Vocabulary

You will see these words in your reading and on your tests. Find their meanings at the back of this book.

- convert
- expand

Graphic Organizer

Draw the section diagram. Use it to organize and write down information as you read.

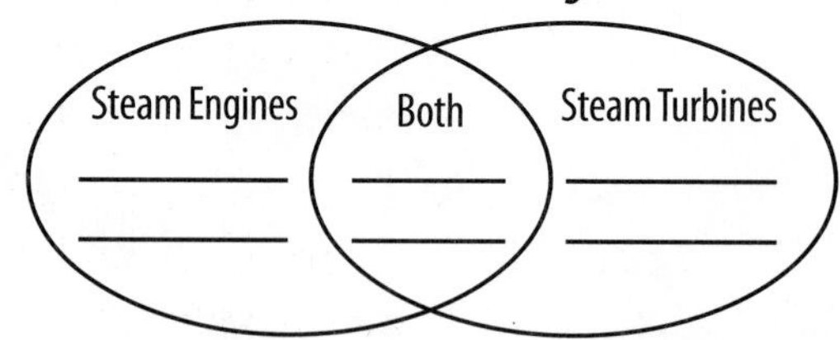

Go to **glencoe.com** to this book's OLC for a downloadable graphic organizer and more.

TECHNOLOGY STANDARDS

STL 3 Relationships & Connections

STL 7 Influence on History

STL 11 Design Process

STL 16 Energy & Power Technologies

ACADEMIC STANDARDS

Science

NSES Content Standard E Understandings about science and technology

Social Studies

NCSS 2 Time, continuity, and change

STL *National Standards for Technological Literacy*

NCTM *National Council of Teachers of Mathematics*

NCTE *National Council of Teachers of English*

NSES *National Science Education Standards*

NCSS *National Council for the Social Studies*

Engines and Motors

How did earlier humans get power for transportation?

Engines and motors produce the power we need to carry people and goods. Until about 300 hundred years ago, there were no engines. The only forms of transportation power came directly from nature, such as wind, flowing water, and muscle power from people and animals. Without power for transportation, how would you get to school or visit far-away relatives or friends? How would food and other products get to our stores?

As You Read

Contrast What is the difference between a steam engine and a steam turbine?

An engine is a power source that uses combustion with air to produce motion. The word *motor* is more general. A motor converts any form of energy into motion. Some examples are air motors and electric motors. People often use the words *engine* and *motor* interchangeably. You can say that a car is powered by an engine or a motor. However, there are no electric engines, only electric motors.

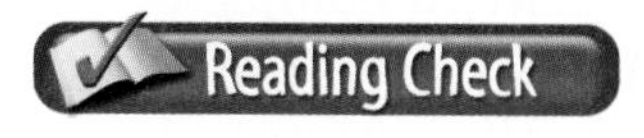

Compare What is the difference between an engine and a motor?

Steam Engines

How do steam engines work?

All steam engines are **external combustion engines**. See **Figure 21.1**. *External* means the power source is outside the engine. *Combustion* refers to burning. Steam engines use the heat from burning coal or wood to change water into steam. The fire is under a boiler outside the engine, so the power source of the engine is external.

Pistons and Flywheels

Steam engines have a piston that moves up and down. A **piston** is a plug that just barely fits inside a closed cylinder. Expanding steam from a boiler pushes on the piston and causes it to move inside the cylinder. The piston's up-and-down movement turns a circular flywheel.

Imagine This...

Personal Rocket Power

Imagine soaring through the air with your own rocket pack. Two companies in Mexico and Colorado are in the early stages of producing rocket packs. Each weighs over 100 pounds and can propel the user to speeds over 60 mph. But each only lasts for 30 seconds and costs over $100,000. *Why do you think the flight time for rocket packs is so short?*

Go to **glencoe.com** to this book's OLC for answers and to learn more about different rocket packs.

Figure 21.1 **Steam Engines**

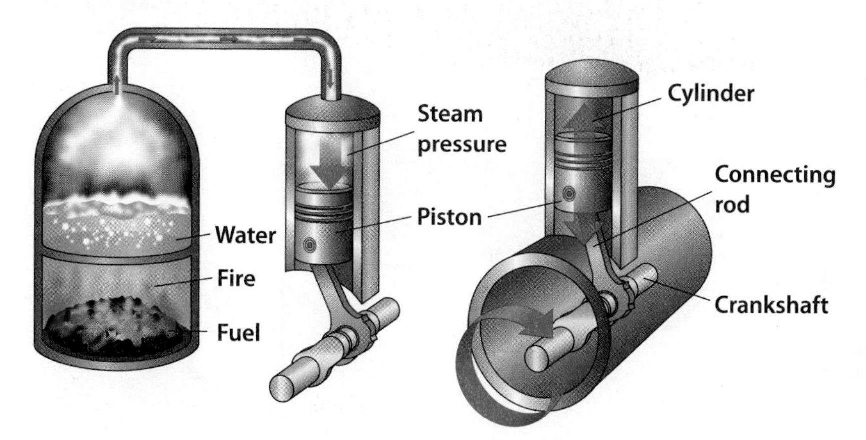

Blowing Off Steam Steam engines (left) produce power when steam pressure pushes on a piston similar to the one shown at the right. *What three types of fuels have been commonly used with steam engines?*

By connecting the spinning flywheel to a vehicle's wheels, steam was used to power land transportation. The first steam engine was built in England in 1712. It operated pumps to remove water from coal mines. This is a good example of how engineers can adapt a technology that was developed for one setting and then use it for another setting.

Gathering Steam Huge locomotives like this one once crisscrossed the country. They were powered by steam engines. *What noise did steam trains make when they ran?*

Almost any type of fuel can be used to produce steam, and many types of engine designs are available. However, steam is dangerous. Also, boilers require a lot of maintenance and space, and so they are used mainly for large vehicles.

Transportation Power

Steam engines powered huge locomotives and ships that transported passengers and cargo. The engines made a loud chuffing noise and produced clouds of steam. Steam engines were also used in some early cars like the Stanley Steamer.

Figure 21.2 Steam Turbines

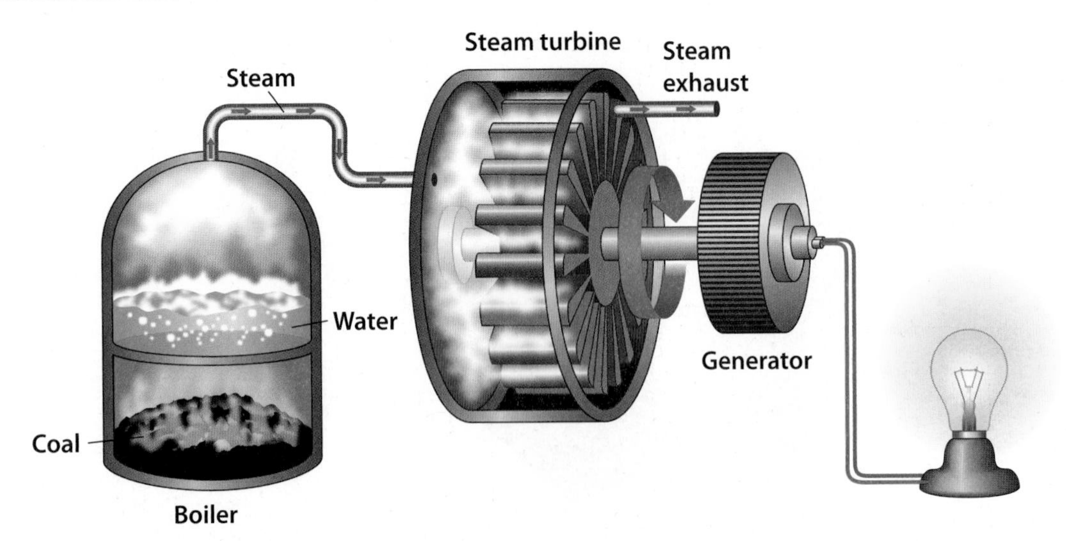

Generating Electricity In one type of electrical power plant, burning coal heats water and makes steam. The steam pressure spins a turbine that is connected to a generator (alternator). The generator produces electricity. *What do steam engines and steam turbines have in common?*

In modern-day America, steam engines are not used very much. Their efficiency is low, and they do not produce much power for their size. People's needs have changed. However, you can better understand today's technologies by studying those used in the past.

Describe How does an external combustion engine work?

Round and Round Steam turning a turbine is similar to blown air turning a pinwheel. *Why are steam turbines external combustion engines?*

Steam Turbines

What are steam turbines?

Steam turbines operate from steam pressure, just like steam engines. That is where the similarity ends, however. Steam engines produce their power by pistons moving up and down. Steam turbines develop power from spinning disks. The two kinds of power sources are very different.

A **turbine** is a continually spinning disk that resembles a pinwheel. Blow on a pinwheel, and it spins. You could call the pinwheel a "breath turbine" because your breath makes it spin. Steam from a boiler spins steam turbines, as shown in **Figure 21.2**. Steam turbines power oceangoing ships and are also used in electrical plants to produce electricity.

After You Read **Self-Check**

1. Identify where and when the first steam engine was built and its purpose.
2. Explain why steam engines are considered *external* combustion engines.
3. Compare a steam turbine to a steam engine.

Think

4. List some safety problems that you think early steam engines might have caused.

Practice Academic Skills

Social Studies

5. Look up improvements made on steam engines by James Watt (1736–1819). Make sketches illustrating his changes and write captions explaining them.

STEM **Mathematics**

6. Jaden wants a new mp3 player. There is one on sale for $95.99 that comes with a 20% mail-in rebate. How much will the player cost Jaden after the rebate?

Math Concept **Percents** A percent is a part of the whole. You can find the price after a discount in two ways.

1. Find 20% of $95.99, and then subtract that amount from the selling price.
2. You can also subtract the discount percent from 100, and then calculate that percentage of the selling price.

For help, go to **glencoe.com** to this book's OLC and find the Math Handbook.

section

21.2 Internal Combustion Engines

Reading Guide

Before You Read **Connect** Where does burning take place in a car's engine?

Content Vocabulary

- internal combustion engine
- reciprocating motion
- rotary motion
- crankshaft
- ignition system
- emissions
- ethanol
- maintenance
- jet engine
- thrust
- propellant

Academic Vocabulary

- transfer
- regulate

Graphic Organizer

Draw the section diagram. Use it to organize and write down information as you read.

Four-Stroke Engine Cycles

Go to **glencoe.com** to this book's OLC for a downloadable graphic organizer and more.

TECHNOLOGY STANDARDS

STL 5 Environmental Effects

STL 16 Energy & Power Technologies

STL 18 Transportation Technologies

ACADEMIC STANDARDS

Mathematics

NCTM Algebra Use mathematical models to represent and understand quantitative relationships.

Science

NSES Content Standard F Science and technology in society

STL *National Standards for Technological Literacy*

NCTM *National Council of Teachers of Mathematics*

NCTE *National Council of Teachers of English*

NSES *National Science Education Standards*

NCSS *National Council for the Social Studies*

Internal Combustion

What is an internal combustion engine?

Did you know that there is a fire inside most car engines? You cannot see the flames because they are deep inside. This internal fire makes these engines **internal combustion engines**. Gasoline, diesel, and rocket engines are all this kind of engine.

Most engines we use create power from a piston sliding inside a cylinder. Fuel and air are placed inside the cylinder. See **Figure 21.3**. An electric spark causes the mixture to ignite, burn, and build up high pressure quickly.

As You Read

Predict What kinds of vehicles use internal combustion engines?

Engine Cycles

How are engine cycles like other types of cycles?

When you pedal a bicycle, your legs go up and down, repeating the motions over and over again. One motion of your leg makes a downward stroke. Lifting your leg creates an upward stroke. Your legs make two strokes before repeating the same motions. We could say that your bicycle is operated by a two-stroke human power plant.

Four-Stroke Cycles

Much like your legs pedaling a bicycle, the pistons inside an engine move up and down. The most popular type of engine is the four-stroke cycle engine. The pistons make four strokes before they repeat themselves. See **Figure 21.4**. These strokes are the intake, compression, power, and exhaust strokes. Most cars and lawn mowers have four-stroke cycle engines.

- **Intake stroke**—The piston moves down. The intake valve is open and fuel and air flow into the cylinder.
- **Compression stroke**—The intake valve closes, and the piston moves up. It squeezes the air-fuel mixture to about ⅛ of its original volume in the top of the cylinder.
- **Power stroke**—An electric spark from a spark plug ignites the mixture. The fuel and air mixture burns very rapidly and increases the pressure inside the cylinder. This pressure forces the piston down.
- **Exhaust stroke**—The exhaust valve opens. The piston moves up and pushes out the exhaust gases. The cycle repeats.

Internal Combustion

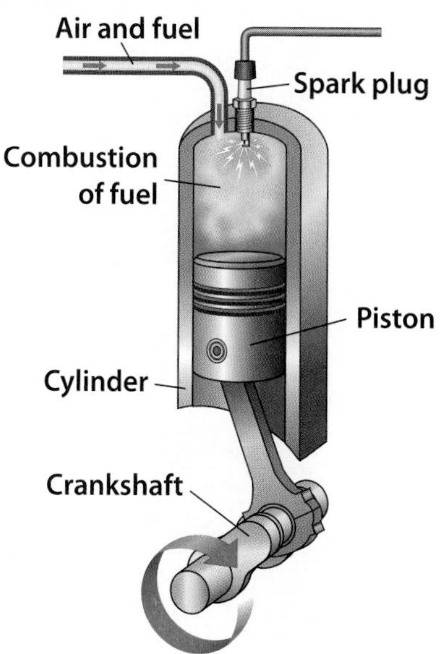

Explosive Force In a typical engine, fuel is ignited in the cylinder. The explosive force from the burning fuel moves the piston. *What keeps the explosive force from leaking around the outside of the piston?*

Figure 21.4 **Four-Stroke Cycle**

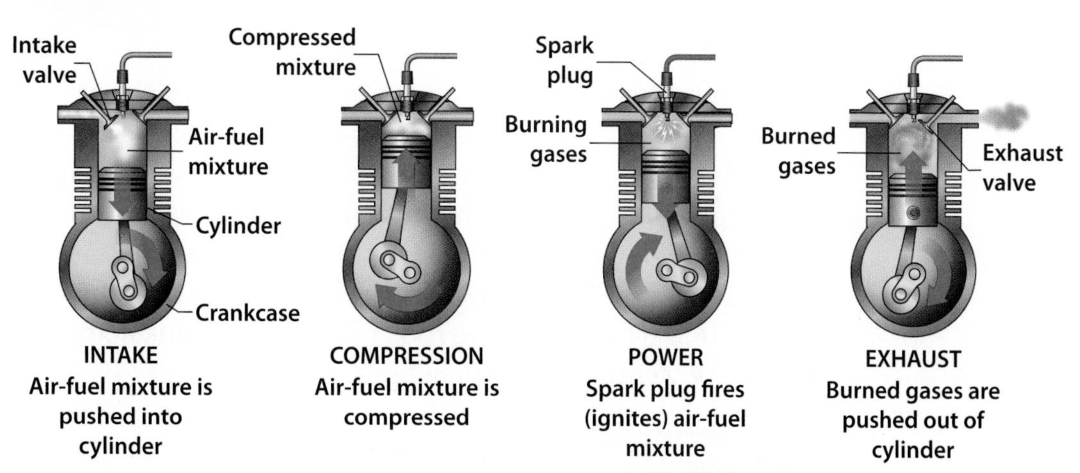

Different Strokes This is what happens during each stroke of a four-stroke cycle gasoline engine. *How many strokes do your legs make when you ride a bicycle?*

Stroke Power
Two-stroke engines power chain-saws, string trimmers, and some motorcycles. *Which is more efficient: a two-stroke cycle engine or a four-stroke cycle engine?*

Two-Stroke Cycles

Some small gasoline engines operate with two strokes. The intake and compression strokes are combined, as are the power and exhaust strokes. Such engines operate on a two-stroke cycle. The piston makes two strokes before it begins to repeat itself. These engines power some off-road motorcycles, chain saws, and other devices. Two-stroke cycle engines are less efficient and emit more pollutants. In two-stroke engines, the fuel and oil are mixed together.

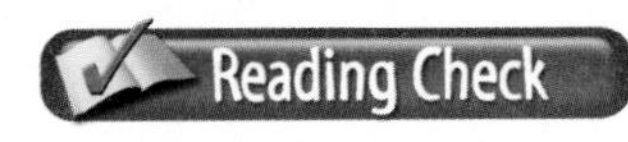

Summarize What are engine cycles?

Gasoline Engines

How do gasoline engines produce power?

There are more gasoline engines in the world than any other type. They start easily, are inexpensive to make, and can be made in almost any size. Automobile engines operate with a four-stroke cycle. Many modern engines have four cylinders, but others may have six or eight cylinders.

Pistons

The piston in a gasoline engine moves only up and down. This up-and-down, straight-line motion is called **reciprocating motion**. Unless we travel by pogo stick, reciprocating motion cannot be used for transportation. It must be converted to **rotary motion**, or circular motion. All of our transportation vehicles and methods, except rocket-powered spacecraft, require rotary motion to turn wheels or propellers.

Crankshaft

Reciprocating motion is changed to rotary motion by a **crankshaft**. See **Figure 21.5**. The pedals on your bicycle are also attached to a crankshaft. The crankshaft converts your reciprocating leg motion to the circular motion of the wheels. Your legs move up and down, but the wheels rotate to move you forward.

An automobile has a crankshaft inside the engine. Each piston is joined to it by a connecting rod. The rotating crankshaft transfers power to the driving wheels.

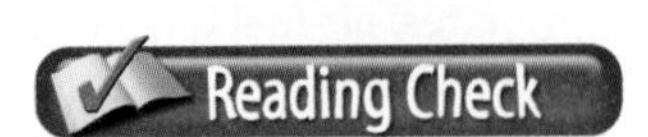

Explain What is the purpose of a crankshaft on a bicycle?

Diesel Engines

How are diesel engines different from gasoline engines?

Diesel engines power trucks, buses, locomotives, ships and some automobiles. They operate smoothly with heavy loads, which would cause a gasoline engine to stall. Diesel engines last longer and require less maintenance.

Parts of a Diesel Engine

The internal parts of a diesel engine are like the parts inside a gasoline engine. Diesels have pistons, cylinders, and a crankshaft. They come in four-stroke and two-stroke cycle versions, which are a bit more efficient and use less fuel than four-stroke engines.

The major differences between diesel and gasoline engines are in the diesel's fuel system and **ignition system**. Since diesel fuel cannot be easily ignited with a spark plug, diesel engines use hot air.

Diesel Engine Cycles

The engine's four strokes are the same as those in a gasoline engine. However, there are some operating differences.

- **Intake stroke**—Only air enters the cylinder. It is not an air-fuel mixture.
- **Compression stroke**—The air is squeezed to about 1/22 of its original volume, causing its temperature to rise to about 1,000°F.
- **Power stroke**—Diesel fuel is squirted directly into the cylinder. The high air temperature ignites the fuel immediately. Pressure builds up very quickly and pushes the piston down.
- **Exhaust stroke**—The piston moves up. Burned gases are pushed out.

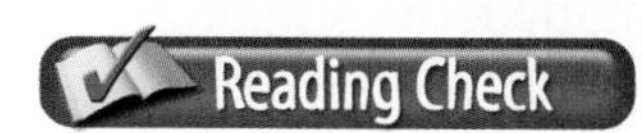

Contrast How are gasoline and diesel engines different?

Crankshaft

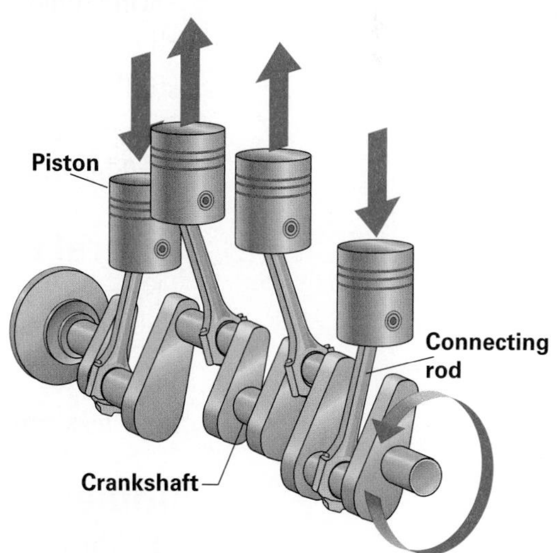

Change in Motion A crankshaft is a strong, heavy piece of metal that converts reciprocating motion to rotary motion. *Where have you seen something that works like a crankshaft?*

Emissions and the Environment

Why does the government control emissions from cars?

When an engine burns fuel, it creates **emissions**. Emissions are gases released into the air because of combustion. A small part of this gas contains pollutants, which are harmful to the environment.

Ethics in Action

Fuel Economy

One important thing to consider when choosing a car is how much fuel it uses, or its "fuel economy." This is measured in miles per gallon. If your car gets more miles per gallon, you spend less on gas and cause less pollution.

Extreme Efficiency The average new vehicle sold in the United States gets less than 30 miles per gallon. However, a French team has built an experimental car called *Microjoule* that can travel 10,705 miles per gallon!

English Language Arts/Writing

Higher Standards Many people want to make fuel economy laws stricter. They want new automobiles to get at least 35 miles per gallon. Others say this would be unfair to automakers and would hurt their business.

1. Research both sides of the issue.
2. Write your own opinion in one paragraph.

Fuel Efficiency

The Environmental Protection Agency of the U.S. government regulates the emissions and miles per gallon for all cars sold in the United States. The emissions must be low to reduce pollution. The miles per gallon must be high to burn less fuel.

Manufacturers are trying to reduce emissions by designing more fuel-efficient automobiles, including electric and hybrid cars (see Section 21.3). They are also trying to reduce harmful emissions by experimenting with cleaner sources of fuel.

Ethanol

Some engines can use **ethanol**, a liquid fuel made from corn. Ethanol burns more cleanly than straight gasoline. Americans usually use ethanol as an "additive," which means that they mix it with gasoline. Cars that can use ethanol are called flex-fuel vehicles. Flex-fuel vehicles in Brazil use ethanol made from sugar.

Natural Gas

Natural gas heats half the homes in America. Scientists are also testing it as a fuel for automobiles. Suppliers compress it to make a liquid. It is compressed natural gas, or CNG. These CNG cars cause less pollution than cars that use straight gasoline.

Biodiesel

Biodiesel is a mixture of soybean oil and diesel fuel. Called B20 because it is 20 percent soybean oil, biodiesel can be used in any diesel engine. Its greenhouse gas emissions are reduced.

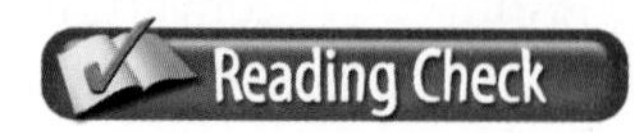

Explain Why is it important to control emissions from automobiles?

Gas Turbine Engines

Why do commercial airplanes use gas turbine engines?

Gas turbine engines power many large airplanes and ships. They are complicated, but they are the most reliable internal combustion engine. A reliable engine requires less **maintenance** because it rarely stops working or breaks.

Advantages and Disadvantages

Gas turbines are smaller and lighter than other engines of the same power rating and have a long engine life. Their biggest disadvantage is high cost. The gas turbine has parts that spin at high speeds and are kept at high temperatures. Such parts must be carefully made from special materials.

Types of Gas Turbine Engines

There are three basic types of gas turbine engines: the turbojet, the turbofan, and the turboshaft (or turboprop). Turbojets and turbofans are also called "jet engines." **Jet engines** push airplanes through the air with a jet of high-pressure exhaust gas. Have you ever blown up a balloon and then let it go? The pressurized air escapes through the end, making the balloon dart around. This force is known as **thrust**.

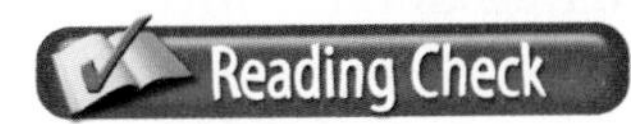

Recall What are three gas turbine engines?

Figure 21.6 Space Shuttle Launch

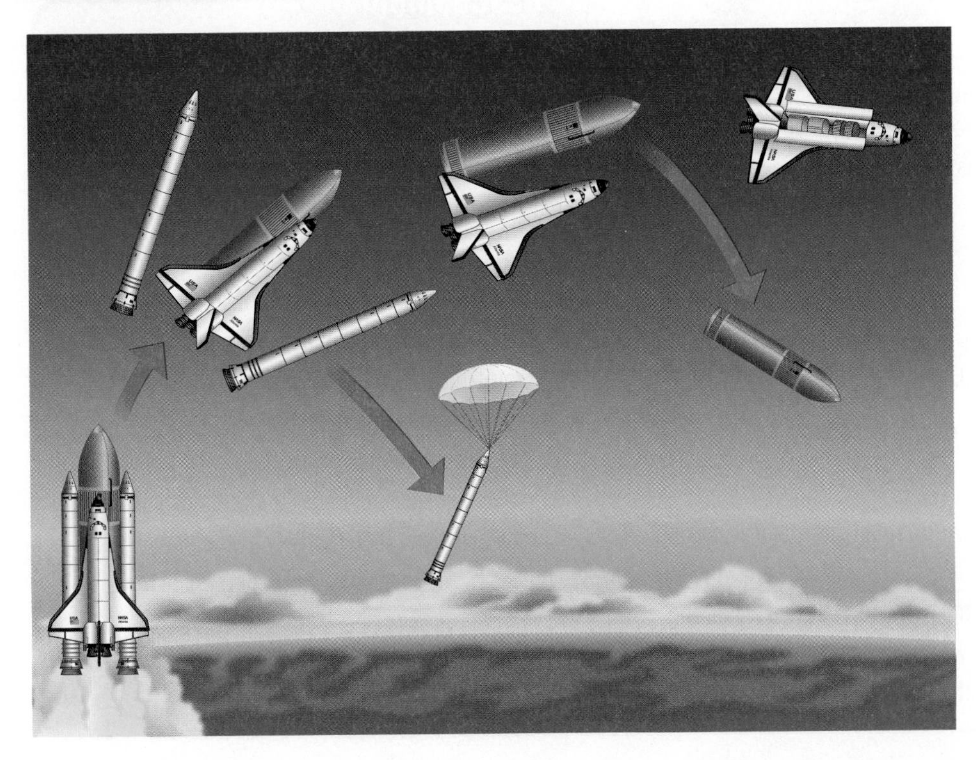

Blast Off! During the launch of the space shuttle, the two solid rocket boosters are released first. The liquid propellants are carried in the big center tank that drops away and is later picked up in the ocean. *Where might the thrust occur in a rocket?*

Academic Connections
Social Studies

Lift Off! America's *Saturn V* rocket put Neil Armstrong and Edwin Aldrin on the moon in 1969. It remains the largest and most powerful production rocket ever built. It was 363 feet tall and developed about 7.5 million pounds of thrust at liftoff.

Apply Do some research at the library on famous astronauts. Write a paragraph describing the effects of the space program on the families of astronauts.

Rocket Engines

Why are rocket engines used in space?

A rocket engine carries its own oxygen for combustion. The oxygen and fuel form a **propellant** producing a high-speed exhaust gas. The gas rushes out the rear to produce thrust. Jet engines depend on oxygen from the air, but there is no air in outer space so only rocket engines can travel in space.

Solid Propellants

The simplest rocket engines use a solid propellant. Solid propellants do not need a special combustion chamber. They are ignited, and then burn. The combustion, however, cannot be stopped once it has begun. The thrust from five powerful rocket engines lifts the space shuttle off the launching pad. Two are solid-propellant rocket engines strapped to each side. See **Figure 21.6** on page 457. They are solid rocket boosters (SRBs) that boost the shuttle into orbit, burning out in two minutes. Then they drop into the ocean.

Liquid Propellants

The three middle engines on the space shuttle are called space shuttle main engines (SSMEs). They do not drop away. They use liquid propellants: liquid oxygen (LOX) and liquid hydrogen (LH2). The propellants are carried in a large external tank covered with insulation. When empty, the tank drops away.

section 21.2 assessment

After You Read Self-Check

1. Explain the difference between a stroke and a cycle.
2. Name the device that turns reciprocating motion into rotary motion.
3. Identify what ignites the fuel inside a diesel engine.

Think

4. The SRBs on the space shuttle use up their propellant in two minutes. Explain why this is good.

Practice Academic Skills

Science

5. Simulate an engine with a small plastic bottle, baking soda, and vinegar. Place the baking soda and vinegar in the bottle. Place the bottle in a basin of water. The acid-base reaction will form a gas that pushes the bottle across the water's surface.

STEM Mathematics

6. Tatyana's car gets 22.5 miles per gallon. Her parents live 144 miles away. If gas costs $2.85 a gallon, how much would it cost her to drive to her parents' home?

Math Concept **Algebra** Write an algebraic equation based on the information in a word problem to solve it. Use x to stand for an unknown quantity.

1. You can write this equation: $x = (144 \div 22.5) \times 2.85$. x stands for the cost of gas.
2. Use the equation to divide the total miles by the miles per gallon to figure out the number of gallons she will need. Then multiply the number of gallons by the cost per gallon to determine the cost.

For help, go to **glencoe.com** to this book's OLC and find the Math Handbook.

section 21.3

Electric Motors

Reading Guide

Before You Read **Preview** What vehicles use electric motors?

Content Vocabulary
- hybrid
- fuel cell

Academic Vocabulary
- environment
- alternative

Graphic Organizer
Draw the section diagram. Use it to organize and write down information as you read.

Hybrid Automobiles

Go to **glencoe.com** to this book's OLC for a downloadable graphic organizer and more.

TECHNOLOGY STANDARDS

STL 4 Cultural, Social, Economic & Political Effects
STL 5 Environmental Effects
STL 6 Role of Society
STL 12 Use & Maintenance
STL 16 Energy & Power Technologies

ACADEMIC STANDARDS

Social Studies
NCSS Content Standard 8 Science, technology, and society
Science
NSES Content Standard F Science and technology in society

STL *National Standards for Technological Literacy*
NCTM *National Council of Teachers of Mathematics*
NCTE *National Council of Teachers of English*
NSES *National Science Education Standards*
NCSS *National Council for the Social Studies*

Electric Transportation

Many transportation vehicles use electric motors. Subways and electric trains are two examples. Trains that use electric motors help reduce air pollution and are quieter than diesel-powered trains or buses. Other electric-powered transportation devices are:

- Segway transporters
- Elevators
- Escalators
- Amusement park rides

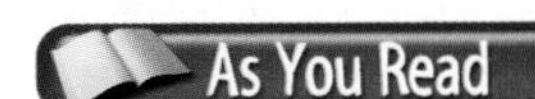
As You Read

Think What are the advantages and disadvantages of hybrid and fuel-cell cars?

Dean Kamen

Inventor of Segway Transporter

Inventor Dean Kamen was born in New York in 1951. While still in college, he invented the wearable infusion pump, a medical device that led to the first insulin pump for diabetics. By age 30, Kamen founded DEKA Research and Development Corporation to be able to further generate inventions in the worlds of science and technology.

Kamen holds more than 440 U.S. and foreign patents. He is perhaps best known as the inventor of the Segway Human Transporter. Introduced in 2001, the two-wheeled, battery-operated vehicle uses sophisticated computer programming. The rider stands while operating the machine. Gyroscopes help the rider maintain balance.

Inspiration Kamen's Segway actually grew out of a prior invention, the Independence IBOT ™ Mobility System. After seeing a man in a wheelchair struggling to get over a curb, Kamen and his team created a self-balancing mobility device. This device allows users to climb stairs and maneuver difficult surfaces, such as rocky areas and sand.

English Language Arts/Writing Write an article for your school's newspaper about what you see as the future of the Segway.

Go to **glencoe.com** to this book's OLC to learn about young innovators in technology.

Locomotives

What engines do locomotives use?

One common type of locomotive uses a diesel-electric drive system. A diesel engine turns a generator to produce electricity. The electricity is sent by wires to electric motors directly connected to the driving wheels. Diesel-electric locomotives can develop over 6,000 horsepower and pull 200 railroad cars.

All-Electric Locomotives

Another type of locomotive is all electric. It has no diesel engine. It usually draws its electrical power from an overhead cable. The stainless steel *Acela Express* is a high-speed all-electric locomotive used for intercity passenger travel. It moves along the popular Washington-to-New York-to-Boston corridor at speeds over 150 mph. Its travel time between Washington, D.C. and New York City is under 2½ hours.

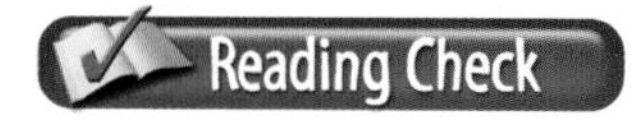

Discuss How are electric motors used to power locomotives?

Electric Cars

Will we ever drive cars powered by electricity?

For almost 100 years, automobile companies have experimented with electrically powered cars. They are hoping to find a technology that does less harm to the environment. Their biggest problem has been the distance you can drive on a single battery charge. It tends to be around 50 miles, which means you can drive only about 25 miles away from a power source.

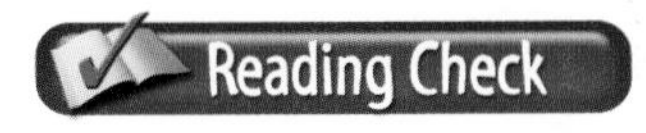

Predict How do you think hybrid automobiles are powered?

Hybrid Automobiles

Although there are few all-electric cars currently available, hybrids are extremely popular with the general public. A **hybrid** is a combination of different elements. The hybrid car combines an electric motor with a gasoline engine.

A hybrid's electric motor is used at low speeds around town and limits exhaust emissions. The small gasoline engine is used for higher speeds on freeways and on the open highway where emissions do not tend to collect. The gasoline engine also operates a generator that in turn recharges the batteries that power the electric motor.

Pure Electricity The Tesla Roadster looks, feels, and drives like many other high-end sports cars, but is a 100% electric vehicle. *Why are companies and consumers interested in developing electric cars?*

EcoTech

All Aboard

The East Japan Railway introduced a hybrid train in 2007. It has a diesel engine, two electric engines under each car, and batteries on the roof. It is 20 percent more fuel efficient than the previous generation of trains and 60 percent cleaner.

Try This Take public transportation, ride your bike, or walk instead of riding in an automobile.

Hybrid cars get very good mileage, often between 45 and 50 miles per gallon of gasoline. They produce only about half the greenhouse gas emissions of an ordinary car that gets about 30 miles per gallon or less.

Fuel-Cell Automobiles

Fuel cells offer another alternative to gasoline-powered transportation. A **fuel cell** combines hydrogen with oxygen to create electricity. The astronauts who operate the space shuttle use three of these fuel cells when travelling in space. For use in transportation, fuel-cell electricity is converted to mechanical power. Some cities are testing fuel-cell-powered buses.

Fuel cells do not pollute the atmosphere because their only emission is water. They also do not have to be recharged, just refilled with hydrogen. In the future, you might drive a car powered by fuel cells. So far, however, they are only experimental. There are none available to the public, partly because they are so expensive to manufacture.

Section 21.3 Assessment

After You Read Self-Check

1. Describe how a diesel-electric locomotive works.
2. Discuss hybrid automobiles and how they work.
3. Identify the emission a fuel cell produces.

Think

4. Electric trains are sometimes designed to draw power from overhead wires or from a protected high-voltage rail on the ground. They have two electrical pickup locations. Explain why you think this is so.

Practice Academic Skills

Social Studies

5. The city of Curitiba, Brazil, has one of the most successful mass transportation systems in the world. Consisting entirely of buses, the system is convenient, efficient, and affordable. More than 85% of the population uses it. As a result, Curitiba has the lowest rates of pollution and gas consumption in Brazil. Do some research on Curitiba's bus system at the library or on the Internet. Write a few paragraphs describing its design and how it works.

STEM Science

6. While the concept of using fuel cells in cars is relatively new, fuel cell technology has been around for some time. Fuel cells have been used in the space program for many years. Research the development of fuel cells and their uses. Write a few paragraphs about what you find. Include a discussion of the advantages and disadvantages of fuel cells.

Exploring Careers in Technology

Ben Hunter

MECHANICAL ENGINEER, MENTOR

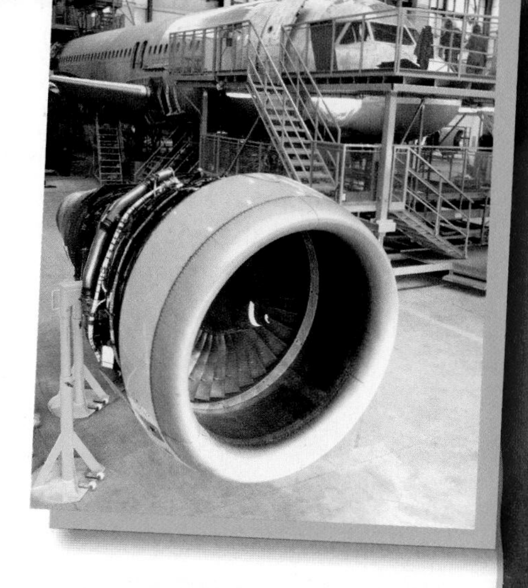

Q: *What got you interested in mechanical engineering?*

A: I began college majoring in aerospace engineering. I've always had a curiosity about how things worked and a love for airplanes. My dad is an air-traffic-control specialist, and my grandfather was a pilot for American Airlines, so aerospace is in the family. I decided to double-major in mechanical engineering. It offered opportunities that otherwise would not be available.

Q: *What do you do for your company?*

A: I am a product development engineer in my company's Aircraft Engine Systems Group. When a company decides to design a new jet turbine engine, we provide components that make that engine work. It's my job to take one of those components from concept, through design, to manufacture, and finally testing and certification. I also mentor the on-the-job work experience of our engineering interns.

Q: *What advice do you have for young people who are considering an engineering career?*

A: Get involved with clubs for hands-on experience in different kinds of projects. Maybe your school has a math or science club or even an engineering club. Of all the classes, math is the most important. It provides the foundation for what you will study in high school and college—courses on differential equations and thermodynamics.

English Language Arts/Writing

Hands-On Experience List the skills you might develop by joining a math, science, or engineering club. How are these skills similar or different?

1. Talk to a teacher, use the Internet, or check the message boards at your school to find out what kinds of clubs are available.
2. Using a spreadsheet program, make a chart of the different clubs and what skills and experiences you can gain from them.
3. How could joining each club affect your future career choices? Write your answers in one paragraph.

Go to **glencoe.com** to this book's OLC to learn more about this career.

Real-World Skills

Problem-solving, observation, speaking, listening

Academics and Education

Mathematics, science, English language arts

Career Outlook

Growth as fast as average for the next ten years

Source: *Occupational Outlook Handbook*

chapter 21 Review and Assessment

Chapter Summary

Section 21.1 Much of our useful power comes from engines. Engines use fuel as an energy source. The world's first engine was a steam engine used to pump water from coal mines. Steam engines are external combustion engines.

Section 21.2 Gasoline engines are internal combustion engines. They use pistons that slide up and down inside a cylinder. Each up or down motion of the piston is a stroke. This motion is converted to rotary motion by a crankshaft. Diesel engines are best suited for heavy work. Rocket engines carry oxygen as well as fuel so they can travel in space.

Section 21.3 Many trains are powered by electric motors. Some locomotives use diesel engines to turn generators that produce electricity. The hybrid car runs on an electric motor and a gasoline engine. Fuel cell cars convert hydrogen to electricity. Their emission is water.

Review Content Vocabulary and Academic Vocabulary

1. On a sheet of paper, use each of these terms and words in a written sentence.

Content Vocabulary

- external combustion engine
- piston
- turbine
- internal combustion engine
- reciprocating motion
- rotary motion
- crankshaft
- ignition system
- emissions
- ethanol
- maintenance
- jet engine
- thrust
- propellant
- hybrid
- fuel cell

Academic Vocabulary

- convert
- expand
- transfer
- regulate
- environment
- alternative

Review Key Concepts

2. **Explain** how an external combustion engine works.
3. **Describe** the difference between a steam engine and a steam turbine.
4. **Explain** how an internal combustion engine works.
5. **Describe** a four-stroke and a two-stroke engine cycle.
6. **Explain** the purpose of a crankshaft.
7. **Identify** differences among various engines.
8. **Describe** how electric motors are used to power locomotives.
9. **Explain** how a hybrid automobile is powered.
10. **List** the advantages and disadvantages of hybrid and fuel cell cars.

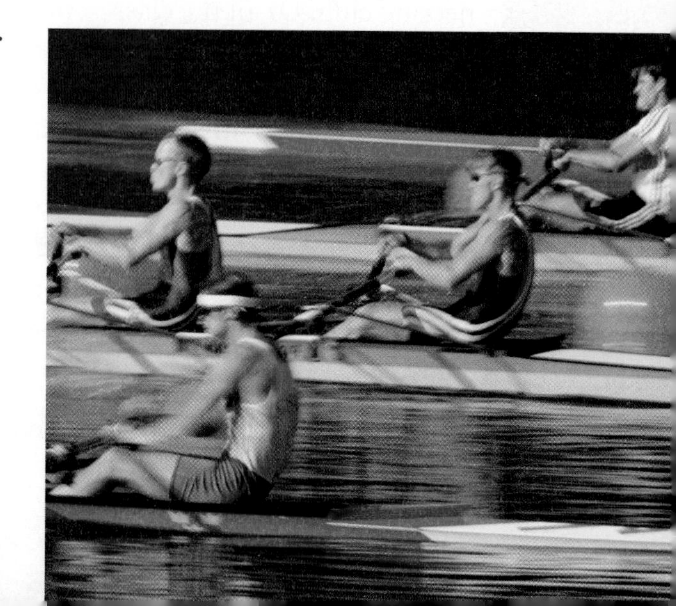

Real-World Skills

11. Human-Powered Transportation The most environmentally friendly types of transportation run off of human-power, such as a bicycle. Research other forms of human-powered transportation and write a few paragraphs describing what you find. Discuss the pros and cons.

STEM **Technology Skill**

12. Space Travel The materials that go into space have to be able to withstand the environment of space. There are many different suppliers of such materials.

a. Conduct research to determine what materials are sent into space.

b. Locate some companies that supply materials to the space program. Look for companies in your state that supply such materials. Write a summary of what you find.

Academic Skills

Social Studies

13. You can find engines and motors almost everywhere. Make a list of every motor you see. Examples might include lawn equipment, appliances, and toys.

STEM **Mathematics**

14. Heather is traveling in Europe and rents a car. The speeds are all in kilometers per hour, not miles per hour. How fast will she be going, in miles per hour, at 120 kilometers per hour? Round your answer to the nearest whole number.

Math Concept **Unit Conversion** One mile per hour (mph) equals 1.61 kilometer per hour (kph). To convert from mph to kph, multiply by 1.61. Divide by that amount to convert from kph to mph.

WINNING EVENTS

Automotive Engineer

Situation You and your team will design and build an electrically powered all-terrain vehicle. Develop sketches and a working, battery-powered model of the vehicle. Your teacher will determine the model's dimensions.

Activity Design the safest, most reliable, and fastest vehicle you can. Make a series of sketches. Then make a working model and conduct "field" tests. Make changes as needed. Race your vehicles against others.

Evaluation Sketches will be evaluated based on how well they communicate your ideas. Models will be evaluated based on these criteria:

- Safety, reliability
- Utility, speed

Go to **glencoe.com** to this book's OLC for information about TSA events.

Standardized Test Practice

Directions Choose the letter of the best answer. Write the letter on a separate piece of paper.

1. What is 53.8% of $42,540?

   A $22,886.52
   B $32,437.08
   C $21,436.52
   D $21,270.00

2. Reciprocating motion is circular.

   T
   F

Test-Taking Tip Read each test question carefully, no matter what kind of test you take.

chapter 21

TECHNOLOGY LAB

Build a Steam Turbine

In a steam power plant, steam is directed toward a fan-shaped turbine. The pressure from the steam causes the turbine to spin. The turbine's rotary motion turns an alternator that generates electricity for use in your home and school.

Tools and Materials

- ✓ Copper tube, 1 inch in diameter and 4 inches long
- ✓ Two corks that fit tightly into the ends of the copper tube
- ✓ Aluminum beverage can
- ✓ Plastic bead
- ✓ Small nail
- ✓ Two metal clothes hangers
- ✓ Propane torch
- ✓ Scissors
- ✓ Razor knife
- ✓ Heavy wire cutters
- ✓ Metal file
- ✓ Safety glasses or goggles
- ✓ Stopwatch

Set Your Goal

For this activity, you will build a steam generator that will turn a fan-shaped turbine.

Know the Criteria and Constraints

In this lab, you will:

1. Build turbine blades that are uniform in size and shape.
2. Keep a record of turbine speeds and how different designs affect speed.

Design Your Project

Follow these steps to complete this lab.

1. Review the drawing of the completed steam turbine.
2. Cut out an octagon-shaped disk from the aluminum can using scissors and the template in picture 1.
   - Punch a hole in the center of the aluminum disc with a nail.
   - Bend the blades. This will be your spinning turbine disk.
   - Color one blade with a permanent marker to help you judge how fast the disk spins.
3. Use the razor knife to cut a notch in one cork. The notch will direct steam from the tube to the turbine disk causing the disk to spin.

SAFETY

Tool Safety

In this activity, you will be using dangerous tools, such as a propane torch. Be sure to always follow appropriate safety procedures and rules so you and your classmates do not get hurt.

Procedure

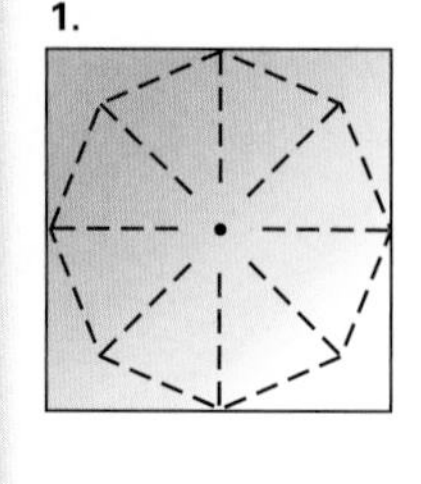

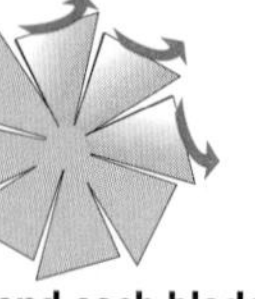
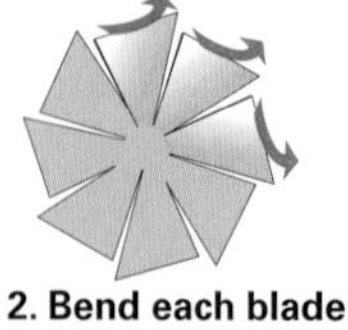

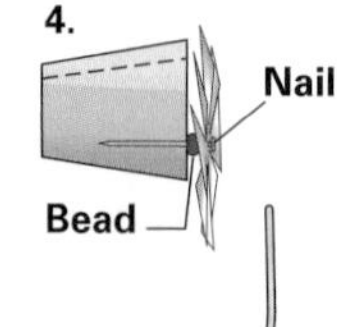

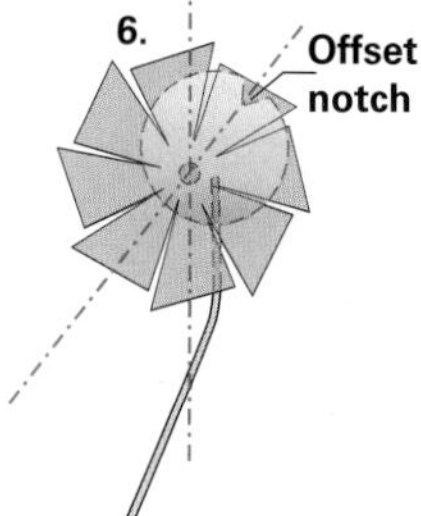

4. Slide the turbine disk onto the nail. Then slide on the plastic bead, which will act as a bearing. Push the nail into the notched cork at a spot across from the notch.
5. Place the cork in each end of the copper tube. Your copper tube is now a completed boiler.
6. Use the heavy wire cutters to make two supports out of clothes hanger wire.
   - Sharpen one end of each piece of wire with a metal file.
   - Then stick the sharpened end of each wire support into the edge of a cork. Your steam turbine is now ready to operate.

Completed Steam Turbine

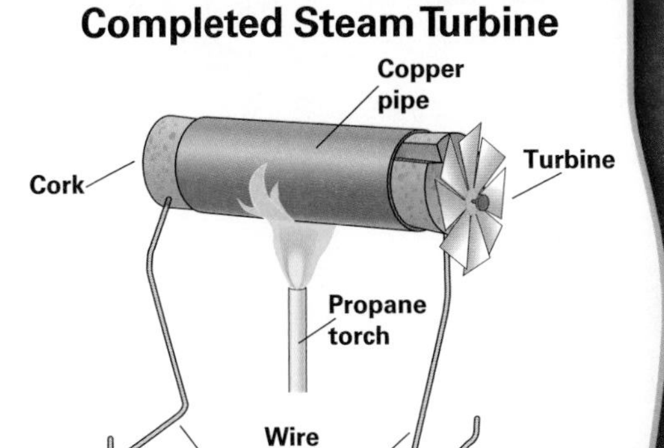

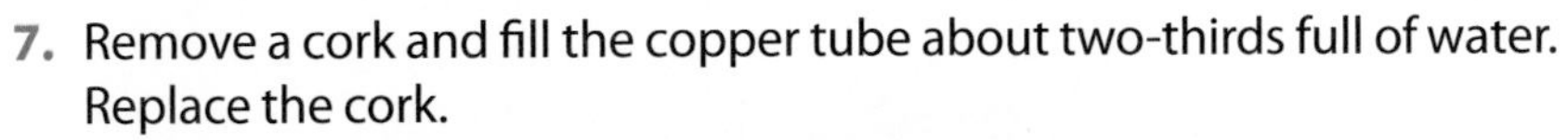

7. Remove a cork and fill the copper tube about two-thirds full of water. Replace the cork.

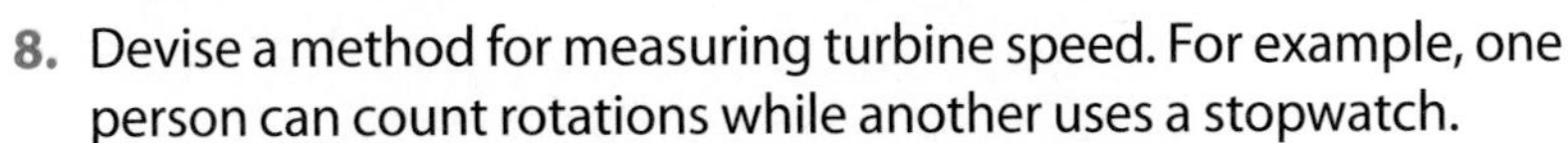

8. Devise a method for measuring turbine speed. For example, one person can count rotations while another uses a stopwatch.
9. Put on the safety glasses, then light the propane torch.
   - Carefully and evenly heat the copper tube.
   - Steam will come out through the notch and will spin the turbine.
10. Measure and record turbine speed using the stopwatch.
11. Try twisting the turbine blades to greater or lesser angles. What effect, if any, does this have on their rotating speed?

Evaluate Your Results

After you complete the lab, answer these questions on a separate piece of paper.

1. Could you tell how fast the disk was rotating? Would it spin faster if the boiler was two-thirds full of water or only one-third full? Why?
2. Would the disk spin faster or slower if you cut a smaller notch in the cork? Why?

Academic Skills Required to Complete Lab

Tasks	English Language Arts	Math	Science	Social Studies
Review safety rules.	✓			
Build steam turbine.	✓	✓	✓	
Devise method for measuring turbine speed.	✓		✓	
Measure and record turbine speed.		✓	✓	
Modify turbine blades and repeat.		✓	✓	

chapter 22

Transportation Systems

Sections

What You'll Learn

- **Name** the different types of land transportation.
- **Summarize** the purpose of transportation subsystems.
- **Examine** the purpose of a transmission.
- **Tell** how oceans and inland waterways are used for transportation.
- **List** the different types of ships.
- **Define** the concept of intermodal transportation.
- **Discuss** ways in which air and space transportation are used.
- **Identify** different types of aircraft and spacecraft.
- **Explain** how an airplane is lifted into the air.

Explore the Photo 

A Pleasant Trip After travelers leave a harbor or airport, they might take a bus, taxi, or commuter plane to their next destination. *A passenger tells the pilot he or she has had a nice trip. Is this part of the transportation system?*

Build a Rubber-Band-Powered Vehicle

At the end of this chapter, you will make a simple land transportation vehicle powered by a rubber band. Get a head start by using this checklist to prepare for the Technology Lab.

PROJECT CHECKLIST

- ✓ Study the designs of different land vehicles at the library or on the Internet.
- ✓ Ask your teacher about what materials you will need for the project.
- ✓ Review your classroom's procedures and safety rules.

section

22.1 Land Transportation

Reading Guide

Before You Read **Preview** What are some examples of land transportation vehicles?

Content Vocabulary

- mass transportation
- transmission
- driving wheel
- front-wheel drive
- rear-wheel drive
- four-wheel drive
- tractor-trailer
- bullet train
- maglev train

Academic Vocabulary

You will see these words in your reading and on your tests. Find their meanings at the back of this book.

- role
- interact

Graphic Organizer

Draw the section diagram. Use it to organize and write down information as you read.

Impacts of Land Transportation

	Automobiles	Mass Transportation
Positive		
Negative		

Go to **glencoe.com** to this book's OLC for a downloadable graphic organizer and more.

TECHNOLOGY STANDARDS

STL 4 Cultural, Social, Economic & Political Effects

STL 12 Use & Maintenance

STL 18 Transportation Technologies

ACADEMIC STANDARDS

Science

NSES Content Standard E Understandings about science and technology

NSES Content Standard F Science and technology in society

STL *National Standards for Technological Literacy*

NCTM *National Council of Teachers of Mathematics*

NCTE *National Council of Teachers of English*

NSES *National Science Education Standards*

NCSS *National Council for the Social Studies*

Transportation Systems

What is mass transportation?

As You Read

Connect What different kinds of subsystems do land transportation vehicles have?

A transportation system is a way of moving people or products from place to place. Transportation systems have inputs, processes, outputs, and feedback. For example, inputs to a city bus system include bus drivers and fuel. Processes include driving the bus and loading passengers. The output is arrival at scheduled stops. Feedback includes comments from satisfied customers.

Transportation systems are interrelated. Each system depends on the other systems. Buses and cars, for example, take passengers to airports and ship docks. Transportation systems are part of the larger technological, social, and environmental systems in our world.

Land Transportation

When you travel in a car, bus, or train, you are using a land transportation vehicle. Land transportation also includes travel by bicycle, motorcycle, and subway. Automobiles are an important part of our land transportation system. However, you need more than just a car to get from place to place. Roads, bridges, and service centers are just a few of the subsystems within a land transportation system that allow you to use your family car. About half of all the world's automobiles are used in the United States.

Mass Transportation

Mass transportation moves many people at one time and is available to the general public. However, it is expensive to develop mass transportation systems that can serve a large country. That is one reason why the automobile has become an important part of our land transportation system. Automobiles are for personal transportation, not mass transportation. Mass transportation is sometimes slower and less convenient. However, people who use mass transportation do not have to worry about parking their cars or paying for gas, parking tickets, or car insurance.

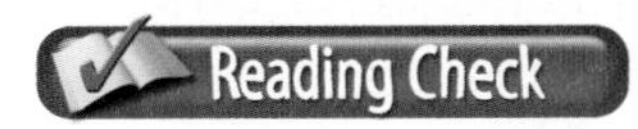

Recall What are types of land transportation?

Automobiles

What does "four-wheel drive" mean?

Modern automobiles are quite different from early models. However, your family car has at least two things in common with those early models. Both have transmissions, and nearly all have front-mounted engines.

Transmission

A car's **transmission** contains gears and other parts that transfer power from the engine to the axles and wheels. The gears work on the same principles as gears on a bicycle.

Mass Transportation Tokyo, the capital of Japan, has a reliable and clean rapid transit system with trains that run above and below ground. More than 7 million people use Tokyo's trains each day. *Why is it easier to build an efficient mass transportation system in a city?*

BusinessWeek Tech NEWS

Moving More People, Making Less Impact

The nonprofit Design Trust for Public Space called for the "greening" of taxicabs. In New York some 13,000 gas guzzlers jam the already congested streets. At an exhibit in 2007, several eco-friendly prototypes were introduced, including a boxy one by Vehicle Production Group. It stands six feet tall, allows for wheelchair access, and uses less gas.

Critical Thinking *List some other ways urban areas can improve their environmental footprint?*

Go to **glencoe.com** to this book's OLC to read more about this news.

When you pedal up a hill, you shift into low gear because it takes more effort to pedal up the hill. The rear wheel moves more slowly. On level ground, it takes less effort to pedal, and the rear wheel of the bicycle turns more quickly. See **Figure 22.1**. A car's engine operates best if you use a low gear while climbing a steep hill or when starting from a stop. Using a high gear is best when driving on a flat road.

Driving Wheels

The power from your legs is transferred to your bicycle's rear wheel with a chain. The rear wheel is a bicycle's **driving wheel**. A car does not use a chain. Instead, it transfers power with one or two metal shafts called "drive shafts." Most cars transfer power to the front wheels and are known as **front-wheel drive** cars. Some cars transfer power to the rear wheels. They are **rear-wheel drive** cars. Some send power to all four wheels. These cars are **four-wheel drive**, or all-wheel drive, cars.

Other Subsystems

An automobile has many subsystems. Besides the transmission and driving wheels, some subsystems include those that provide structure and support (frame), propulsion (engine), and guidance (steering wheel). Others are safety related (air bags). Some cars have mapping computers that connect to GPS satellites.

Many of these subsystems are manufactured by outside suppliers. The anti-lock braking system (ABS), for example, is not usually made by the same company that makes the car. The CD player, tires, and windshield may also be manufactured by a different company.

Figure 22.1 **Transmission**

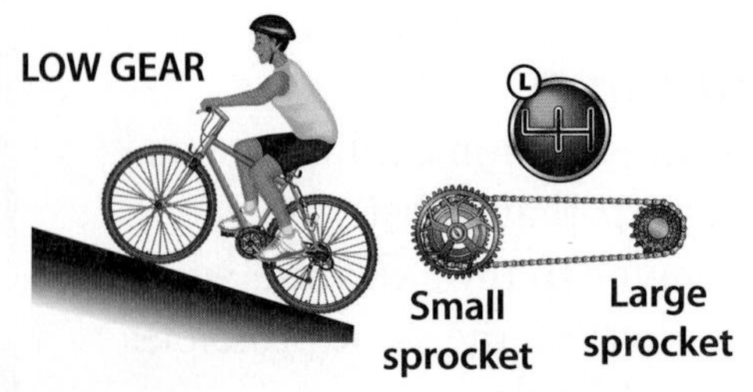

Shifting Gears A car's transmission works just like the gears in a bicycle. High gear combines the larger crank sprocket with the smallest sprocket at the rear wheel. *What combination of sprockets does low gear use?*

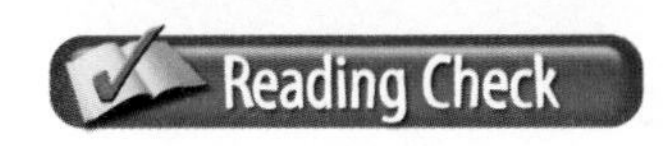

Summarize What is the transmission in an automobile?

Buses

How did school buses help promote better schools?

Buses usually carry 30 or more passengers and are used for mass transportation between cities. This is called "intercity transportation." Buses are also used for transportation within cities and for school transportation. School systems in the United States alone use over 400,000 buses.

School Buses

School buses were being used as early as 1920. They made it possible to gather students together from small rural schools into one larger school with improved facilities. Millions of students ride safe and sturdy buses to school each day. Buses are made from strong steel to meet federal manufacturing requirements. Seats and other support systems are specially designed for safety. Many use diesel fuel, which would not burn as easily as gasoline if an unexpected leak should occur.

Intercity Buses

Intercity buses are also powered by diesel engines. These types of buses carry up to 64 seated passengers. They are generally less expensive for passengers to ride than trains or airplanes. They also make stops at smaller communities that are not served by trains and airplanes.

The Black Box

Many cars today have data recorders, or "black boxes," as part of their control systems. The black box monitors the release of safety air bags in an accident. It also records the driving actions of the driver.

Caught in the Act If an accident happens, the police and insurance companies can use the black box to see what the driver was doing. If the driver was speeding or doing anything wrong, the accident could be considered his or her fault.

English Language Arts/Writing

Big Brother Is Watching People might be more likely to obey traffic laws if they know their actions are being recorded. But is this an invasion of privacy?

1. George Orwell, the author of the novel *1984*, once said, "Men are only so good as their technical developments allow them to be." What did he mean?
2. Do you agree with the quote? Write a one paragraph response.

Urban Buses

Urban buses can carry more people than intercity buses because they allow some passengers to stand. Use of urban buses eases traffic congestion and saves fuel. They use about one-third as much fuel per passenger as do automobiles. However, urban buses account for only 15 percent of all passenger miles traveled in the United States.

Summarize Why are school buses safe?

Figure 22.2 **Wind Deflectors**

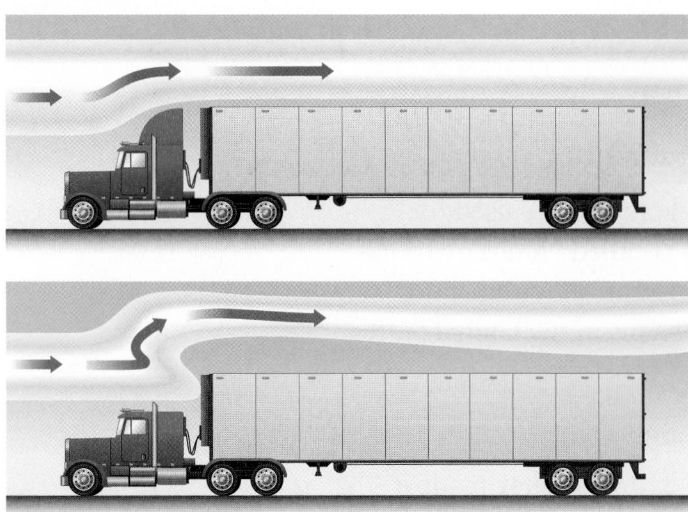

Air Resistance A wind deflector on top of the truck cab can save fuel costs each year. *Why?*

Trucks

What is the purpose of a truck's wind deflector?

American cities rely on trucks to supply them with food, fuel, furniture, and other products. Trucks play an important role in transportation. They go directly from the supply location to the customer. Most trains, airplanes, and ships do not do that.

Types of Trucks

We use hundreds of different kinds of trucks. Most are diesel powered, but gasoline engines are also used. Some commercial trucks are as small as pickup trucks. Others are as large as the 18-wheel semi-trailer trucks that carry cargo on interstate highways. These large trucks are also called **tractor-trailers**.

There is no such thing as a standard truck, but there are three general types: light duty, medium duty, and heavy duty. Panel and pickup trucks are examples of light-duty trucks. Medium-duty trucks are used locally and include sanitation trucks, soft drink delivery trucks, and heating oil trucks. Heavy-duty trucks carry large loads, and an 18-wheel tractor-trailer is one type.

Wind Deflectors

The flat front and square shape of many trucks present a large surface that the air can press against. This means that the trucks waste fuel. Some manufacturers place a wind deflector on the tractor's roof. A deflector directs the air around the truck, which reduces resistance and saves fuel. See **Figure 22.2**.

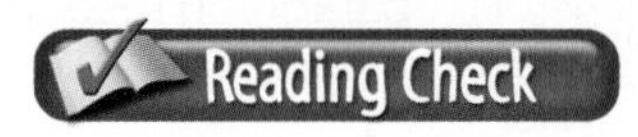

Identify What are the truck types?

Locomotives

What kind of engines replaced steam engines in locomotives?

Railroads earn most of their money by hauling freight. They deliver bulky items like coal and iron ore. They also carry things such as automobiles and television sets. About 10,000 freight trains roll over the tracks each day. Some are over 200 cars long.

Trains also carry passengers, and the busiest lines are on the East Coast. These are operated by AMTRAK, the intercity passenger railroad system. Travel between the terminals at Washington, D.C., and Boston, Massachusetts, including cities along the way, is quite popular. The distances are fairly short, and the trains travel directly from downtown to downtown. However, trains carry less than 1 percent of all U.S. intercity passengers. That is a very low number. In some countries, trains carry up to 50 percent of intercity travelers.

Commuter and subway trains transport workers, tourists, students, shoppers, and others to their daily destinations. They are also part of the mass transportation network.

High-Speed Trains

The newest all-electric locomotives, called **bullet trains**, travel at high speeds. The speed and the pointed shape of the locomotive's nose inspired its name. Japan's *Shinkansen* was the first bullet train in 1964.

The bullet train in the United States is the *Acela Express*, for acceleration and excellence. Several all-electric stainless steel trains entered service in the year 2000. Each six-coach train can carry over 300 passengers at speeds up to 150 mph.

Maglev Trains

Another new kind of train does not roll on wheels. It does not even touch the ground. The train is called a **maglev train**. Maglev stands for "magnetic levitation." The forces of magnetic attraction and repulsion allow the train to float, or levitate, less than one inch above its guideway, or path. The same forces interact to move the trains.

Magnetism The Shanghai Maglev Train carries passengers 22 miles in seven minutes using electromagnetic force. *Since maglev trains float in the air, how do they slow down and then stop?*

Academic Connections
Language Arts

The Language Barrier Large tractor-trailer trucks, which are called "18-wheelers" in the United States, are known as *articulated lorries* in Britain. A "dual-carriage way" in Britain is a *divided highway* in the United States. High-speed, limited-access roads or interstate highways are called "motorways" in Britain.

Apply Use a dictionary to look up the words *tram, omnibus, caravan,* and *the underground.* What terms are used in the United States?

Maglevs are very quiet and produce almost no vibration. However, they are expensive to construct. Only a few experimental ones have been built in the United States. The first maglev to enter commercial service runs on a 22-mile line to the Shanghai airport in China. The train's highest speed during the trip is normally about 260 miles per hour. It was built in Germany and began passenger service in 2004.

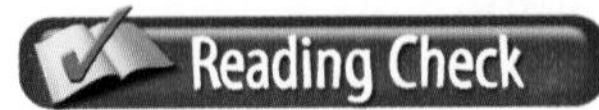

Explain What does the word *maglev* mean?

Pipelines

How is cargo transported through pipelines?

When you turn on a faucet to get a drink of water, you are using a transportation system. Some cargo, like water, oil, and natural gas, travels long distances through pipelines. Pipes may be as small as two inches in diameter or as large as 15 feet in diameter. Most pipelines are buried in the ground.

Pipelines require service facilities such as pumping and control stations. The stations are located along the pipeline and keep the cargo moving. When the cargo is made of particles, such as gravel, it is mixed with liquids to form a "slurry." The pumps then force the material through the pipeline.

section 22.1 assessment

Self-Check

1. Identify the purpose of a transmission.
2. Describe how an intercity bus is used. Explain how this is different from the use of an urban bus.
3. Explain what causes a maglev to "float" above its guideway.

Think

4. Some people think that high-speed trains instead of airplanes should be used for distances less than 200 miles, especially in highly populated areas. Discuss why you think this might be desirable.

Practice Academic Skills

Social Studies

5. Use the Internet to learn about bullet trains used in Japan, England, France, Germany, and Spain. Report your findings to the class.

STEM **Mathematics**

6. A tractor-trailer is hauling 32 television sets. Each TV weighs 112 pounds. What is the total weight of the cargo after eight of the TVs are delivered?

Math Concept **Two-Step Problems** Problems that require two calculations to solve take some extra concentration.

1. The first step is to determine the number of TVs that are left in the truck.
2. After you determine the number of TVs, use multiplication to determine the total weight.

For help, go to **glencoe.com** to this book's OLC and find the Math Handbook.

section

22.2

Water Transportation

Reading Guide

Before You Read **Connect** What are the different types of water transportation?

Content Vocabulary

- navigable waterway
- displacement
- supertanker
- intermodal transportation
- containership

Academic Vocabulary

- overseas
- via

Graphic Organizer

Draw the section diagram. Use it to organize and write down information as you read.

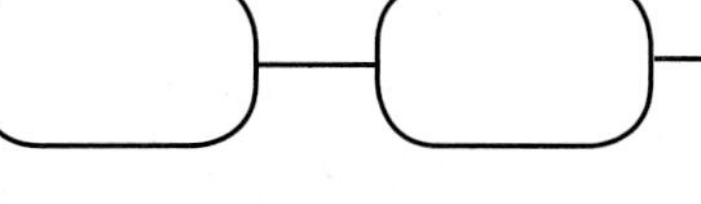

Go to **glencoe.com** to this book's OLC for a downloadable graphic organizer and more.

TECHNOLOGY STANDARDS

STL 3 Relationships & Connections

STL 4 Cultural, Social, Economic & Political Effects

STL 18 Transportation Technologies

ACADEMIC STANDARDS

Social Studies

NCSS Content Standard 2 Time, continuity, and change

Science

NSES Content Standard F Science and technology in society

STL *National Standards for Technological Literacy*

NCTM *National Council of Teachers of Mathematics*

NCTE *National Council of Teachers of English*

NSES *National Science Education Standards*

NCSS *National Council for the Social Studies*

Waterways

What is a navigable waterway?

Water has provided transportation routes for centuries. Oceans, rivers, lakes, and other navigable waterways have made natural routes between cities, states, countries, and continents. A **navigable waterway** is a lake or river that is deep and wide enough to allow ships and boats to pass. The five Great Lakes are navigable waterways, as are many rivers, such as the Mississippi and Ohio.

Engineers can also construct canals through which boats and ships can pass. The Suez Canal in Egypt connects the Mediterranean Sea with the Red Sea, allowing ships to carry cargo back and forth between Europe and Asia.

As You Read

Predict How does intermodal transportation help move people and cargo?

It usually costs less to transport goods by water than by rail, highway, or air. Whenever possible, people try to save as much money as they can by transporting products the cheapest way.

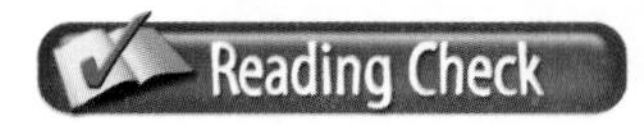

Explain How are navigable waterways used for transportation?

Boats and Ships

What is the difference between a boat and a ship?

A small, open vessel is called a "boat." A large, deep-water vessel is called a ship. There are about 24,000 ships in the world.

There are three general types of water transportation vessels. Passenger vessels that carry people are one type. Cargo ships are another type. They transport oil, grain, iron ore, automobiles, and many other products. Specialty craft include every other type, such as river barges used for transporting coal and other goods, tugboats for pulling large ships into dock, and icebreakers.

Overseas Cargo

Ships deliver most of the overseas cargo leaving or arriving in the United States. To transport by ship, we need docks with special loading and unloading equipment. We need properly trained people to operate the ship. We also need good communications for weather data and other information. They are all necessary for the entire system to operate efficiently.

Luxury Ships Large luxury ships like the *Navigator of the Sea* are used as vacation cruise liners. *How do such large ships maneuver into a tight dock space?*

Heavy Ships

For many centuries, people used sailing ships to haul cargo and passengers. The ships of the 1800s had displacements of about 1,200 tons. **Displacement** is a measure of how much water a ship and its cargo push aside as the ship floats. It is an indication of the ship's size. Today an average cargo ship might have a displacement of 21,000 tons.

A heavy ship is difficult to move, so today's ships are pushed by powerful engines. Ocean liners and cargo ships often use gas turbine or diesel engines.

Used with permission of Royal Caribbean Cruises Ltd.

Ivan Getting

Inventor of GPS

Ivan Getting, who was born in New York City in 1912, was a physicist and electrical engineer. His early contributions to the world of technology included research on the use of radar, ballistic missiles, and microwave tracking systems during World War II. In 1960, he founded the Aerospace Corporation.

Getting is best known for his original ideas about the Global Positioning System (GPS). In the 1970s, Bradford Parkinson of the U.S. Air Force was appointed to run the GPS program for the U.S. Department of Defense using Getting's ideas. By 1978, the NAVSTAR GPS was functional and accurate within three meters.

Map to Reward GPS uses satellites orbiting above the earth for mapping and navigation. In 2003, the year Getting died, he and Parkinson were honored with the Charles Stark Draper prize of $500,000 for their invention.

English Language Arts/Writing Describe in a short essay the importance to the 21st Century of Getting's GPS invention.

Go to **glencoe.com** to this book's OLC to learn about young innovators in technology.

It is not unusual for a large modern ship to displace 100,000 tons. That means it is about 80 times bigger than a ship of the 1800s. Very large ships called **supertankers** transport oil across oceans in storage tanks. Their displacements are as high as 500,000 tons. Does this tell you why they are called *super*tankers?

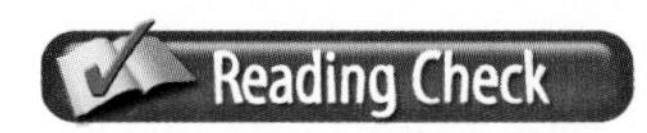

Recall What are the different kinds of water transportation vessels?

Intermodal Transportation

Can modes of transportation be linked to carry people or products?

Can you think of a time when you used more than one mode (form) of transportation in a single journey? When two or more modes of transportation are used together to move people or cargo more efficiently, it is called **intermodal transportation**.

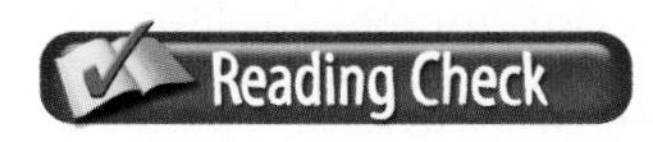

Predict What is intermodal transportation?

Moving People

To understand how intermodal transportation helps move people, imagine that you are going to take a cruise. To get to the city where you will board ship, you need to take an airplane.

EcoTech

Ecotourism

Tourism can result in gaudy hotels, create overpriced restaurants, and generate trash. It can displace and inconvenience local people. Ecotourism includes tourism practices that try to preserve the environment and improve the well-being of the locals.

Try This Next time you travel, consider the effect you have on the places you visit.

In order to reach the airport, you might take a bus to a subway station, and then travel to the airport via subway train. You might also drive your own car or take an airport shuttle.

Different systems have been organized to work together to help you. There is a bus stop near the subway station, a subway stop at the airport, and a taxi to take you from the airport in the coastal city to the cruise ship.

Moving Cargo

Cargo is moved most efficiently when it is packed into large containers. When a product travels overseas, the containers are loaded on ships called **containerships**. Loading and unloading is easy because the containers are usually the same size and the same shape.

Containers can then be loaded onto a train. The products do not have to be unpacked and then repacked. Some containers are designed as truck semi-trailers. They ride on trains to a terminal, where they are then attached to truck tractors. They continue their journey on the highway. Because intermodal transportation is so efficient, it saves time and money.

section 22.2 assessment

After You Read Self-Check

1. Identify the difference between a boat and a ship.
2. Name the three general types of water transportation vessels.
3. Explain why intermodal transportation is so efficient.

Think

4. Describe how you think engine-powered ships affected society when they were first introduced.

Practice Academic Skills

Social Studies

5. Look up information and write a short report on how a particular transportation system is a part of the technological, social, and environmental systems that make up our world.

STEM Mathematics

6. Xavier is going to take his boat across a lake that is four miles wide. A bandstand is directly across the lake. Three miles to the bandstand's right is a dock for boats with a playground. If Xavier decides to go on his boat to the playground, what distance will he have to travel across the lake?

Math Concept **Pythagorean Theorem** The Pythagorean theorem states that the sum of the squares of the lengths of the two shorter sides of a right triangle equals the square of the length of the third side, called the "hypotenuse." This formula allows you to determine the unknown length of a side if you know the other two lengths.

1. Draw a diagram of the possible paths he could take. Label the distances that are given.
2. Use the Pythagorean theorem to determine the distance to the playground.

For help, go to **glencoe.com** to this book's OLC and find the Math Handbook.

section 22.3

Air and Space Transportation

Reading Guide

Before You Read **Connect** What are the main types of air transportation vehicles?

Content Vocabulary
- lift
- commercial airplane
- jumbo jet
- helicopter
- lighter-than-air craft
- payload

Academic Vocabulary
- complex
- approach

Graphic Organizer
Draw the section diagram. Use it to organize and write down information as you read.

Air and Space Transportation

K	W	L

Go to **glencoe.com** to this book's OLC for a downloadable graphic organizer and more.

TECHNOLOGY STANDARDS

STL 3 Relationships & Connections

STL 4 Cultural, Social, Economic & Political Effects

STL 12 Use & Maintenance

ACADEMIC STANDARDS

Science

NSES Content Standard E Understandings about science and technology

NSES Content Standard D Earth in the solar system

STL *National Standards for Technological Literacy*

NCTM *National Council of Teachers of Mathematics*

NCTE *National Council of Teachers of English*

NSES *National Science Education Standards*

NCSS *National Council for the Social Studies*

Air Transportation

What is a jumbo jet?

Transportation by air takes place in airplanes, helicopters, and lighter-than-air craft. Hang gliders and sailplanes are used for recreation. Military airplanes are used for defense. Can you think of other vehicles for air transportation?

Airplanes are the most important part of our air transportation system. However, many other components are necessary for safe air travel. For example, we need airports, training programs, and radar. Many airplanes are in the air at the same time. This is why air travel is our most **complex** transportation system and is regulated by the government.

As You Read

Compare What are some of the uses for air and space transportation?

Welcome Aboard This Boeing 777 aircraft can carry up to 500 passengers from San Francisco to Tokyo at speeds up to .84 Mach. *What components are necessary for safe air travel?*

Airplanes

Did you ever wonder how something as heavy as an airplane could get off the ground? The secret is in the shape of the airplane's wing. As the airplane **approaches** the runway, it gathers speed and air rushes over the wing.

The shape of the wing causes the air to travel faster over its upper surface. See **Figure 22.3**. This reduces air pressure above the wing. It also helps increase the pressure on the wing's lower surface, pushing it upward and creating **lift**. Almost like magic, the airplane rises.

Commercial Airplanes

Many important airplanes were built and flown after the Wright brothers' first flight in 1903. One of the most important airplanes was the 1935 DC-3, which had two gasoline engines. It was the first **commercial airplane** as it made a profit by carrying just passengers.

The Boeing 707 came out in 1958 and carried 179 passengers, a huge number at the time. In following years the number of airline passengers increased so much that manufacturers decided to build **jumbo jets**, very large jet airplanes that can carry about 500 passengers. The first was the Boeing 747.

Because jumbo jets have such powerful engines, they can lift more weight than other airplanes. This allows them to carry a lot of fuel. As a result, they can stay up in the air for a long time. Jumbo jets can fly non-stop from Cincinnati to London and from Detroit to Tokyo.

Smaller Jets

Smaller jet aircraft, like the two-engine Canadair RJ-200, are used on shorter flights between smaller cities. They use less fuel and do not need a long runway to take off.

Figure 22.3 Lift

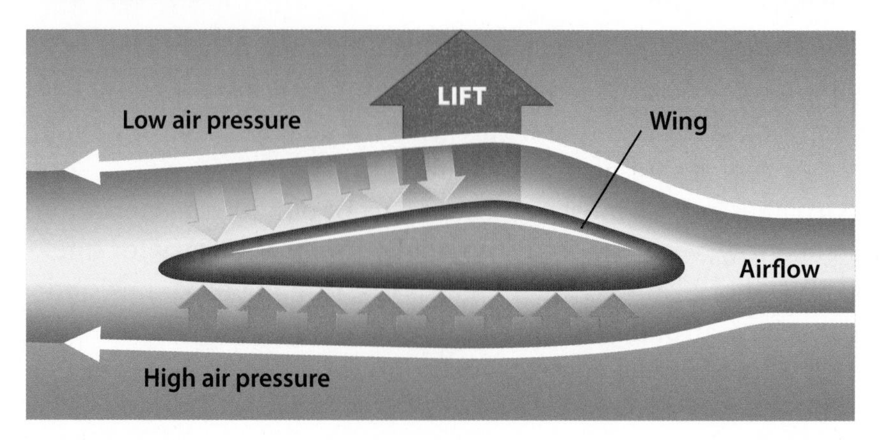

The Secret of Flight The lower air pressure above the wing helps create lift. *What happens to the air pressure if the plane starts to slow down?*

Cargo Planes

Some airplanes carry only cargo. However, even the biggest airplane can carry only a fraction of what a ship or train can haul. A Boeing 747 can carry 100 tons of cargo. Transport by air can be very expensive. Therefore only lightweight items, such as mail and electronics, are usually shipped by air.

Helicopters

A **helicopter** is an aircraft with one or two rotors that allow it to lift straight up. The twirling blades of the rotor create lift. Some helicopters also have small tail rotors to keep them from spinning out of control.

Helicopters can be as small as a one-person machine or as large as a cargo-carrying helicopter capable of lifting ten tons. Passenger-carrying helicopters connect some large cities with major airports. Some helicopters deliver parts to construction sites. Others are used to check on traffic or to transport people to hospitals.

Lighter-than-Air Craft

Lighter-than-air craft include aircraft such as dirigibles, blimps, and airships. Helium lifts them into the air, and gasoline engines turn propellers to move them forward. The engines are located in gondolas, or cars, that are suspended from the craft. The passenger compartment is located in a separate gondola.

A hot-air balloon is also a lighter-than-air craft. It uses large torches to heat the air inside a huge nylon bag. Hot air weighs less than cooler air, so the balloon rises. The pilot controls the balloon by ascending or descending into wind currents.

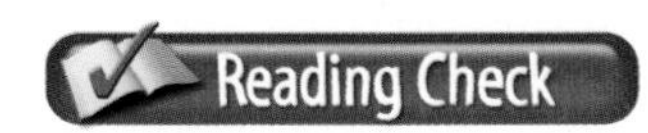

Explain How is an airplane able to lift up into the air?

Space Vehicles

What does NASA do?

The National Aeronautics and Space Administration (NASA) is responsible for regulating and directing the entire U.S. space program. This includes the space shuttles and vehicles for exploration.

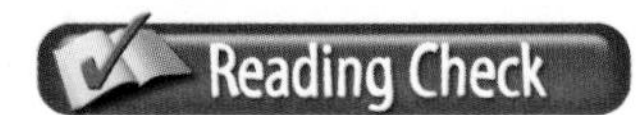

Identify What do the letters NASA stand for?

Work Space This space shuttle is in orbit 115 miles above the earth. *What usually happens during a typical space shuttle mission?*

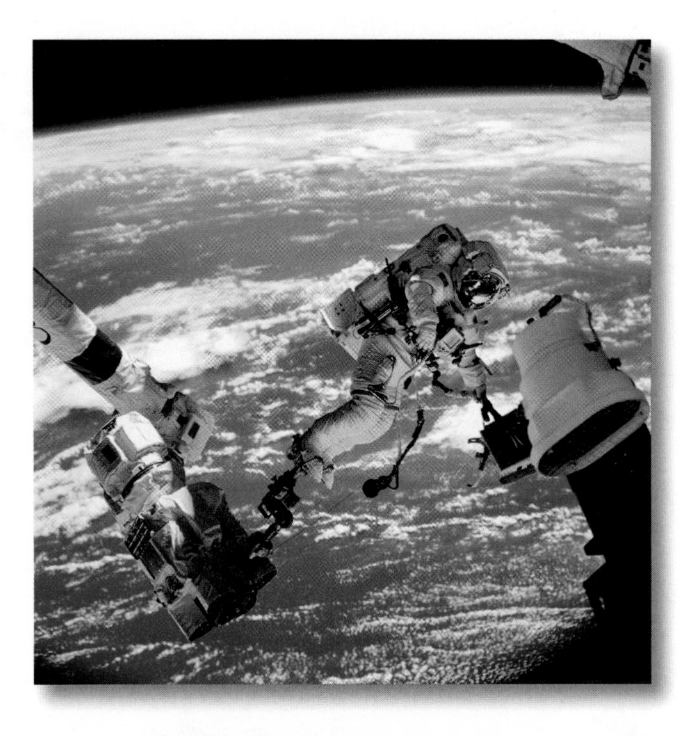

Back to the Moon

Imagine flying to the moon in a space shuttle. NASA is preparing to return to the moon with a project called *Constellation*. To get there, they are working on designing a new type of rocket that combines the best features of the space shuttle and the *Saturn V*, which first took astronauts to the moon. The rocket is named *Ares 1*. Engineers hope to have it flying by 2014. *What do you think are some of the criteria and constraints for designing the Ares I?*

Go to **glencoe.com** to this book's OLC for answers and to learn more about the *Ares 1*.

Space Shuttles

The first space shuttle that went into orbit was the *Columbia* in 1981. On a typical mission, a space shuttle with four astronauts orbits 115 miles above the earth. Each flight carries cargo, called a **payload**, in the large cargo bay. The payload can weigh up to 65,000 pounds.

Shuttle speed is about 17,000 miles per hour. Once in orbit, an astronaut opens the cargo bay doors, which may contain a communications satellite. The astronaut controls a 50-foot mechanical arm that grabs the satellite and releases it a safe distance from the shuttle. The arm can also grab satellites already in orbit.

Shuttle Missions

The space shuttle can also carry a complete scientific laboratory in the cargo bay. However, many laboratory experiments are now done on the International Space Station (ISS). A shuttle mission can also include carrying supplies to the ISS.

To return to earth, the astronauts fire small rockets to slow down the space shuttle. It re-enters the atmosphere and glides toward a landing strip. Space shuttle astronauts have a very dangerous job, and several have been lost during in-flight failures. For example, seven astronauts on the *Challenger* in 1986 and another seven on the *Columbia* in 2003 perished during their missions.

Spacecraft can also travel without astronauts. These ships consist of hollow containers mounted on booster rockets. The hollow containers usually carry satellites for communication or scientific research. Once a spacecraft is in orbit, its container opens, and the satellite is put into position. The satellite then begins orbiting on its own.

section 22.3 assessment

Self-Check

1. Name the type of jet that carries the most passengers.
2. List several jobs that helicopters perform.
3. Explain the letters *NASA*.

Think

4. Explain the value you think the space program has for the development of technology for use on Earth.

Practice Academic Skills

English Language Arts/Writing

5. Do research on new rockets designed to take people to the moon and the types of missions planned for them. Imagine you are chosen to be an astronaut on one of these missions. Write a short story describing your experience going to the moon.

Science

6. The space shuttle is the first reusable spacecraft that NASA has used. Some feel that the space shuttle program is dated and becoming obsolete. Research new space shuttle designs. Write a few paragraphs about the designs you find and explain their advantages over the current space shuttle design.

Exploring Careers in Technology

Ronnie Coursey

AIR-TRAFFIC CONTROLLER

Q: *What is a typical day like at your job?*
A: The typical day begins with information briefings on air-space conditions, weather, traffic flow control, and equipment status. Then I receive a position-specific briefing before assuming control of a sector. A controller ensures that all aircraft are separated and controlled in a safe, orderly, and expeditious manner. Air-traffic controllers are limited to two hours at a time on each position. We rotate control positions often and receive a new briefing each time.

Q: *What kind of training and education did you need to obtain your job?*
A: I received most of my basic air-traffic training in the U.S. Army. Also, the math and science courses I took in high school were essential.

Q: *What do you like most about your job?*
A: No two days are alike. Air traffic is very dynamic, because the number and types of aircraft, weather, turbulence, and destinations are always changing. This job never gets boring!

Q: *How did you get interested in your job?*
A: I joined the Army because I wanted to have the opportunity to train for a good career and receive educational benefits while serving my country. I looked for a career that was fun, challenging, and offered good pay in a civilian life, and I found it.

English Language Arts/Writing

On-the-Job Communication In order to succeed in most careers, it is important to have good communication skills.

1. Interview three adults. You can interview anyone who works full-time, including an air-traffic controller.
2. Ask him or her to explain how they use communication skills on the job, and to give you some examples.
3. Summarize your findings in a one-page report.

Go to **glencoe.com** to this book's OLC to learn more about this career.

Real-World Skills

Speaking, listening, organization, problem-solving

Academics and Education

Science, mathematics, English language arts, social studies

Career Outlook

Growth as fast as average for the next ten years

Source: *Occupational Outlook Handbook*

chapter 22 Review and Assessment

Chapter Summary

Section 22.1 About half of all the world's automobiles are used in the United States. The car's transmission contains gears for transferring power. Buses are an important part of mass transportation. Some larger trucks use deflectors to reduce air resistance. A train that floats over a guideway is called a "maglev."

Section 22.2 There are about 24,000 large deep-water vessels in the world. Displacement is the measure of how much water a vessel moves aside as it floats. Intermodal transportation combines several different kinds. Cargo is often placed in containers for efficiency.

Section 22.3 Jet airplanes servicing smaller airports do not need a long runway and use less fuel. Helicopters can carry cargo, and some can easily lift ten tons. The force that allows a plane to rise in the air is called "lift." During missions, space shuttles orbit about 115 miles above the earth.

Review Content Vocabulary and Academic Vocabulary

1. On a sheet of paper, use each of these terms and words in a written sentence.

Content Vocabulary
- mass transportation
- transmission
- driving wheel
- front-wheel drive
- rear-wheel drive
- four-wheel drive
- tractor-trailer
- bullet train
- maglev train
- navigable waterway
- displacement
- supertanker
- intermodal transportation
- containership
- lift
- commercial airplane
- jumbo jet
- helicopter
- lighter-than-air craft
- payload

Academic Vocabulary
- role
- interact
- overseas
- via
- complex
- approach

Review Key Concepts

2. **Describe** the different types of land transportation.
3. **Explain** the purpose of transportation subsystems.
4. **Explain** the purpose of a transmission.
5. **Describe** how oceans and inland waterways are used for transportation.
6. **Identify** the different types of ships.
7. **Explain** the concept of intermodal transportation.
8. **Discuss** ways in which air and space transportation are used.
9. **Identify** types of aircraft and spacecraft.
10. **Explain** how a plane is lifted into the air.

Real-World Skills

11. Driving Wheels A car's driving wheels can be in the front or in the rear. Other cars have all four wheels as driving wheels. Research what configuration your family's car has. Write a few paragraphs describing the car, and the advantages and disadvantages of the different types of driving wheels.

Technology Skill

12. Lighter than Air Aircraft that fly because they are lighter than air include blimps, zeppelins, dirigibles, and more.

- **a.** Obtain a helium-filled balloon and see how much weight it will lift by attaching weights to its string.
- **b.** Based on what you determine in step one, figure out how many balloons it would take to lift you. If you are not sure of your weight, weigh yourself.

Academic Skills

Social Studies

13. When deciding how to transport goods, people base their decision on price, safety, and speed. Research different shipping methods. Write a few paragraphs comparing the different ways goods can be shipped. Include a discussion about why some modes are preferred.

STEM Mathematics

14. The Mach number is the speed of something, relative to the speed of sound. The speed of sound is called Mach 1. If something is traveling at Mach 1.2, it is going 1.2 *times* the speed of sound. If the speed of sound is 735 mph, how fast is a plane traveling if its speed is Mach 1.8?

Math Concept **Operations** If a problem uses the word *times*, it is indicating that multiplication should be used. To determine the speed of sound based on a Mach number, simply multiply the given speed of sound by the Mach number.

WINNING EVENTS

TECHNOLOGY STUDENT ASSOCIATION

Aerospace Engineer

Situation You and your team will design and build a glider. Your teacher will determine the glider's minimum and maximum dimensions. You will launch the glider by an elastic band.

Activity Design the glider model within the given constraints and criteria. Make a series of sketches. Then make a working model of your glider and conduct flight tests. Make changes as needed. Finally, conduct a series of five final flight tests, recording the times aloft.

Evaluation Time aloft—total number of seconds of five flights

Go to **glencoe.com** to this book's OLC for information about TSA events.

Standardized Test Practice

Directions Choose the letter of the best answer. Write the letter on a separate piece of paper.

1. What is the cost of a $56.50 computer program if it is discounted 20%?
   - **A** $22.80
   - **C** $45.20
   - **B** $56.30
   - **D** $11.30
2. The lift of a wing is a result of the high pressure generated on top of the wing.
   - **T**
   - **F**

Test-Taking Tip Learn ahead of time the kind of test you will be taking, when and where the test will be, and what materials to bring.

Build a Rubber-Band-Powered Vehicle

Modern land vehicles have many basic parts that are the same as those used hundreds of years ago. Most have wheels, a body, and a way to provide power. You can learn more about land vehicles by making your own.

Tools and Materials

- ✓ One wooden block, 1 inch thick, 2 inches wide, and 6 inches long
- ✓ Two wooden ready-made wheels, about 2¼ inches in diameter
- ✓ Two wooden ready-made wheels about 1 inch in diameter
- ✓ One dowel rod, ⅜ inch in diameter and about 3 or 4 inches long
- ✓ Three heavy rubber bands, ¼ inch wide and about 2 inches in diameter
- ✓ Staples
- ✓ About 18 inches of fishing line
- ✓ Punch
- ✓ Wire cutters
- ✓ Hand drill with ⅜-inch drill bit
- ✓ Hammer
- ✓ Tape measure, 25-feet long
- ✓ Saw

Set Your Goal

For this activity, you will make a simple land vehicle powered by a rubber band. You will then experiment with the design to see if you can make the vehicle go faster or farther.

Know the Criteria and Constraints

In this lab, you will:

1. Follow the basic design in the picture for your first vehicle.
2. Keep a record of all speed and distance measurements.
3. Submit sketches of your design improvements.

Design Your Project

Follow these steps to complete this lab.

1. Look at the drawing of the completed vehicle on page 489.
2. Build the vehicle.
   - Drill a ⅜-inch hole at one end of the body, as depicted in picture 1.
   - Use the saw to cut notches at the front and rear.
   - Put the dowel rod through the hole.
   - Cut it to length and attach the two 2¼-inch wheels.

Body Detail

1.

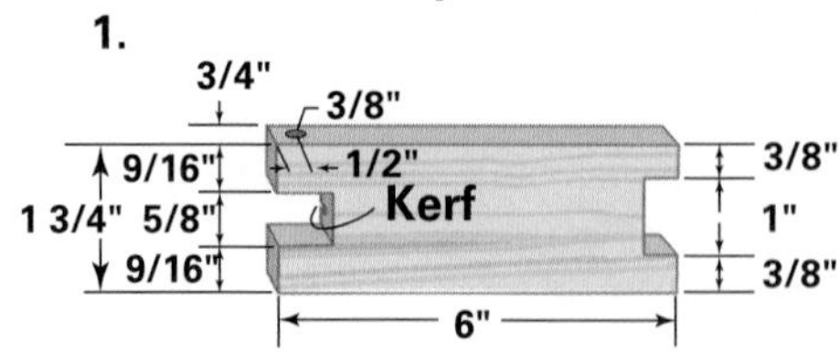

2. Assembly of front roller

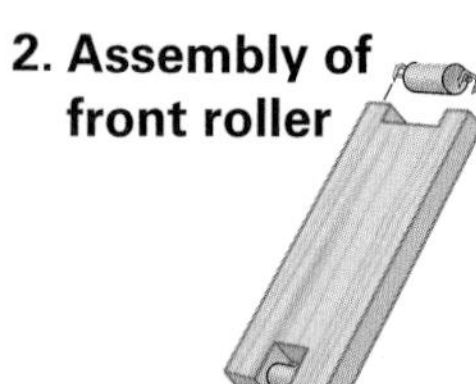

3. Attaching rubber band

4. Marking center of axle

5. Winding wheels of the vehicle in preparation for operation

- Place a rubber band around the outside of each wheel. These wheels will be your car's drive wheels, providing its driving power.
- Mark the center of the dowel rod with a punch and hammer in a small nail. Leave about ¼ inch sticking out.
- Attach the front wheels with small nails. Make sure all four wheels can spin freely.
- Make the front roller from a ⅛-inch-long section of the dowel rod.
- Hammer small nails into each end and cut off the nail heads.
- Attach the roller to the body with staples, as shown in picture 2. Do not drive the staples too deep. The roller must turn easily.
- Attach a rubber band to the bottom of the body by hammering in a staple, as depicted in picture 3.
- Tie one end of the fishing line to the rubber band. Pass the line over the front roller and under the body of the vehicle.
- Tie the other end of the fishing line to the small nail on the dowel rod connecting the large wheels.
- To provide power for your car, wind the larger wheels backward, as in picture 5. Continue until all the fishing line is wound onto the axle and the rubber band is stretched over the front roller.

Completed Vehicle

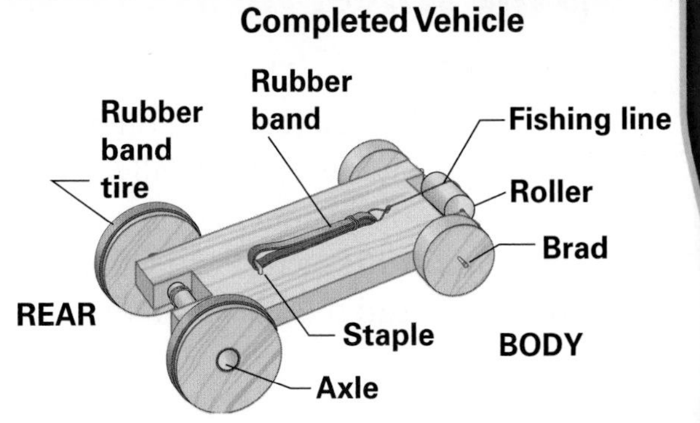

3. Place your vehicle on a smooth floor and let go. It will quickly accelerate and then coast to a stop.
   - Use the 25-foot tape measure to measure how far it traveled.
   - Use a stopwatch to measure its speed.
   - Repeat a few times to get averages of speed and distance.
4. Sketch several improvements for the vehicle and try them out.

SAFETY

Reminder

In this activity, you will be using tools and materials. Always follow appropriate safety procedures and rules so you and your classmates do not get hurt.

Evaluate Your Results

After you complete the lab, answer these questions on a separate piece of paper.

1. Did your vehicle travel as fast as you expected it would?
2. How did you change your vehicle to make it go faster or farther?
3. What would happen if you used bigger or smaller drive wheels?

Academic Skills Required to Complete Lab

Tasks	English Language Arts	Math	Science	Social Studies
Build car.	✓	✓	✓	
Test car; measure time distance.		✓	✓	
Retest car several times to obtain average distance and time.		✓	✓	
Make changes to car and retest.		✓	✓	

Technology Time Machine

As the Wheel Turns

Play the Game This time machine will travel to the past to show you how the invention of the wheel dramatically changed transportation. To operate the time machine, you must know the secret code word. To discover the code, read the clues, and then answer the questions.

Clue 1

3500 B.C.E. The earliest evidence of the wheel comes from the area between the Tigris and Euphrates Rivers in what is now known as Iraq. There, the ancient Sumerians formed wheels from three planks of wood that were bound with leather ties held in place by some copper nails. A hole in the middle held an axle.

Clue 2

2000 B.C.E. Use of the wheel for organized warfare created a demand for something that weighed less and moved faster. Spokes eliminated most of the heavy wood in the center of the wheel. Sculptures from Chinese tombs from the 13th century B.C.E. show spokes on chariot wheels.

Clue 3

1800s Heavy railroad cars required sturdy metal wheels that ran along a track. Locomotives in the United States used large driving wheels to help increase speeds to about 30 mph. Some of these wheels were more than six feet in diameter.

Clue 4

1845 The wheels on horse-drawn vehicles were covered with a solid, rubber tire. However, solid rubber was hard on roadways and did not produce a comfortable ride. In England, Robert W. Thompson patented the first pneumatic (air-filled) tire for use on carriages.

Clue 5

1888 John Dunlop of Ireland patented a pneumatic tire for bicycles and formed a company to manufacture the tires.

Clue 6

1900 At the turn of the century, the use of pneumatic tires spread to automobiles. Separate inner tubes held the air, and the outer covering was made from rubber-coated cotton cloth.

Clue 7

1954 The tubeless tire was introduced. These tires permitted higher speeds by reducing surface friction.

Clue 8

2001 Dean Kamen designed the battery-powered Segway Human Transporter. Its two wheels are made of plastic and are surrounded by air-filled tires. Each wheel is powered by its own motor and controlled by a computer. Kamen hopes it will eventually replace automobiles in crowded cities.

Crack the Code

On a piece of paper, write the answers to these questions:

1. What is another term for an air-filled tire?
2. Which part of the tire was made from rubber-coated cotton cloth?
3. Name the material from which the earliest wheel was made.
4. In which country was the first pneumatic tire patented?
5. Wheels on horse-drawn vehicles were covered with this material.

Now write down the first letter of each answer. Put them together to discover the secret code word!

Engines and motors produce this to provide transportation.

unit 7 Thematic Project

Advancing Transportation

Getting from Place to Place In Unit 7, you learned about the types of engines that make vehicles go. You also looked at transportation systems people use to move in neighborhoods, travel on water, fly through the air, and even rocket into space! The way people get from place to place affects their lives. Think about your own community. If you live in a city, do you have an easy way to get to the country? If you live in the country, do you have an easy way to get to the city? Do you live near an airport? How difficult is it to visit relatives who live in another city?

Design for the Future Think about how to improve transportation systems in your community. For example, if you live near a river that is upstream from a city, a system of hovercraft ports every few miles could reduce traffic.

This Project In this project, you will research and design a transportation system to meet the future needs of your community.

Your Project

- Choose a mode of transportation.
- Complete the following tasks:
  1. Design a new system of transportation for your community. Include time schedules, seating capacities, and safety procedures.
  2. Design it to use a sustainable fuel.
  3. Build a model of the transportation system.
- Write a report.
- Create a presentation with posters, video, or presentation software.
- Present your findings to the class.

Tools and Materials

- ✓ Computer
- ✓ Internet access
- ✓ Trade magazines
- ✓ Word-processing software
- ✓ Presentation software
- ✓ Posterboard
- ✓ Colored markers

The Academic Skills You'll Use

- Communicate effectively.
- Speak clearly and concisely.
- Use correct spelling and grammar when taking notes or writing presentations.
- Think about transportation technology design.

English Language Arts

NCTE 4 Use written language to communicate effectively.

NCTE 12 Use language to accomplish individual purposes.

Social Studies

NCSS 8 Science, Technology, and Society

Step 1 Choose Your Topic

You can choose any mode of transportation that interests you. Examples might include:

- Submarines
- Helicopters
- Monorails
- Escalators
- Rocketships
- Hydrofoils
- Ferries
- Aerial Trams

Choose something you have used before!

Step 2 Do Your Research

Research your project. Your fact finding may include a combination of any of these ideas:

- Talk to or write to someone in the transportation industry.
- Study plans for similar systems.
- What do old and new articles in libraries and Web sites say about your topic?
- Take a ride on the transportation mode of your choice and write your observations.

Use the library, the Internet, and magazines.

Step 3 Explore Your Community

- Visit a power plant that generates power from hydroelectric, solar, or wind power.
- Ask local people what kind of innovations they would like to see in local transportation.

Thank everyone who gives you information!

Step 4 Create Your Project

Your project should include:

- 1 research project (design, blueprint, or model)
- 1 report
- 1 presentation

Project Checklist

	Objectives for Your Project
Visual	✓ Make a poster or slide presentation to illustrate your project.
Presentation	✓ Make a presentation to your class and discuss what you have learned.
	✓ Turn in research and names from your interview to your teacher.

Step 5 Evaluate Your Presentation

In your report and presentation, did you remember to:

- Demonstrate your research and preparation?
- Present an imaginative solution to a problem?
- Label your designs or test your model?
- Use facts and evidence to back your ideas?
- Engage your audience?
- Speak clearly?

Rubrics Go to **glencoe.com** to the book's OLC for a printable evaluation form and your academic assessment form.

GLOBAL TECHNOLOGY

High-Speed Rail

France has been a world leader in high-speed public transportation since the 1950s. Later the first TGV (*Train à Grande Vitesse*) was built in 1981. Passengers could ride from Paris to Lyon at 186 miles per hour! Today the railway system is so expansive and effective that some airline services have been discontinued. Double-decker cars have been added to the train to handle the many passengers.

Critical Thinking *Do you think the TGV could succeed in most U.S. regions? Why or why not?*

French	
hello	bonjour
goodbye	au revoir
How are you?	Comment allez-vous?
thank you	merci
You're welcome	de rien

Go to **glencoe.com** to the book's OLC to learn more and to find resources from **The Discovery Channel.**

STEM Handbook

Science, **T**echnology, **E**ngineering, and **M**athematics offer broad career pathways into the high-growth industries competing in the global marketplace. The jobs in these areas are offered by businesses on the frontline of technological discovery. With a solid background in STEM, you'll be following a career path that will offer the opportunity to participate in the development of high-demand and emerging technologies.

With advances in science driving innovations in technology, materials and procedures researched in the lab are now the backbone of many new processes. Science is used in the design of cell phone circuits, the construction of surfboards, and the processing of food products. While science begins with careful observation, research, and experimentation, it involves more than lab work. Some opportunities will take you into the field. You might also be involved in quality control or technical writing. The range of science career opportunities is extremely large.

Technology joins science skills with those in engineering and math. It is essential in managing communication and information systems. In manufacturing, technology is used to focus human resources and materials in the production of products ranging from electronic devices to footwear. If you want a career that combines science, math, and engineering, consider a career in technology. In that field, you'll work closely with individuals who have a hands-on knowledge of these subjects. As part of their team, you'll help make the decisions needed to bring a product from design through manufacturing to market.

A career in engineering will involve you in the technological innovations that flow from discoveries in science and math. You'll be incorporating these new discoveries in products that expand our technological reach in many areas, including communications. If you think that engineering is unexciting, name a product or service of interest to you that was introduced within the last year. Then identify the main benefit of that product or service. That benefit probably resulted from an engineering effort that built on developments in science, technology, and math. A career in engineering could place you in a position to participate in the development of such products and services.

Mathematics input is essential to create and develop products—whether the product is real, such as a DVD, or virtual, such as an online video game. If you want to be a valued member of the teams providing the math input for the development of products, you will need solid math skills. You'll need these skills for high-level problem solving and innovation. As you advance along the career pathway, you'll be challenged to apply mathematics to real-world problems. There is a wide variety of math careers from which to choose.

SCIENCE

Machines

Although machines make work easier, they do not actually reduce the amount of work that has to be done. How, then, do machines help us? Recall the formula for work:

$$\text{Work} = \text{force} \times \text{distance}$$

In this equation, if you increase the force, you can decrease the distance and still do the same amount of work. If you increase the distance, you can decrease the force. Machines help us by increasing distance, increasing force, or changing the direction of force. Complicated machines are combinations of a few basic machines. These basic machines are shown and described in the illustrations below.

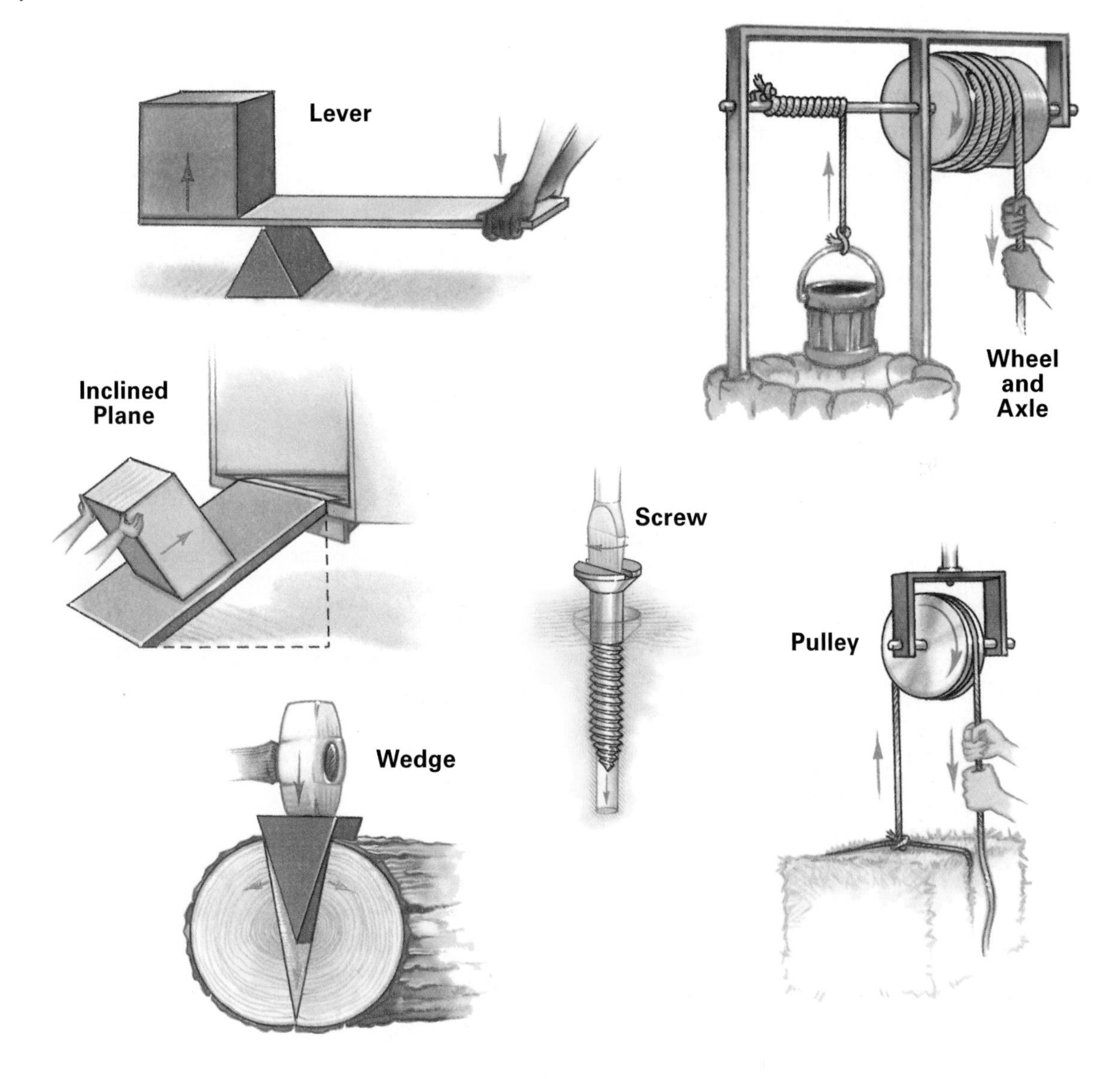

Mechanical Advantage

Some machines are more helpful than others. The measure by which a machine increases force is known as its mechanical advantage (MA). This can be calculated by the formula:

$$MA = \frac{F_r}{F_e}$$

In this equation, F_r stands for resistance force (load). This is the force applied by the machine. For example, a screwdriver exerts a force on a screw. F_e stands for effort force, the force applied to the machine. An example is the force you apply to the screwdriver. If a small-effort force results in a large-resistance force, the machine has a high mechanical advantage. The greater the mechanical advantage, the more helpful is the machine.

For machines that increase distance, we can calculate velocity ratio. The velocity ratio is the distance the load moves divided by the distance the effort moves.

Efficiency

Efficiency compares the work a machine can do with the effort put into the machine. The energy output of a machine is always less than the energy input. That's because some of the energy input is used to overcome friction between moving parts. Efficiency is usually stated as a percentage. Machines with a higher percentage are more efficient.

$$\text{Efficiency} = \frac{\text{work output}}{\text{work input} \times 100\%}$$

What if you applied all of a machine's energy output to the machine itself? Is it possible to build a perpetual motion machine—a machine that will run forever? Unfortunately, no. All machines do need a continuous outside supply of energy. If no energy is brought into the system, the machine will eventually stop. Also, all machines use more energy than they output. No machine is 100 percent efficient.

Power

Power is work done within a certain period of time. To calculate power, determine the amount of work done. Then divide the amount of work by the time it took to do the work.

$$\text{Power} = \frac{\text{work}}{\text{time}}$$

Gears and Cams

Gears are intermeshing toothed wheels and bars that transmit force and motion. They can alter the force's size and the motion's speed and direction. A cam is an offset wheel. Cams are connected to rods. A second rod, or follower, rests on the top of the cam. As the cam turns, the top rod moves up-and-down.

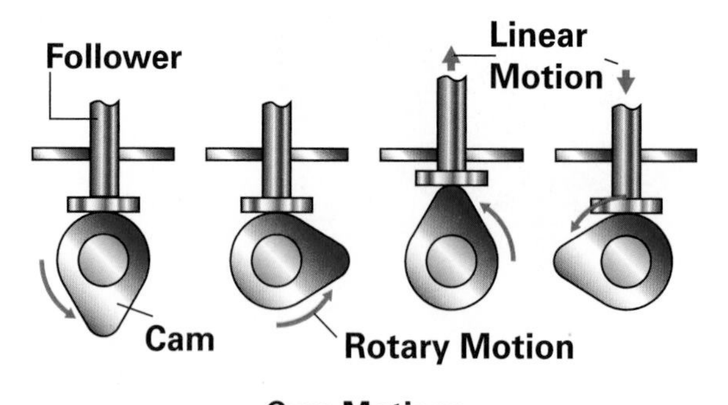

Cam Motions

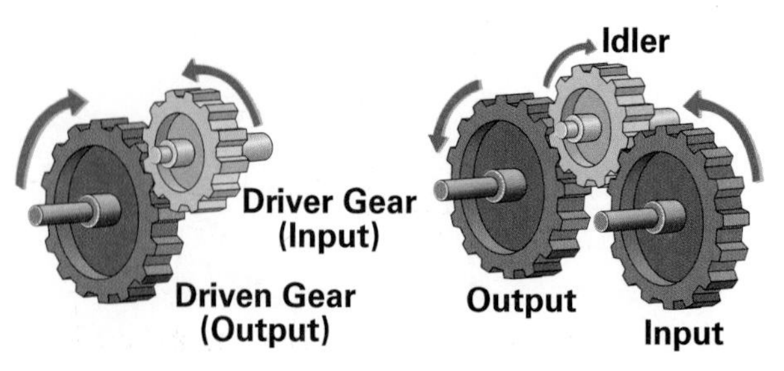

Gears

Forces and Motion

A force is a push or a pull. The force of gravity keeps your feet on the ground. Motion takes place when a force causes an object to move. Forces can act from a distance, without objects touching one another. For example, Earth's gravity affects the moon, even though the moon is nearly 250,000 miles away. The moon's gravity also affects the earth, causing the tides.

The "weightlessness" of astronauts as they orbit the earth in the space shuttle is not the result of their distance from the earth. The astronauts float because they are falling toward the earth at the same rate as the shuttle. However, the shuttle is also moving forward at a fast speed. This forward motion, combined with the downward pull of gravity, keeps the shuttle in a circular path around the earth.

Gravity

Gravity is a force that pulls objects toward each other. Gravity increases as mass (amount of matter) increases. Gravity decreases as distance increases.

Weight is the amount of force exerted on matter by gravity. While mass remains constant, weight varies, depending on the amount of gravitational attraction. For example, gravity on the moon is only one-sixth of that on Earth.

Terminal Velocity

Gravity accelerates a falling object at the rate of 32.2 feet per second, per second. But an object won't just keep falling faster and faster because of air resistance. As the object falls faster, air resistance increases. Eventually, the air resistance equals the force of gravity. The object then falls at a steady speed. This speed is the object's terminal velocity.

Pressure

Pressure is the weight (force) acting on a unit of area. When your textbook is flat on your desk, it exerts pressure on the desktop. The book covers an area of approximately 80 square inches. It weighs about 2.75 pounds. To calculate how much pressure your book exerts on the desktop, use this formula:

$$\text{Pressure} = \frac{\text{force}}{\text{area on which the force acts}}$$

The book exerts 0.034 pounds of pressure per square inch.

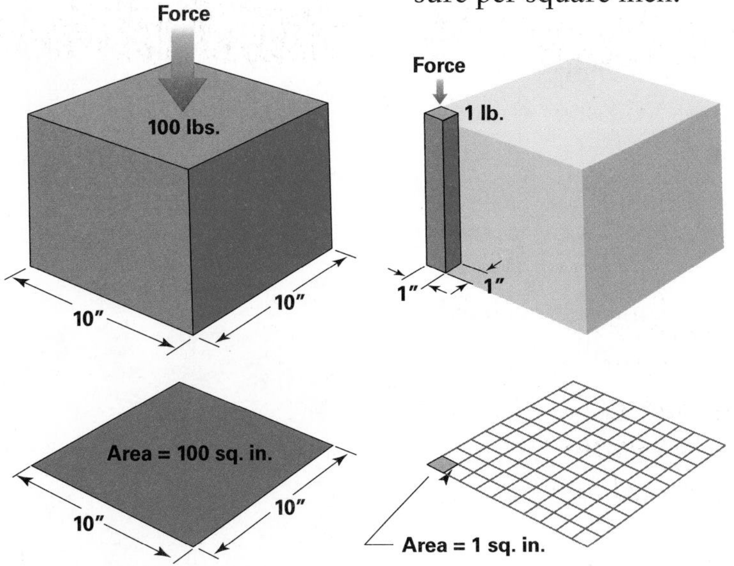

Newton's Laws of Motion

In the 17th century, the English scientist Isaac Newton described three laws of motion.

Newton's First Law Sometimes called the "law of inertia," Newton's first law states that an object at rest will stay at rest unless a force acts on it. If an object is in motion, it will stay in motion unless a force acts on it. For example, have you ever noticed that when riding in a car that stops quickly, your body will continue to move forward and tighten up against the seatbelt? This is because your body is trying to stay in motion, but the force of the seatbelt is stopping you. The car, too, is trying to stay in motion, but the force on the brakes is slowing it down.

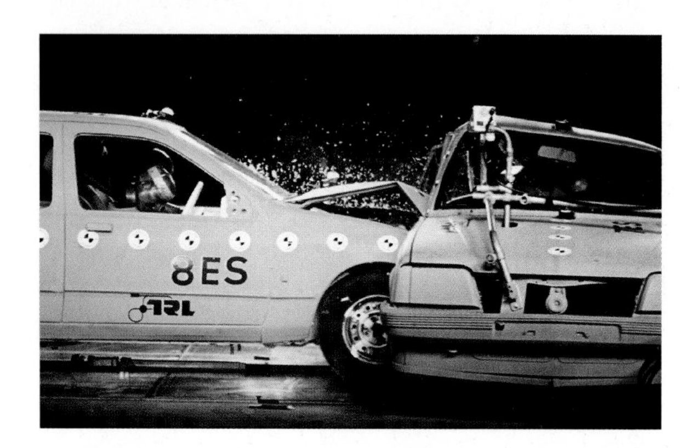

Newton's Second Law When a force acts on an object, the result is a change in speed or direction. The object will start to move, speed up, slow down, or change direction. The greater the force, the greater the change. For example, if you and a friend push on a stalled car, the car will move. Once the car begins to move, the harder you push, the faster the car will go.

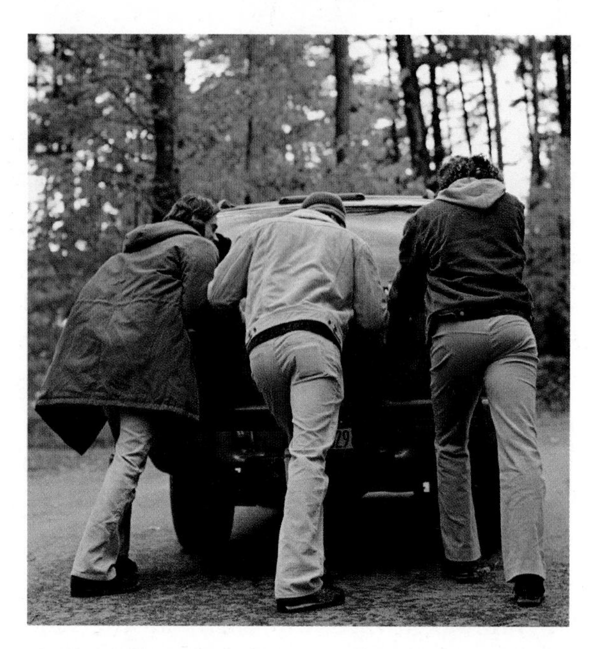

Newton's Third Law For every action, there is an equal and opposite reaction. For example, when you bounce on a trampoline, you exert a downward force on the trampoline. The trampoline then exerts an equal force upward, sending you up into the air.

Speed, Velocity, and Acceleration

Speed, velocity, and acceleration are related, but each is distinct. Speed is a measure of how fast an object is moving.

Velocity is the speed of an object in a certain direction. The velocity of a car on the highway might be 65 miles per hour. If the car moves onto an exit ramp, there is a change in velocity.

A change in velocity is called acceleration. The greater the mass of an object, the greater the force needed to accelerate it. If the change is a decrease in speed, it is negative acceleration, or deceleration.

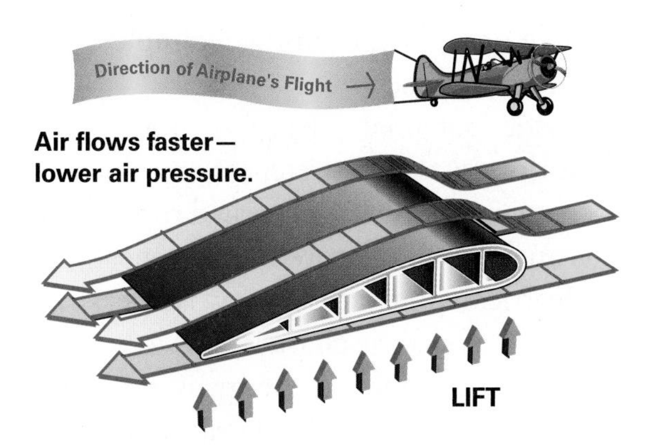

Forces and Fluids

Hydraulics is a branch of science that deals with the motion of liquids. Aerodynamics deals with the motion of gases. Liquids and gases are fluids—substances that can flow. The properties of fluids make them well suited to use for certain technological applications.

- When a force is applied to a confined fluid, the pressure is transmitted throughout the fluid (Pascal's principle).
- Pressure increases with depth.
- A fast-moving fluid has a lower pressure than a slow-moving fluid (Bernoulli's principle). That's why airplanes can fly. The air moves faster above the wings of an airplane than below them. The difference in air pressure enables the airplane to stay aloft.

Force, Mass, and Acceleration

Force (a push or pull) causes a mass to accelerate (change speed and/or direction). The formula for finding force is:

Force = mass × acceleration

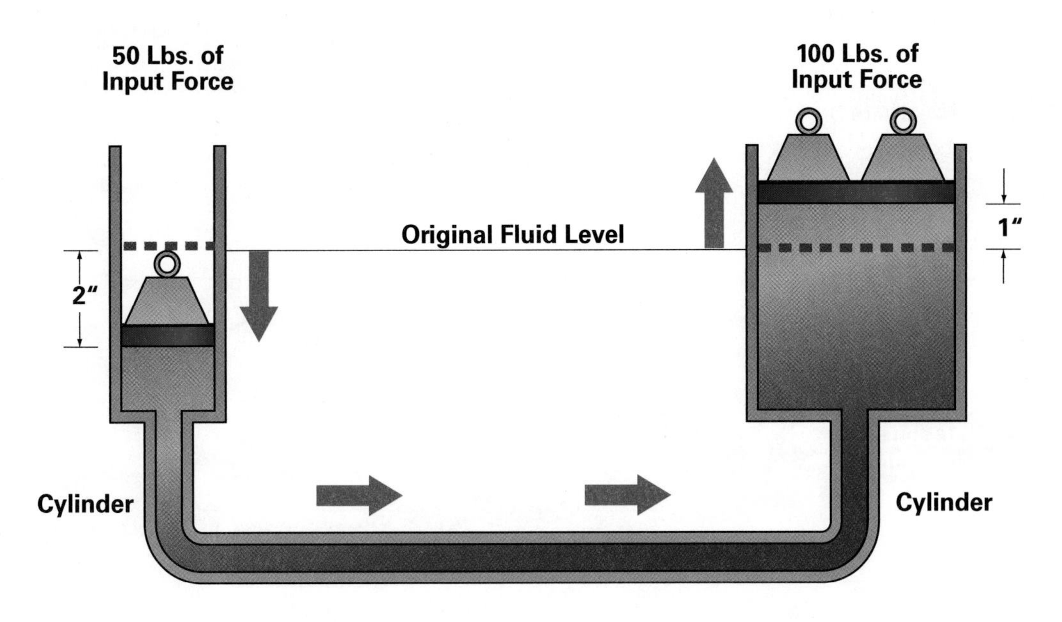

Energy and Work

Potential and Kinetic Energy

Energy is the ability to do work. Although it can take many forms, there really are only two kinds of energy: kinetic and potential. Kinetic energy is energy in motion. Potential energy is energy at rest.

Consider what happens when you use a hammer to drive a nail. As you lift the hammer, you do work. You exert force that moves an object over a distance. Some of your energy is transferred to the hammer. That energy is stored in the hammer as potential energy. When you move the hammer towards the nail, the potential energy becomes kinetic energy. When the hammer head hits the nail, energy is transferred from the hammer to the nail. The hammer stops, but the nail moves.

Types of Potential Energy

Potential energy is the energy an object has because of its position or condition.

- Gravitational potential energy is stored when work lifts an object against the force of gravity, as in the example of lifting the hammer.
- Elastic potential energy is stored when work twists or stretches an object or changes its shape. A stretched rubber band is an example of stored elastic energy.
- Electrical potential energy is the stored energy of electric charges.
- Magnetic potential energy is the stored energy of a piece of iron near a magnet.

Measuring Kinetic Energy

The kinetic energy (ek) of a moving object is found by using the formula:

$$E_k = \frac{1}{2}mv^2$$

In this formula, m is the mass of the object, in kilograms, and v is the velocity in meters per second.

This formula shows that kinetic energy increases with the mass and speed of a moving object. An object with twice the mass has twice the kinetic energy. However, an object with twice the speed has four times the kinetic energy. That's why automobile accidents at high speeds cause more damage. There is more kinetic energy.

Conservation of Energy

Energy does not disappear. The total amount of energy stays the same. This principle is known as the conservation of energy. It states that energy cannot be destroyed. However, energy can change form. It can move from object to object.

Forms and Uses of Energy

Technology converts energy from one form to another to accomplish work. Here are a few examples.

- Chemical energy stored in a flashlight battery is converted into light and heat energy.
- Mechanical energy (movement of objects) is converted to electrical energy by generators.
- Heat energy from the earth is converted to electrical energy in geothermal power plants.
- Light energy is converted to electrical energy in a solar cell.
- Sound (a type of mechanical energy) is converted to electrical energy in a telephone transmitter. At the other end of the line, it is converted back to sound.
- Electrical energy is converted to sound and light in a television set.
- Nuclear energy is released when the nuclei of atoms are split. This splitting, or fission, yields large amounts of heat energy that nuclear power plants convert to electricity. Nuclear energy is the strongest known force. If you could release the nuclear energy in 1 kilogram [2.2 pounds] of coal, you'd have the energy obtained from burning 3 million kilograms [6.6 million pounds] of coal.

Work

When a force acts on an object and moves that object in the same direction as the force, then work has been done. Work involves a transfer of energy. For example, energy from gasoline or diesel fuel is transferred through a car's engine to the wheels. The car moves.

It's important to remember that work involves both force and motion. If applying a force does not result in motion, then no work has been done. Also, the motion must be in the direction of the force.

Measuring Work

The amount of work done is found by multiplying force times distance.

$$W = f \times d$$

In the metric system, force can be expressed in newtons and distance in meters. The unit for work is the newton-meter, or joule. One joule is equal to the work done when a force of one newton moves something a distance of one meter in the direction of the force.

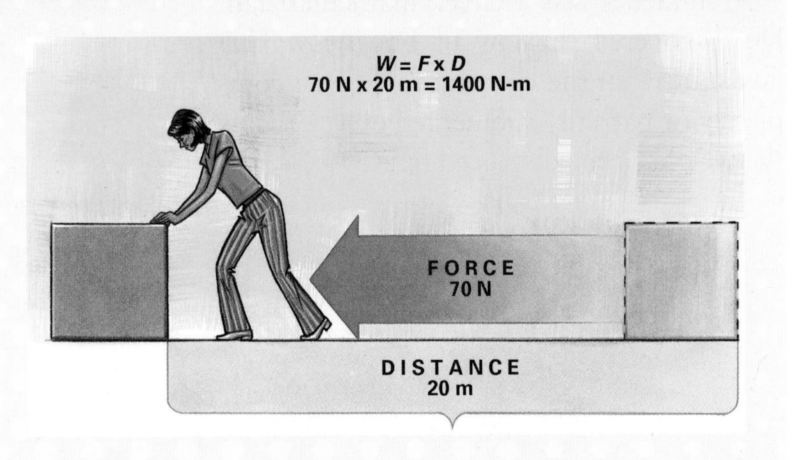

Matter

Our world consists of energy and matter. Energy is the ability to do work. Matter is the "stuff" that things are made of.

An atom is the smallest unit of matter, so small that trillions of atoms fit on a pencil point. An atom has a positively charged core called the nucleus and one or more negatively charged electrons. These orbit the nucleus. Electrons respond to nearby atoms to form bonds.

For example, atoms may donate electrons to neighboring atoms. The hardness of quartz comes from this kind of bond. Plastics share electrons in covalent bonds. Metals have their own form of electron sharing, known as the metallic bond. All these bonds involve the attraction of electrons to neighboring atoms. They are chemical bonds.

Mass and Density

Mass is the measure of the amount of matter in an object. Density is a measure of mass per unit of volume. Put simply, if mass is a measure of how much "stuff" there is in an object, density is a measure of how tightly that "stuff" is packed.

Archimedes' Principle

Archimedes was a Greek mathematician. He discovered the law of buoyancy. This law states that the buoyant force on a completely or partially submerged object equals the weight of the displaced fluid.

Elements

An element is a substance containing only one kind of atom. One kind of atom differs from another according to its number of protons. An atom of hydrogen (H) has one proton. Helium (He) has two protons.

Compounds

Elements can exist in pure form, such as pure oxygen. More often, elements combine to form a substance that is quite different from its ingredients. For example, hydrogen and oxygen, both gases, combine to form water, a liquid. A substance that combines two or more elements in specific proportions is called a compound.

Elements can combine in many ways—forming different kinds of chemical bonds and different arrangements of atoms. The specific proportions determine how two or more elements will bond and how the resulting compound will behave. Two atoms of hydrogen combine with one atom of oxygen to form water (H_2O). Two hydrogen atoms combine with two oxygen atoms to form hydrogen peroxide (H_2O_2). This is a bleaching agent that has been used as rocket fuel.

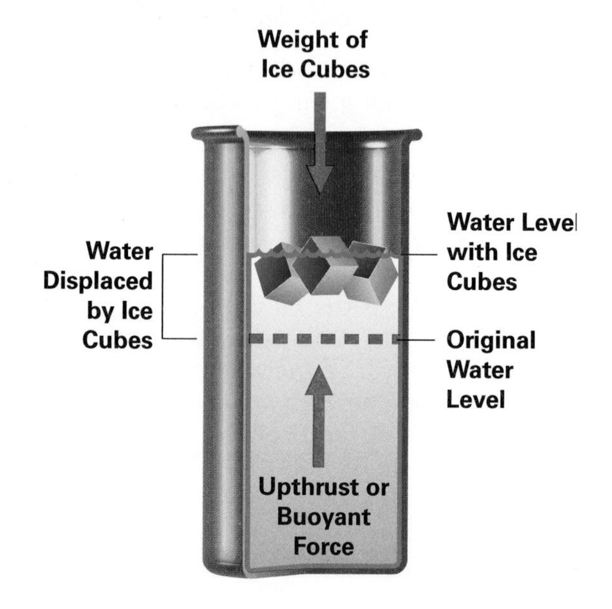

Mixtures

A mixture is a combination of elements or compounds that does not involve a chemical bond. In other words, the ingredients hold onto their atomic identity. As a result, the ingredients of a mixture can be separated and recovered by ordinary mechanical means. These include filtering, settling, and distilling.

The ingredients in a mixture generally keep their characteristics. Sugar that is dissolved in water, for example, is both wet and sweet even though the sugar disappears from sight.

An alloy is a mixture made by mixing two or more pure metals or by mixing metals with nonmetals. The mixture is stronger than pure metal. For example, mixing small amounts of carbon with iron makes steel. Differences between compounds and mixtures include the following:

- Mixtures can be separated mechanically. Separating a compound requires breaking chemical bonds.
- In a mixture, ingredients keep their original properties.
- In a mixture, the proportion of ingredients does not have to be exact. Brass, for example, may have varying amounts of copper and zinc.

Properties of Materials

Designers, engineers, and builders must know the properties of materials in order to select the right ones for the job. Materials have many properties. The following explanations and drawings describe common properties.

- Sensory properties are those we can see, hear, smell, taste, or feel. Colors may be used to please the eye or to grab attention. Materials for blankets and infants' clothes are chosen for their softness. Scents are added to cleaning products to emit fresh, clean odors.
- Optical properties determine how a material reacts to light. Window glass is transparent—light passes through it and you can see through it clearly. Frosted glass is translucent—light passes through it, but you cannot see through it. Opaque materials do not allow light to pass through. Some opaque materials, such as aluminum foil, also reflect light.
- Thermal properties determine how a material reacts to heat. Most metals are good heat conductors. This is why pots and pans are frequently made of metal. Fiberglass and Styrofoam™ resist the flow of heat. Therefore, these materials are commonly used to insulate structures and products.
- Electrical properties determine whether a material is a conductor, an insulator, or something in between (a semiconductor). Copper is a good electrical conductor, so it is frequently used in electrical wiring. Rubber blocks the flow of electricity. It is frequently used to cover the copper wire in electrical cords for safety.
- Magnetic properties determine how a material reacts to a magnet.
- Chemical properties deal with things such as whether a material will rust or whether it can dissolve other materials.
- Mechanical properties are those that describe how a material reacts to forces.

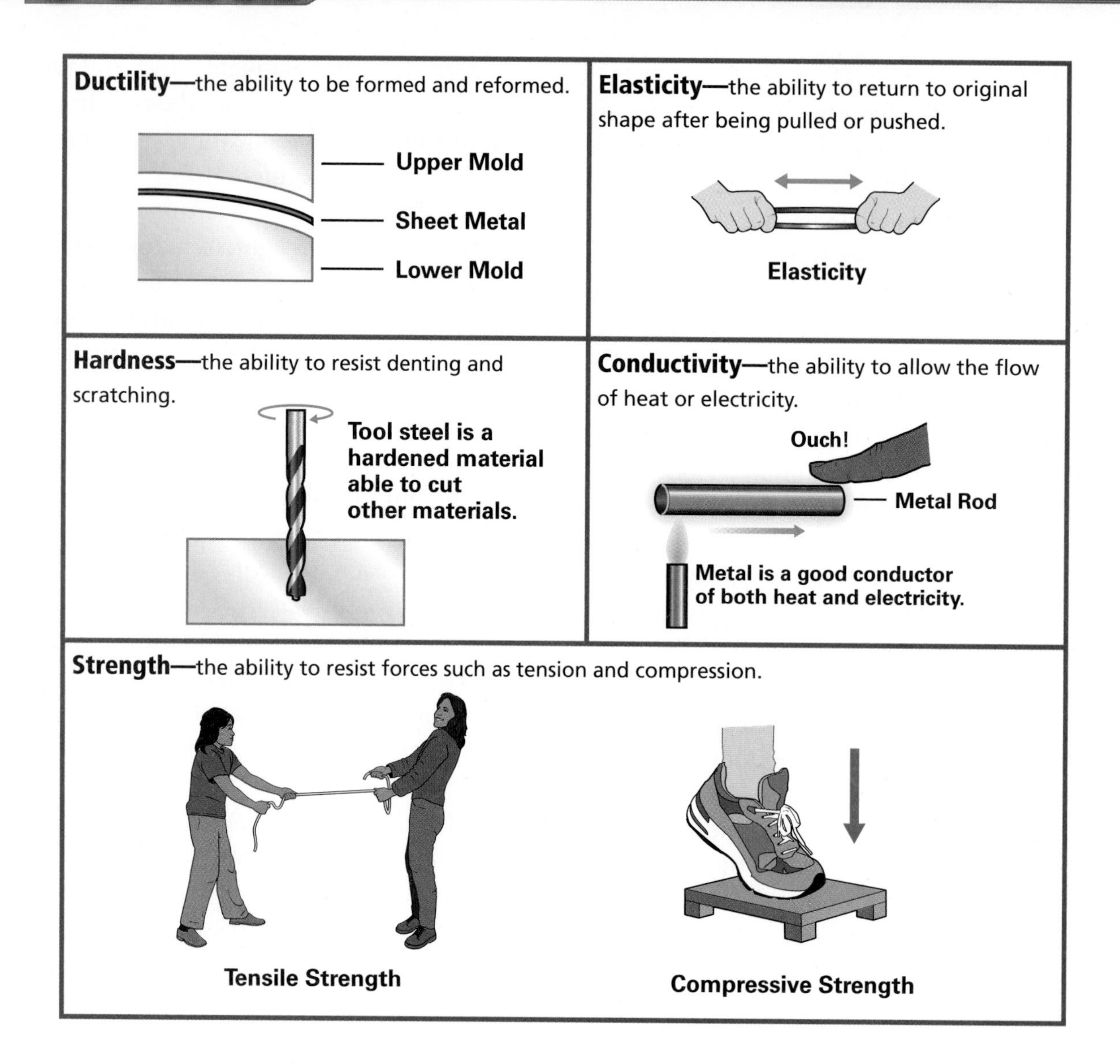

Electricity and Magnetism

Electricity

Electricity exists because most elementary particles of matter (such as electrons and protons) have a negative or positive charge. Like charges (two positives or two negatives) repel each other. Unlike charges attract. This electrical force is responsible for many natural phenomena. Technology has found many uses for it.

Electricity can be classified as static or current. In static electricity, the charges are "at rest." They remain in certain positions on objects. Static electric charges can build up when two different materials rub together, as when you walk on a carpet. If you then touch metal, the built-up charges suddenly flow away. They are no longer static, and you feel a shock. Lightning is another example of the sudden discharge (flowing away) of static electricity.

A flow of electricity is a current. A material through which current flows easily is called a conductor. A material through which almost no electricity can flow is called an insulator. Some materials can behave either as conductors or insulators, depending on what other materials are mixed with them. These are called semiconductors. They can be used to control the flow of electricity. For example, semiconductors are used to make computer chips.

Series and Parallel Circuits

The path electricity takes as it flows is called a circuit.

In a series circuit, individual components are connected end to end to form a single path for current flow. Series circuits have two major disadvantages. First, when connected in series, each circuit has to have its own switch and protective device. Second, if one component is open, the entire circuit is disabled.

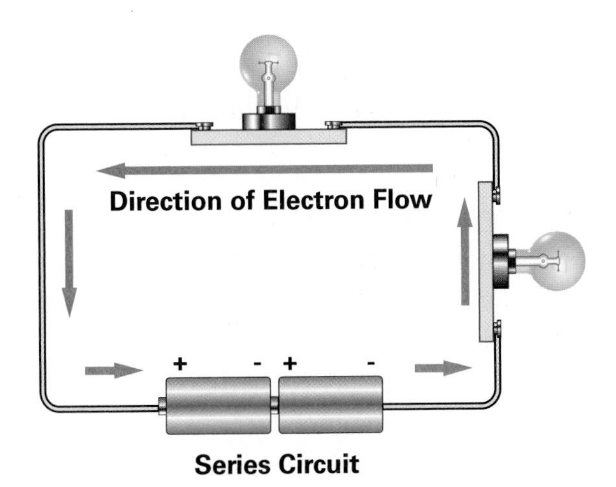

Series Circuit

In a parallel circuit, two or more loads are connected in separate branches. In most cases the parallel circuits are connected in series with a common switch and protective device. Equal voltage is applied to each branch of a parallel circuit. Current flow divides as it reaches the parallel path. The amount of current flowing through each branch depends on the resistance in that path only.

A flashlight has a simple circuit. Turning on the switch completes the circuit. This allows electricity to flow from the battery, through the switch, then through the bulb and back to the battery.

Parallel Circuit

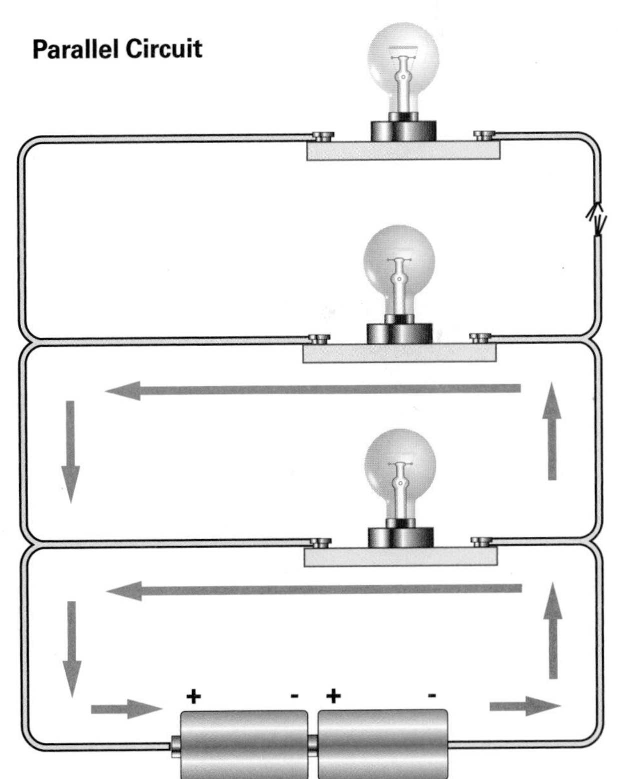

Ohm's Law

Current is the amount of charge that flows past a point in a circuit during a given time. Current is measured in amperes (A).

For a current to exist in a conductor, there must be an electromotive force (emf), or potential difference, between the two ends of a conductor. This electromotive force is measured in volts (V).

Even conductors resist the flow of current somewhat. The greater the resistance, the less current that flows. Resistance is measured in ohms (Ω).

The voltage, amperage, or resistance of an electrical circuit can be calculated by using Ohm's law (named for German physicist George S. Ohm). This law states that electric current equals the ratio of voltage and resistance.

$$I = E/R$$

In this formula, *I* stands for the intensity of the current, measured in amperes; *E* stands for electromotive force, measured in volts; and *R* stands for resistance, measured in ohms.

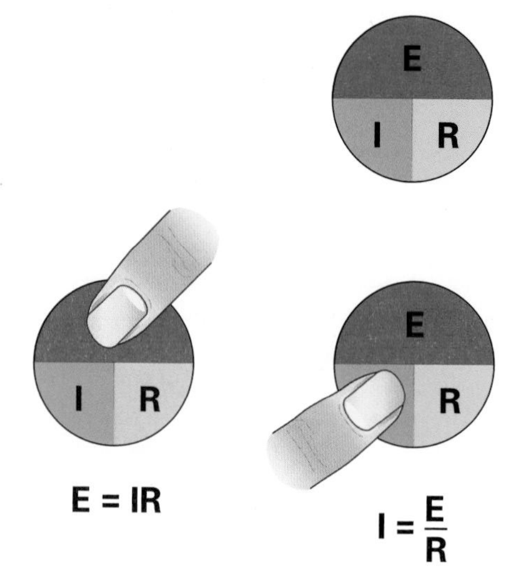

Magnetism

A pole is an area where magnetic density is concentrated. All magnets have two poles. These are designated as the north and south poles. The north pole of one magnet will attract the south pole of another magnet, while it will repel the other magnet's north pole.

Around the magnet is an invisible force field. When certain materials, such as iron, enter this force field, they become magnets, at least temporarily.

The planet Earth is a magnet with a force field extending into space. The force field, called the magnetosphere, protects us from harmful solar radiation. The poles of the earth magnet are not in exactly the same place as the geographic poles.

There is a scientific explanation for the action of magnets. Atoms produce magnetic fields. The magnetic field around a single atom is very weak. If the magnetic fields of a group of atoms align (point in the same direction), the force is much stronger. Such a group of atoms is called a domain. If the domains in a material align, that material becomes a magnet.

Electromagnetism

Magnets are not the only source of magnetic fields. Electric currents also produce magnetic fields. Magnetism produced by electricity is called electromagnetism. A magnet created by electricity is called an electromagnet. Electromagnets are very useful in many devices. Unlike other magnets, they are easily turned on and off. Their strength can be controlled by controlling the strength of the current.

Uses for Electricity and Magnetism

The relationship of electricity to magnetism has led to many developments in technology.

- Electromagnets make possible the magnetic levitation in maglev trains.
- A magnetic resonance imaging (MRI) machine creates a strong magnetic field that affects the atoms in the human body. The machine detects the patterns of energy absorption.
- Generators in power plants use magnets to convert the energy of motion (from steam, falling water, or wind) into electricity.
- Electric motors use magnets to change electricity into motion.
- Computers, VCRs, and tape players use magnetic tapes and disks to record and play back.

Light

Light is a form of energy. Sometimes it behaves like waves, sometimes like a stream of particles. The study of light has helped us develop technologies that make use of light as energy.

Properties of Light

Light travels in waves. The behavior of light waves is similar to the behavior of sound or water waves. For example, light waves bend as they pass through a small opening and spread out on the other side of the opening. This behavior is called diffraction. However, unlike water or sound waves, light waves can travel through a vacuum (an absence of air).

The Electromagnetic Spectrum

Visible light is only a part of the electromagnetic spectrum. This spectrum comprises a range of wavelengths that also includes radio and television waves, radar (microwaves), infrared rays, ultraviolet rays, X-rays, and gamma rays.

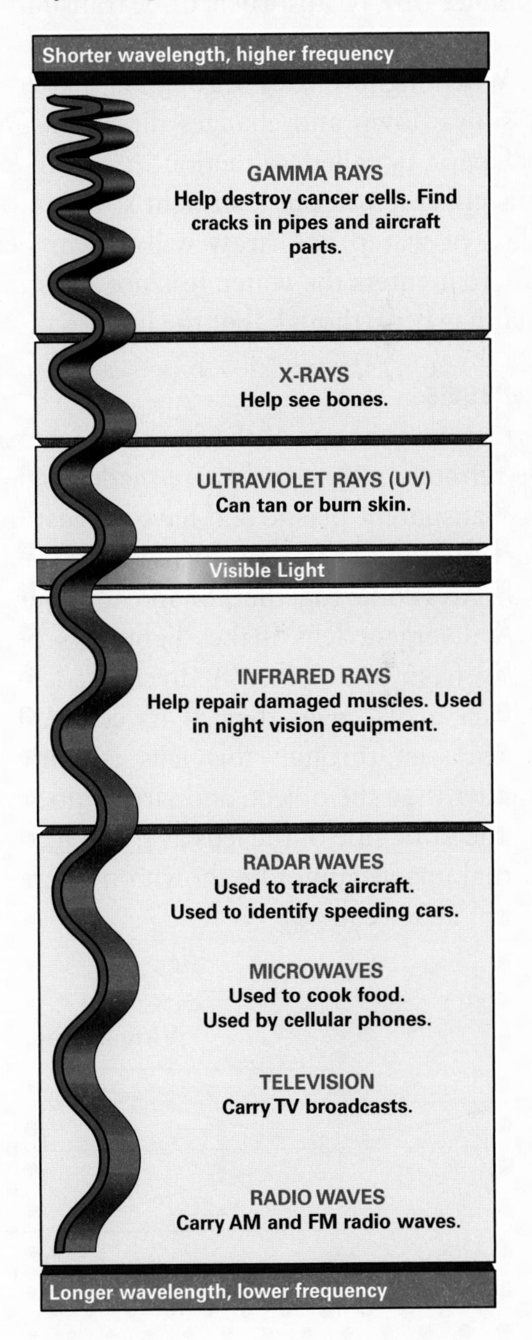

The speed of light is 300,000 kilometers (186,000 miles) per second in a vacuum. It travels at slower speeds through air, water, glass, or other transparent materials. When moving through an empty vacuum, light travels in a straight line. When light meets an object, the light may reflect (bounce off), be absorbed, or be transmitted (travel through).

When light passes through a material, it slows down and changes direction. This behavior is called refraction. You can see this effect if you put a straight straw into a glass of water. The straw will appear bent where it enters the water. It is not the straw that has bent, though, but the light.

Lenses

Lenses make use of the fact that light can be refracted. Most lenses are made of glass or transparent plastic and have at least one curved side. They can focus light (make light rays come together) or spread it out.

A divergent lens makes light rays bend away from each other. A divergent lens is thicker at its edges than at its center. The image seen through this lens is upright, smaller than the object, and virtual (located on the same side of the lens as the object). A virtual image cannot be shown on a screen. A real image can.

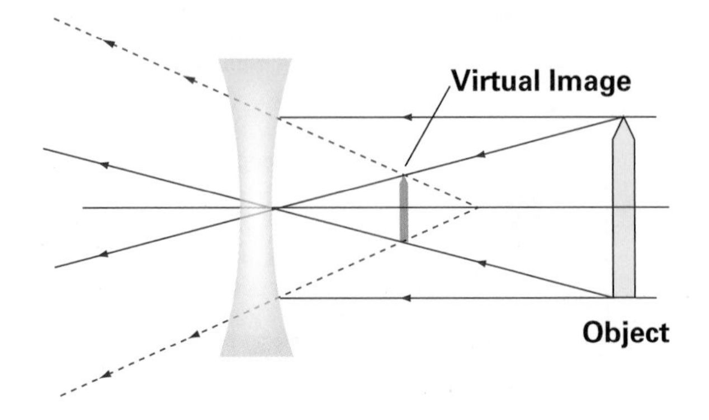

A convergent lens bends light rays toward each other. A convergent lens is thicker at the center than at the edge. The image formed by a convergent lens may be real or virtual, depending on the object's position in relation to the lens.

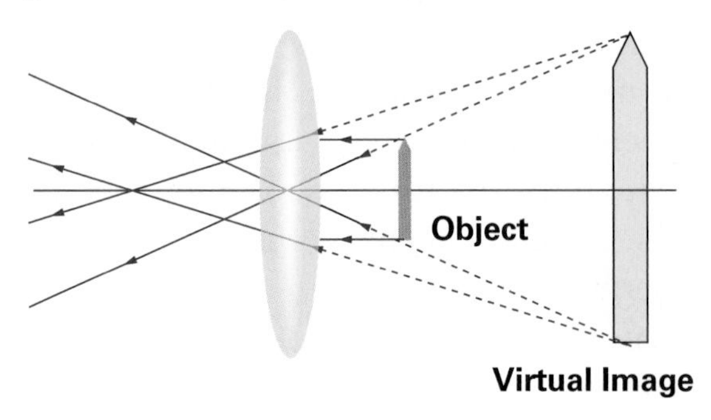

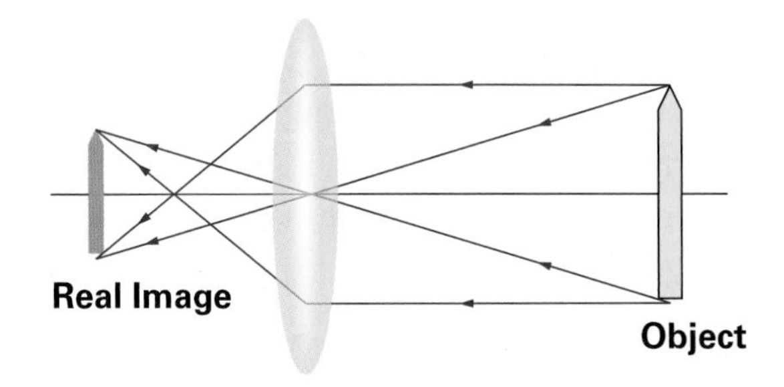

Uses of Light

Light finds various uses in communication and energy production.

- Pulses of light can be transmitted through fiber-optic cables. These pulses can carry telephone conversations, TV signals, or computer data.
- Photography (discussed in Chapter 7) is the control and recording of light. In video photography, light is recorded as magnetic patterns.
- Photovoltaic cells turn light energy into electrical energy. Some experimental vehicles are solar-powered.
- Lasers produce a very precise beam of light. The light waves are "in step" with each other. Laser uses include surgery and reading compact discs.

Cells

All living things are made of cells. Most cells are microscopic. An egg is an example of a very large cell.

Creatures such as bacteria consist of only one cell. Others are made up of many kinds of cells that have specialized functions. In the human body, for example, hair cells differ from liver cells. Both differ from muscle cells. In a plant, cells that make up the roots differ from the cells in the stalk.

Despite their variety, all cells share certain basic traits. Cells live by taking in food, converting it into energy, and eliminating waste. Cells reproduce by dividing in two. To carry out these functions, cells regulate their chemical reactions.

The chemical reactions that occur inside a cell depend on long, chainlike molecules called proteins. Even the simplest cell contains many different kinds of proteins. Each has a highly specific job to perform. DNA, for example, is a complicated combination of proteins. A technologist can cause a cell to produce something or stop producing something. The tool to make it happen would be a protein.

Plant and animal cells have certain features in common. A double-layered cell membrane made of a lipid (fat) keeps the watery contents of a cell in. It keeps almost everything else out. The membrane is dotted with proteins that act as gatekeepers. If a substance pushes against a gatekeeper protein and offers the right chemical fit, a passage opens. This is how a cell lets food in and discharges waste.

- Cytoplasm is a gel-like substance that fills the cell. It provides a medium for internal transport of materials.
- The nucleus is the core of the cell, containing its chromosomes. Chromosomes are made of DNA. This provides "templates" for assembling the right chemicals in the right way to carry on the life of the cell. Certain single-celled organisms, such as bacteria, do not have a nucleus.
- Mitochondria are structures (organelles) that provide energy to the cell. They break down nutrients with enzymes.
- Vacuoles are pockets for storing nutrients. Some animal cells do not have vacuoles. They may have other structures for storing materials until they are needed.
- Plant cells have two features not found in animal cells: chloroplasts and a thick cell wall.
- Chloroplasts create energy for plant cells. They use sunlight to convert water and carbon dioxide into simple sugars.
- A thick, fibrous cell wall surrounds the membrane of a plant cell. This provides rigidity to the overall structure of a plant.

TECHNOLOGY

Technology consists of processes and knowledge that people use to extend human abilities and to satisfy human needs and wants. The chapters of this textbook discuss the nature of technology, its effect on our lives, and the ways technology is used in the designed world.

Technology can be classified into six broad areas.

Communication

Energy & Power

Manufacturing

Transportation

Construction

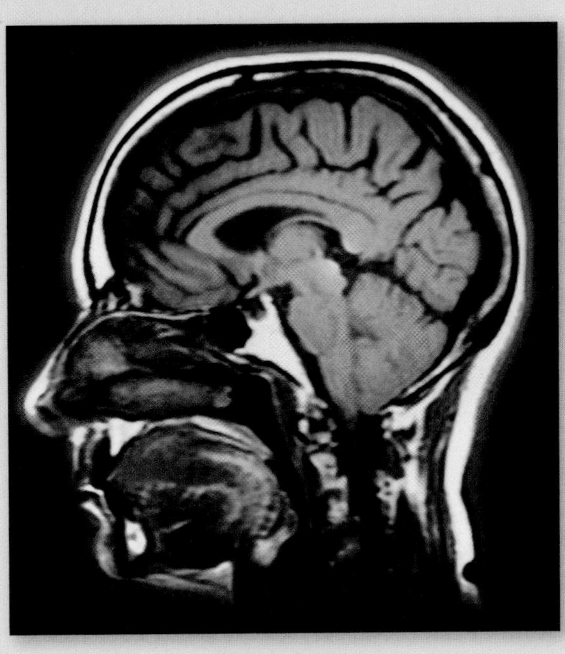
Bio-Related

ENGINEERING

Design and Drawing for Engineering

Suppose you received the following instructions: "Go north on Route 6. Turn right at Hendricks and park by the pool." These instructions sound clear enough, don't they? But we all know that even well-chosen words might not be enough. The drawing below, on the other hand, is much clearer.

If you want to be sure someone understands your message, use a picture. As the old saying goes, "A picture is worth a thousand words."

Drawing pictures is a way of "thinking on paper." Pictures aid problem solving. If you're not sure whether the jewelry box you're making would look better with one drawer or two, sketch different designs until you find one you like.

This section of the STEM Handbook will help you use a few graphic methods to communicate and work out solutions to problems. It is not a course in mechanical drawing. However, it will help prepare you for activities in this student text.

If you have access to a CAD system, it is the tool of choice for drafting. Can a computer draw by itself? No. Will it turn a poor student into a good student? No. So what good is it? Perhaps the best thing about CAD is that it allows the user to modify a drawing, just as word processing allows the editing of text. You can cut and paste and save and recall files. You can quickly perform functions that might otherwise take much time and effort.

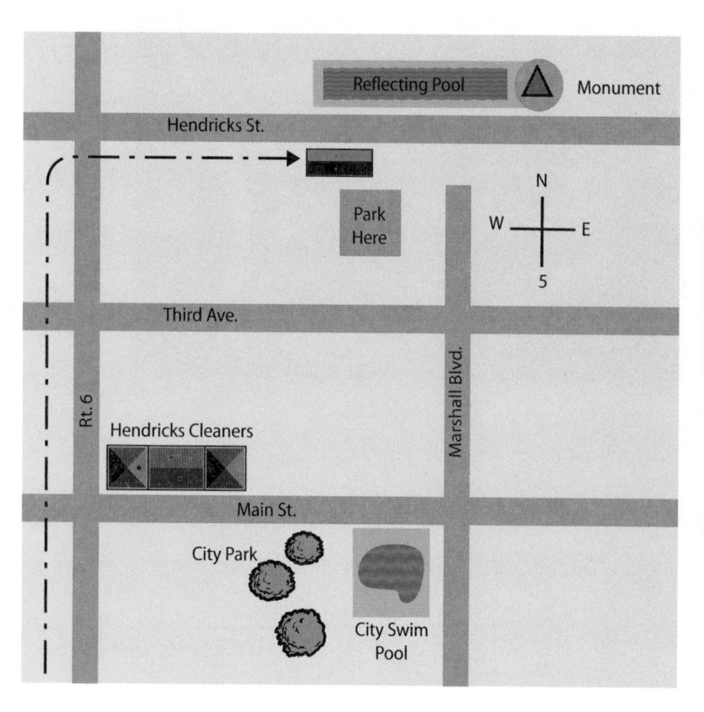

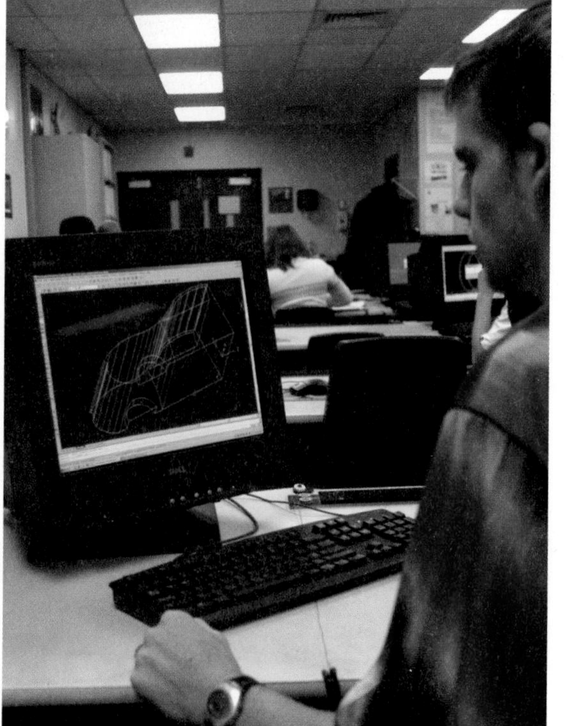

Sketching

The best way to express yourself graphically when time is short is by sketching. A single sketch can sometimes clarify complex thoughts. You can learn to draw good sketches by practicing.

Sketching Tools

Pencils are classed by the hardness of their lead. Pencils with hard leads are used when accuracy is needed, such as for charts and diagrams. Medium pencils are used for technical drawing and sketching. Soft pencils are used for doing general artwork and quick sketches.

Use a soft pencil. Hold it about 1½ inches from the point. Work big. Don't make small, cramped pictures. A good sketch should not look stiff and "perfect" like a mechanical drawing. The lines should be free and loose.

When sketching straight lines, spot your beginning and end points. Keep your eye on the point toward which you are drawing (A). Draw vertical lines downward (B).

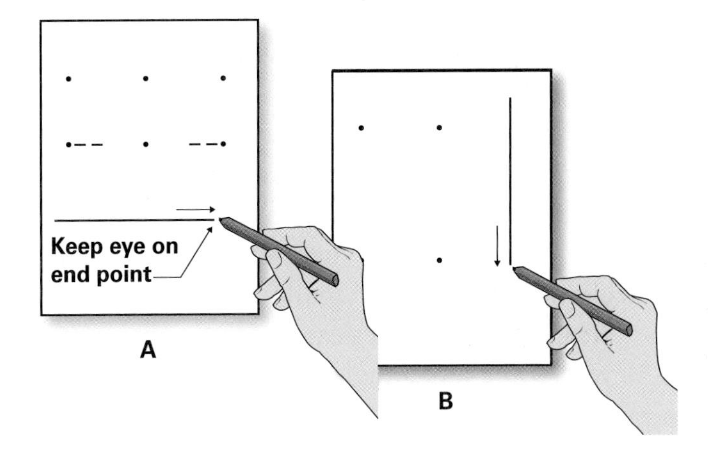

If an inclined line is to be nearly vertical, draw it downward (C). If it is to be nearly horizontal, draw it to the right (D).

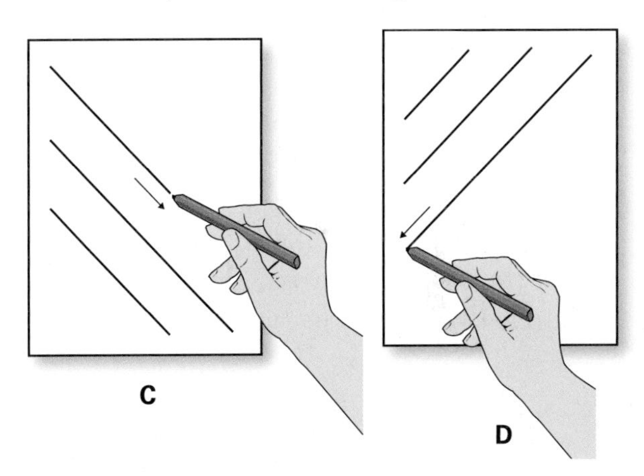

Inclined lines can also be drawn as if they were vertical or horizontal by simply turning the paper (E).

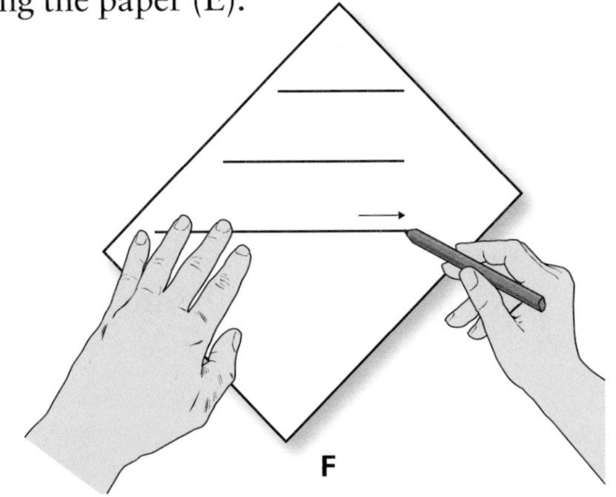

Is It a Good Design?

Before you begin to design a product and during each design stage, ask yourself these questions:

- Is it appropriate to the need you are trying to fulfill?
- Do people really need it or think they need it? (If they don't, they won't use it.)
- Can it be made under the present conditions? (For example, if you have only a week to make it, can you make it in that time?)
- How much does it cost to make? (If it's too expensive, you'll have trouble marketing it.)
- Is it durable? Will it last?
- Is it attractive?

Sketching a Circle

Circles are hard for many people to draw. Many times they are lopsided. There are tricks you can use, however, that make drawing circles easier.

First, lightly sketch a square that is the diameter of your circle (1). Then mark the midpoints of the sides. Next, draw diagonal lines (2) and mark where the line of the circle will cross them. Finally, sketch in the circle (3). (Oval shapes can be drawn the same way, using a rectangle instead of a square.)

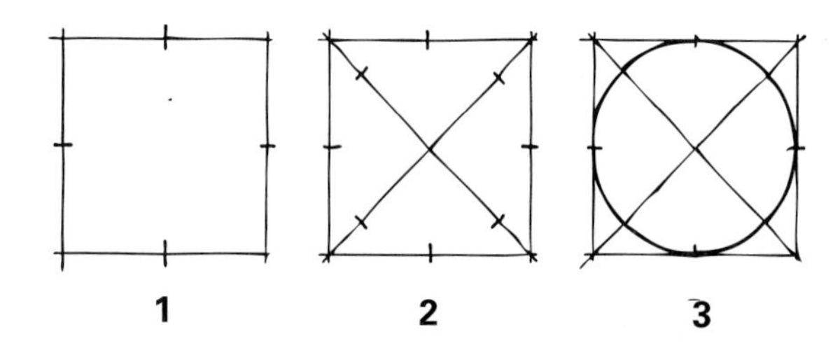

You can also draw a circle with the aid of centerlines. First, sketch two centerlines (1). Then add light radial lines, or "spokes," between them (2). Make marks where the line of the circle will cross. Finally, sketch in the circle (3).

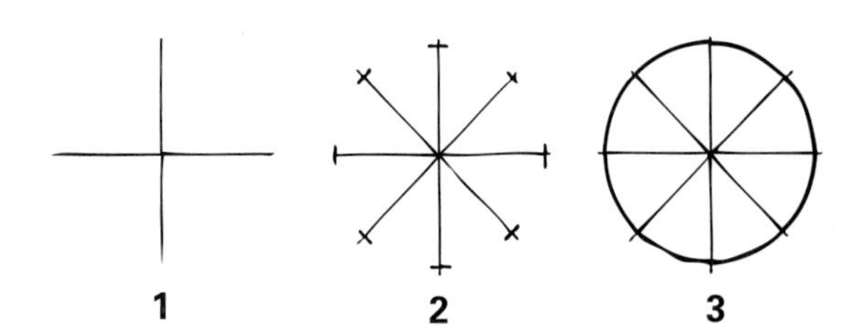

Arcs can be sketched using the same general methods as for circles. First, lightly sketch a square of the same size as the arc. Draw a diagonal line, and mark off the size of the arc. Finally, sketch in the arc itself.

Building from Basic Shapes

The square, the rectangle, the circle, the cylinder, the triangle, and the cube can be used to draw other objects. Identify which of them have been used to help create the objects shown here.

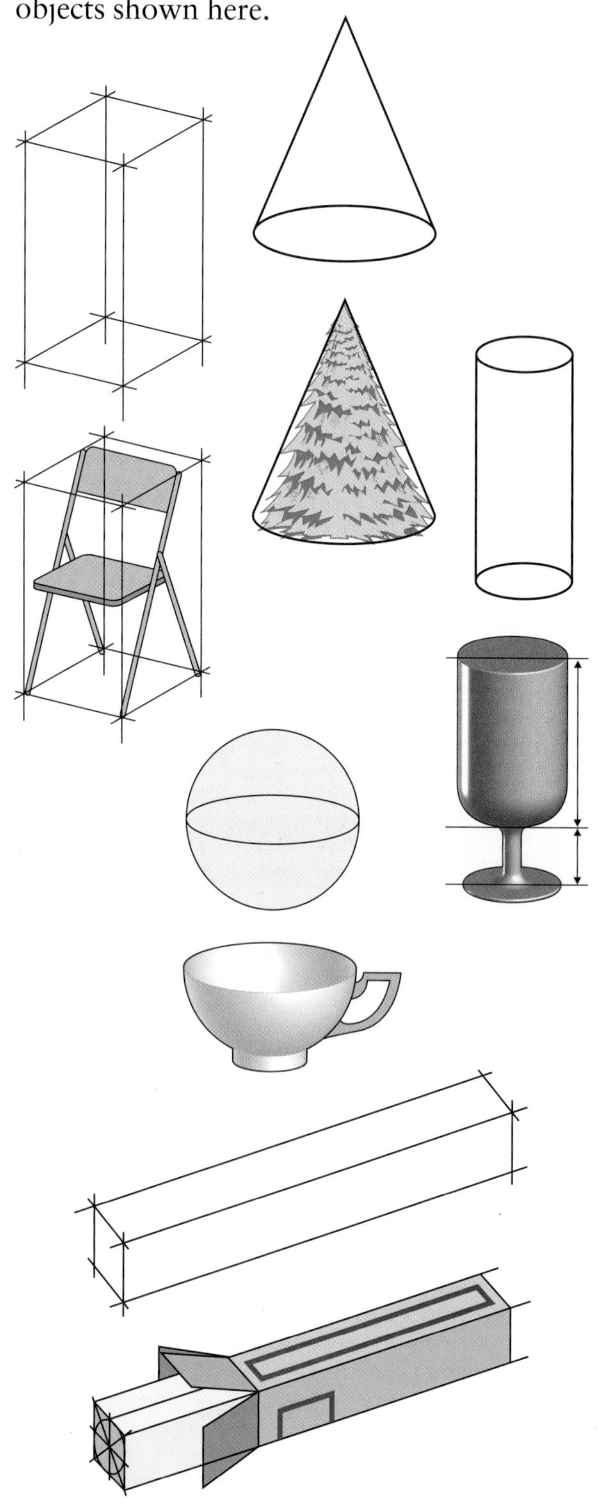

Using a Protractor

The protractor is used for measuring different angles.

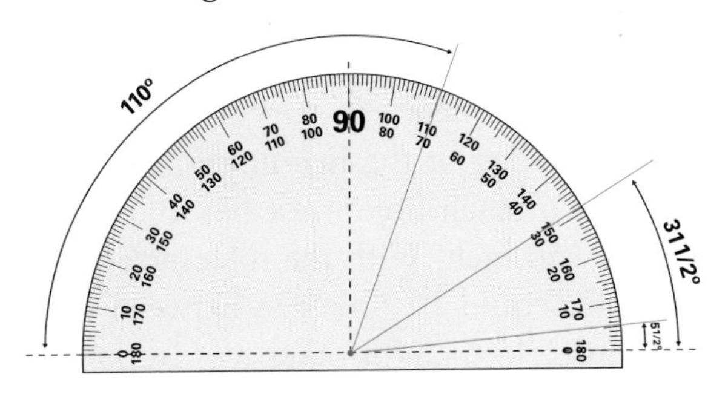

Drawing to Scale

Some objects cannot be drawn full-size. They are too big or too small. For this reason they are drawn smaller or larger than actual size. A building, for instance, may be shown on a map 1/96 size. Drawing objects larger or smaller than actual size, but in the correct proportions, is called "drawing to scale."

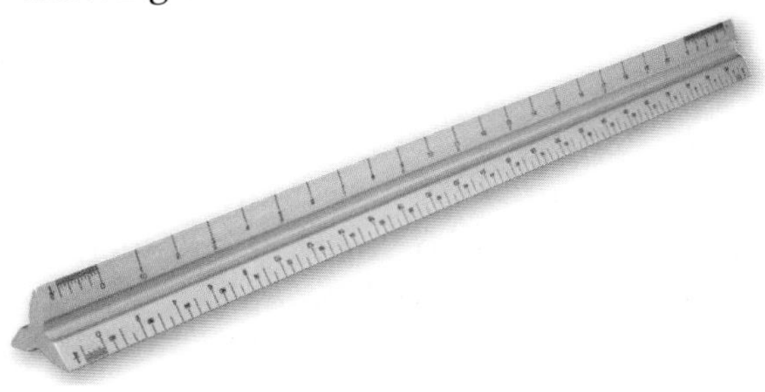

The scale to which a drawing is made should be indicated on the drawing. For example, "Scale: Half size" or "Scale: 1 in. = 1 ft."

Tools called scales are used by architects and engineers to quickly calculate scale reductions or enlargements. The architect's scale includes reduced-size scales in which fractions of an inch represent feet.

Dimensioning

Dimensions on a drawing give sizes. They are important when you or someone else must make an object from a drawing. There are many lines and methods used in dimensioning.

You will need to be able to recognize the various kinds of lines. The most important lines are object lines, hidden lines, dimension lines, extension lines, and leader lines.

- Object lines show the visible lines of the object.
- Hidden lines show where a line would be if you could see through the object.
- Dimension lines usually have the dimension written in the center. They end in arrows.
- Extension lines extend from the object to the dimension line and a little beyond it. They show the boundaries of the area being measured.
- Leader lines are usually used to give information or to dimension interior details, like holes. They lead from the dimension to the part referred to. They end in an arrow that touches the part.

All dimension figures should align with the bottom of the drawing. Never place a dimension directly on the object itself unless there is a good reason for it. The largest dimension goes on the "outside" of any shorter dimensions. Diameters of round objects are indicated with a symbol, which means "diameter." It is written before the dimension number. The symbol for a radius is R. It, too, is written before the number.

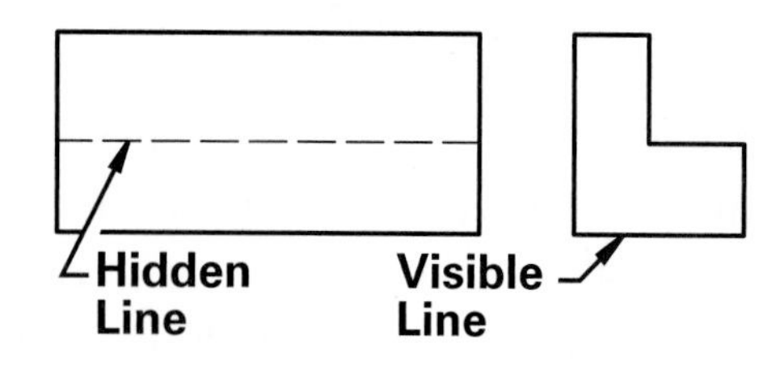

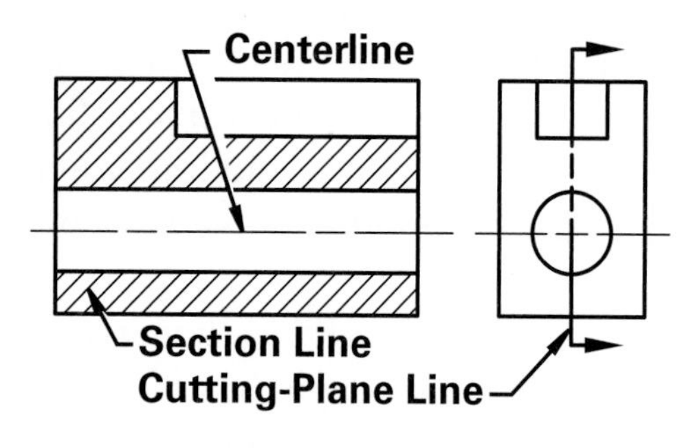

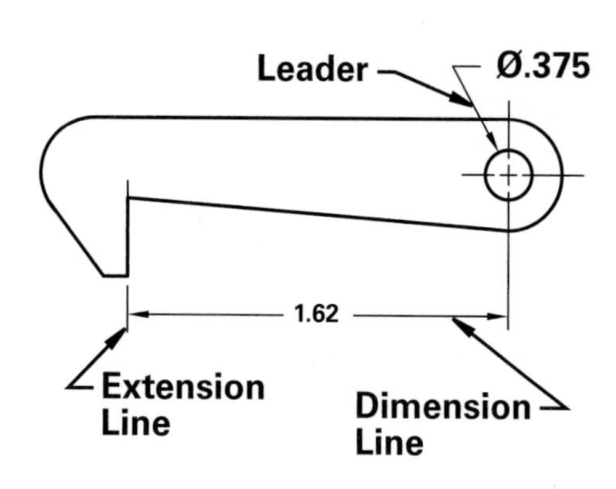

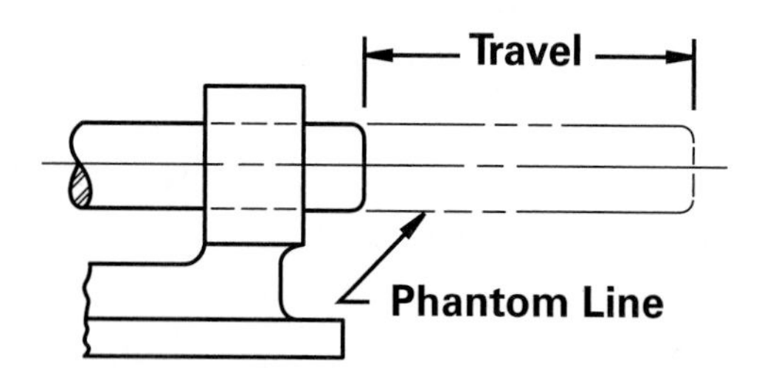

A dimension on a drawing is exact. It is usually not possible to make an object that is exact to within thousandths of an inch. For that reason dimensions are often given tolerances. A tolerance is the amount a given measurement can vary. For example, in the drawing of the five holes below, the diameter of each hole must be drilled to a size of .250 inch. With the tolerance given, the holes could be any size between .255 inch and .245 inch. In other words, there is a tolerance of plus or minus .005 (five-thousandths) of an inch.

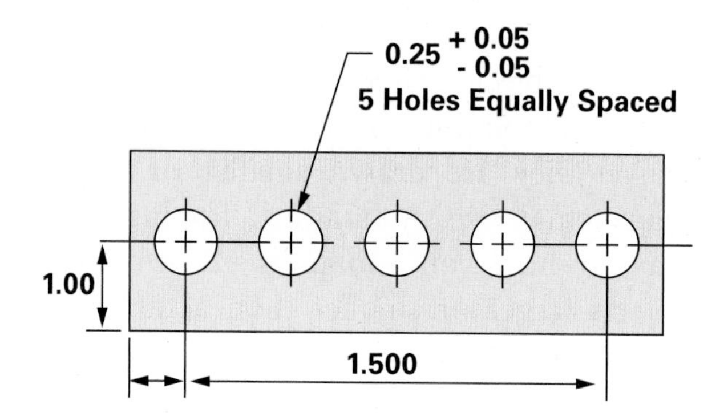

Working Drawings

Working drawings, or multiview drawings, show how to make a product. All dimensions are given. The object is shown from as many angles as needed to indicate its width, height, and depth.

Working drawings show "head-on" views of the top, front, back, sides, and bottom of an object. Ordinarily, only three views are shown—top, front, and right side.

They are produced by means of orthographic projection. "Ortho" means straight or at a right angle. In orthographic projection the pictures (views) of the object are of surfaces at right angles to one another.

Working drawings may require dimensioning. Most are drawn to scale. Drawing A shows the steps to follow in making a three-view working drawing.

A

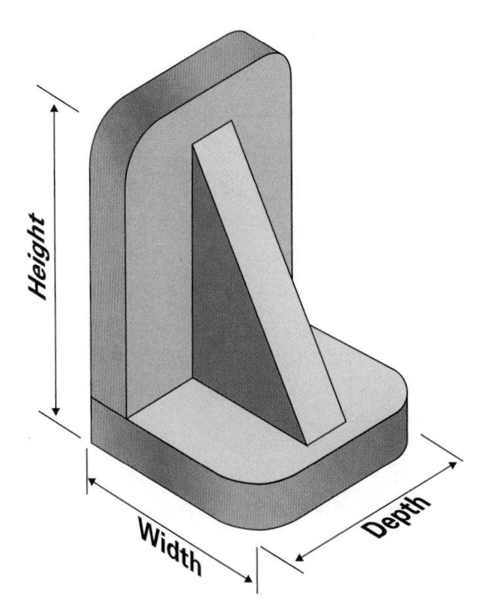

1. Block in boxes for views in proportion.

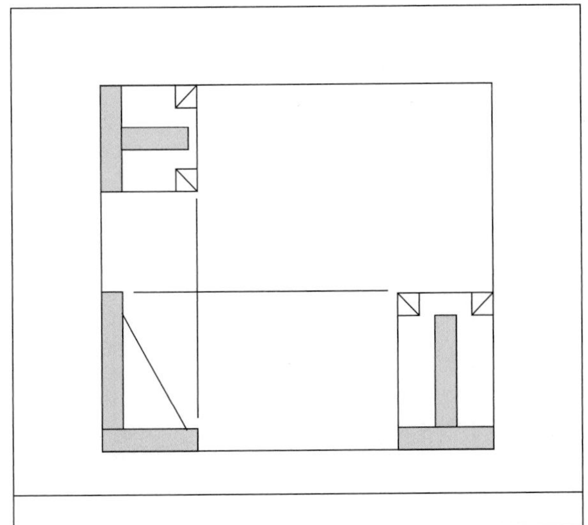

2. Block in construction for arcs, circles, etc.

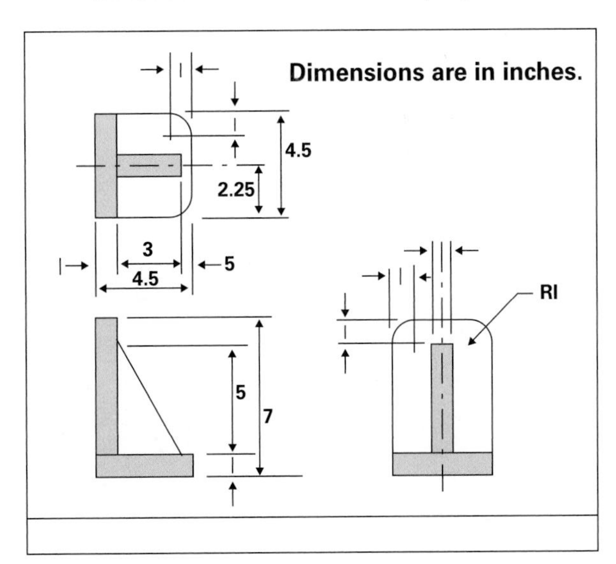

3. Complete details.
4. Add dimensions.

Orthographic projection is used to make a working drawing. Orthographic projection is the technique of showing several surfaces of an object "head on." Because many objects have six sides, as many as six views are possible. Usually, however, only three views are shown. These show the front, top, and right side.

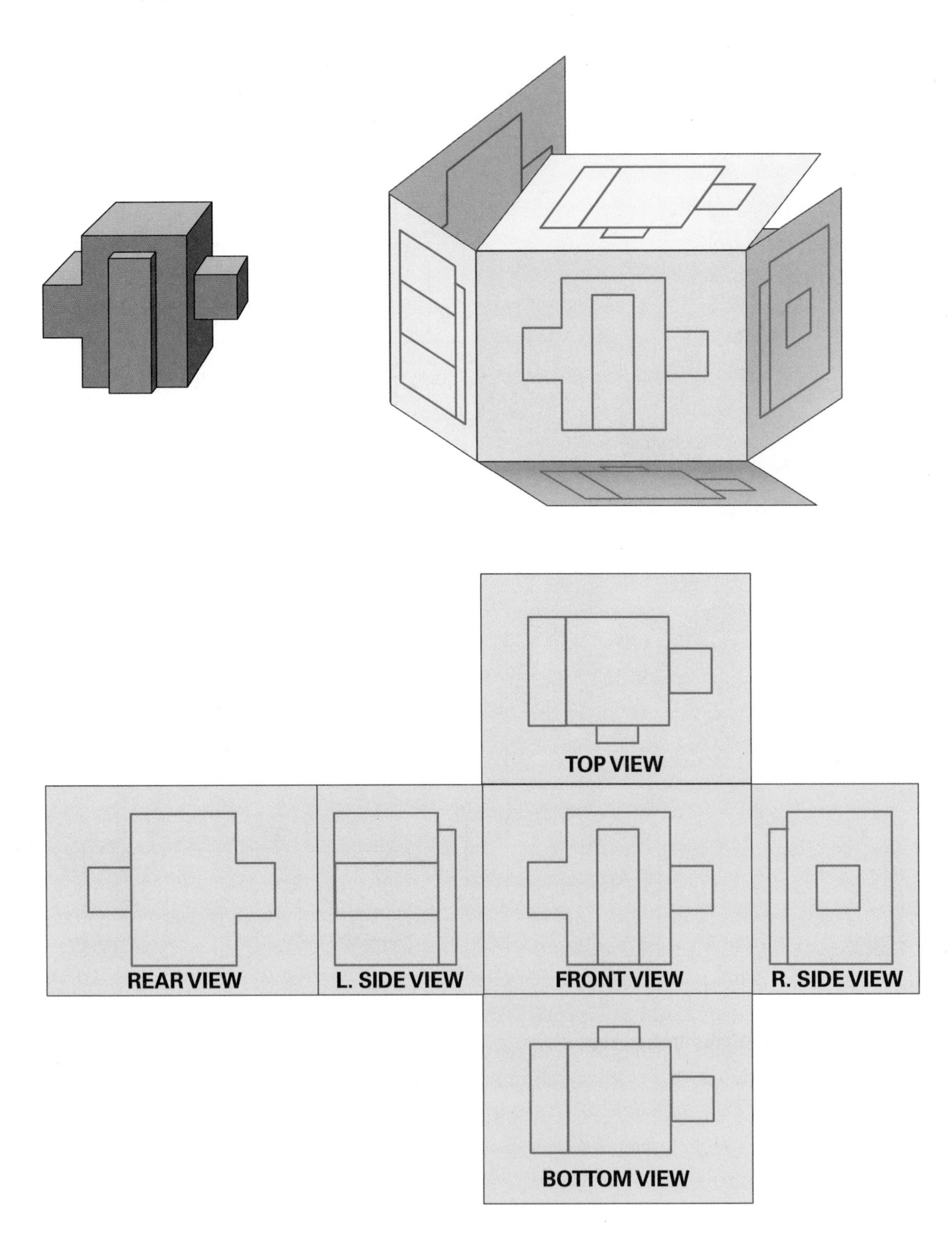

Engineering Design

Engineering Design Process

These are the main steps in the engineering design process:

1. Define the problem.
2. Brainstorm, research, and generate ideas.
3. Identify criteria and specify constraints.
4. Develop and propose designs and choose among alternative solutions.
5. Implement the proposed solution.
6. Make a model or prototype.
7. Evaluate the solution and any of its consequences.
8. Refine the design.
9. Create the final design.
10. Communicate the processes and results.

Defining the Problem

The design problem must be clearly stated before design work begins. It must include all of the important design requirements. You may need to clarify and restate the design problem. The problem will determine the design.

Checking the Design

Defining and refining the design is just the first part. A good design depends on the proper relationship of several elements. This requires careful planning. A good design depends on good planning. Check the design at each stage to ensure that the design will work.

Elements of Design

The elements of design are the main parts of the design. These elements include:

- **Line** This can be thin, thick, straight, or curved, among other forms.
- **Shape** Is it a rectangle, a triangle, or a circle?
- **Mass** What is the size of the object?
- **Color** This element can be used to highlight function or add interest.
- **Pattern** What is the appearance of the object's surface? Is it plain or does it carry a design?

Principles of Design

The effectiveness of a design can be judged against certain principles. These principles include:

- **Balance** Pleasing integration of various elements.
- **Proportion** Proper relation of parts to one another.
- **Harmony** Pleasing arrangement.
- **Unity** Ordering of parts to create an undivided total effect.

MATHEMATICS

Linear Measurement

A linear measurement is one made along a line. To make a linear measurement, use a rule or a tape measure.

Working with Fractions and Decimals

In the customary system of measurement, you'll often be working with fractions, especially when measuring materials. Most customary rules are divided into inches. A one-foot rule has 12 inches. A yardstick has 36 inches. The inches are divided into smaller parts (fractions). Typically, there are marks to show $\frac{1}{2}$ inch, $\frac{1}{4}$ inch, $\frac{1}{8}$ inch, and sometimes $\frac{1}{16}$ inch.

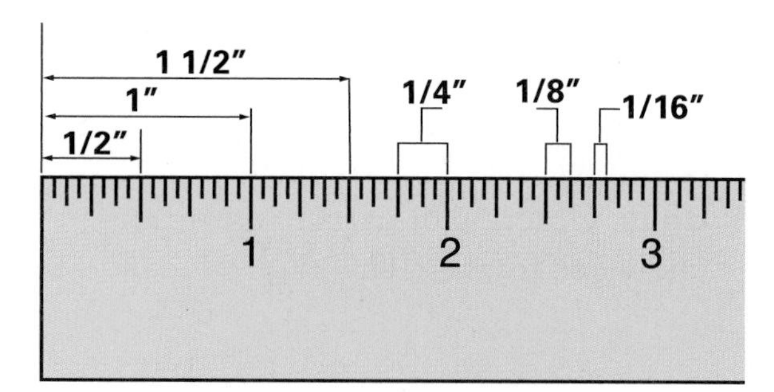

Adding Fractions

If the denominators are the same, add the numerators. Simplify the sum if necessary.

Example:

$$\frac{1}{8} + \frac{1}{8} = \frac{2}{8}$$
$$\frac{2}{8} = \frac{1}{4}$$

If the denominators are different, you must rename the fractions with a common denominator before you can add them.

Example:

$$\frac{1}{4} + \frac{3}{8}$$
$\frac{1}{4}$ is equal to $\frac{2}{8}$
$$\frac{2}{8} + \frac{3}{8} = \frac{5}{8}$$

If the numbers are mixed, add the fractions. (If necessary, rename them with a common denominator first.) Add the whole numbers. Rename and simplify.

Example:

$$1\,\frac{3}{4} + 2\,\frac{1}{2}$$
$2\,\frac{1}{2}$ is equal to $2\,\frac{2}{4}$
$$1\,\frac{3}{4} + 2\,\frac{2}{4} = 3\,\frac{5}{4}$$
$3\,\frac{5}{4}$ is simplified to $4\,\frac{1}{4}$

Subtracting Fractions

If the denominators are the same, subtract the numerators.

Example:

$$\frac{3}{8} - \frac{2}{8} = \frac{1}{8}$$

If the denominators are different, you must rename the fractions with a common denominator before you can subtract them. Simplify if necessary.

Example:

$$\frac{5}{16} - \frac{1}{8}$$
$\frac{1}{8}$ is equal to $\frac{2}{16}$
$$\frac{5}{16} - \frac{2}{16} = \frac{3}{16}$$

If the numbers are mixed, subtract the fractions. (If necessary, rename them with a common denominator first.) Subtract the whole numbers. Rename and simplify.

Example:

$$4\,\frac{1}{8} - 2\,\frac{1}{16}$$
$\frac{1}{8}$ is equal to $\frac{2}{16}$
$$4\,\frac{2}{16} - 2\,\frac{1}{16} = 2\,\frac{1}{16}$$

Multiplying Fractions

Multiply the numerators and then multiply the denominators. Simplify if necessary.

Example:

$$2/16 \times 1/2 = 2/32$$
$$2/32 = 1/16$$

If the numbers are mixed, rename them as improper fractions. Then multiply the fractions. Rename and simplify as needed.

Example: Suppose you have 3¼ pounds of cement. You need to use ⅔ of it. How many pounds would that be?

$$2/3 \times 3\ 1/4$$
$$2/3 \times 13/4 = 26/12$$
$$26/12 = 2\ 1/6$$

Dividing Fractions

To divide by a fraction, multiply by its multiplicative inverse. In other words, invert the fraction and then multiply. Simplify if necessary.

Example: You have a collection of CDs. Each one, in its case, is ⅜" thick. If you make a CD holder that is 6" wide, how many of your CDs will fit in it?

$$6 \div 3/8$$
$$6 \times 8/3 = 48/3$$
$$48/3 = 16$$

Converting Fractions to Decimals

To convert fractions to decimals, divide the numerator by the denominator.

Example: To convert the fraction ⅜, divide 3 by 8.

$$3 \div 8 = .375$$

Converting Decimals to Fractions

To convert decimals to fractions, write the decimal as a fraction and simplify.

Example 1:

$$.25 = 25/100$$
$$25/100 = 1/4$$

Example 2:

$$.375 = 375/1000$$
$$375/1000 = 3/8$$

The greatest common factor of the two numbers is 125. Dividing 375 by 125 = 3. Dividing 1,000 by 125 = 8.

Angles

An angle is the figure formed when two lines or two surfaces originate at the same point. Angles are measured in degrees. The following drawings show some common angles.

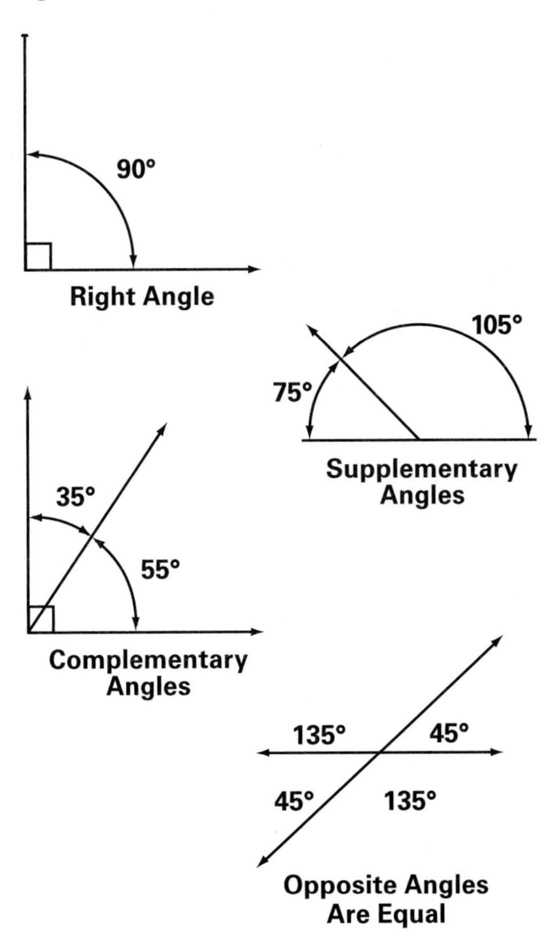

Polygons

A polygon is a closed figure with straight sides. It is classified by the number of sides. In a regular polygon all sides are equal in length and all angles are equal.

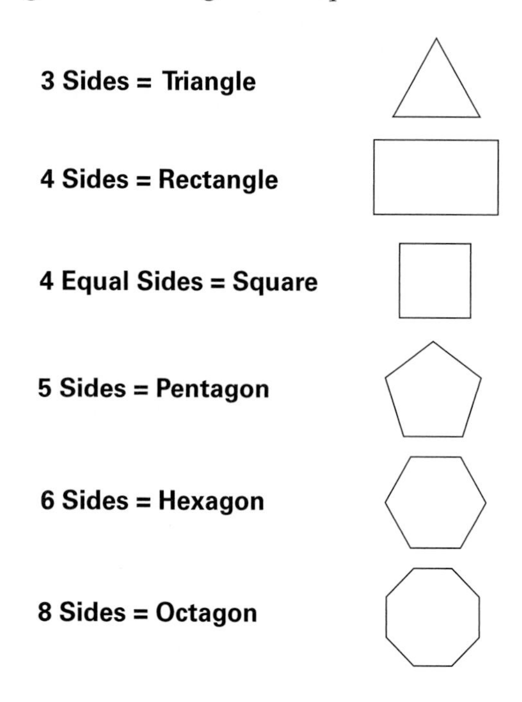

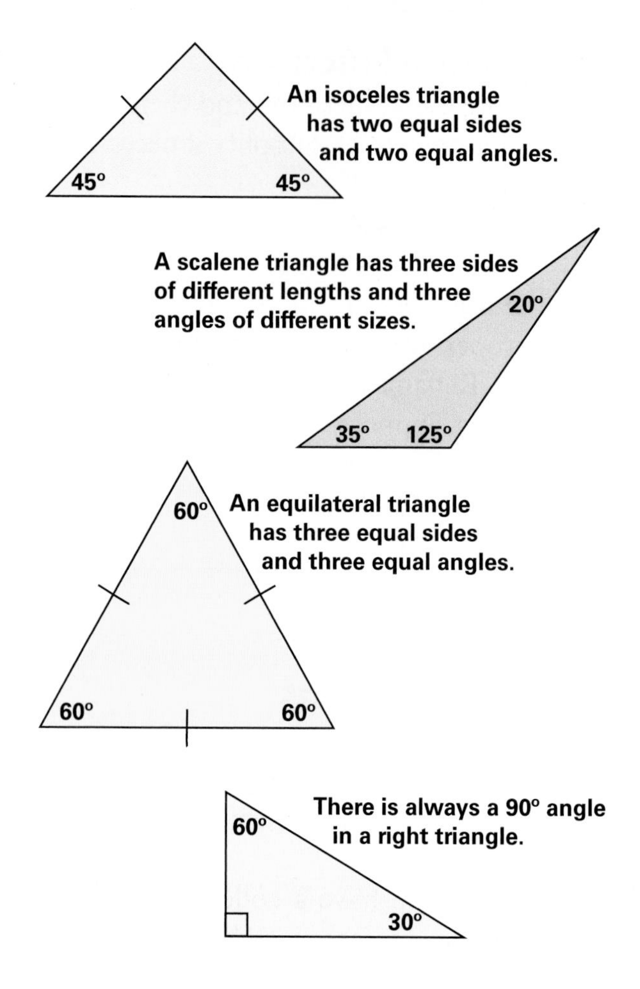

Types of Triangles

Triangles are classified according to the length of their sides.

- **Isosceles** A triangle having two equal sides.
- **Equilateral** A triangle having all sides equal.
- **Scalene** A triangle having three sides of unequal length.

Triangles are also classified according to the size of their largest internal angle. The right triangle has one internal angle of 90°, which is a right angle.

Rectangles and Triangles

A rectangle is a four-sided shape with four right angles (90°). The diagonals of a rectangle are always equal. They divide the rectangle into two right triangles. A right triangle has three sides. The longest side is called the "hypotenuse." These facts are important when laying out a perfect rectangle. You can use it when constructing a simple box or when laying out a large building.

$A^2 + B^2 = C^2$

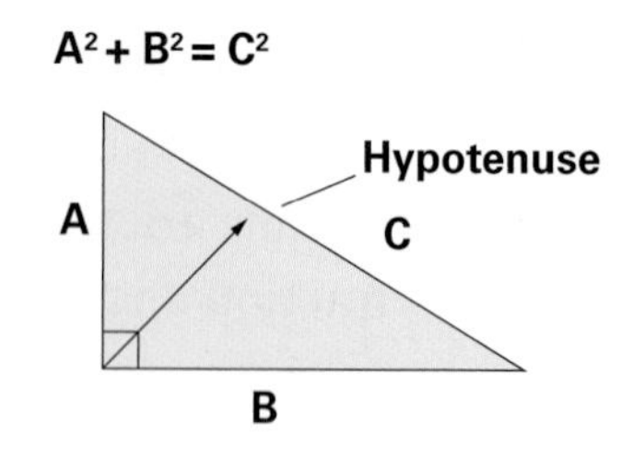

Using Right Triangles

If you know the length of two sides of a right triangle you can figure the length of the third side by using the equation:

$$a^2 + b^2 = c^2$$

c = the hypotenuse, the side opposite the right angle

a and b = the two other sides

This equation is called the Pythagorean theorem. Here's a practical application. Let's say you want to build a storage shed that is 6' × 8'.

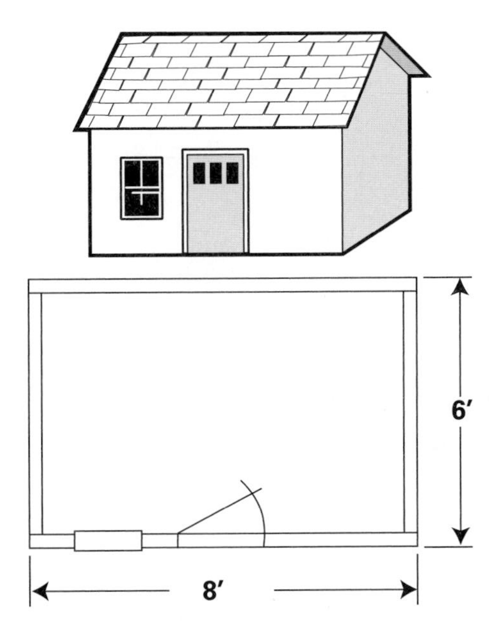

1. First establish the 8' side of the shed. Place a stake in the ground at each end of this side.

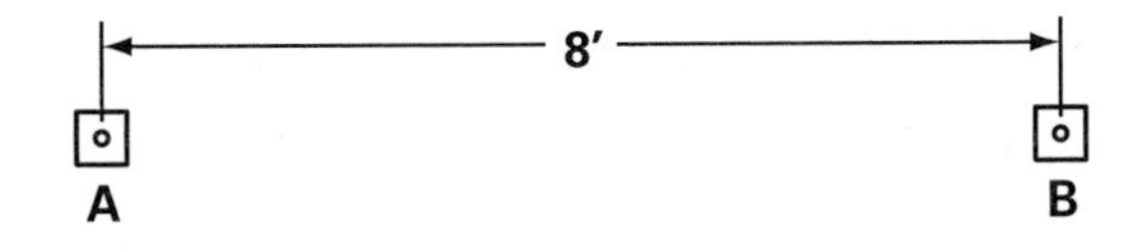

2. Using a calculator, enter 6. Then push the x^2 key. If you do not have that key, multiply 6 × 6. The result should be 36. Write down the answer or store it in the calculator's memory.
3. Repeat Step 2 using the 8' dimension.
4. Add the answers from Steps 2 and 3. It should be 100.
5. With the answer of 100 still showing, press the square root key. The answer should be 10. This is the length, in feet, of the diagonal for your shed.
6. From the stakes you placed in Step 1, and using two tape measures, measure 6' from one stake and 10' from the other stake. Where the two measures intersect is where one of the back corners of the shed is located.

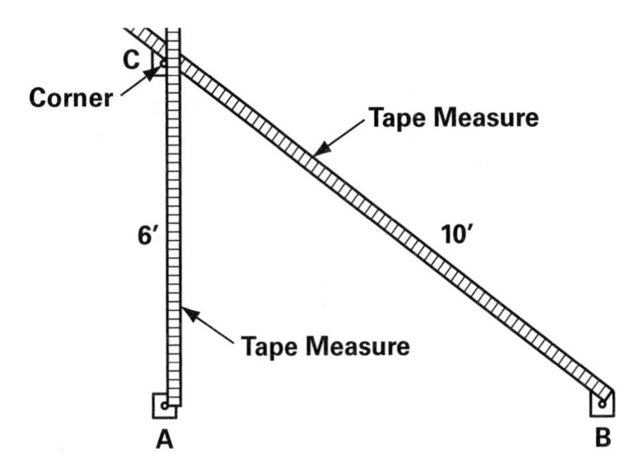

7. Repeat Step 6 from the other stake on the 8' line. Because diagonals of a rectangle are equal, you will have the layout of a perfect rectangle for your shed.

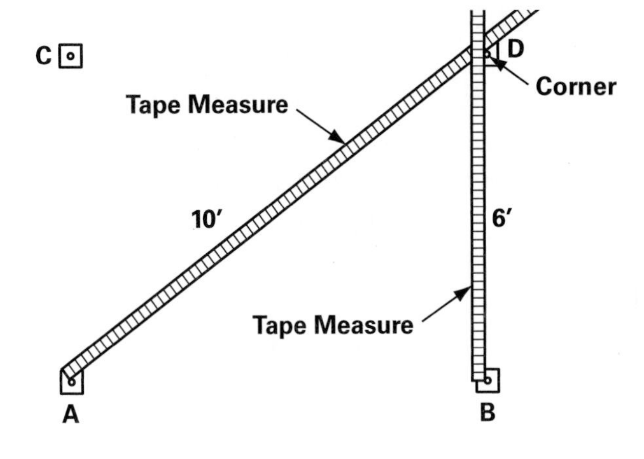

Surface Measurements

Surface measurements measure either the distance around a shape or the area that the shape covers.

Perimeter of a Rectangle or Square

The perimeter of a rectangle or square is the sum of all the sides. Since rectangles have two pairs of equal sides, you can find the perimeter (P) by adding twice the length (l) plus twice the width (w).

$$P = 2l + 2w$$

This formula works also for a square, but there's a simpler way to find the perimeter of a square. All four sides of a square are of equal length. Multiply one side by 4.

Area of a Rectangle or Square

To find the area (A) of a rectangle or a square, multiply one side (length, or l) by an adjacent side (width, or w). **Note:** Area is measured in square units.

$$A = lw$$

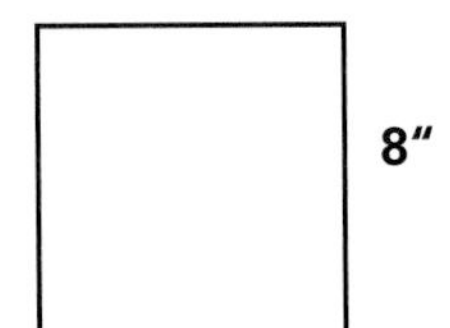

Circumference of a Circle

The circumference is the distance around a circle. To find the circumference (C) of a circle, multiply the diameter (d) by the value of π, which is 3.14.

$$C = \pi d$$

The diameter of a circle is equal to twice its radius. You can find the circumference if you multiply twice the radius (r) by π.

$$C = 2\pi r$$

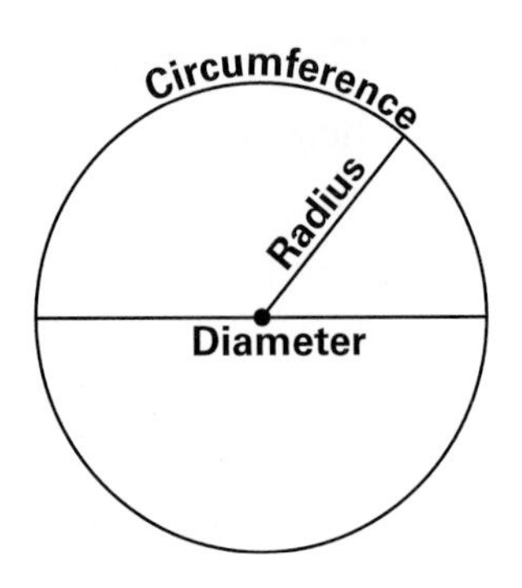

Area of a Circle

To find the area (A) of a circle, multiply π by the radius (r) squared.

$$A = \pi r^2$$

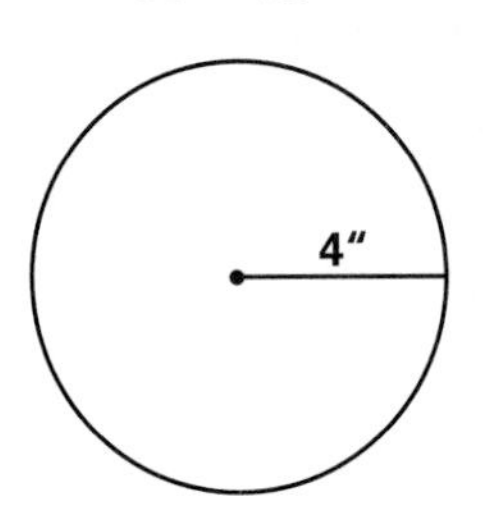

Perimeter of a Triangle

To find the perimeter of a triangle, add the lengths of the three sides.

$$P = a + b + c$$

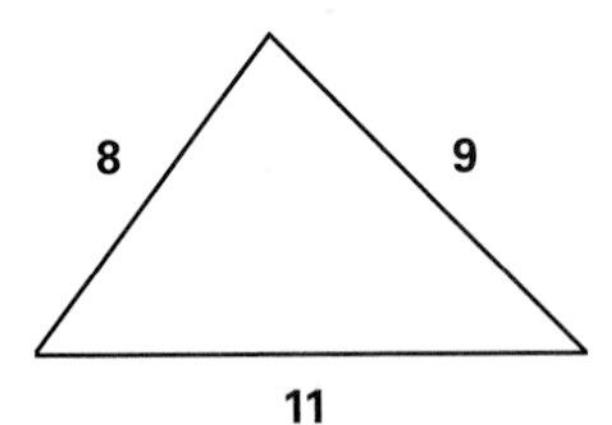

Area of a Triangle

The area (A) of a triangle is one-half the product of the base (b) and the height (h).

$$A = \frac{1}{2} bh$$

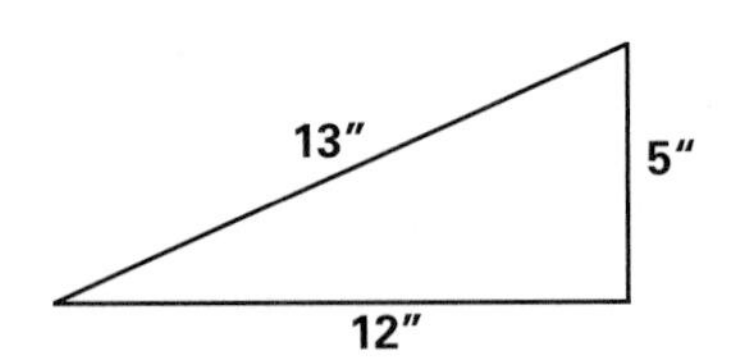

Volume Measurements

Volume measurements measure the space inside three-dimensional shapes. Volume is measured in cubic units.

Volume of a Rectangular Prism

To find the volume (V) of a rectangular prism, multiply the length (l), the width (w), and the height (h).

$$V = lwh$$

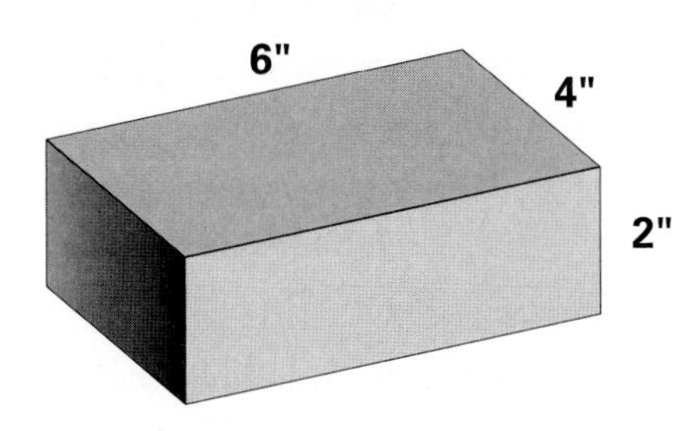

Volume of a Cylinder

The base of a cylinder is a circle. To find the volume of a cylinder, multiply the area of the circular base by the height. (Remember, to find the area of a circle, multiply π times the radius squared.)

$$V = \pi r^2 h$$

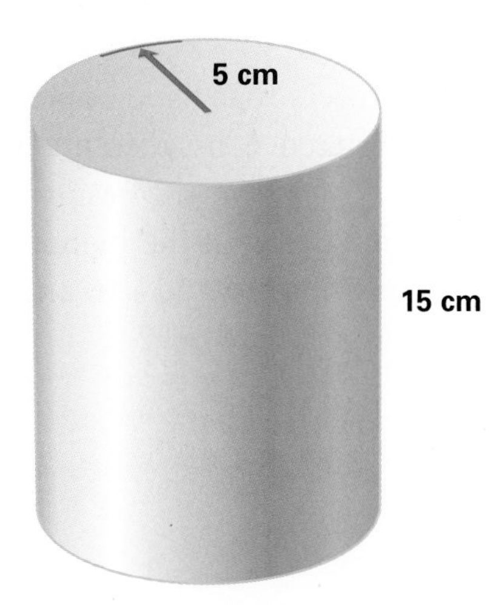

Ratios

A ratio is a way to compare two numbers, using division. For example, directions for mixing concrete may call for 120 pounds of sand and aggregate and 20 pounds of portland cement. The ratio of sand-aggregate to cement can be expressed as 120:20 or $^{120}/_{20}$.

The ratio from this mix yields about one cubic yard of concrete. To find the right amounts to make 5 cubic yards of concrete, multiply both terms of the ratio by 5.

$$120 \times 5 = 600 \text{ pounds sand-aggregate}$$
$$20 \times 5 = 100 \text{ pounds cement}$$

To make a smaller batch—say, one-tenth of a cubic yard—divide both terms of the ratio in the same manner.

$$120 \div 10 = 12 \text{ pounds sand-aggregate}$$
$$20 \div 10 = 2 \text{ pounds cement}$$

The three ratios presented here are equivalent ratios. This is easy to see when you work the ratios as division problems.

$$120 \div 20 = 6$$
$$600 \div 100 = 6$$
$$12 \div 2 = 6$$

It's useful to reduce equivalent ratios to their simplest form. This means dividing until there is no number except one that goes into both terms evenly. Both terms of the ratio $^{12}/_{2}$ can be divided by 2. To find the simplest form of $^{12}/_{2}$, divide:

$$12 \div 2 = 6$$
$$2 \div 2 = 1$$

The simplest form of $^{12}/_{2}$ is $^{6}/_{1}$. Knowing the simplest form makes for the easiest conversions. With the concrete mix, we can say that a batch of any size needs a ratio of six parts sand-aggregate to one part cement.

Proportion

Proportion is the relation of one part to another. A proportion is an equality of two ratios. A proportion might be expressed as *a* is to *b* as *c* is to *d*.

Scale

A scale is the ratio of the number 1 to some larger number. Generally, the larger the second number, the smaller the scale. For example, if 1" on a drawing is the equivalent of 50", that would be expressed as 1:50.

Working with Percentages

A percent is a ratio that compares a number to 100. For example, if 80 percent of household waste is sent to landfills, that means 80 out of every 100 pounds goes to a landfill.

Equations involving percents include the percentage, the base, and the rate. If you know two of these, you can find the third.

Rate × base = percentage
Percentage ÷ base = rate
Percentage ÷ rate = base

Integers

An integer is any of the whole numbers, as shown here on a number line. A fraction, since it is not a whole number, is not an integer. Integers can be positive or negative. Integers to the left of zero on the number line are negative. Integers to the right of zero are positive.

Negative numbers may represent values such as temperatures below zero or a project's costs figured against expected income. Knowing how to solve equations that have negative numbers helps in figuring out the net effect of opposing forces. Examples would include torque versus rolling distance and thrust versus drag.

To make calculations with negative numbers, it is useful to refer to absolute value. The absolute value of an integer is the integer with its plus or minus sign taken away. Absolute value indicates how many steps an integer is away from zero. It is symbolized by a vertical bar before and after the integer. The absolute value of -3 is |3| and the absolute value of +8 is |8|.

To add integers that are both positive or both negative, add their absolute values and give the sum the same sign as the integers. For example, to add +8 and +4:

+8 + +4
|8| + |4| = |12|
+8 + +4 = +12

To add two negative numbers, -5 and -3:

-5 + -3
|5| + |3| = |8|
-5 + -3 = -8

To add integers that have opposite signs, find their difference in absolute value by subtracting. Then give the result the same sign as the integer with the higher absolute value. For example, to add +16 and -2:

+16 + -2
|16| – |2| = |14|

-6 -5 -4 -3 -2 -1 0 1 2 3 4 5 6

After finding a difference in absolute value of |14|, we next wanted to attach a plus or minus sign to the answer. We attached a plus because the integer with the higher absolute value, 16, was positive.

As another example, let's add -9 and +4:

$$-9 + +4$$
$$|9| - |4| = |5|$$
$$-9 + +4 = -5$$

Given integers with opposite signs, we subtracted and found a difference in absolute value of |5|. To decide whether the answer was plus or minus, we looked to the integer with the higher absolute value, -9.

Cartesian Coordinates

If you overlay two number lines perpendicular to each other so that they intersect at 0, the result is a Cartesian coordinate system. The usefulness of a coordinate system lies in its ability to express a series of related numbers as a graphic. For example, the work life of a machine tool might be represented as a line that slopes gently and then falls off sharply. This line would indicate that the tool has passed a critical point of wear.

The intersection marked zero is called the origin. The horizontal and vertical number lines are labeled x and y, respectively. To plot the points that will form a line or curve, use ordered pairs of coordinates such as (3, 1). The first coordinate, called the abscissa, locates the point horizontally (on the x-axis). The second coordinate, called the ordinate, locates the point vertically (on the y-axis).

In the example, the x-axis might represent the days of operation. The y-axis might show how far from nominal (ideal) the tool is in a particular dimension. Let's say design specifications allow the tool to vary from nominal by plus or minus one unit. At first the tool measures consistently as 1 unit above nominal. By day 5, the tool has worn down to the nominal measurement (5, 0). On day 6, the tool is still within 1 unit of specification (6, -1), but by day 7 the tool has worn out (7, -4).

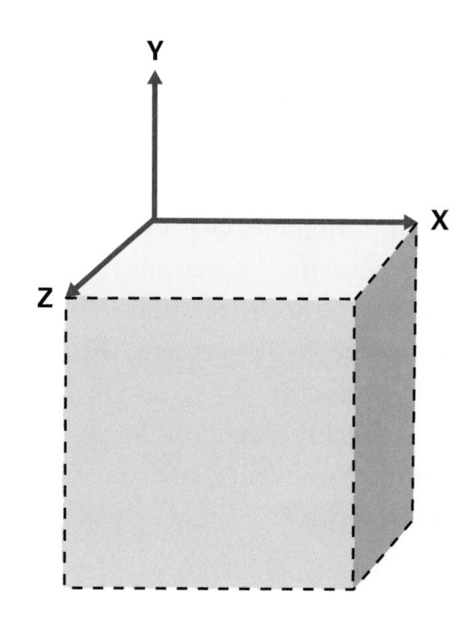

Probability

The probability that a game die will come up 5 on any given roll is ⅙. Rolling a 5 is one outcome among six possible outcomes. The expression ⅙ is equivalent to .17 and may be stated as "a probability of 17 percent." A formula for finding the probability (P) of a specific outcome is

$$P = s/t$$

In this formula, s is a specific outcome (rolling a 5) and t is the total number of possible outcomes (the six sides of a die).

The probability formula relies on the assumption that all possible outcomes are equally likely. If the die is weighted on one side, then a factor other than chance will have a decisive effect. Suppose you are calculating the likelihood of rain tomorrow. The nearness of a massive storm would be a factor other than chance affecting tomorrow's weather. Probability formulas will tell you only about the effect of chance on outcomes.

If you flip a true coin five times and it comes up tails every time, people may say that your chance of coming up heads on the next flip has improved. This idea is incorrect. For any one flip of the coin, the probability of coming up heads is always ½, or 50 percent.

Looking at a record of many coin flips, you will find the number of heads to be around 50 percent. As the number of coin flips increases, the percentage for heads will get closer to 50 percent. Probability is a weak predictor for a single event. It becomes powerful when you are looking at a large number of events, such as estimating traffic over a bridge.

Statistics

Statistics is the analysis of data, usually in search of patterns that show trends, cause-effect relationships, or other useful factors. Data refers to numerical measurements. Examples include temperatures, copper prices, and real estate values.

Statistics typically relies on sampling. This is the collecting of a small number of measurements to represent the whole. Television ratings, for example, typically rely on a sample of about one thousand households. Statisticians have ways of confirming that a sample is valid. This is important. The alternative—taking all possible measurements—would be extremely expensive.

The use of statistics is particularly effective in two areas: identifying a norm within the data set and identifying change. Global warming and increased productivity in agriculture are examples of change identified by comparing previous data with current data. Such comparisons do not necessarily predict the future. However, they do give a basis for making plans.

Identifying the norm helps us understand what is typical in a data set. Statistics uses three concepts for finding mid-range value that is meaningful in a given application. These three concepts include mean, mode, and median.

Mean

The statistical mean, sometimes called the arithmetic average, is the sum of all the values in a data set divided by the number of items in the data set. Using the table of ABC Company sick days as an example, we find the statistical mean to be:

$$51 \div 10 = 5.1$$

The mean gives a representative view when values are well distributed from one end of the range to the other.

Mode

The mode is the value that appears most often in a data set. In the ABC Company data, the mode is 3. Mode is a choice where in-between values obtained by averaging don't make sense, as in the number of children in a household.

Median

The median is the middle number in a data set when the data is arranged in numerical order. In the ABC Company example, the data set would be arranged as follows: 2 2 3 3 3 5 6 8 9 10. Since the data set has an even number of items, average the two middle numbers (3 and 5) and fix the median at 4. The median is the most useful when a few very high or very low values in the data would tilt the average in a misleading way.

ABC Company Workers Out Sick on Each of the Last Ten Workdays	
Workday	**Number of Workers Sick**
1	2
2	3
3	2
4	5
5	9
6	10
7	6
8	3
9	3
10	8

Using a Calculator

Calculators can help you do math operations faster. Of course, you still need to understand math to know which operations to do. Here are some tips.

- Estimate the answer before you work the problem on the calculator. After you've used the calculator, compare the answer with your estimate. If the two differ, there has been an error. Work the problem again.
- After you have entered the number on the calculator, check the display. If you've made an error, press the "clear entry" (CE) key to remove the number.
- You do not need to enter zero before a decimal point. You do not need to enter final zeroes after a decimal point. Enter the number 0.327000 as .327.
- Remember that the answer displayed on the calculator may not be in its final form. For example, suppose you were solving the problem: $45 is what percent of $90? On a simple calculator, the answer would appear as .5. However, you should write this answer as 50 percent. (Recall that a percent is a ratio that compares a number to 100. The answer .5 is 50⁄100, or 50 percent.)

Metric System of Measurement

Most countries use the metric system. The United States still uses the customary, or standard, system. However, many industries in the United States use the metric system. It is a good idea to learn to use both systems. The modernized metric system is known as the International System of Units. This is sometimes shortened to SI (for Système International d'Unités). It uses seven base units. In everyday life, only four units are in common use:

- The meter, the unit of length, is a little longer than a yard (39.37 inches).
- The kilogram, the unit of weight (mass), is a little more than two pounds (2.2 pounds).
- The liter, the unit of liquid capacity or volume, is a little more than a quart (about 1.06 quarts).
- The degree Celsius is the unit for measuring temperature.

Common Metric Units (SI base units are indicated by *)		
Unit	**Symbol**	**Quantity**
ampere*	A	Electric current
candela*	cd	Luminous intensity
degree Celsius	°C	Temperature
kilogram*	kg	Weight (mass)
kelvin*	K	Thermodynamic temperature (for scientific use)
liter	L	Liquid capacity (volume of fluids)
meter*	m	Length
mole*	mol	Amount of substance
second*	s	Time
volt	V	Electric potential
watt	W	Power

The metric system is a decimal system that uses seven base units. All units larger and smaller than the base units are based on multiples of ten, with no fractions. To indicate these larger and smaller units, prefixes are added to the term for the base unit. For example, larger and smaller units of length are indicated by adding such prefixes as kilo-, centi-, and milli- to the word meter. A kilometer is 1,000 times larger than a meter. A centimeter is 100 times smaller than a meter (one-hundredth of a meter). A millimeter is 1000 times smaller than a meter (one-thousandth of a meter). These three prefixes are the most common. They are used for nearly all units of measurement. Both the prefixes and the names of the units can be shortened by using symbols.

Metric-customary conversions are shown in the table on page 531.

Common Metric Prefixes		
Prefix	**Symbol**	**Meaning**
nano	n	One-billionth of
micro	μ	One-millionth of
milli	m	One-thousandth of
centi	c	One-hundreth of
kilo	k	One thousand times
mega	M	One million times
giga	G	One billion times

Metric-Customary Conversions

When you want to convert:		Multiply by:	To find:
Length	inches	2.54	centimeters
	centimeters	0.39	inches
	feet	0.30	meters
	meters	3.28	feet
	yards	0.91	meters
	meters	1.09	yards
	miles	1.61	kilometers
	kilometers	0.62	miles
Mass and Weight	ounces	28.35	grams
	grams	0.04	ounces
	pounds	0.45	kilograms
	kilograms	2.20	pounds
	tons (short)	0.91	tonnes (metric tons)
	pounds	4.45	newtons
	newtons	0.23	pounds
Volume	cubic inches	16.39	cubic centimeters
	cubic centimeters	0.06	cubic inches
	cubic feet	0.03	cubic meters
	cubic meters	35.31	cubic feet
	cubic miles	4.17	cubic kilometers
	cubic kilometers	0.24	cubic miles
	liters	1.06	quarts
	liters	0.26	gallons
	gallons	3.785	liters
Area	square inches	6.45	square centimeters
	square centimeters	0.16	square inches
	square feet	0.09	square meters
	square meters	10.76	square feet
	square miles	2.59	square kilometers
	square kilometers	0.39	square miles
	hectares	2.47	acres
	acres	0.40	hectares
Temperature	Fahrenheit	5/9 (°F - 32)	Celsius
	Celsius	9/5 °C + 32	Fahrenheit

°F: 210, 200, 190, 180, 170, 160, 150, 140, 130, 120, 110, 100, 90, 80, 70, 60, 50, 40, 30, 20, 10, 0, -10

°C: 100, 90, 80, 70, 60, 50, 40, 30, 20, 10, 0, -10, -20

Weight as measured in standard Earth gravity

Academic Vocabulary Glossary

accurate Free from error, especially as the result of care, p. 104
achieve To attain a desired end or aim, p. 33
acquire To come into possession or control of something, p. 362
adequate Sufficient for a specific requirement, p. 325
allocate To apportion for a specific purpose or to particular persons or things, p. 426
alternative One of two or more choices or courses of action, p. 462
analyze Distinguishing the parts of something in order to discover its true nature, p. 220
approach The taking of preliminary steps toward a particular purpose, p. 482
area A geographic region, p. 114
associate To connect one thing with another in the mind, p. 406
attitude A feeling or emotion toward a fact or state, p. 58
automate To operate by use of a self-acting or self-regulating mechanism, p. 370
automatic Done by machine, p. 11
aware Having or showing realization, perception, or knowledge, p. 151

benefit To be useful or profitable, p. 432

community A body of persons of common and especially professional interests scattered through a larger society, p. 405
compare Examine and note the similarities or differences of, p. 174
compensate To offset an undesired effect, p. 249
complex A whole structure (such as a building) made up of interconnected or related structures, p. 481
component A constituent part; ingredient, p. 210
concept An idea generalized from particular instances, p. 361
conduct To direct or take part in an operation or management, p. 297
consider To think about carefully, p. 100
consist To be composed or make up of, p. 429
constant Something invariable or unchanging, p. 140
contact To get in communication with, p. 367
control To exercise power or influence over something, p. 424
convert To change from one form or function to another, p. 449
convince Bring to belief, consent, or a course of action, p. 318
create To make or bring into existence something new, p. 62
crucial Of extreme importance, p. 351

demonstrate To prove or make clear by reasoning or evidence, p. 249
determine To settle or decide by choice of alternatives or possibilities, p. 257
device A piece of equipment, p. 9
distinct Distinguishable to the eye or mind as discrete; separate, p. 31
distribute To give out or deliver, p. 344

element One of the factors determining the outcome of a process, p. 119
enable To make possible, practical, or easy, p. 270
enhance To increase or improve in value, quality, desirability, or attractiveness, p. 36
ensure To make sure, certain, or safe, p. 295
environment The aggregate of social and cultural conditions that influence the life of an individual or community, p. 461
equip To furnish appropriate provisions for service or action, p. 254
error Something produced by mistake, p. 217

Academic Vocabulary Glossary

establish To put on a firm basis; set up, p. 342

evaluate To determine the significance, worth, or condition of, usually by careful appraisal and study, p. 79

expand To increase the extent, number, volume, or scope of, p. 449

expert Someone with special knowledge or ability, p. 304

F

factor An active contributor to the production of a result, p. 192

feature To have or present something or someone as an important element, p. 258

formula A general fact, rule, or principle expressed in mathematical symbols, p. 269

function The action for which a person or thing is specially fitted or used or for which a thing exists, p. 212

G

generate To bring into existence, p. 167

goal The end toward which effort is directed, p. 142

identify To establish the identity of someone or something, p. 238

impact A significant or major effect, p. 38

income A gain or recurrent benefit, usually measured in money, that derives from capital or labor, p. 316

indicate To point out or point to, p. 272

individual Something or someone existing as a distinct entity, p. 91w

integral Essential to completeness, p. 422

interact To act upon one another, p. 475

internal Of, relating to, or occurring on the inside of an organized structure, p. 299

invest To commit money in order to earn a financial return, p. 201

item An object of attention, concern, or interest, p. 351

journal A record of current transactions, p. 228

labor Human activity that provides the goods or services in an economy; or workers available for employment, p. 29

link On a Web-site page, an identifier that permits connection to another Web-site page or element; also known as hyperlink, p. 274

maintain Keep in an existing state, p. 387

method A way, technique, or process of or for doing something, p. 200

monitor To watch, keep track of, or check, p. 313

network A usually informally interconnected group or association of persons, often within professions, p. 171

objective Something toward which effort is directed, p. 395

obtain Gain or attain, usually by planned action or effort, p. 60

overseas Of or relating to movement, transport, or communication over the sea, p. 478

percent A value determined on the basis of a whole divided into 100 equal parts, p. 148

period A portion of time determined by some recurring phenomenon, p. 341

potential Possibility for development, p. 368

predict To foretell on the basis of observation, experience, or scientific reason, p. 15

principle A comprehensive and fundamental law, doctrine, or assumption, p. 318
process A continuous operation, p. 198
project To plan, figure, or estimate for the future, p. 7
publication A published work such as a magazine or journal, p. 239
purchase To obtain by paying money or its equivalent, p. 345
purpose The goal or intended outcome of something, p. 231

range To change or differ within limits, p. 73
recall To remember, p. 25
region A broad geographic area distinguished by similar features, p. 119
regulate To bring under the control of law or constituted authority, p. 456
relationship The state of being related or interrelated, p. 94
require To call for as suitable or appropriate, p. 390
response To show a reaction, p. 197
restrict To confine within bounds, p. 409
role A function or part performed in a particular operation or process, p. 474
route An established or selected course, travel, or action, p. 396

S

secure Free from risk or loss, p. 411
seek To make an attempt, p. 291
sequence A continuous or connected series, p. 76
series A number of things or events of the same class coming one after another in spatial or temporal succession, p. 168
similar Having characteristics in common; strictly comparable, p. 49
source One that provides information, p. 324
specific Applying to, characterized by, or distinguishing something particular, special, or unique, p. 99
structure Something arranged in a definite pattern of organization, p. 104
survey To query in order to collect data for the analysis of some aspect of a group or area, p. 12

target To set as a goal or mark, p. 366
technical Of or relating to proficiency in a practical skill, p. 144
technique A method of accomplishing an desired aim, p. 48
technology A manner of accomplishing a task using technical processes, methods, or knowledge, p. 188
theory A belief, policy, or procedure proposed or followed as the basis of action, p. 73
tradition An established or customary pattern of thought, action, or behavior, p. 16
transfer To convey from one person, place, or situation to another, p. 454
transmit To send or convey from one person or place to another, p. 162

ultimate Last in a progression or series, p. 53
undergo Submit to, p. 252
unique Distinctively characteristic, p. 433

vary To make differences between items, p. 163
vehicle A means of carrying or transporting something, p. 136
via By way of, p. 480
visual Capable of being seen; visible, p. 303
volume Mass or the representation of mass, p. 114

abrading Changing the shape of a material by rubbing off small pieces, such as with sandpaper, p. 347

abutment A structure that supports the end of a bridge or dam, p. 428

AC Alternating current; electrical flow that constantly changes direction, p. 162

acid rain A weak sulfuric acid created when sulfur dioxide in the air mixes with rain and oxygen, p. 148

added value The increase in how much a material is worth after it has been processed into a finished product or a part for a finished product, p. 345

advertising Making a public announcement that a product is available for sale, p. 369

agroforestry Turning forests into controlled environments dedicated to the replacement of trees, p. 326

AI Artificial intelligence; computer program that can solve problems and make decisions ordinarily handled by humans, p. 218

alloy A material made by mixing a metal with other metals or materials, p. 63

alternating current See *AC*, p. 162

amperage (AM-purr-ag) The electrical flow, p. 163

analog signal An electrical signal that changes continuously, p. 173

animation Creating a series of slightly varying drawings or models so that they appear to move and change when the sequence is shown, p. 268

antibiotic In medicine, a substance that kills bacteria, p. 301

aperture (APP-ih-chur) The opening that controls the amount of light that enters a camera, p. 249

aquaculture Growing fish, shellfish, or plants in artificial water ecosystems, p. 326

artificial ecosystem A human-made, controlled environment built to support humans, plants, or animals, p. 325

assembly drawing A drawing that shows how to put parts together to make an item, p. 99

assembly line Series of work stations at which individual steps in the assembly of a product are carried out as the product is moved along, p. 343

atom The smallest particle of an element that still retains the properties of the element, p. 161

audio The recording and reproduction of sound, p. 267

balance A state of steadiness or stability; one of the principles of design, p. 74

beam bridge A bridge that rests on girders laid across the span and is frequently supported by piers partway along the span, p. 429

beankstalk Principle The rule that states systems, processes, and products should not grow beyond an optimal or ideal size, p. 37

binary code An electronic code that a computer can understand and which is based on the binary number system, p. 212

biometrics The science of measuring an individual's unique features, p. 200

bionics Creating replacements for human body parts, p. 304

bioremediation Using bacteria and other organisms to clean up contaminated land and water, p. 324

biosynthesis Making chemicals using biological processes, p. 324

brainstorming A group problem-solving technique in which members call out possible solutions, p. 79

browser Software program that provides access to the World Wide Web, p. 274

building code A rule used to control how structures are built, p. 395

building site Location for construction of a building, p. 406

bullet train High-speed, all-electric locomotive with a bullet-shaped nose, p. 475

cable-stayed bridge A bridge supported by inclined cables connected to towers, p. 428

CAD Computer-aided drafting or computer-aided design; the use of a computer system in place of mechanical drawing tools to create technical drawings and/or design an object, p. 95

calorie The measure of energy in food, p. 136

CAM Computer-aided manufacturing; a system that uses computers to operate the machinery in a factory, p. 365

cantilever bridge (CAN-tih-lee-vur) A bridge made of beams supported by the ground at only the ends; the beams meet in the middle of the bridge, p. 429

capital Accumulated wealth, which may be money, credit, or property, p. 29

CCD Charge-coupled device; a special microchip inside a digital camera that converts light into an electrical signal, p. 251

central processing unit See *CPU*, p. 211

ceramic Material made from nonmetallic minerals that are fused together with heat, p. 64

CIM Computer-integrated manufacturing; the use of one computer system to control the design, manufacturing, and business functions of a company, p. 365

circuit The pathway that electricity flows along; often includes parts such as a wire and a device to which the electricity is being delivered, p. 162

cloning Producing an identical copy of an individual plant or animal, p. 321

closed-loop system A system that has a way of controlling or measuring its product, p. 33

CMOS A special microchip inside a digital camera that converts light into an electrical signal, p. 251

CNC Computerized numerical control; machine tool operation controlled by commands from a computer, p. 365

combining Joining materials together, p. 49

commercial airplane An airplane that carries passengers or freight in order to make money, p. 482

commission A payment made to a salesperson or agent for business they have done, p. 352

communication technology The transfer of messages among people and/or machines through the use of technology, p. 188

composite A material made by combining two or more other materials, p. 64

computer-aided drafting See *CAD*, p. 95

computerized axial tomography See *CT scan*, p. 299

computerized numerical control See *CNC*, p. 365

computer virus A set of destructive instructions that "infects" a computer system and can cause damage, p. 213

concrete A mixture of cement, sand, stones, and water that hardens into a construction material, p. 394

conditioning Changing the inner structure of a material, p. 50

conductor Material that allows electricity to flow easily through it, p. 166

conservative design Design strategy used to be sure structures can bear more weight than required under normal conditions, p. 396

constraint A restriction on a product, p. 36

containership Large ocean-going vessel designed to carry specially designed containers filled with cargo, p. 480

copy In printing, a graphic message ready for reproduction, p. 237

CPU Central processing unit; the part of a computer that processes information; the "brain" of the computer, p. 211

craft A skilled occupation, usually done with the hands, such as carpentry or sewing, p. 341

crane In construction, a large machine used to lift heavy loads by means of a hook attached to cables, p. 424

crankshaft The part of an engine that changes the reciprocating motion of the pistons to the rotary motion that turns the wheels, p. 454

criteria Standards that a product must meet in order to be accepted, p. 36

CT scan Computerized axial tomography; an X-ray image enhanced by a computer and shown on its screen, p. 299

curtain wall Exterior wall on a skyscraper that does not help support the building, p. 393

DC Direct current; the one-directional flow of electrons, p. 162

dehydrate To remove moisture, p. 317

deoxyribonucleic acid See DNA, p. 318

desktop publishing The use of desktop computers and small printers for publishing, p. 239

developer In photography, a chemical used during processing to reveal the latent image, p. 256

digital compression Reducing the size of a digital file by removing bits of data that can be recreated later, p. 270

digital signal An electrical signal having distinct values, p. 173

dimension A size or location of object parts indicated on a technical drawing, p. 92

direct current See *DC*, p. 162

director The person in charge of instructing the performers and guiding the camera work during a video production, p. 272

displacement A measure of the weight of the water that is moved out of the way by a floating ship and its cargo, p. 478

distributed computing Network of computers that researches and analyzes data during computer downtime, p. 220

division of labor System of organizing manufacturing by giving separate tasks to separate workers or groups of workers, p. 343

DNA Deoxyribonucleic acid; the molecules in a gene that carry genetic information, p. 318

drafting The techniques used to make drawings that describe the size, shape, and structure of objects, p. 89

driving wheel A wheel that transmits motion from one part of a vehicle to another, p. 472

drywall In construction, the inside covering of walls and ceilings that is made from plaster and sturdy paper, p. 414

durable good Manufactured product that lasts three or more years, p. 341

dynamic digital printing The use of printing machines that print directly from a computer file instead of a printing plate, p. 232

editing Cutting and arranging material in order to decide its final sequence and content in an audio or video production, p. 273

efficiency The ability to achieve a desired result with little effort and waste, p. 145

electricity The movement of electrons from one atom to another, p. 161

electromagnetic carrier wave See *electromagnetic wave*, p. 194

electromagnetic wave Wave of electromagnetic energy used to carry an electronic signal through the atmosphere; also referred to as an *electromagnetic carrier wave*, p. 171

electronic device A device that changes one form of energy, such as sound, into an electrical signal that can be transmitted by a sender through a channel to a receiver, p. 170

e-manufacturing Using electronic information in the manufacture of a product, p. 366

emissions Gases released into the air as a result of combustion, p. 455

embankment The main part of a dam that holds back the water, p. 431

endoscope In medicine, a small, flexible instrument, having a camera attached, that is inserted into a patient through an incision or by some other means, p. 300

energy The capacity or ability to do work, p. 134

energy conservation The management and efficient use of energy sources, p. 150

engineering Designing products or structures so that they are sound, and deciding how they should be made and what materials should be used, p. 11

ergonomics The design of equipment and environments to promote human safety, health, and well-being; also called *human factors engineering*, p. 295

ethanol A liquid fuel made from corn, p. 456

excavation In construction, a large hole, p. 424

expert system Form of artificial intelligence in which information from experts is collected and stored in a computer's memory, p. 218

external combustion engine An engine that burns fuel to create energy; its power source is outside the engine, p. 449

feedback The part of a closed-loop system that provides control or measurement of the product, p. 34

fertilizer A chemical compound used to restore nutrients to the soil, p. 315

fiber optic cable In communication, thin, flexible, glass strands that transmit light over great distances in order to carry information, p. 172

finishing In manufacturing, the last step in making a product; used to improve appearance, p. 51

flexography A relief printing process that uses a raised, rubber printing plate, p. 229

focus In photography, a term used to describe the sharpness of an image, p. 249

footing The bottom part of a foundation, made of hardened concrete and located under the foundation wall, p. 410

forming In manufacturing, changing the shape of a material, p. 48

fossil fuel Fuel produced from the fossils of long-dead plants and animals, p. 137

foundation The part of a house that rests on the ground and supports the upper structure, p. 410

four-color process printing Combining magenta, cyan, yellow, and black to produce all the other colors used on printed materials, p. 239

four-wheel drive Transfer of power from the vehicle's engine to both front and rear wheels, p. 472

frequency The number of cycles or changes in direction of alternating current; measured in hertz, or cycles per second, p. 163

front-wheel drive Transfer of power from a vehicle's engine to the front wheels, p. 472

fuel cell A device that converts the chemical energy of a fuel, such as hydrogen, into electrical energy, p. 462

gable roof A roof with two sloping sides that meet at the ridge and form a triangular shape at either end, p. 411

gene In biotechnology, the factor in cells that carries heredity, p. 318

genetic engineering Altering or combining genetic material in order to treat a disease or modify body characteristics, p. 303

genetic testing Evaluation of a person's genes to discover if the person is at risk for a particular disease, p. 298

geothermal energy Heat energy produced under the earth's crust, p. 141

graphic communication Sending and receiving messages using visual images and printed words or symbols, p. 193

gravure printing (grah-VYOOR) A printing process in which letters and designs are etched or scratched into a metal plate; ink fills these grooves and is then transferred to paper, p. 230

greenhouse effect Increase in the temperature of the earth's atmosphere caused by a rise in carbon dioxide, p.149

halftone In printing, a photograph reproduced using a series of dots, p. 238

hand tool Tool powered by human muscle, p. 54

helicopter An aircraft that is lifted straight up by one or two rotors, p. 483

horsepower A measurement of power based on lifting, pounds one foot in one second, p. 143

HTML Hypertext markup language; the code in which information on the World Wide Web is written, p. 274

human factors engineering The design of equipment and environments to promote human safety, health, and well-being; also called *ergonomics*, p. 73

humanities School subjects having to do with cultural knowledge, such as language and history, p. 116

hybrid In biotechnology, an organism bred from two different species, breeds, or varieties; in manufacturing, a manufactured product made from a mixture of different elements, p. 314, p. 461

hydraulic power Fluid power produced by putting a liquid under pressure, p. 144

hydroelectric power Electricity generated by turbines that are propelled by flowing water, p. 141

hydroponics Growing plants in nutrient solutions without soil, p. 325

hypertext markup language See *HTML*, p. 274

hypothesis An explanation for something that is used as the basis for further investigation, p. 113

ignition system In transportation, an engine system that starts the fuel burning, p. 455

immunization In medicine, a process for making the body resistant to disease, usually by vaccination, p. 294

implant In medicine, an electronic device inserted into the body, p. 303

Industrial Revolution Social and economic changes in Great Britain, Europe, and the United States that began around 1750 and resulted from making products in factories, p. 341

Information Age Period beginning around 1900 in which human activities focused on the creation, processing, and distribution of information, p. 118

ink-jet printing Printing process that uses spray guns to spray ink on the printing surface, p. 235

innovation (in-noh-VAY-shun) The process of modifying an existing product or system to improve it, p. 73

inorganic material Material that comes from minerals that were never alive, p. 60

input Whatever resources that are put into a system, p. 32

insulation Material used to keep heat or cold from entering or leaving a building, p. 413

insulator Material that resists the flow of electricity, p. 166

integrated circuit A tiny chip of semiconducting material, such as silicon, that contains many of the electrical circuits needed to operate an electronic device, p. 174

intermodal transportation Using two or more forms of transportation to move people or cargo more efficiently, p. 479

internal combustion engine In transportation, an engine in which the fuel is burned inside, p. 452

invention Turning ideas and imagination into new products and systems, p. 73

irradiation In biotechnology, treating products with radiation to destroy pathogens, p. 291

irrigation Bringing a supply of water to crops, p. 315

isometric drawing A pictorial drawing that shows three sides of an object that has been rotated thirty degrees and tilted forward thirty degrees, p. 92

jet engine A type of gas turbine engine that pushes a vehicle forward, as hot air and exhaust are shot out the back of the engine, p. 457

jumbo jet A very large airplane that carries several hundred passengers at one time, p. 482

just-in-time delivery In manufacturing, a method of scheduling the arrival of materials at the time they are needed so that storage is not necessary, p. 351

laser A very powerful, narrow beam of light in which all the light rays have the same wavelength, p. 172

laser surgery In medicine, surgery done with a laser beam instead of a scalpel, p. 302

latent image In photography, the invisible image produced on exposed film, p. 252

layout Arrangement of elements on a page to be printed, p. 238

lens In photography, a piece of glass used to focus and magnify light, p. 249

letterpress printing A printing process that uses raised letters, symbols, or designs that are inked and then pressed against paper, p. 229

lift The upward movement of an airplane resulting from reduced pressure above the wing and increased pressure below it, p. 482

lighter-than-air craft An aircraft lifted by being filled with a gas that is lighter than air, p. 483

line art In printing, drawings and other art elements made of solid lines and shapes, p. 238

lithography A printing process that is based on the principle that oil and water do not mix, p. 230

load The output force of a power system, p. 144

lumber In construction, pieces of wood cut from logs into convenient shapes, p. 393

machine A tool with a power system that takes advantage of certain scientific laws that enable the tool to work better, p. 27

machine tool In manufacturing, a machine used for shaping or finishing metals and other materials, p. 16

machine-to-machine communication The transfer of messages from one machine, usually a computer, to another, p. 198

maglev train Train that is levitated and propelled by the use of electromagnets, p. 475

magnetic resonance imaging See *MRI*, p. 299

maintenance The process of inspecting and servicing a system on a regular basis to enable it to continue functioning properly, to extend its life, or to upgrade its capability, p. 457

marketing Telling potential customers about products and services in such a way as to make them eager to buy, p. 352

market research The process of getting people's opinions about a product so that a company knows what changes to make or whether to sell the product, p. 350

mass transportation Transportation that moves many people at one time and is available to the general public, p. 471

measuring tool Tool used to identify size, shape, weight, distance, density, or volume, p. 53

mechanical property Way in which a material reacts to a force, p. 61

mode A way of doing something, p. 196

model Replica of a proposed product that looks real but does not work, p. 102

monoculture farming In biotechnology, raising only one crop or one species of plant, p. 316

mortar In construction, a mixture similar to concrete used to fasten concrete blocks or bricks together, p. 410

Motion Capture The process whereby an actor performs physical actions that are recorded by sensors attached to his or her body, p. 274

MRI Magnetic resonance imaging; in medicine, the use of a magnetic field to create an image of body structures, p. 299

multimedia A combination of different media into one presentation to create a more enriched and entertaining message, p. 266

multiview drawing A technical drawing that shows an object from several different views, p. 91

nanotechnology The science of working with the atoms or molecules of materials to develop very small machines, p. 15

National Institute of Occupational Safety and Health See *NIOSH*, p. 367

navigable waterway A body of water that is deep and wide enough to allow boats and ships to pass, p. 477

negative In photography, an image produced on exposed film after processing; light areas appear dark and dark areas appear clear, p. 256

NIOSH National Institute of Occupational Safety and Health; the government agency that approves for use protective equipment, such as safety glasses, p. 367

non-durable good A manufactured product that lasts less than three years, p. 341

oblique drawing A pictorial drawing that shows one side of the object face-on, without any distortion, and the other sides at an angle, p. 93

Occupational Safety and Health Administration See *OSHA* , p. 367

Ohm's law Scientific law expressing the relationship among electrical amperage, voltage, and resistance; it takes one volt to force one ampere of current through a resistance of one ohm, p. 164

open-loop system A system with no way of controlling or measuring its product, p. 32

operating system The program a computer follows that tells it how to operate, p. 212

optimization In manufacturing, creating the most effective and functional product, system, or process, p. 36

organic material Material that comes from something that is or was once alive, p. 60

OSHA Occupational Safety and Health Administration; the government agency that sets safety rules and checks to make sure the rules are being followed, p. 367

outlet works Section of a dam with gates that allow water to flow through, p. 431

output What a system produces, p. 33

parallel circuit An electrical circuit having multiple pathways carrying current to individual devices; if one device stops working, the others are not affected, p. 169

pasteurization Heat treatment used to destroy pathogens in food, p. 291

pathogen Organism that causes disease, p. 291

payload Another term for cargo, p. 484

personal privacy The right of individuals to keep certain information away from public view, p. 200

perspective drawing A realistic pictorial drawing in which receding parallel lines come together at a vanishing point, p. 94

pharming In biotechnology, using genetically modified organisms to produce medicines, p. 324

photosite In photography, a tiny light-sensitive cell that converts light into an electrical charge, p. 253

pictorial drawing A drawing that shows a three-dimensional object realistically, p. 92
pier In construction, a concrete column used to add support to a foundation, p. 424
pile In construction, a large shaft driven deep into the soil to support a structure, p. 424
piston A plug that slides inside a cylinder of an engine, p. 449
pixel One of many tiny dots of light used to create a video image; acronym for *picture element,* p. 267
pneumatic power Fluid power produced by putting a gas under pressure, p. 144
portable electric tool Small, portable tool powered by electricity, p. 54
power A measure of work done over a certain period of time when energy is converted from one form to another or transferred from one place to another, p. 142
primary tool A basic tool that is hand held and muscle powered, p. 27
print A printed copy of a graphic message, p. 237
problem statement A statement that clearly defines a problem to be solved, p. 77
process The conversion of a system's input into a useful product, or output, p. 33
producer The person responsible for an entire audio or video production, p. 271
profit Money left over after all the bills for making a product are paid, p. 352
program A set of instructions that a computer follows, p. 212
propellant In transportation, a fuel mixture that causes an explosive thrust, p. 458
proportion The correct relationship between sizes and quantities in a design, p. 74
prototype In manufacturing, a working model of a proposed product, p. 102

quality assurance The process of inspecting products to make sure they meet all standards that have been set, p. 351

R&D See *research and development*, p. 361
RAM Random access memory; memory in a computer's CPU that stores data temporarily, p. 211
random access memory See *RAM*, p. 211
rapid prototyping Using CAD and a special machine to make a three-dimensional model of an object, p. 102
read-only memory See *ROM*, p. 211
rear-wheel drive Transfer of power from a vehicle's engine to its rear wheels, p. 472
reciprocating motion (ree-SIP-roh-kay-ting) In an engine, up-and-down or back-and forth motion that occurs in a straight line, p. 454
recycle To reuse all or part of materials, such as metal, glass, paper, and plastics, p. 151
research and development (R&D) Manufacturing department that searches for and develops new products and processes, p. 361
residential building (rezz ih DEN shul) A building in which people live, p. 405
resin (REZZ-in) A chemical compound used to make plastics, p. 347
resistance Anything that opposes or slows the flow of electrical current, p. 163
resource Something that supplies help or aid to a system; can be a source of information, capital, supply, or support, p. 25
retailer A merchant who buys products from a wholesaler and sells them to consumers, p. 370
robotics The technology involved in building and using industrial robots, p. 366
ROM Read-only memory; permanent memory in a computer's CPU that cannot be deleted or changed, p. 211
rotary motion Circular motion, p. 454

sanitation In biotechnology, removal of waste products or contaminants that could cause disease, p. 293

scale drawing Drawing of an object that is not true size but that is in the correct proportions, p. 91

schedule In manufacturing or construction, a plan that includes a list of the work to be done and when it should be finished, p. 364

schematic diagram Drawing that shows the circuits and components of electrical and electronic systems, p. 100

science Knowledge covering general truths or laws that explain how something happens, p. 11

scientific law A theory that has been proven true so often that it is accepted as fact, p. 113

scientific management In manufacturing, the system of developing standard ways of doing particular jobs, p. 342

scientific theory Scientific conclusion carefully developed through experimentation, p. 113

script The written version of an audio or video production that contains a list of characters and their dialogue, p. 272

search engine Software program that helps users find information on the Internet, p. 275

section drawing A drawing that shows the interior of an object, p. 94

semiconductor Material that can be used as either a conductor of electricity or an insulator, p. 167

sensory property Property of a material that we register with our senses, such as taste or texture, p. 61

separating In manufacturing, removing pieces of a material, p. 47

series circuit A single pathway that electrical current flows through to more than one device; if one device along the path stops working, they all stop, p. 168

serigraphy (sih-RIG-rah-fe) A printing process that uses a printing plate made of an open screen of silk, nylon, or metal mesh, p. 232

shadowing program School program in which students spend time in a work environment, p. 120

sheathing In construction, a layer of material between house framing and the outer covering, p. 411

shield In construction, a metal tube that fits inside a tunnel to support the walls, p. 430

shutter A covering that opens to let light into a camera, p. 249

skill The combination of knowledge and practice that enables a person to do something well, p. 25

skyscraper A very tall building, p. 386

solar cell A device that converts sunlight into electrical energy, p. 140

solar heating system A system in which energy from the sun is used to heat a building, p. 139

sound waves Vibrations traveling through air, water, or some other medium that can be perceived by the human ear, p. 194

span Distance between bridge supports; the entire length of a bridge, p. 428

spillway Portion of a dam that allows excess water to spill over the dam, p. 432

standard In manufacturing, a rule or guideline for making a product, p. 366

static electricity A buildup of electrons that are not in motion and are seeking to discharge, p. 162

stick construction Method of building construction in which lightweight pieces of wood or steel are used, p. 386

studs In construction, pieces of lumber, usually 2 × 4s used for the framework of walls, p. 411

subfloor In building construction, the first layer of flooring, usually made of plywood, p. 411

subgrade In construction, the layer of soil beneath a roadway, p. 427

subsystem A system that is part of another, larger system, p. 31

superconductor Material that has no measurable resistance to electricity, p. 167

supertanker A very large ship that is fitted with tanks for carrying oil across oceans, p. 479

supplier Person or company that provides one or more of the materials or parts for a manufactured product, p. 350

surfacing Applying a material to a roadway that provides a smooth surface, p. 387

suspension bridge A bridge that hangs from large cables across a wide span, p. 428

system A group of parts that work together in an organized way to complete a task, p. 31

technical drawing To accurately represent the size, shape, and structure of objects; also called *mechanical drawing,* p. 88

technologically literate Term used to describe someone who is informed about technology and feels comfortable with it, p. 9

technology The practical use of human knowledge to extend human abilities and to satisfy human needs and wants, p. 7

telecommunication Communicating over a distance, p. 195

telemedicine Using communication technology to transmit medical information and advice over a distance, p. 304

thrust The high pressure that pushes a jet engine forward, p. 457

time and motion study In manufacturing, an investigation into the best ways of doing a job; things such as working conditions, wasted time, and unnecessary movement are considered, p. 342

tolerance Respect for others' differences, p. 201

tool An instrument or apparatus that increases a person's ability to do work, p. 27

tractor-trailer A two-part truck that includes a tractor and a trailer that holds the load, p. 474

trade-off A compromise in which one thing is given up in order to gain something else, p. 37

transformer Device used to change electricity from one voltage to another, p. 163

transgenic organism In biotechnology, an organism into which genes from another organism have been transplanted, p. 319

transistor Small device made of a semiconducting material and used to control electric current, p. 174

transmission Vehicle system that contains gears and other parts that transfer power from the engine to the axles and wheels, p. 471

transmitter The part of a communication system that changes the sender's message into electrical impulses and sends it through the channel to the receiver, p. 172

troubleshooting A problem-solving method used to identify a malfunction in a system, p. 366

truss In construction, a prefabricated triangular framework that supports a roof, p. 411

truss bridge A bridge made from steel beams fastened together in triangular shapes, p. 429

turbine (TUR-bin) A disk or wheel that changes the energy of moving gases or liquids into rotary motion, p. 451

ultrasound In medicine, the use of sound waves to create an image of internal body structures on a computer screen, p. 299

uniform resource locator See *URL*, p. 274

unity A state in which all parts of a design work together, p. 75

URL Uniform resource locator; an address on the World Wide Web, p. 274

vaccine In medicine, a preparation containing dead or weakened pathogens used to stimulate a person's immune system, p. 294

video The recording and reproduction of moving images, p. 267

virtual factory A three-dimensional computer model of a factory, p. 363

visualization software CAD software used to create virtual models, p. 104

voltage The pressure needed to push electricity through a circuit, p. 163

wattage A measure of electric power; calculated by multiplying the amperes used times the voltage of the circuit, p. 165

wholesaler A merchant who buys large quantities from a manufacturer and sells smaller quantities to retailers, p. 370

wi-fi A local wireless connection to your network components and the Internet, p. 220

WiMAX A long distance wireless connection to your network components and the Internet, p. 220

wind farm A large collection of windmills located in an area that has fairly constant winds, p. 140

working drawing One of a set of technical drawings needed to make a product or structure, p. 99

xerography (zee-RAH-graph-fe) A printing process that transfers negatively charged toner to positively charged paper, p. 233

Index

Index

F

J

K

L

M

N

T

Photo Credits

Lindsey Parnaby/epa/Corbis, **Cover;** Marc Hill/Alamy, **v, 4–5, 18;** James Leynse/Corbis, Getty Images Royalty Free, **vi, 27, 97; vii, 148, 194(r), 197, 201, 202(2), 235, 264–265, 278, 288–289;** Issei Kato/Reuters/Corbis, **viii, 267;** SuperStock Royalty Free, **ix, 28(bl), 28(tm), 114, 132–133, 136, 138(2), 139, 140, 154, 215, 257, 262, 320, 361;** China Images/Alamy, **xi, 475;** Fridmar Damm/Corbis, **2–3;** Firefly Productions/Corbis, **7;** Corbis/Bettmann, **8, 294(b), 376(tl), 377(ml), 377(tl), 377(ml);** Scott Speakes/Corbis RF, **15;** Krista Kennell/ZUMA/Corbis, **17;** Bojan Breceij/Corbis, **22–23, 40;** Arthur Tilley/Getty Images, **25;** Jupiter Images, **26;** Masterfile Royalty Free, **28(bl), 28(br), 28(tl), 130–131, 165, 338–339, 354;** Brian Sytnyk/Masterfile, **28(tr);** Michael Pole/Corbis, **34;** Ted Soqui/Corbis, **36;** Klaus Hackenberg/zefa/Corbis, **39;** Gail Mooney/Masterfile, **42;** Bill Miles/Corbis, **44–45, 66;** SuperStock, **50, 92, 95, 105, 173, 180, 208–209, 222, 450, 451;** Time & Life Pictures/Getty Images, **61;** Digital Stock/Corbis, **65, 200, 221;** Image Source/Corbis, **70–71, 82;** Will & Deni McIntyre/Corbis, **73;** Robin Sachs/PhotoEdit, **79;** John Madere/Corbis, **81;** Dennis MacDonald/age fotostock, **84, 387;** Charles Gupton/Corbis, **85;** Handout–Lower Manhattan Developement Corporation/Reuters/Corbis, **86–87, 106;** Age Fotostock Royalty Free, **110–111, 113, 122, 336–337;** Mike Blake/Corbis, **117;** Corbis, **121;** Kevin Dodge/Masterfile, **124;** Peter Griffith/Masterfile, **125;** The Bridgeman Art Library/Getty Images, **126(bl);** Dmitri Kessel/Getty Images, **126(br);** Steve Vidler/SuperStock, **127(tl);** Eyewire/Corbis, **127(tr);** Big Cheese Photo/Jupiter Images, **128, 182, 284, 334, 378, 442, 482;** Bob Rowan/Progressive Image/Corbis, **146(r);** Momatiuk–Eastcott/Corbis, **148;** David R. Frazier/Photo Edit, **150;** David Young–Wolff/Photo Edit, **151;** Skycam/Corbis, **153;** Roger Ressmeyer/Corbis, **158–159, 176;** Ron Stroud/Masterfile, **162;** World History Archive/age fotostock, **167;** Image100/Corbis, **175;** Time Life Pictures/Getty Images, **180(r);** SuperStock, Inc., **181(tr);** Radius Images/Alamy, **184–185;** Corbis Royalty Free, **186–187, 204, 206;** AFP/Getty Images, **191, 207, 218, 277, 423, 444–445;** First Light/Getty Images, **193;** Klaus Lahnstein/Getty Images, **194(l);** Thinkstock/Corbis, **203;** PhotoDisc, **212;** Jim Craigmyle/Corbis, **214;** Michael Newman/PhotoEdit, **232;** Workbook Stock Image/Jupiter Images, **249(l);** Anton J. Geisser/age fotostock, **249(r);** Image 100/Corbis, **259;** John Cumming/Getty Images, **263;** Issei Kato Reuters Corbis, **207;** Dennis Hallinan/Alamy, **270;** Warner Bros./Getty Images, **272;** Getty Images Entertainment, **273;** Paolo Patrizi/Alamy, **274;** Philippe Wojazer/Reuters/Corbis, **275;** Digital Vision/Getty Images, **277, 327;** Image Source, **282(tr);** George S. de Blonsky/Alamy, **286–287;** Javier Larrea/age footstock, **291;** Frans Lanting/Corbis, **294(t);** Jeff Greenberg/PhotoEdit, **295(b), 405;** Jupiter Images RF, **295(t), 386(b);** Brand X Pictures/age fotostock RF, **298;** Stock Connection Image/Jupiter Images, **299(b);** Javier Larrea/age fotostock RF, **300, 358–359, 372;** epa/Corbis, **302;** Creatas/SuperStock, **305;** Andy Sacks/Getty Images, **310–311, 328;** Russ Munn/AGStockUSA, **313(b),** SuperStock, Inc./SuperStock, **313(t), 324;** Bryan Mullennix/age fotostock RF **315(b);** Frank Lukasseck/age fotostock, **315(t);** Bruce Forster/Getty Images, **326;** Ben Weaver/Getty Images, **330;** Index Stock Imagery/Jupiter Images, **332;** Stan Waxman/Getty Images, **332(bl);** Lester V. Bergman/Corbis, **333(tr);** Raymond Forbes/age fotostock, **341;** Bill Aron/PhotoEdit, **346;** Courtesy of Apple, **350;** Alvis Upitis/Superstock, **353;** Culver Pictures, Inc. / SuperStock, **362;** Oxford Scientific Image/Jupiter Images, **366;** Susan Van Etten/PhotoEdit, **369;** Creasource/Corbis, **371;** GoodShoot/SuperStock, **380–381, 398;** ARCO/J. Hildebrandt/age fotostock, **382–383, 398;** Bettmann/Corbis, **386(t);** Hubertus Kanus/SuperStock, **390;** Bernard Bisson/Sygma/Corbis, **392;** Ken Davies/Masterfile, **393(b);** David R. Frazier/PhotoEdit, **394(t);** Tim Street–Porter/Beateworks/Corbis, **395;** Construction Photography/Corbis, **397;** Bill Frymire/Masterfile, **402–403, 416;** George Hunter/Getty Images, **406;** Peter Widmann/Alamy, **407(b);** vario images GmbH & Co. KG/Alamy, **407(t);** Philippe Eranian/Corbis Sygma, **411;** Aardvark/Alamy, **413(b),** Stockbyte/SuperStock, **413(t), 463;** David R. Frazier Photolibrary, **414;** Getty Images/Photodisc, **415;** D. Hurst/Alamy, **418;** FAN Travelstock/Jupiter Images, **420–421, 436;** Peter Mackinven/Index Stock, **428;** Agence Image/Age Fotostock, **432(b);** David Rae Morris/epa/Corbis, **432(t);** Courtesy of NASA, **434;** Travelpix Ltd./Getty Images, **435;** Image Source Pink/SuperStock, **440(bl);** Dennis Cox/Getty Images, **440(tl);** age fotostock, **440(tr), 471;** Joel W. Rogers/Corbis, **446–447;** Richard Hutchings/Photo Edit, **451;** Dennis MacDonald/PhotoEdit, **454;** Rob Howard/Corbis, **460;** Car Culture/Corbis, **461;** Joel W. Rogers/Corbis, **464;** John Warden/SuperStock, **468–469, 486;** Used with permission of Royal Caribbean Cruises Ltd., **478;** Thinkstock Images/Jupiter Images, **479;** Courtesy of American Airlines, **482;** Bill Howe/Alamy, **483;** Donald C. Johnson/Corbis, **485;** Michael Macor/San Francisco Chronicle/Corbis **491(b).**